1999 GEOLOGIC TIME SCALE

CENOZOIC

Period		Epoch		Age	Picks (Ma)
QUATERNARY		HOLOCENE			0.01
		PLEISTOCENE		CALABRIAN	1.8
TERTIARY	NEOGENE	PLIOCENE	L	PIACENZIAN	3.6
			E	ZANCLEAN	5.3
		MIOCENE	L	MESSINIAN	7.1
				TORTONIAN	11.2
			M	SERRAVALLIAN	14.8
				LANGHIAN	16.4
			E	BURDIGALIAN	20.5
				AQUITANIAN	23.8
	PALEOGENE	OLIGOCENE	L	CHATTIAN	28.5
			E	RUPELIAN	33.7
		EOCENE	L	PRIABONIAN	37.0
			M	BARTONIAN	41.3
				LUTETIAN	49.0
			E	YPRESIAN	54.8
		PALEOCENE	L	THANETIAN	57.9
				SELANDIAN	61.0
			E	DANIAN	65.0

MESOZOIC

Period	Epoch		Age	Picks (Ma)	Uncert. (m.y.)
CRETACEOUS				65	±2
	LATE		MAASTRICHTIAN	71.3	±1
			CAMPANIAN	83.5	±1
			SANTONIAN	85.8	±1
			CONIACIAN	89.0	±1
			TURONIAN	93.5	±4
			CENOMANIAN	99.0	±1
	EARLY		ALBIAN	112	±2
			APTIAN	121	±3
		NEOCOMIAN	BARREMIAN	127	±3
			HAUTERIVIAN	132	±4
			VALANGINIAN	137	±4
			BERRIASIAN	144	±5
JURASSIC	LATE		TITHONIAN	151	±6
			KIMMERIDGIAN	154	±7
			OXFORDIAN	159	±7
	MIDDLE		CALLOVIAN	164	±8
			BATHONIAN	169	±8
			BAJOCIAN	176	±8
			AALENIAN	180	±8
	EARLY		TOARCIAN	190	±8
			PLIENSBACHIAN	195	±8
			SINEMURIAN	202	±8
			HETTANGIAN	206	±8
TRIASSIC	LATE		RHAETIAN	210	±8
			NORIAN	221	±9
			CARNIAN	227	±9
	MIDDLE		LADINIAN	234	±9
			ANISIAN	242	±9
	EARLY		OLENEKIAN	245	±9
			INDUAN	248	±10

Magnetic polarity: RAPID POLARITY CHANGES

PALEOZOIC

Period		Epoch	Age	Picks (Ma)
PERMIAN				248
		L	TATARIAN	252
			UFIMIAN-KAZANIAN	256
			KUNGURIAN	260
		E	ARTINSKIAN	269
			SAKMARIAN	282
			ASSELIAN	290
CARBONIFEROUS	PENNSYLVANIAN	L	GZELIAN (S.)	296
			KASIMOVIAN	303
			MOSCOVIAN (W.)	311
			BASHKIRIAN (N.)	323
	MISSISSIPPIAN	E	SERPUKHOVIAN	327
			VISEAN	342
			TOURNAISIAN	354
DEVONIAN		L	FAMENNIAN	364
			FRASNIAN	370
		M	GIVETIAN	380
			EIFELIAN	391
			EMSIAN	400
			PRAGHIAN	412
			LOCKHOVIAN	417
SILURIAN		L	PRIDOLIAN	419
			LUDLOVIAN	423
			WENLOCKIAN	428
		E	LLANDOVERIAN	443
ORDOVICIAN		L	ASHGILLIAN	449
			CARADOCIAN	458
		M	LLANDEILIAN	464
			LLANVIRNIAN	470
		E	ARENIGIAN	485
			TREMADOCIAN	490
CAMBRIAN*		D	SUNWAPTAN*	495
			STEPTOEAN*	500
		C	MARJUMAN*	506
			DELAMARAN*	512
		B	DYERAN*	516
			MONTEZUMAN*	520
		A		543

PRECAMBRIAN

Eon	Era	Bdy. Ages (Ma)
PROTEROZOIC		543
	LATE	900
	MIDDLE	1600
	EARLY	2500
ARCHEAN	LATE	3000
	MIDDLE	3400
	EARLY	3800?

GEOLOGICAL SOCIETY OF AMERICA

*International ages have not been established. These are regional (Laurentian) only.

Sources for nomenclature and ages: Primarily from Gradstein, F., and Ogg, J., 1996, *Episodes*, v. 19, nos. 1 & 2; Gradstein, F., et al., 1995, SEPM Special Pub. 54, p. 95–128; Berggren, W. A., et al., 1995, SEPM Special Pub. 54, p. 129–212; Cambrian and basal Ordovician ages adapted from Landing, E., 1998, *Canadian Journal of Earth Sciences*, v. 35, p. 329–338; and Davidek, K., et al., 1998, *Geological Magazine*, v. 135, p. 305–309. Cambrian age names from Palmer, A. R., 1998, *Canadian Journal of Earth Sciences*, v. 35, p. 323–328.

The History and Sedimentology of Ancient Reef Systems

TOPICS IN GEOBIOLOGY

Series Editors: **Neil H. Landman,** American Museum of Natural History, New York, NY
Douglas S. Jones, University of Florida, Gainesville, FL

Current volumes in this series

Volume 4 THE GREAT AMERICAN BIOTIC INTERCHANGE
Edited by Francis G. Stehli and S. David Webb

Volume 5 MAGNETITE BIOMINERALIZATION AND MAGNETORECEPTION IN ORGANISMS
A New Biomagnetism
Edited by Joseph L. Kirschvink, Douglas S. Jones, and Bruce J. McFadden

Volume 6 *NAUTILUS*
The Biology and Paleobiology of a Living Fossil
Edited by W. Bruce Saunders and Neil H. Landman

Volume 7 HETEROCHRONY IN EVOLUTION
A Multidisciplinary Approach
Edited by Michael L. McKinney

Volume 8 GALÁPAGOS MARINE INVERTEBRATES
Taxonomy, Biogeography, and Evolution in Darwin's Islands
Edited by Matthew J. James

Volume 9 TAPHONOMY
Releasing the Data Locked in the Fossil Record
Edited by Peter A. Allison and Derek E. G. Briggs

Volume 10 ORIGIN AND EARLY EVOLUTION OF THE METAZOA
Edited by Jere H. Lipps and Philip W. Signor

Volume 11 ORGANIC GEOCHEMISTRY
Principles and Applications
Edited by Michael H. Engel and Stephen A. Macko

Volume 12 THE TERTIARY RECORD OF RODENTS IN NORTH AMERICA
William Korth

Volume 13 AMMONOID PALEOBIOGRAPHY
Edited by Neil H. Landman, Kazushige Tanabe, and Richard Arnold Davis

Volume 14 NEOGENE PALEONTOLOGY OF THE MANONGA VALLEY, TANZANIA
A Window into the Evolutionary History of East Africa
Edited by Terry Harrison

Volume 15 ENVIRONMENTAL MICROPALEONTOLOGY
The Application of Microfossils to Environmental Geology
Edited by Ronald E. Martin

Volume 16 PALEOBIOGEOGRAPHY
Bruce S. Lieberman

Volume 17 THE HISTORY AND SEDIMENTOLOGY OF ANCIENT REEF SYSTEMS
Edited by George D. Stanley, Jr.

Volume 18 EOCENE BIODIVERSITY
Unusual Occurrences and Rarely Sampled Habitats
Edited by Gregg F. Gunnell

The History and Sedimentology of Ancient Reef Systems

Edited by

George D. Stanley, Jr.
University of Montana
Missoula, Montana

Springer Science+Business Media, LLC

Library of Congress Cataloging-in-Publication Data

Stanley, George D.
The history and sedimentology of ancient reef systems/George D. Stanley, Jr.
p. cm. — (Topics in geobiology; v. 17)
Includes bibliographical references and index.
ISBN 978-1-4613-5446-8 ISBN 978-1-4615-1219-6 (eBook)
DOI 10.1007/978-1-4615-1219-6

1. Coral reefs and islands. I. Title. II. Series.

QE565 .S73 2001
551.42′4—dc21

00-046613

The reef system of living corals and other organisms from Chumbe Island, Zanzibar, Tanzania. Photograph by J. E. N. Veron, Australian Institute of Marine Science.

ISBN 978-1-4613-5446-8

Originally published by Kluwer Academic / Plenum Publishers, New York in 2001
Softcover reprint of the hardcover 1st edition 2001

http://www.wkap.nl/

10 9 8 7 6 5 4 3 2 1

A C.I.P. record for this book is available from the Library of Congress.

Contributors

Randolph B. Burke North Dakota Geological Survey, Bismarck, North Dakota 58505-0840

Paul Copper Department of Earth Sciences, Laurentian University, Sudbury, Ontario P3E 2C6, Canada

Erik Flügel Institute of Paleontology, University Erlangen-Nürnberg, Erlangen, Germany D-91054

Ivan P. Gill Department of Geology and Geophysics, University of New Orleans, New Orleans, Louisiana, 70148

Pamela Hallock Department of Marine Sciences, University of South Florida, St. Petersburg, Florida 33701

Dennis K. Hubbard Department of Geology, Oberlin College, Oberlin, Ohio 44074

Claudia C. Johnson Department of Geological Sciences, Indiana University, Bloomington, Indiana 47405

Erle G. Kauffman Department of Geological Sciences, Indiana University, Bloomington, Indiana 47405

Wolfgang Kiessling Department of Geophysical Sciences, University of Chicago, Chicago, Illinois, 60637

Reinhold R. Leinfelder Institute for Palaeontology and Historical Geology, University of München, D-80333 München, Germany

Norman D. Newell Paleontology Division (Invertebrates), The American Museum of Natural History, New York, New York 10024

Baba Senowbari-Daryan Institute of Paleontology, University Erlangen-Nürnberg, D-91054 Erlangen, Germany

George D. Stanley, Jr. Department of Geology, The University of Montana, Missoula, Montana 59812

Gregory E. Webb School of Natural Resource Sciences, Queensland University of Technology, Brisbane, QLD 4000, Australia

Andrey Yu-Zhuravlev Paleontological Institute, Academy of Russian Sciences, Moscow, 117337 Russia

Foreword

Reef. The very word conjures up restful, tropical images wrapped in warmth and displayed in shimmering colors. So it is with this volume: a series of reef images, past and present, brought to life by thoughtful and probing analysis and revealed through well-written and beautifully illustrated chapters.

Ever since Darwin reported on his investigations of far-flung reefs, these structures have held an unending fascination for earth scientists. The fact that their fossil predecessors contain a wealth of information about past life and can be filled with a king's ransom of hydrocarbons means that they always will be subjects of legitimate scientific study. This volume likewise is a gold mine of up-to-date information, a report card on the status of reefs and reef research at the beginning of the millennium. Although the perspective is geological, the chapters are multidisciplinary, and so the volume is a treasure trove for biologists, ecologists, and oceanographers. It brings together chapters by the experts who have for many years sought to understand the nature of reefs in different time periods. While it might have suffered from the lack of a centralist view, it does not, because each author has clearly tried to place temporally distinctive reefs in a larger global context. Thus the authors rarely just report on "their" reefs, but to our benefit, cannot resist comparing and contrasting reefs with one another and wresting as much paleoceanographic information out of their data as possible.

One particular phrase from this volume that sticks in my mind is that reefs are a "plethora of paradoxes." This truism is illustrated again and again as different reefs are chronicled. It is clear that if we want to read the valuable information that they contain, we must do so carefully using all the tools at our disposal and with an open mind. Some of the most engaging parts of the volume occur when different authors interpret reefs in the same period in earth history in different ways. Likewise, it is particularly compelling to read the view of ongoing reef studies by a pioneering reef worker.

In short, this volume is at the same time a wonderful source of basic information and a grand overview of modern and ancient reef systems for students of all ages. Although you may begin in a restful frame of mind, when you have read it all, there will be much to ponder, not only about the fossil record, but also about what reefs are telling us of our modern world.

Noel P. James
Kingston, Ontario

Preface

Reefs are complex physical, chemical, and biological systems—a tropical phenomenon, quantitatively and qualitatively important, and yet so extremely fragile. Ironically, while fragile ecologically, reefs are among the most enduring and robust of Earth's ecosystems. Both direct and indirect biological, physical, and chemical processes have interacted with reefs to result in the development of positive and negative feedback loops with the planet. During a span of more than 600 million years, reef ecosystems have witnessed a number of profound changes in composition and paleoecological structure. The integration of many facts, ideas, and hypotheses gleaned from the study of modern and ancient reefs allows us to recognize a number of critical turning points in their evolution. Some of these coincide with episodes of reef crises, mass extinction, and global change. An analysis of reefs through time reveals that this ecosystem has experienced a number of major reorganizations.

Reefs systems also are a sedimentological phenomenon, not only because of the volume of carbonate rock produced, but also because of the way that the carbonate interacts with Earth's biosphere and atmosphere, particularly with regard to global climatic change. In addition, many ancient fossil reefs, with their thick and well-developed carbonate rocks, have come to be appreciated for vast petroleum reserves they contain. Research on reef diagenesis and cements has direct relevance not only to the evolution of the biota, but also to petroleum potential as well.

The chapters of this volume contain contributions from an international group of specialists. They address some important themes in both modern and ancient reef systems. Some chapters contain "snapshots" of reefs of particular intervals, while others touch on relevant themes of both modern and ancient reefs—themes that weave their way through reefs of all ages. It was my pleasure as editor of the volume to work with these authors to produce this volume, which I trust will illuminate a wide spectrum of reef research to a diverse readership.

I open this volume with an introduction to both modern and ancient reefs and reef ecosystems. This chapter also is intended as a basic introduction for students, general geologists, and professionals or others who may be uninitiated to reefs and reef ecosystems. The chapter addresses the living coral reef ecosystem, stressing among other relevant factors the importance of ecological and physical interactions between the organisms and their environment. The chapter also addresses mass extinctions and provides a general overview of the history of reefs.

Databases have proved their relevancy in analyzing trends in the fossil record and reefs are no exception. In Chapter 2, Wolfgang Kiessling summarizes major trends in reefs of the Phanerozoic. His summary is part of a large German research project called the "PaleoReef Database"—a paleogeographically controlled database of 2,700 Phanerozoic reefs based on critical time slices. The database examines fluctuating reef characteristics such as reef abundance, reef size, bathymetry, and biotic diversity through time. These are combined with a set of paleogeographic maps plotting ancient reef distributions.

Paul Copper examines the Precambrian through mid-Paleozoic history of reef ecosystems in Chapter 3. After reviewing some salient aspects of reefs and their definitions, Copper enters the Archean to Neoproterozoic "prelude." The lion's share of this chapter, however, deals with the Precambrian history—a topic largely neglected in many overviews of reef evolution. This vast interval of geologic time, largely before metazoans, was dominated by cyanobacterial (microbial) communities that Copper refers to as "chloroxybacteria," because they were responsible for increasing atmospheric oxygen in our hydrosphere and our atmosphere through photosynthesis. The latest part of the Precambrian, the Neoproterozoic interval (1000–544 Ma), was marked by a curious decline of microbial communities. Copper reviews a body of current literature, searching for possible explanations for this decline. His list includes global warming, changes in carbon dioxide levels, and flux in oceanic pH. Copper provides insight into the late Neoproterozoic glaciation and speculates on the evolution of single-celled to multicellular eukaryotes and their importance in later calcifying reef biotas of the Paleozoic.

Continuing with the theme of early Phanerozoic reef ecosystems, one of the obvious changes during the Cambrian history was the evolution of the first calcifying metazoans. The names of some of the players—receptaculids, pharetronid sponges, coralomorphs, hyoliths, chancelloriids—which sound unfamiliar to most readers, remind us how far-removed these reefs were from those that came later. These reefs were a consortium of archeocyaths, calcimicrobes, and other organisms. In Chapter 4, Andrey Zhuravlev summarizes these metazoan reefs of the Cambrian in time and space, examining their paleoecology and evolution. Zhuravlev attempts to dispel many myths about archeocyath ecosystems with a wealth of information he presents on the spatial and temporal distributions of Cambrian reefs. Although archeocyaths functioned as primary builders, the author does not support the idea of previous workers, namely, that these sponges were photosymbiotic and lived in oligotrophic settings. In this chapter, we also learn the latest ideas on what caused the dramatic and pronounced Cambrian mass extinctions which had disastrous effects on reef ecosystems.

Throughout geologic history, great quantities of calcium carbonate are precipitated in reefs. This carbonate is most noticeable to geologists when it appears in the form of shells and massive skeletons of reef organisms, such as corals and stromatoporoid sponges. Such types of carbonates are termed "enzymatically secreted," meaning that they are controlled directly by the organisms. However, a relatively unappreciated but significant quantity of biological carbonate is not secreted directly by the organisms; rather, it is biologically "induced." Such carbonate is called "nonenzymatic" and is the subject of Chapter 5, by Gregory E. Webb. Webb explains not only how nonenzymatic reef carbonates exerted important controls on the evolution of reef ecosystems, but he also presents the proposition that no reef makes sense without consideration of this phenomenon. Reeflike features can be produced largely, if not entirely, by biologically induced, nonenzymatic precipitation that comes to be related with calcimicrobes and microbiolites, as well as with a variety of microbial crusts, rinds, and many types of reef cements. Not only do these make up a significant percent of the total carbonate in many Proterozoic to Phanerozoic reefs, but biologically induced carbonates have significance in explaining fundamental aspects of the evolution of reef ecosystems. They may in fact be tied with Earth's global history of long-term change.

Norman D. Newell provides a brief review in Chapter 6 of what is deemed to be the most famous of all fossil reefs—the Permian Reef Complex of west Texas and eastern New Mexico. What geologic experience is complete without a field trip to these carbonate rocks making up the Guadalupe Mountains and the surrounding area? Newell, one of the pioneers in this field, shares his insight. Since first proposed as a model for modern counterparts, this reef has weathered considerable debate and continues to stand as perhaps the most powerful reef model available to geologists. With insight derived from half a century of his own research, Newell reviews the early history and geology of the Guadalupes, including the early petroleum-related research that paved the way toward recognition of these important Permian carbonates. Recapping the history of the Capitan reef, Newell provides a succinct summary of the latest findings, based on both paleontologic and stratigraphic studies. He also touches on the issue of the great Permian mass extinction.

The great mass extinction at the end of Permian time decimated the luxurious reefs of that time and irrevocably changed reef structure and composition, ushering in the reef ecosystems of the early Mesozoic. Global perturbations characterized the Early Triassic marine environments and it was not until much later that new reef ecosystems emerged. By Middle Triassic time, reefs reappeared but their composition and paleoecological structure were substantially different from Permian predecessors. These reefs again changed from the Middle to Late Triassic age. Some of the best records of these first Mesozoic reefs are found in mountain ranges stretching from central Europe through the Himalayan Mountains and eastward to Papua New Guinea, forming what is known as the backbone of Eurasia. They are the remains of a once great east–west seaway called the Tethys. Reefs from this region are the subject of Chapter 7, by Erik Flügel and Baba Senowbari-Daryan. Their chapter is extracted from a larger German priority program on Reef Evolution. Based on detailed studies of some of the best reefs of the Tethys, Flügel and Senowbari-Daryan review the principal reef builders of this interval. They examine the distributional patterns and discuss how reef ecosystems changed during the Triassic. Among the organisms discussed are segmented and nonsegmented sponges, the problematic organism "*Tubiphytes*," and scleractinian corals. Corals and sponges became especially important as reef builders during the latest Triassic interval.

Following the major end-Triassic mass extinction and the Early Jurassic recovery, Middle to Late Jurassic reefs proliferated across vast regions of the shallow Tethys, and some of these reefs reached diversity levels comparable to those of modern reefs. In Chapter 8, Reinhold Leinfelder draws from his extensive research experience on the Jurassic reefs to provide a lucid and detailed summary of the state of knowledge of Jurassic reef ecosystems. This chapter provides an understanding of the evolution, distribution, and general types of reefs that developed in the Tethys. Leinfelder marshals evidence from plate tectonics, paleoecology, sedimentation, and stratigraphy to address what he considers to be the principal factors controlling reef ecosystems during this important period of the Mesozoic. Leinfelder summarizes the chapter by providing a simplified model for Jurassic reef development, relating to this model the factors controlling reefs. These include water depth, nature of the shelf slope, nutrients, water circulation, and so forth. He notes how closely Jurassic reefs were tied to sea-level changes, sedimentation style, and climatic regimes.

Leinfelder also explores reef bathymetry and the photosymbiosis issues, making a comparison between corals of the Jurassic and modern-day zooxanthellate species. While some Jurassic corals show sizes and growth rates comparable to those of living zooxanthellate species, other taxa appear to have grown more slowly and may not have been as well adapted to oligotrophic shallow-water reef settings as are modern corals. Some elements of the biotas, such as lithistid demosponges, have no relevant counterparts among living reefs.

As we understand more about the oceanographic and climatic nature of the icehouse world inhabited by living coral reefs, climate models have been developed. These models led us to understand more about how the hydrosphere interfaces with other parts of planet Earth, including how heat is transferred within the tropics. The Cretaceous world presented some global situations very different from today, as discussed in Chapter 9, by Claudia Johnson and Erle Kauffman. In the Late Cretaceous the Caribbean and much of the tropics were basking in a major global greenhouse climate. Reefs had been claimed by rudistid bivalves and mollusks but earlier had been inhabited by scleractinian corals. Earlier in the Cretaceous, they began constructing bafflestone mounds. Later, they invaded carbonate platforms where they created impressive reeflike features characterized by topographic relief. By the Late Cretaceous, rudist bivalves were producing framework, as well as other features associated with reefs.

Johnson and Kauffman claim that not only temperature, but salinity, oceanographic circulation, and oxygen concentrations changed during Cretaceous time. Based on considerable field experience, these authors present an impressive number of stratigraphic sections in the Caribbean. They also outline plate tectonic models for the mid- to latest-Cretaceous Caribbean. Johnson and Kauffman also explore causes accounting for the dramatic Cretaceous changeover from coral to rudistids—a perennial subject of discussions and vigorous debate. Johnson and Kauffman summarize some of the latest hypotheses to explain this remarkable event, extracting data from diversity trends and paleogeographic patterns among rudistids.

Interlocking frameworks of calcified organisms have become not only the hallmark of reefs but also an integral part of the definition. Dennis Hubbard, Randolph Burke, and Ivan Gill address this subject in Chapter 10. For students of modern and ancient reefs, the recognition and classification of those structures have relied heavily on the presence of in-place, interlocking, calcified organisms—framework. This chapter addresses framework and its relevance in our records of living, Holocene, and ancient reefs. These authors bring a unique perspective to bear on the problem. They recognize in-place or primary framework and differentiate it from secondary framework. Their data come from several decades of coring efforts on Holocene reefs, especially those of the Caribbean region. Results beg the question, "Where's the reef" and the dogma of relying on pervasive framework and *in situ* calcifying organisms to define these reefs. Bioerosion, destruction by storms, hurricanes, and a multifarious interplay of taphonomic and diagenetic factors are thought to be responsible for the structure we have come to call ancient reefs. Are those impressive biological structures we observed while snorkeling or scuba diving on a reef really representative of what we find in the rock record? Hubbard and coauthors explore this theme, while integrating data from paleoecology, taphonomy, and sedimentology.

Pamela Hallock completes the volume by addressing modern reefs in Chapter 11. This chapter addresses coral reefs as carbonate sinks and trophic resources. Providing an overview focusing on the carbonate sediments, reef organisms, nutrients, and global change, Hallock focuses on carbonate sediments, their constituents, and the classification of reefs. Aimed at the basics, Hallock's chapter serves as an excellent review of the biological and sedimentological nature of living reefs and the complex interactions between the reef organisms and the sediments. She also emphasizes salient aspects of nutrients in reefs. Included also in this chapter are Hallock's findings on the impact of humans on coral reefs and global change. This is relevant, especially with regard to the rising incidence of devastating black-band disease, ozone depletion, biologically damaging ultraviolet radiation, and other adverse factors working against reefs, making the 21st century look rather bleak for coral reefs.

In summary, these 11 chapters afford an intellectually stimulating look at what may be the oldest ecosystems on Earth, an ecosystem that has been a source of debate for over a century. Controversy not withstanding, I hope you will find this volume to be an invaluable resource for your professional and classroom needs, or simply an up-to-date source to satisfy your own curiosity about one of Earth's most fascinating ecological and geologic systems.

George D. Stanley, Jr.
Missoula, Montana

Contents

Chapter 1

Introduction to Reef Ecosystems and Their Evolution

GEORGE D. STANLEY, JR.

1. Introduction to Reefs

Living reefs are geologically and biologically a conspicuous ecosystem and one undeniably important both quantitatively and qualitatively. Reefs are restricted to tropical and subtropical settings primarily on eastern trunks of continents or western parts of oceans, and today range between 20° to 30° north and

GEORGE D. STANLEY, JR. • Department of Geology, The University of Montana, Missoula, Montana 59812.

The History and Sedimentology of Ancient Reef Systems, edited by George D. Stanley, Jr. Kluwer Academic/Plenum Publishers, New York, 2001.

south of the equator. Indeed, the very word "reef" usually invokes images of warm trade winds, swaying palm trees, tropical coral seas with coral sand islands, and the crash of the surf on a rocky edifice. Living reefs, and the rocky ramparts they produce, are sizable. Today they cover about 15% of the total amount of shallow sea floor of ocean basins with a surface area of over 600,000 km^2. Reefs require warm, well-lighted, shallow water marine conditions and can be found growing on top of volcanic islands and on the shallow shelf adjacent to continents. Ancient reefs are well known but they do not always resemble living counterparts so closely.

A peasant, living in land-locked southeast Asia, once inquired, "What is a reef?" The response was that it is a kind of living undersea mountain that grew up to reach the sea surface. What a simple and clear mental picture! The effects of this living underwater mountain in altering the physical environment, of course, can be profound. Witness the growth of the Great Barrier Reef with an incredible diversity of microbes, algae, plants, and vertebrate and invertebrate life. This ribbon of life runs roughly 1,200 km offshore the eastern edge of the Australian continent. A variety of unique subenvironments can be delineated on reefs and they include a narrow reef crest and reef flat on which waves swell and storms crash, while out in front a forereef zone descends into deeper water, and immediately landward a calm backreef zone develops with a more laterally extensive shallow lagoon. The shallowest reef crest absorbs the brunt of waves and storms, protecting the coastline. The amount of broken forereef debris, thrown up around the reef and the slumping of giant reef blocks into deep water seaward of the reef, attest to its ability to resist the great power of surge, waves, and storms. The upward growth of the calcified surfaces of the reef alters not only the entire ecological setting of a region but the physical environment as well.

Reefs, both living and ancient, are without a doubt of great importance to humans. Complex physical and biological environments on living reefs allow the growth of fish, mollusks, and other kinds of marine life that are of great economic importance. Over 100 island countries in the tropics are supported by fishing, aquaculture, and the harvesting of pearls and other products from the sea, many of which revolve around reefs. Important pharmaceutical products derived from reef organisms are being discovered and marketed to help cure disease and medical problems. Corals are even known to provide a bone substitute that can be grafted in the human body to replace lost bone tissue. The porous and broken debris characterizing ancient reefs and related carbonate rocks is host to nearly half the world's petroleum resources; in addition, fossil reefs and carbonate rocks are host to some economically important metallic deposits.

In geology reefs are significant sedimentological phenomena. Calcifying plants and animals on reefs promote the production of calcium carbonate in quantities that stagger the imagination. Chave *et al.* (1972), for example, estimated that calcium carbonate or limestone is produced at yearly rates of between 400 and 2,000 tons for each hectare of reef surface exposed on the sea floor. This carbonate has a pronounced effect on the balance of the world's oceans, producing approximately 700 billion kg of carbon each year. Carbonate rock formation represents carbon taken out of the global CO_2 cycle and locked in vast deposits. Limestone might help explain why life can exist on the earth, as Pamela Hallock has discussed (1997). Carbon dioxide is a well-known greenhouse gas, and the amount of CO_2 locked in carbonate rocks could mediate world climate. Currently the reef CO_2 budget is a subject of active research. As Hallock suggests in Chapter 11 (this volume), changes and fluctuations in the amount of limestone locked in geologic deposits through time may have directly or indirectly influenced major events in the history of life of many organisms and plants. The amount of CO_2 in our atmosphere has varied through time and its concentration may have been a major driving force in biological evolution.

A periodical entitled *Coral Reefs* is the official journal of the International Society for Reef Studies and it concerns itself with a broad spectrum of reef sedimentology and biology. Recently Hatcher (1997) addressed how nutrients flow through the coral reef ecosystem. He emphasized that reef ecosystem processes act to "link the physical environment to interacting assemblages of organisms," and he discussed how ecosystem processes affect reefs. Hatcher also discussed ways to study and apply data on ecosystems to coral reefs. Like counterparts in tropical forests, coral reefs follow similar principles of ecology and evolution and reef ecosystems foster the development of diverse plant and animal communities, and such study lends itself well to modeling. In order to understand ancient reef ecosystems and their evolution, we need to better understand the living reef.

1.1. The Living Reef Ecosystem

Living reefs are showy ecosystems with vivid colors, special adaptations, and myriad interacting plants and animals. They are places of high diversity, biological intricacy, and special adaptations. Organisms building the reef can be gregarious or colonial and often are closely packed in three dimensions. Invertebrate–plant–fish communities are adapted to high light intensity and some are autotrophs. Other reef dwellers are heterotrophs. Amid the diversity, colonial scleractinian corals are conspicuous as framework builders. These clonal invertebrates have evolved a symbiosis with one-celled algae, usually the dinoflagellate *Symbiodinium* (see Section 1.2). This constitutes an ecologically and geologically important relationship. Colorful coral polyps are surrounded by batteries of stinging tentacles and they may exude sticky fluids to help capture their food, mostly microscopic swimming plankton. Corals reproduce both sexually and asexually. After a free-swimming stage, sexually produced coral larvae settle down and attach to a hard surface, where they secrete a calcareous skeleton and grow into a sessile adult. Corals increase asexually by budding to produce a variety of morphologies ranging from flat-shaped to hemispherical or branching colonies. While not so obvious to a casual observer, corals are quite aggressive invertebrates. They are scrimmag-

FIGURE 1. Artistic depiction of a living reef community showing the diversity of interacting organisms which populate it. Two massive scleractinian boulder corals (not labeled) dominate. Cutaway view of boring organisms excavating or living inside a coral colony. (a) Parrot fish; (b) fire coral; (c) coral-eating fireworm; (d) flexible sea fan; (e) a calcified green alga, *Halimeda*; (f) moray eel; (g) Christmas-tree worm; (h) rock-boring clam; (i) encrusting coraline algae; (j) sediment-eating sea cucumber; (k) tube sponge; (l) lettuce coral; (m) grazing snail; (n) spiny sea urchin; (o) soft-bodied anemone; (p) staghorn coral; (q) butterfly fish. The guilds represented include constructors (b, p), bafflers (d, k), binders (i, l), destroyers (h, n), and dwellers (e, m). From Stanley, 1992. Reproduced with permission.

ing constantly with nearby corals and other sessile reef organisms for limited resources and space on the reef, and they can extend special tentacles or extrude long mesenterial filaments to attack and kill neighboring sessile organisms. Such competition may help explain in part the distinct zonation observed across reefs and the changes that occur in the composition and growth forms of the reef builders.

In a typical reef ecosystem (Fig. 1), hard corals coexist with a variety of vertebrates, invertebrates, plants, and algae. These include sessile organisms like sponges, soft corals, seafans, and algae, and a host of free-living invertebrates such as starfish, sea urchins, herbivorous and carnivorous snails, and crab and shrimp. Sea cucumbers are holothurian echinoderms and they mostly are deposit feeders. Because they ingest large quantities of mud and organic sediment, which are processed through their guts, sea cucumbers are among the many vagile organisms responsible for altering sediment around a reef.

When these organisms die and decay, they usually are represented only by their tiny internal calcite ossicles scattered in the sediment. This makes their detection in the fossil record difficult.

Considerable primary producers are found in reefs. They include cyanobacteria, algae, and plants. *Halimeda*, one of the many calcified algae, are major sediment producers. Within their complex ecosystems, many reef organisms, including fish, have evolved complex and subtle mutualistic interactions. They have colorful sexual displays and exhibit strong examples of mimicry and highly developed behavioral adaptations. In response to predatory starfish, crabs, coral-eating fireworms, and voracious snails on the reef, potential prey have countered by evolving hard shells, noxious toxins, or stinging tentacles.

Herbivorous and carnivorous fish are important on the reef (Fig. 1). Moray eels hide in the hollows and cavities. Parrot fish use their heavy beaks to rasp off calcareous algae and, in the process, break up considerable quantities of reef rock into sediment. Some species of butterfly fish are corallivores, plucking individual coral polyps, while other species prey on worms. The vast number of herbivorous fish exert strong influences over the reef. In their ceaseless browsing and grazing on microbial turf or on algae, fish assist corals and other sessile-encrusting organisms that otherwise would be crowded out. Although incomplete, the fossil record of herbivorous reef fish extends back over 200 million years. Reef fish must have a major impact on the ecological structure of modern reefs (Wood, 1993, 1999).

The cryptic biota are largely unseen, living mostly under, inside, or actually producing some of the framework of the reef. With reduced pressure from predation, cryptic organisms seek out a life within the reef in settings of low light to total darkness. Included are some bioeroders that bore and excavate solid rock, producing millions of metric tons/km^2 of calcareous sediment with habitat for other elements of the cryptic biota. The passageways and chambers excavated within the framework of the reef are habitat for numerous hidden communities of sessile organisms. Important bioeroders include clionid sponges, sipunculid worms, spiny sea urchins, and rock-boring clams (Fig. 1). While such destructive biological activities can undermine coral colonies, a healthy reef requires biological and physical breakdown.

Biotic diversity is a distinctive feature of all tropical ecosystems (Paulay, 1997) and reefs are no exception. Biólogists who study and inventory the taxonomic diversity of reefs view them as an ecosystem where specializations are the norm and vast numbers of species are packed into a small amount of ecospace. Only a few species dominate. Two contrasting views of the significance of the tropics have emerged: first, that they are special refuges or "museums" of obsolence. Thus, the tropics create the kinds of environments that protect species, allowing their accumulation and survival. Alternately, the second view regards the tropics as a kind of dynamic evolutionary "cradle," where new species frequently arise, evolve, and spread geographically, eventually adapting to more temperate regions. As suggested by the latter view, the tropics may be a life support system for the earth. Further, the tropics could be an important source of evolutionary innovation for life (Jablonski, 1993) and part of an evolutionary "diversity pump."

Nutrients exert important controls on coral reefs today and presumably also in the past. Contrary to popular misconceptions, the clear waters around coral reefs are not places of great nutrient availability. Reefs actually prosper best under conditions of low nutrients. When nutrients in the form of sewage, agrochemicals, and so forth are dumped near or seep through ground water into the marine environment, reefs are affected adversely. The clear blue water that characterizes reefs is largely devoid of nutrients and stands in stark contrast to temperate and polar marine ecosystems whose cloudy green waters contain an abundance of plankton and nutrients. In this respect reefs may be regarded as oases of life amid biological deserts and, with some notable exceptions such as reefs around Oman, they generally cannot tolerate excess nutrients. Numerous cases of reef demise in the fossil record have been linked to overnutrification (Hallock and Schlager, 1986). In addition to nutrients, other controlling factors for reefs include salinity, temperature, and sunlight. Most researchers would rank temperature as a prime control as it is closely linked with the precipitation of $CaCO_3$. Many large, reef-building organisms such as corals were adapted to sunlight, normal salinity, and warm temperatures. Because they can be adversely affected by salinity changes and smothered by land-derived sedimentation, which also introduces nutrients and cuts down ambient light levels, reefs flourish best offshore and away from the influence of rivers and deltas.

Researchers once believed that corals and other organisms lived mostly by plankton feeding, but subsequent field surveys on robust reef ecosystems failed to show how the full nutrient requirements of coral reefs could be met. There seemed to be a lack of primary producers. Then came the realization that vast amounts of nutrients in the form of biomass are locked intracellularly inside the tissues of invertebrates such as corals in the form of endosymbiotic algae. A prime example of this is the symbiotic relationship between corals and one-celled dinoflagellate algae called zooxanthellae. Most zooxanthellate corals calcify much faster than nonzooxanthellate counterparts and this ecological group dominates on modern reefs.

1.2. Algal Symbiosis and Scleractinian Corals

Endosymbiosis is perhaps the most pervasive of reef themes and it was evoked to explain the history of reefs by Cowen (1988) and Talent (1988). However, Wood (1999) offered a different opinion, suggesting that only during select times in reef history has this kind of symbiosis idea been relevant. We know with certainty that photoautotrophy is an ecological characteristic of crucial relevance on the living reef ecosystems and it certainly helps to explain why both modern and most ancient reefs are restricted to warm, tropical to subtropical shallow-water settings. As previously mentioned, the development of symbioses between calcifying metazoans and one-celled zooxanthellae is associated with both rapid calcification and growth to large sizes. The general shape of the corallum of some living corals and the shells and skeletons of other reef-dwelling creatures reflect an adaptation to insolation and ambient light levels. Of course, the resulting morphology often is a kind of compromise with other factors necessary for survival such as feeding, rates of sedimentation, and sediment type. For example, while some corals show a flattening of the corallum in deeper water so as to maximize sunlight, they also must remain a slightly convex shape in order to shed sediment and avoid being smothered.

Rudistid bivalves of the Late Cretaceous reached giant size and show many modifications of the shell, suggesting that photoautotrophy in conjunction with filter feeding was utilized (Kauffman and Johnson, 1988; see also Chapter 9, this volume). Today, the giant so-called Pacific "killer clam" *Tridacna*, a well-known photoautotrophic bivalve, attains the maximum size for any living bivalve. The tissues of these clams are filled with endosymbiotic algae and are exposed to light outside the shell. Ancient reef bivalves living in a manner similar to *Tridacna* are known as far back as the Middle Paleozoic and bivalves shaped like Mexican hats have been discovered from the Upper Triassic of western North America. These Triassic bivalves possessed marginal extensions of the body cavity partitioned into chambers and postulated to have housed symbiotic algae (Yancey and Stanley, 1999).

For corals, the terms *hermatypic* and *ahermatypic* are ingrained deeply in much of the geologic literature. These terms have been firmly rooted since their introduction by John Wells (1933). His original definitions were based on the presence or absence of symbiotic algae (zooxanthellae) and the ability of the corals to build or not to build reefs. By definition, hermatypic corals build reefs or mounds and possess zooxanthellae, while ahermatypes lack these attributes. Several authors (Schumacher and Zibrowius, 1985; Stanley and Cairns, 1988) discussed hermatypic versus ahermatypic and pointed out the problematic nature of these definitions.

The practical distinction between hermatypic and ahermatypic begins to break down when applied strictly to modern corals and reefs. For example, deep- and cold-water corals build large, reeflike thickets and mounds below the photic zone but do not possess symbiotic algae. Also, there are living shallow-water corals that possess symbiotic algae but never build reefs or occur in reef associations. An added problem for paleontologists is that the intracellular zooxanthellae are never preserved in fossil corals, making the distinction difficult to apply. With this in mind, Schumacher and Zibrowius (1985) proposed a more practical approach with division into separate categories. The first category is zooxanthallate versus nonzooxanthellate; the second category is constructional versus nonconstructional. Thus, if a coral is known to possess symbiotic algae, it is called *zooxanthellate*. In a separate category, if a coral takes part in reef construction (i.e., is demonstrated to build a reef), then it is termed *constructional*. While a fossil coral can be considered constructional, the question of whether it possessed zooxanthellae usually must remain hypothetical.

While most living reef corals acquire food as active zooplankton feeders (micropredators), they only partially fulfill their total nutrient requirements in this way. Endosymbiotic algae (zooxanthellae) residing in the endothermal tissues of corals provide a great boost to the energy requirements of their hosts. In corals the density of zooxanthellae reaches over one million algal cells per cubic centimeter of animal tissue (Muller-Parker and D'Elia, 1997) and their biomass may equal or exceed that of the animal tissue. So are they plant or animal?

In a mutualistic relationship, these symbionts fix carbon in the marine environment photosynthetically and translocate the resulting photosynthate to the coral host (see Chapter 11, this volume). This subtle ecological relationship is said to account for the high productivity found on coral reefs. The zooxanthellae take up the CO_2 and nitrogenous wastes of the coral. Since the coral animal lacks gills and excretory organs, the symbiotic zooxanthellae can be envisioned as the equivalent of "lungs" and "kidneys" in more complex animals. The zooxanthellae also are thought to be responsible for the rapid calcification and high growth rates observed among most reef-building corals. For this reason zooxanthellate corals dominate in the shallow, well-lighted portions of the reef, while nonzooxanthellate corals are relegated to deeper-water settings or to low-light cryptic habitats within the reef where competition with faster growing zooxanthellates is minimized.

Rates of calcification among zooxanthellate scleractinian corals and perhaps many other calcifying organisms as well are believed related to the amount and intensity of sunlight. The zooxanthellae are ensured a well-illuminated safe niche within the host's coral tissues and they receive from the host a continuous supply of CO_2 and nutrients. Ecological factors connected with coral–algal symbiosis and nutrient availability have been cited to explain the fact that reef ecosystems are restricted to tropical and subtropical shallow biotopes where ambient sunlight is strong and nutrients are low.

The zooxanthellate coral symbiosis may help explain the geographic distribution of present-day modern reefs that occurs today on the western sides of ocean basins. When we examine patterns of water circulation in ocean basins, we see circulation gyres of cool waters with high nutrients and upwelling along the western sides of continents. Presumably this also occurred in the geologic past.

When did the unique ecological relationship of the coral–zooxanthellate coevolution begin? Scleractinian corals go back to Middle Triassic time, but Stanley (1981) postulated that the coral–zooxanthellae symbiosis did not evolve until some 20 million years after the first appearance of the group. Stanley and Swart (1995) tested this idea by conducting stable isotopic analysis of aragonitic skeletons of Triassic reef corals. They deciphered strong signals in oxygen and carbon isotopes from some Late Triassic scleractinians and the latest Triassic coincided with a surge in the dominance of these corals in reefs of that time. Gautret *et al.* (1997) extracted soluble organic compounds trapped within the skeletal crystallites of modern corals, and differences in composition allowed distinction between zooxanthellate and nonzooxanthellate species. This technique has been applied successfully to corals of the Late Triassic (J. P. Cuif, personal communication, Jan. 1999) and provides support for isotopic conclusions.

Was endosymbiosis a dominant factor in the success of many if not most ancient reef ecosystems? Paleozoic corals and stony sponges grew to large sizes on reefs of the Silurian and Devonian and show apparent light adaptations. Accordingly, there have been debates about whether Paleozoic corals were zooanthellate. If density bands are reliable indicators of growth rates, on average, many Paleozoic corals grew as fast as modern scleractinian reef corals.

2. What Is a Reef?

2.1. The Reef Concept

The historical roots of the term for reef are difficult to trace. The root may be either "rif," an old German or Norse term for a ridge of rock (an obstruction) that lay at or near the surface of the water, or "Er Rif," an Arabic term for "hills" found in shallow waters between Tangiers and Melilla and coined by traders who traveled the north coast of Africa toward the Gibralter Straits. After the earliest understanding of reefs by ancient mariners as rocky substrates that could wreck a ship's hull, more scientific reef concepts and definitions have taken shape. A student may wonder about the debates and discussions concerning what constitutes a reef. After all, reefs and reef ecosystems certainly are readily observable living manifestations that may be characterized, quantified, and modeled. It ought to be easy to compare them to counterparts in the fossil record.

Alas, one need only read the papers, treatises, and book volumes discussing what is or is not a reef to realize the degree of difficulty in reaching a consensus among biologists and geologists. The reef, both modern and ancient, has been investigated, dissected, defined, clarified, and redefined, only to remain rooted in controversy. The reader is referred to Heckel (1974) who has provided one of the best overview papers.

Lowenstam (1950) presented the concept of framework at a very early stage in the study of ancient reefs. He applied this term to the rigid fabrics described in the walls and floors of quarries around the Great Lakes where large, moundlike Silurian reefs are exposed. Later, Dunham (1970) proposed a distinction between stratigraphic and ecological reefs. He emphasized the concept of stratigraphic reef as employed by exploration geologists working with thickened masses of limestone or dolomite. These are important as reservoir rocks for oil and gas. Dunham also included in his definition of reefs the concepts that paleontologists and marine ecologists use. These include "biological" and "ecological " characteristics. The term "stratigraphic" reef is not essentially ecological, but rather descriptive in nature and pertains to a thickened mass of carbonate that differs appreciably from the surrounding rocks. A large number of such deposits yield few or no fossils owing to recrystallization, dolomitization, or other diagenetic processes that have affected the rocks. Another general descriptive term is simply carbonate build-up, which is used similarly without reference to internal composition.

The "ecological" reef concept, on the other hand, is strongly rooted in biological characteristics. It stresses biological diversity and complexity of ecological/paleoecological structure within shallow-water settings. What we observe in the ecological reef is really the complex interplay of physical, chemical, and biological factors. The ecological concept emphasizes the importance of an organically produced framework that is said to impart a degree of rigidity and resistance to waves and storms. For more on these views, the reader is referred to Hubbard and others (see Chapter 10, this volume).

Organic framework and wave resistance are generally thought to be important components of any reef concept. As discussed above, ideas on framework were presented by Lowenstam (1950) during the early formulative stages of reef studies to characterize certain rigid fabrics found in Silurian reefs. Examples of framework are commonly found in quarries, hillsides, and roadcuts. Large, closely packed, and rapidly upward-growing calcified organisms such as corals present spectacular structures attributed to framework. However, Heckel (1974) proposed a different idea. In a largely descriptive classification, he maintained, framework was just one feature in the study and characterization of fossil reefs. Heckel stressed constituent composition (e.g., skeletal grains and lime mud), stratigraphic shape of the structure, and organic composition.

James (1983), like Heckel, took a broadbrush approach in defining reefs. He included the "reef mound" category for a variety of carbonate buildups of calcareous algae, coral, and skeletal sediment and for mounds of lime mud. His view of reefs also included deep-water carbonate buildups and coral thickets (Teichert, 1958; Stanley and Cairns, 1988), which contrast ecologically with shallow-water coral reefs.

The guild concept provides a functional classification for reef ecosystems in the ecological sense (Fagerstrom, 1985, 1987, 1994, 1997). Five reef guilds were proposed by Fagerstrom: constructors, bafflers, binders, destroyers, and dwellers. Guild membership is not mutually exclusive and in fact cases may be overlapping. Although this approach has received criticism (Precht, 1994), it is preferred by many workers because of the way it simplifies the analysis of a reef, allowing organisms to be categorized by their functional–ecological roles in the reef ecosystem. Because it stresses paleoecological function rather than taxonomic classification, some workers feel that functional guild categories greatly simplify analysis of reef ecosystems and facilitate comparisons between reef ecosystems of different ages and compositions. The contructor guild produces the framework. The binder guild contains organisms that encrust and overgrow the constructional framework and help consolidate the reef. Membership in a binder guild could be assigned to a variety of different organisms such as corals, encrusting red alga, bryozoans, or a calcified sponge and the same is true for bafflers, destroyers, and dwellers.

In many fossil reefs, evidence for rigidity and wave resistance is absent and must be postulated from large amounts of talus or broken reef debris that characterizes many fossils reefs. Other reeflike carbonate masses in the fossil record are dolomitized and/or recrystallized and fail to yield any fossils or talus. The Carboniferous Period was an unusual interval in reef history when constructional framework was either absent or uncommon. Carboniferous time was characterized by an abundance of reeflike Waulsortian mounds dominated by a preponderance of mud and possibly cyanobacteria. Other geologic intervals were marked by mounds built by cyanobacteria. While these structures appear very much reeflike, some workers prefer to exclude them from the reef category.

It is known that a kind of coherent, massive, and wave-resistant feature may be imparted through diagenesis or by cementation. A high volume of reef cements characterizes some ancient reefs. Pore space, cavities, and interstices in many carbonates become filled by sediments and cements early in the burial history. This process leads to the consolidation and lithification of many carbonates. Long after burial, cements even can be remobilized. Both dissolution and cementation of reef carbonates have been influenced by the influx of both fresh and marine waters. Interestingly, the formation of many cavity coatings and cements in ancient reefs appears to have been mediated by organisms, especially during microbial activity, a process active deep inside the calcified interstices. Grotzinger and Knoll (1995) believed that voluminous quantities of carbonate cements and microbial crusts formed early in the history of some Precambrian and Paleozoic carbonates. Using an example like the famous Permian reef of west Texas (see Chapter 6, this volume), these authors could find analogies with Precambrian carbonates, recognizing cements and microbial crusts within this Permian reef. Thus some reefs of the Phanerozoic, which lack biological framework, may have close analogies with carbonate deposits of late Precambrian (Proterozoic) age. Biogenic crusts and microbial communities appear to have influenced the formation of reef ecosystems throughout the Phanerozoic (Webb, 1996; Chapter 5, this volume).

When entombed in strata, many reefs and reeflike deposits show pronounced differentiation from the surrounding sedimentary deposits in terms of relief, bedding, and porosity. Cummings and Schrock (1928) and Cummings (1932) introduced the terms, bioherm and biostrome, to help differentiate between more massive, upward doming reef structures and more lenticular, distinctly bedded, reeflike features. In this sense a bioherm is a reef; but because a biostrome lacks relief and framework, many workers do not regard it as a reef. Heckel (1974) used the term "bioherm," with a slightly different connotation, to mean mudmounds formed in the absence of calcifying organisms. Kershaw (1994) proposed a "biostrome" classification for carbonate buildups based on a continuum of changes in composition, thickness, geometry, and internal bedding. Unfortunately this classification has not gained wide acceptance. More quantitative approaches have been developed. For example, Gerhard (1991) proposed three major reef classes based on architectural end points: (1) framework reefs: reefs with topographic relief, characterized by framework organisms, low amounts of internal sediment, and large amounts of cement; (2) biodetrital reefs: reefs with less topographic relief characterized by little *in situ* framework but with extensive bioerosion of living and dead reef skeletons to produce much detritus; and (3) hydrodynamic reefs: reefs characterized by the accumulation of organic reef debris by wave and storm action. The last reef category could provide suitable substrate for settlement and later growth of framework. These categories are not mutually exclusive but rather are considered merely end points in a continuum of change. The subject of framework and the classification of growth fabrics with respect to scleractinian corals was presented by Insalaco (1998). He suggested that the term "growth fabric" be reserved as a descriptive term for *in situ* corals or other organisms within the matrix and that it should be used as an alternative to "framework." Insalaco also distinguished between growth fabrics projected high above the substrate (suprastratal) and growth fabrics that remained mostly buried in the sediments during the life of the organism (constratal). In addition, he proposed a practical descriptive classification of reef fabrics based mostly on the general shapes of *in situ* skeletons. This reef classification parallels that of Embry and Klovan (1971).

Features known as mud mounds are unique to parts of the fossil record but are little known in present-day oceans. They are rather enigmatic deposits, sometimes associated with deeper water. They have distinct topographic relief, are dominated by fine-grained, carbonate mud (micrite), and lack evidence of framework or biotic construction. Many are microbial, and examples are especially common in the Devonian and Carboniferous periods. Based principally on textural features, James and Bourque (1992) divided mud mounds into microbial, skeletal, and mud-dominated types. The reader is referred to Copper (Chapter 3, this volume) for examples of these deposits.

2.2. Problems with the Reef Concept

As we have seen, there are problems with the reef concept that appear to revolve around operational aspects rather than practical applications. We need a common set of rules and definitions to recognize, classify, and understand all reefs. Can geologists, paleontologists, and biologists find common ground within the broad phenomena of reefs? Reefs go through profound changes in their transition from the biosphere into the lithosphere. Viewed within the spectrum of carbonate rock types, sedimentology, and the effects of taphonomy, when is a reef still recognizable as a reef and at what point might it lose this distinction? Some of the central problems revolve around the differing perspectives of biologists, paleobiologists, and stratigraphic geologists. At one end point lie the organisms and the importance of their biological interactions. Sedimentologic composition and stratigraphic appearance of reefs rest at the other.

It is well known that many reefs in the fossil record are highly altered from their original states. Reefs, regardless of age, must have some unifying biological and physical characteristics: (1) shape, growth fabric, and diversity of the calcified organisms; (2) nature and type of the sediments; and (3) nature of the internal cements. Many fossil reef organisms, including framework, are preserved in life positions and seem easy to interpret by homology with modern reefs. Other fossil examples seem to utterly defy this approach. In large part, fossil reef workers may be fettered unnecessarily by an inflexible adherence to the living reef as the prototype with its obviously impressive constructional framework and wave resistance. Many ancient "reefs" simply do not show much evidence of framework. In addition, there appear times in reef history when the constructor guild was unrecognized or absent, so "reefs" may have assumed different characteristics ecologically and sedimentologically. Even in the presence of an original framework, the processes of taphonomy and multifarious changes (physical, chemical, and biological) exert profound influences and may yield a product with little correspondence to the original features.

Reef workers should appreciate the taphonomy of reefs and the nature of the transition from living ecosystems into geologic deposits. Examples are found in a classification of biostromes by Kershaw (1994) and in descriptions of skeletal composition and hydrodynamics of deposits by Heckel (1974). Heckel's classification includes categories such as "organically (?) bound, skeletal-debris reef" and "spar-cemented debris reef." Many workers question how well organic remains or ecological structures of a once-living reef might survive millions of years of burial and how closely ancient reefs might resemble the living structure they once were.

Some studies (Greenstein and Moffat, 1996; Greenstein and Pandolfi, 1997; Pandolfi and Greenstein, 1997) found Pleistocene examples to be reasonably good proxies for their once living counterparts, especially in diversity, biotic composition, and ecology. However, Hubbard's (1997) findings on Holocene framework are different. In this study, extensive quantitative data compiled from drilling and coring Holocene coral reefs at hundreds of sites around the eastern Caribbean revealed an architecture predominantly of debris or something more akin to sediment piles. Clearly the sediment was reef derived, with an overwhelming contribution from corals, but there were considerably fewer in-place and interlocking corals than expected. This led Hubbard (1997) to suggest that "the vast majority of ancient reef deposits are comprised not of in-place, interlocking framework, but rather are loose assemblages with reef-building organisms usually 'floating' in a matrix of reef-derived debris" (p. 43). Thus studies of reef taphonomy are relevant and important to understand and classify ancient reefs. The reader is referred to Hubbard, Burke, and Gill (Chapter 10, this volume) for further discussion of the problem.

Although growth of biological framework on modern shallow-water reefs is rapid, we must realize that it is counterbalanced by highly destructive (recycling) processes of physical, chemical, and biological breakdown. The biodetrital and hydromechanical reef end points proposed by Gerhard (1991) may be relevant in interpreting ancient counterparts. Waves, hurricanes, typhoons, and major storms degrade reefs. Hurricanes not only destroy reef structures but also exert long-term changes on the ecology of the reef (Woodley *et al.*, 1981).

As a reef grows upward, it is constantly degraded by waves, storms, and bioeroders. The bioeroders are important agents that convert coral colonies and other skeletons into calcareous detritus. This "guild" includes a myriad of organisms: vertebrates such as fish, surface invertebrates such as echinoids and sessile infaunal, rock-boring clionid sponges, and lithophagous clams. Internal galleries range from large excavations down to the microscale of fungal and bacterial borings. Even small-scale destroyers can undermine and topple fairly massive coral colonies after sufficient time. Destroyer or bioeroding organisms live on the surface, inside, and around the reef. Through boring, rasping, scraping, and chemical dissolution, destructive organisms remove vast quantities of the internal and external calcified framework of reefs (Glynn, 1997). In low-nutrient, clear-water environments of the tropical reef, the breakdown is intensified when influxes of nutrients kill corals and encourage algal growth.

Such a system is always in a dynamic state of flux with a delicate balance between reef growth and reef destruction. Reefs are greatly influenced by sea-level changes and climate. A reef must create substrate faster than it is being destroyed in order to deposit a recognizable structure. According to Hubbard (1997),

> What we see on the surface of the reef today is only one possible snapshot. The real "reef" of geologists is the temporal integration of all the snapshots over time and the underlying control is the series of processes that take place between the snapshots.

Geologic studies confirm the magnitude and multifarious nature of the processes. Figure 2 summarizes the development of the reef concept and some of the types of reefs that have been discussed.

3. Ancient Reef Ecosystems

3.1. How Do Reef Ecosystems Evolve and Are There Any Common Patterns?

From the perspective of geology and the history of life, one cannot help but be impressed by the profound and global nature of the changes that have affected the reef ecosystem during its more than one billion-year tenure on Earth. As discussed below, these changes were closely tied to a number of mass extinctions. Mass extinctions punctuate Earth history and serve as high-level mechanisms of change. The disruption of life sometimes results in complete collapse of a long-lived reef ecosystem and consequently opens up possibility for wholesale restructuring of new ecosystems.

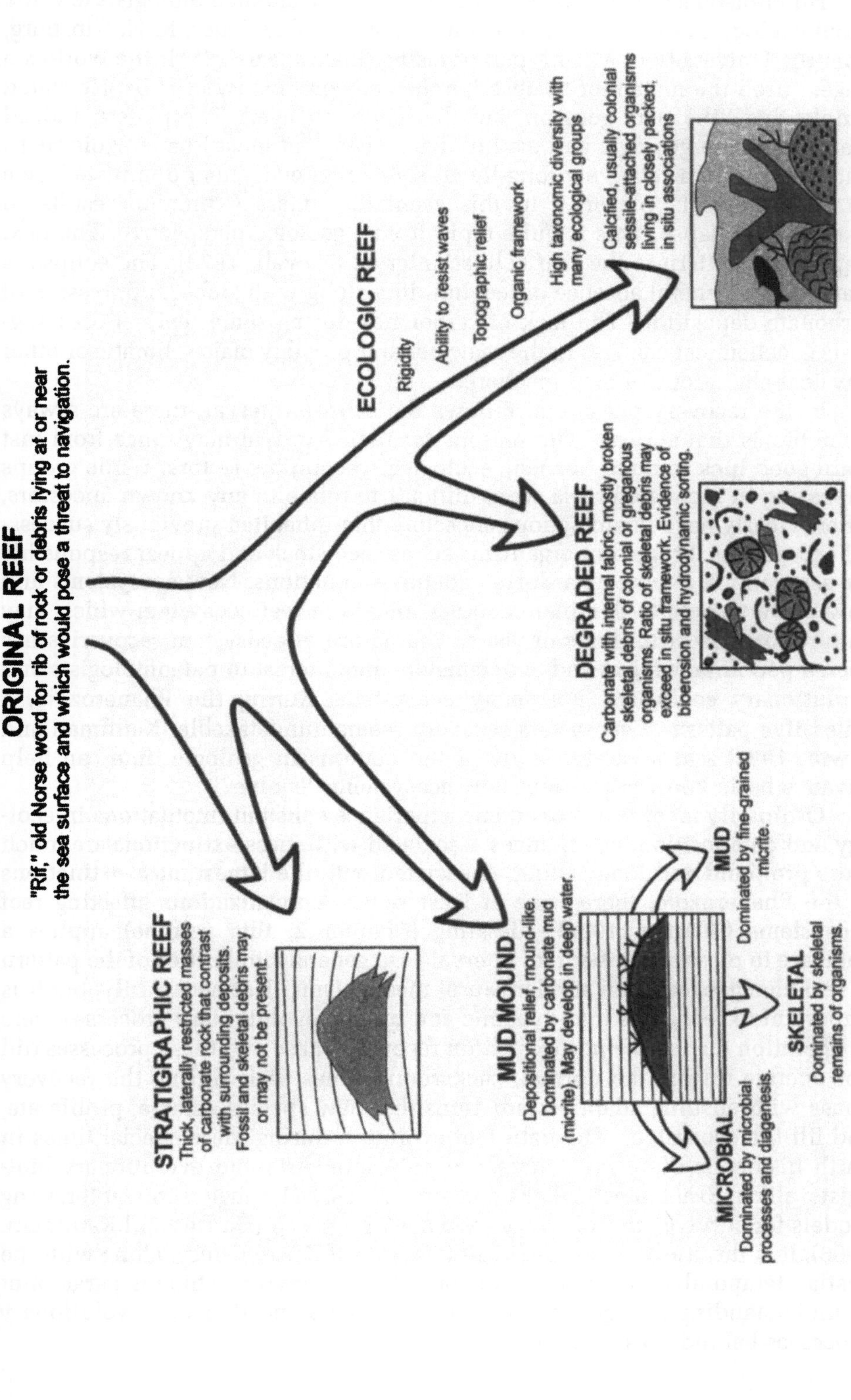

FIGURE 2. Development of reef concept from the original meaning to some of the commonly used examples today.

Hutchinson's (1978) classic book on ecology stimulated biologists to think about ecology as the theater and evolution as the play. Later, R. N. Ginsburg, from the University of Miami, paraphrasing Shakespeare's "All the world's a stage," used the metaphor to describe the geologic history of reefs. He stated simply that "the play goes on, but the actors change." The pattern indeed seems to fit the geologic and evolutionary history of most Phanerozoic reefs. They start with a relatively long-lived, stable reef ecosystem dominated by a set of ecological "players." In this scenario a mass extinction results in ecosystem collapse. This is quite rapid from a geologic perspective. The next step in the pattern is the reef eclipse interval (Newell, 1972). The eclipse is marked by a general absence of reef building along with global suppression of carbonate deposition. The lack of carbonates during some reef eclipses suggests geochemical changes in the seawater and possibly major climatic or other physical alterations of the biosphere.

In the recovery phase that follows the eclipse interval, there are always some biotas that survive. The reasons for their survival may range from just plain good luck to more intrinsic ecological/geographic factors. While groups appearing in recovered reefs seem difficult to relate to any known ancestors, others clearly have unambiguous ancestors that inhabited previously successful ecosystems. Progenitor organisms act as seed stock and appear responsible for generations of new taxa during adaptive radiations. New ecosystems and new players come to dominate another long-term reef ecosystem which may endure for tens of millions of years. The nature of ecosystem recoveries has been a perennial subject and is of considerable interest to paleontologists and evolutionary ecologists. For many ecosystems during the Phanerozoic, a reiterative pattern of ecosystem recovery seems unmistakable (Kauffman and Erwin, 1995) and a careful study of the patterns in geologic time can help advance basic knowledge about how ecosystems "evolve."

Ordinarily most reef ecosystems experience constant fluctuations in ecology and composition, but changes associated with mass extinctions are much more profound and long lasting. Associated with the largest mass extinctions of the Phanerozoic, there were at least seven reorganizations affecting reef ecosystems (Stanley, 1992). Kiessling (Chapter 2, this volume) applies a database to recognize other reef intervals. Do general similarities of the pattern of reef changeover mean similar causal mechanisms? Not necessarily, but it is important to recognize that dynamic and unique evolutionary processes were in operation during the postextinction recovery phase and these processes did not operate during the normal, background times. It is during the recovery phase with ensuing adaptive radiations that new species emerge, proliferate, and fill the ecosystem. The nature of evolution during these special times in Earth history is of special interest to paleontologists and evolutionary biologists alike. Exact mechanisms remain unclear, although some interesting models for survival and recovery have been proposed (Harries and Kauffman, 1996). It is the study of these critical intervals of Earth history, along with the spatial, temporal, and ecological history of the survivors, which is paramount to understanding and quantifying the remarkable and dynamic evolutionary processes behind these reorganizations.

The "eclipse interval," when flourishing ecosystems seem virtually to vanish from the face of the earth, is particularly interesting. If we continue to follow the Shakespearean metaphor in our comparison of the evolution of reef ecosystems with a long-running theatrical production, then the play is staged in different acts, usually with "intermission time." However, with respect to the reef eclipses, often lasting from a few to tens of millions of years, there certainly were lengthy intermissions! When the curtains rise, we find new actors have been cast. Some of the newly emergent biotas are morphologically and ecologically similar enough to those of preceding ecosystems, especially their functional adaptations, to infer that they evolved from some previous ancestors, ancestors that managed to survive. Yet in many postextinction intervals, researchers are at a great loss to discover any record of survivors. Where did such biotas reside during the eclipse intervals and why does such a crucial part of their evolutionary history seem to be missing from the fossil record?

One of the prime examples of this kind of pattern is the reemergence of reef ecosystems after the Permo-Triassic mass extinction, which included the lengthy Early Triassic eclipse (Stanley, 1988, 1992; Flügel, 1994). Some of the organisms appearing in the first reefs of the Middle Triassic seem to be derived from Permian ancestors, yet there is no record of any intermediate species that lived during the postextinction Early Triassic eclipse, which lasted some 20 million years (Flügel and Senowbari-Daryan, Chapter 7, this volume). Are such biotas really missing or are their habitats undiscovered? One possibility is that Permian organisms survived in small and isolated refugia. Were these refugia present in the deep sea or were they located on isolated islands as proposed by Stanley (1988, 1996)? The origins of scleractinian corals are a current subject of discussions (Stanley and Fautin, 2001). Based on molecular data, Romano and Palumbi (1996) suggested that ancestors of scleractinians could extend back 300 million years. Ezaki (1997) proposed that a Permian coral, *Numidiaphyllum*, was actually a scleractinian ancestor but considerable controversy surrounds this interpretation. Another possibility is that the ancestors of scleractinian corals were soft-bodied anemone-like forms that did not build any kind of calcareous skeletons until later in Triassic time. The records of such "naked corals" would be geologically "censored."

A number of overview papers published on ancient reefs and reef ecosystems (Newell, 1971, 1972; Copper, 1974; Laporte, 1974; Fagerstrom, 1987; Stanley and Fagerstrom, 1988; Stanley, 1992) have emphasized the importance of mass extinctions in reshaping ancient reef ecosystems. After decades of study, a reasonably good fossil record has been integrated with the history of reefs. Next we will briefly examine this history to search for common threads in ecosystem evolution.

3.2. The First Reefs

The so-called "stromatolite" buildups of the late Precambrian were the earliest reef ecosystem. These upward-growing calcified masses often coalesced to reach as much as 50 m in thickness. Such structures, produced by microbial communities, existed long before the advent of metazoans. Some were several kilometers long and produced pinnacles and barrierlike features. They occurred in shallow-water platforms bordering ancient continents during late Archean and Proterozoic time and were the earliest calcifying organisms. Cyanobacterial sheaths trapped sediment by a slimelike substance and these microbes responded by growing toward sunlight, producing finely laminated, upward-doming sedimentary features. These features have been called the first reefs. Such microbial ecosystems became particularly widespread in late Proterozoic time, especially near the Proterozoic–Phanerozoic boundary, at a time when calcified metazoans were just evolving. Precambrian ecosystems of this type are discussed by Copper (Chapter 3, this volume). Webb (Chapter 5, this volume) offers insight into the production of "biologically mediated" carbonate during this time.

Many workers attribute the great decrease in these calcifying microbes to the evolution of grazing and browsing hard-shelled invertebrates near the dawn of the Cambrian period or perhaps some soft-bodied grazers in the Neoproterozoic. However, the real explanation may not be so simple. Cyanobacteria in the marine environment create calcified structures by forming minute carbonate crystals. Workers such as Knoll *et al.* (1993) postulated that important chemical changes in ancient seawater during the late Precambrian altered saturation levels, which inhibited formation of the carbonate crystals necessary for calcimicrobes to calcify. The reduction in carbonate crystals was brought about by the evolution of diverse and abundant calcite-secreting metazoans at or near the Cambrian–Proterozoic transition. At that time, a global mass extinction took place and this extinction was accompanied by global cooling and glaciation. Regardless of the theory one chooses to accept, this time signaled major changes in the marine environment and ecosystems.

3.3. The First Metazoan Reefs

The first metazoan reefs of the Paleozoic took shape near the start of the Cambrian period as part of the evolutionary "big bang," an amazing adaptive radiation of metazoan life that began filling the oceans and was not completed until Ordovician time. A major event for reef ecosystems was the evolution of the first calcified organisms (i.e., those with the first hard shells or skeletons). Although the first metazoan reefs did not appear until the early Cambrian, the ancestors of calcifying organisms date from the latest Precambrian and earliest Cambrian time. The first reef ecosystems of the Phanerozoic were composed of archeocyathids, a group of calcified sponges that made their debut in the Early Cambrian (Zhuravlev and Wood, 1995; Riding and Zhuravlev, 1995). Archeocyaths joined an ecosystem already dominated by calcifying microbes, the cyanobacteria (Zhuravlev and Wood, 1995; Riding and Zhuravlev, 1995) and a variety of calcified reef dwellers, including trilobites and the first corals, called coralomorphs. Was this assemblage comparable to modern reef ecosys-

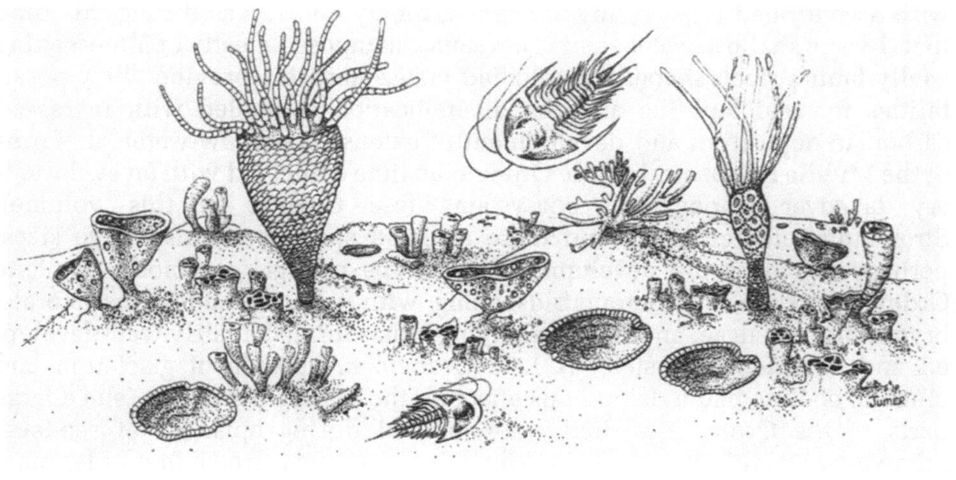

FIGURE 3. An archeocyathid ecosystem from the Early Cambrian. These include vase-shaped, upright, and encrusting archeocyathids that filled many guilds. Armor-plated and short-stalked echinoderms as well as trilobites are depicted. From Stanley (1992). Reproduced with permission.

tems? Many aspects of their composition and ecological structure differed from modern counterparts (Fig. 3). This ecosystem is considered the first "metazoan" reef, but it probably appeared different from modern reef ecosystems principally because it lacked large-scale constructional organisms and was dominated by Precambrian holdovers, the calcifying microbes (Fig. 3). It also contained many marine organisms and groups of organism now extinct. Most archeocyaths were relatively small in size and probably lived without benefit of zooxanthellae symbiosis (Wood, 1993). The reader is referred to Zhuravlev (Chapter 4, this volume) for details of these earliest metazoan reefs.

The duration of the first archeocyathid–microbial ecosystem was short. Near the close of the Early Cambrian, it collapsed in a mass extinction or more likely through a series of extinctions. Anoxia has been implicated as a possible cause along with global cooling, rise in oxygen levels, nutrient increase, and drops in sea level. Although lithistid sponges appeared in deeper-water settings of the Middle Cambrian, most of the remaining Cambrian plus Early Ordovician time interval marked a lengthy reef eclipse. During this eclipse, few ecosystems resembling reefs are known, but calcifying microbes, sponges, and some problematic organisms seem to have survived. This interval stands as the longest-lasting eclipse in the history of reef ecosystems.

3.4. Mid-Paleozoic Reef Expansion and Collapse

The evolution and importance of stromatoporoids, bryozoans, tabulate and rugose corals, crinoids, and calcified algae in the Ordovican coincided with a continued burst of invertebrate diversity and it was during this time that diverse shallow-water shelf ecosystems, composed mostly of filter-feeding shelly faunas, took shape. For calcified metazoans, this provided new possibilities for reef building and the diversification coincided with increased carbonate deposition and development of extensive shallow-water platforms of the Middle Paleozoic. Middle Ordovician time coincided with an evolutionary takeover among reef ecosystems (see Chapter 3, this volume). Stromatoporoids emerged, and these calcified sponges attained large sizes, perhaps signaling the development of a type of algal symbiosis. By Late Ordovician time, stromatoporoids along with tabulate and rugose corals, bryozoans, red algae, and other calcifying biotas formed a fairly well-developed and complex ecosystem. A Late Ordovician episode of glaciation and climatic cooling had a deleterious effect on the marine ecosystem, since large parts of the Ordovician shelf were drained during episodes of sea-level "drawdown." Although the end-Ordovician was considered a first-order mass extinction, surprisingly it had only secondary effects on the reef ecosystem and was not characterized by as severe a disruption or collapse as subsequent reef ecosystems.

The emergence of the first major reef ecosystems of the Silurian signaled the longest-standing, most continuous episode in the history of reefs, lasting some 75 million years. The coral–stromatoporoid–red algal reef ecosystems of this interval included some of the most spectacular reefs and huge shelf areas developed. Bioherms, biostromes, and major barrier reefs are known (Copper, 1996; Riding, 1981). Although we envision this ecosystem on the broadbrush scale of whole geologic periods, Brunton *et al.* (1997) distinguished at least eight global phases or episodes of reef building in the Silurian at positions on the inner and outer shelf. Many but not all Silurian reef ecosystems were dominated by corals and stromatoporoids. Bryozoan, cyanobacterial, and siliceous sponge bioherms also are known in the Silurian. These included well-studied examples preserved in the Great Lakes region, England, the island of Gotland, and in Siberia. Brunton concluded that climate was the major control on Silurian reefs and that climate functioned in conjunction with sea-level change. Also important to reefs were changes in nutrients and salinity-driven bottom water that periodically invaded shallow shelves.

The domination of stromatoporoids continued into the Devonian with large-scale constructional guilds composed of tabulate corals, rugose corals, red algae, and stromatoporoids (Fig. 4). These ecosystems yielded impressive structures such as barrier reefs, forming what may be the largest reefs ever produced. Complex paleoecological associations and vertical zonation characterized these reefs (Fig. 4). These Devonian reefs later became important reservoirs for oil and gas. Among the best studied is the "Great Barrier Reef" of the Canning Basin, western Australia (Playford, 1980), which is estimated to have measured over 300 km long and 50 km wide when living.

The stromatoporoid–red algae–coral ecosystem was maintained for a lengthy interval of geologic time. An ecosystem collapse near the end of the Devonian period has been termed the Frasnian–Famennian crises (McGhee,

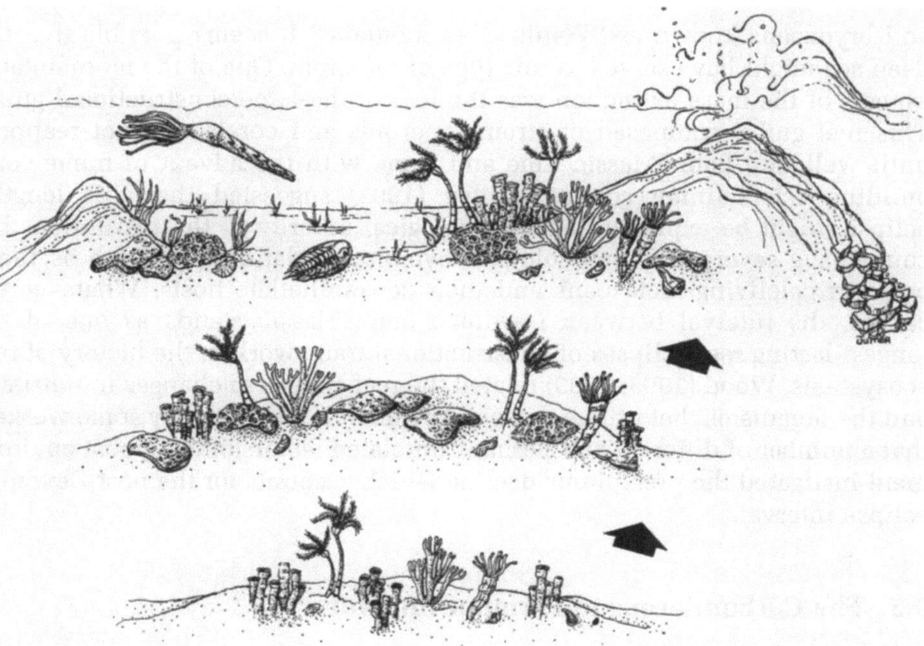

FIGURE 4. Paleoecological changes in a Devonian reef ecosystem of corals, stromatoporoids, red algae, crinoids, and a straight, nautiloid cephalopod. Lower sketch starts with a pioneering fauna of rugose and tabulate corals and crinoids, which is succeeded by an intermediate paleocommunity of more diverse organisms (center) that forms more of a topographic mound, and finally to a climax-type paleocommunity, characterized by large stromatoporoids at the reef crest on which waves break, and a lagoon and backreef paleocommunity. Arrows indicate successive stages of development. Based on Copper (1974). Reproduced with permission from Stanley (1992).

1996). It signaled the end of this long-lived mid-Paleozoic reef ecosystem. As discussed by Copper (1994b, 1996, Chapter 3, this volume) the demise of large, calcified reef-building organisms including stromatoporoids and corals has been associated with sea-level drop, global cooling of Gondwanaland, and anoxia, but extraterrestrial impacts and iridium anomalies also have been discussed. Just before the end of the Devonian, metazoan reefs vanished. The end of the Devonian thus was characterized by a resurgence of calcifying microbes (cyanobacteria) that proliferated in shallow-water carbonate settings. As discussed earlier, some workers consider these microbes "disaster taxa." However, some sponges (including stromatoporoids) also survived in isolated refugia and reappeared in Paleozoic reefs along with bryozoans and some rugosan and tabulate corals. One of the great mysteries of paleontology is the question of just where geographically these organisms were when they weathered the post-Devonian crises. The postextinction eclipse interval lasted some 27 million years from the latest Devonian into Carboniferous time. The Carboniferous interval contained some deep-water mud mounds with corals and bryozoans known as "Waulsortian mounds." It seems possible that the deep sea might have served as one type of refugium. One of the more notable aspects of the mass extinction was the loss of the large constructional guild. This reef guild, composed of stromatoporoids and corals, did not reappear until well into Late Triassic time and came with the advent of framework-building scleractinian corals. Stanley (1992) suggested that this lengthy eclipse might be explained by the ecological severity of the disruption that could have severed subtle ecological symbiotic relationships such as those between calcifying metazoans and their zooxanthellate hosts. Whatever the causes, the interval between Devonian and Triassic stands as one of the longest-lasting reef eclipses of constructional framework in the history of reef ecosystems. Wood (1993, 1999) related the reef eclipse to changes in nutrients and the organisms that utilize those nutrients. It is suggested by some workers that a number of different and perhaps unrelated factors of the global environment instigated the precipitous decline which accounts for the post-Devonian eclipse interval.

3.5. The Carboniferous to Permian Interval

Much of the Carboniferous was a time of climatic changes and cooling events. Although there were some coral patch reefs, the Carboniferous was generally a noncoral and nonreef interval characterized by abundant level-bottom ecosystems of diverse calcified microbes, algae, calcifying sponges (including chaetetid sponges), fenestrate bryozoans, brachiopods, crinoids, and abundant problematical organisms, including one called *Tubiphytes*. Enigmatic (probably deeper water) mud mounds and shallow-water, mound-like skeletal accumulations have been described as reef mounds or patch reefs (West, 1988). A notable exception is found in thick Early Carboniferous carbonate rocks of the Akiyoshi Limestone in Japan (Sugiyama and Nagai, 1994). Stromatolites, chaetetid sponges, rugose corals, and bryozoans dominated in what some workers regard as "framework" reefs. Similar associations also are found in Derbyshire, England (Fagerstrom, 1987).

By Permian time, the continents had assembled into the one-world continent of Pangea (Fig. 5), and it is postulated that this plate tectonic consolidation had major effects on climate and ocean circulation (Parrish, 1993) and in turn affected the robustness and distribution of reefs and carbonate buildups. Recessed within the continent of Pangea, the great tropical seaway called the Tethys developed. Like the Indo-Pacific today, the Tethyan seaway was the center of Mesozoic reef building and it continued to be the focus of high diversity during and after the tectonic breakup of Pangea.

New reef ecosystems took shape during the Early to Late Permian (Flügel and Stanley, 1984; Flügel, 1994) in the Tethys and tropical to subtropical regions around Pangea. Widely distributed reef occurrences have been investigated in Russia, China, Greece, the Arabian Peninsula, and west Texas. At

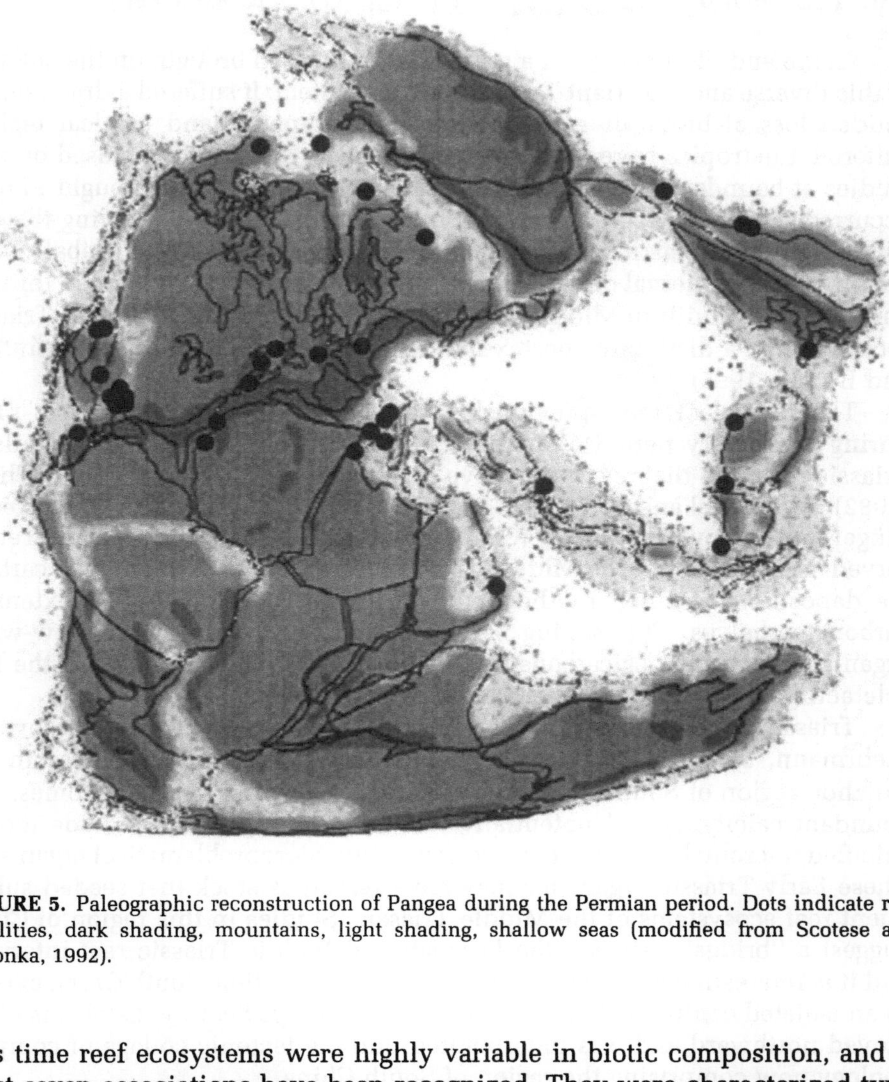

FIGURE 5. Paleographic reconstruction of Pangea during the Permian period. Dots indicate reef localities, dark shading, mountains, light shading, shallow seas (modified from Scotese and Golonka, 1992).

this time reef ecosystems were highly variable in biotic composition, and at least seven associations have been recognized. They were characterized typically by stabilizing and baffling organisms such as calcified algae, calcisponges consisting of both nonchambered "inozoan" and chambered "sphinctozoan" types, bryozoans, and a variety of problematical organisms, including the tiny, tubelike, branching *Tubiphytes*. As in the Carboniferous interval, bryozoans and noncolonial organisms dominated. Most Permian reef ecosystems lacked many large-scale, heavily calcified reef organisms (Fagerstrom, 1987). The Late Permian reef of west Texas, constitutes one of the best-studied examples of this age (see Chapter 6, this volume). The origin of this reef has been widely discussed, as has the issue of framework (Fagerstrom and Weidlich, 1999).

3.6. The Permo-Triassic Eclipse and the Triassic Recovery

At the end of the Permian a great mass extinction brought on the collapse of this diverse and luxuriant Tethyan reef ecosystem. It suffered a dramatically sudden loss of biotic diversity. While both temperate and tropical regions suffered, the tropics were more severely affected (Erwin, 1993). Based on new studies at boundary sections in China, the mass extinction is thought to have occurred rapidly in geologic terms (Bowring *et al.*, 1998). Following the end of the Permian, reefs mostly dissipated and they were marked globally by a lengthy eclipse interval of some 12–14 million years, which extended through Early Triassic and into Middle Triassic time. In some places, Lower Triassic rocks contain a distinctive postextinction calcified, microbial biota (Schubert and Bottjer, 1995).

The revival of reefs previously extinguished in the Tethys took place during a recovery period after the start of Middle Triassic time. Details of Triassic reefs and their evolution have been reviewed by Stanley (1988), Flügel (1982), and Flügel and Senowbari-Daryan (1996). The reader also is referred to Flügel and Senowbari-Daryan (Chapter 7, this volume). The western Tethys served as the center of the Middle Triassic recovery. After the hiatus, carbonate deposition resumed in the Tethys with the development of extensive carbonate shelves. This setting became populated by diverse shallow-water organisms including calcareous algae and sponges, hydrozoans, and the first scleractinian corals (Senowbari-Daryan *et al.*, 1993).

Triassic rocks in South China have provided exciting new discoveries (Lehrmann, 1999). Thick, relatively pure Lower Triassic carbonates in the Guizhou region of South China contain small, constructional biostromes. The abundant calcifying and potentially reef-building organisms include mostly calcified microbial communities and some tiny microproblematical organisms. These Early Triassic organisms may have served as stock that seeded subsequent reef ecosystems of the Middle Triassic. Studies in this region of China suggest a "bridge" between the Permian and Middle Triassic reef intervals, and it is interesting to note that during Early Triassic time South China existed as an isolated continental block in the eastern Tethys Seaway. Later this block moved northward and was incorporated into the tectonic collage of complex geology now comprising the region of South China.

Thick carbonates and reefs of the Middle Triassic Tethys region contained some of the same organisms as those of the Permian: calcisponges, calcareous algae, and bryozoans with reef dwellers consisting of *"Tubiphytes"* and a host of microproblematical organisms. Some stromatolites and "algal crusts" also are important in reefs of that age. According to some workers, during the Permian large constructional framework was notably absent and reefs of the Middle Triassic contained mostly baffling and binding guilds. Some new types of organisms also appeared, such as spongiomorphs and the first scleractinian corals. Scleractinians seem unrelated to most Paleozoic corals and may have evolved from soft-bodied anemone-like ancestors that developed the ability to calcify and secrete hard skeletons (Stanley, 1988, Stanley and Fantin, 2001).

Did Middle Triassic reefs contain any holdover taxa from the Permian? According to Flügel (1994; Flügel and Senowbari-Daryan, Chapter 7, this volume), they did not, but instead consisted of many newly evolved taxa, which were morphologically similar to those of the Permian. Flügel and his co-workers emphasize that reef organisms most similar to those of the Permian do not make their appearances until Late Triassic time!

While our interpretations of the Triassic recovery are based largely on the Tethys model, it is constructive also to look outside this well-studied region. The surprising discovery of Early Triassic reefal activity from South China is especially interesting, but more geographically distant displaced terranes of western North America are demonstrating an important but neglected record of Permian and Triassic reefbuilding (Coney *et al.*, 1980). Research on fossils from these terranes indicates their roles as possible refugia, and some volcanic terranes with thick carbonate records could have existed either as isolated islands near the western shores of Pangea or as great expanses of the Panthalassia Ocean (Soja, 1996). Reefs and carbonates grew along the flanks of many islands, and later, after the volcanoes subsided under the sea, they continued to grow upward, keeping pace with subsidence and sea-level rise, to produce impressive reef complexes. Because of dynamic and ceaseless processes of seafloor spreading and plate tectonic subduction, all the ancient seafloor containing the volcanic islands has been swept clean. Vestiges are only preserved as terranes of the American Cordillera. One of the most famous of these volcanic carbonate complexes is a Triassic reef studied by Reid (1985). It was found to contain many holdover taxa previously known from the Permian, including sponges and "phylloid" algae.

The transition from Middle to Late Triassic time witnessed an increase in reef development within vast carbonate complexes in the Tethys seaway. There was a turnover in composition among both chambered and non-chambered calcified sponges, calcified algae, and other organisms including scleractinian corals. Scleractinians were still present but were neither volumetrically nor ecologically important as reef constructors. This type of ecosystem continued into Late Triassic time, disrupted by a smaller mass extinction some 12–17 million years before the end of the Triassic. Sometime either between the Carnian and Norian stages or during early Norian time, a major Triassic reorganization took place among reefs of the Tethys (see Chapter 7, this volume). This reorganization has been associated with mass extinction. The reorganization was marked by a paleoecological shift in guild structure within reef ecosystems of the Tethys. This changeover segued into the rise of corals and the revival of the long-absent constructor guild.

Changes in the latest Triassic included the increased importance of reef-building scleractinian corals which caused a revival of the constructor guild, as previously discussed, which had been absent since late Devonian time. Colonial scleractinians increased both in size and volumetric importance in thick carbonate complexes of the Tethys (Chapter 7, this volume). With the revival of the constructor guild, reef ecosystems quickly took on an ecology more like that seen in modern reefs, but curiously, many Permian holdovers are present. This series of events has been termed the modernization of reefs (Stanley, 1988, 1992). Stanley and Swart (1995) speculated that corals coevolved symbiosis with zooxanthellae during the Late Triassic. After studying stable isotopes of carbon and oxygen in some typical corals of the Tethys, the isotopic signatures indicative of zooxanthellae were found in the calcified skeletons of Late Triassic corals from the Tethys. These investigations led to the conclusion that corals had acquired rapid growth rates and reef-building potential, like modern species, by this time in their evolutionary history. In addition to corals, red algae, chambered sponges, hydrozoans, and disjectoporoid sponges participated in reef building in a variety of settings within carbonate platforms of the Tethys.

3.7. The End-Triassic Collapse and Jurassic Reef Ecosystems

The end of the Triassic was marked by a disruption and rapid collapse of a luxuriant but short-lived coral and of sponge-dominated reefs of the Tethys (Stanley, 1988). The destruction of reefs was the result of a first-order mass extinction event (Hallam, 1990) affecting a broad spectrum of marine and terrestrial ecosystems. Anoxia and sea-level changes have been blamed for this event (Hallam and Goodfellow, 1990), but climatic change also seems implicated. Other causes including the impact of meteorites or comets have been proposed. Whatever the cause, destruction was swift and recovery delayed.

Much of Early Jurassic time occupied an eclipse interval of some 6 to 8 million years. Like previous examples, this interval was marked by a near global reduction in carbonate deposition and a virtual absence of reef building. The environmental perturbation seems to have affected the Tethys more severely than the Pacific islands now represented by Cordilleran terranes of North America. A well-developed Early Jurassic reef in the Cordilleran terrane of Stikinia, British Columbia, Canada is perhaps the earliest example yet known (Stanley and McRoberts, 1993; Stanley and Beauvais, 1994). This reef developed in a volcanic island-arc setting and was dominated by a constructional framework of scleractinian corals (Fig. 6). Surprisingly, the chief constructor was a Triassic species from Tethys that had been thought extinct since the Late Triassic. The details of the growth and succession of this reef are presented in Fig. 7.

The Jurassic recovery was slow and marked by an ebb in diversity among corals and other marine faunas. A local perturbation of the marine environment in the Tethys was followed by community reorganizations near the close of the Early Jurassic period (Beauvais, 1984; Hallam, 1996). By Middle Jurassic time, reef ecosystems in the Tethys had revived and expanded. Corals dominated once again and they underwent a major adaptive radiation along with sponges and calcareous algae. Reef ecosystems became ecologically complex and varied in composition and their ecological structure was controlled by water depth and sedimentation. A variety of different reef types existed in different water depths and positions on the shelf (Chapter 8, this volume).

FIGURE 6. The Telkwa reef, a steep-sided Early Jurassic reef from the volcanic island arc, Stikine terrane of central British Columbia. The light-colored limestone is composed of reef-building corals, while the darker rock is volcanic or volcaniclastic.

Despite a minor setback during the transition from Middle to Late Jurassic time, reefs continued to be robust and proliferated in the Tethys. Many were dominated by scleractinian corals that were taxonomically different from those of the Triassic. They coexisted with calcareous algae, large calcareous sponges, and other organisms. Also important during Late Jurassic time were deeper-water reef mounds, dominated primarily by siliceous sponges and calcified microbes. Warming climate and flooding of extensive shallow-shelf areas enhanced reef development in the Jurassic.

3.8. Cretaceous Reefs and the Rise of the Rudists

The expansion of reef ecosystems within the warm-water tropical Tethys continued into Early Cretaceous time with a full complement of reef guilds. Despite a small mass extinction marking the end of the Jurassic, reef ecosystems continued to develop. Corals and other reef organisms had been joined already by a specially adapted group of gregarious bivalves known as rudistids. By the end of Early Cretaceous time, rudistids were proliferating in reef

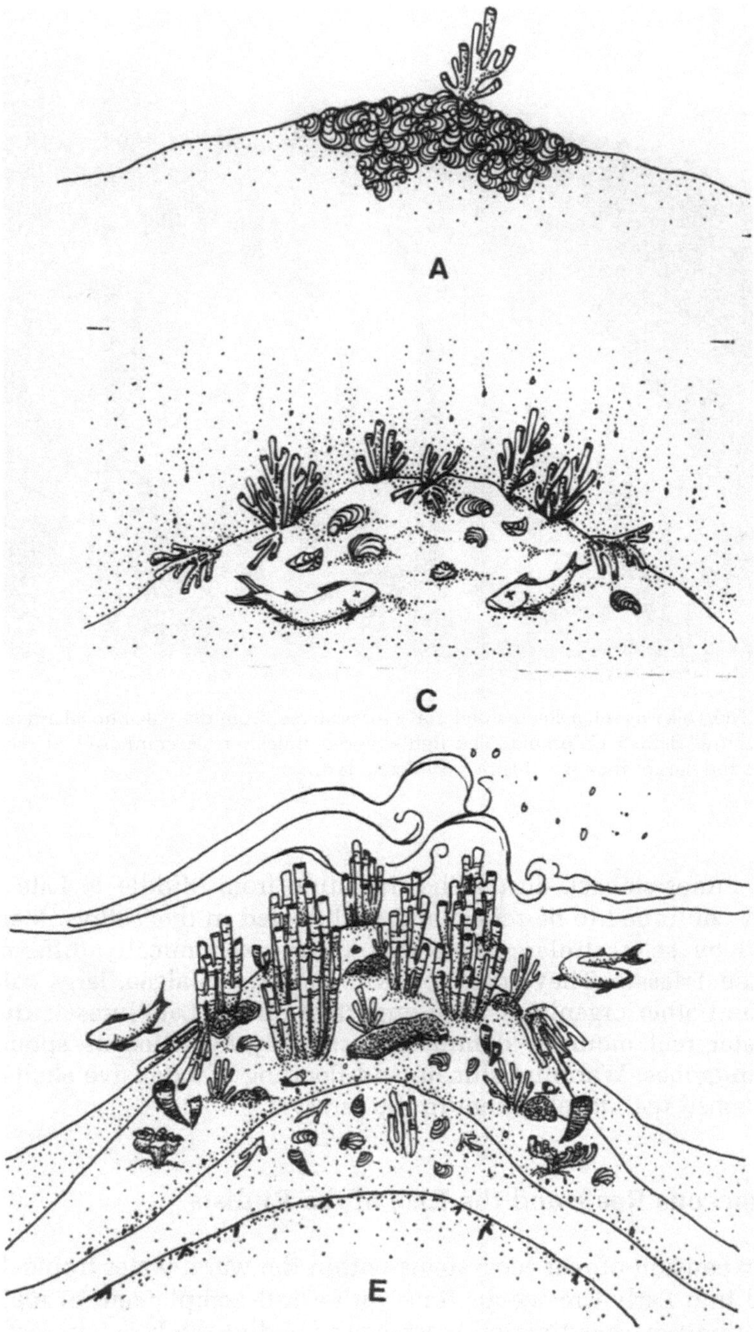

FIGURE 7. The successional history of the Telkwa reef (see Fig. 6). This reef grew in a volcanic island setting in the ancient Pacific Ocean. (A) Gregareous bivalves formed a hard substrate on a sand pile. (B) Colonies of branching corals grew on the bivalve substrate. (C) A volcanic ash eruption killed the reef. (D) Reef was recolonized by large branching corals and other invertebrates.

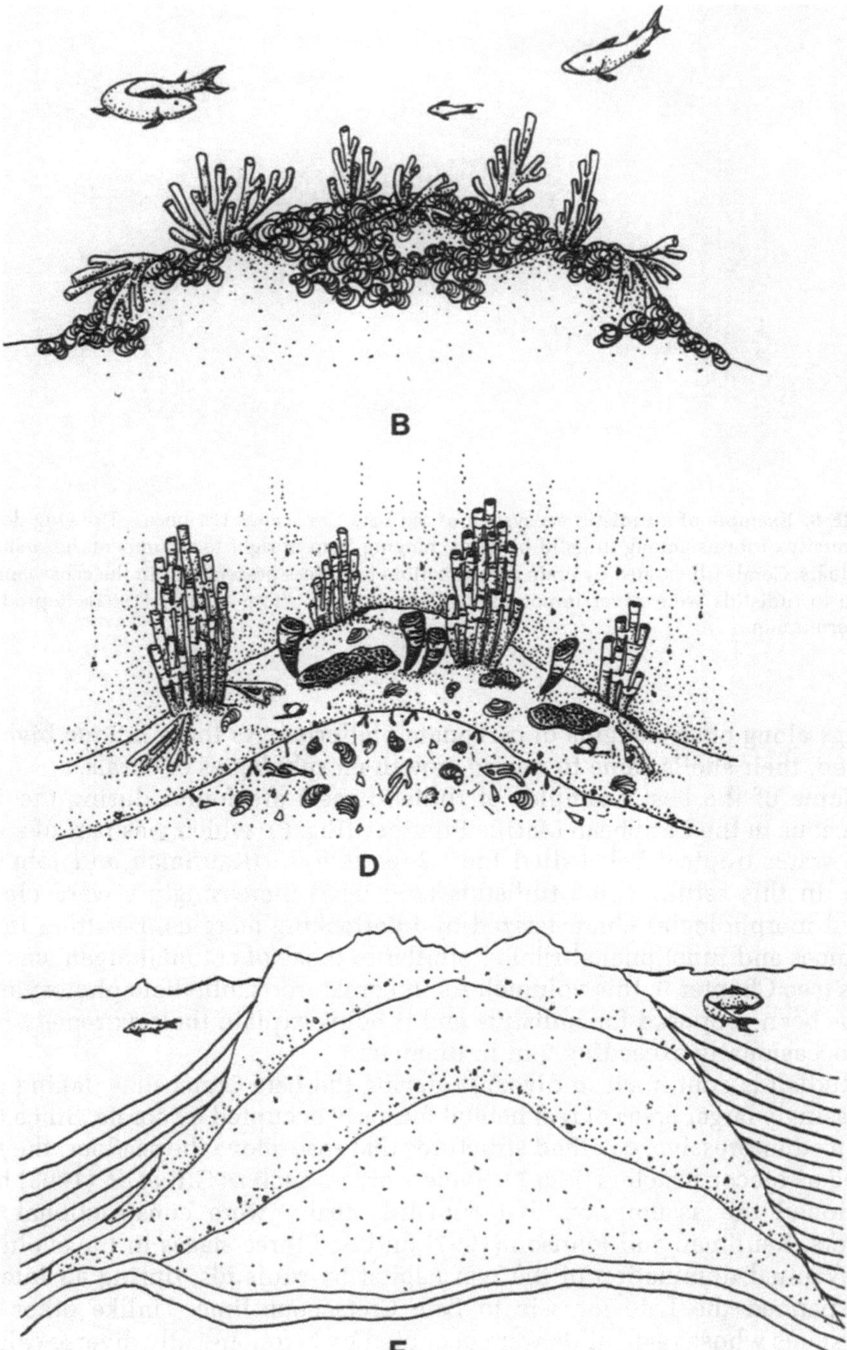

(E) Reef grew up to the shallow wave zone and produced a wave resistant structure composed primarily of corals. (F) Much later after the reef died, the limestone was exposed, eroded, and karstified, still preserving the massive limestone and the flanking beds, composed of volcaniclastic sediment. (Reproduced with permission of the Natural Research Council of Canada.)

FIGURE 8. Example of a rudistid ecosystem of the Late Cretaceous Caribbean. Drawing depicts the diversity of forms among rudistid bivalves, ranging from upright to recumbent, horn-shaped individuals. Corals (illustrated by brain corals, bottom left) were present within the ecosystem but relative to rudistids were never important reef components. From Stanley (1992). Reproduced with permission.

settings along open margins of carbonate platforms. As these unique bivalves evolved, their shells came to mimic growth morphologies of corals.

Some of the best examples of rudistid reefs are found during the Late Cretaceous in the Caribbean Marine Province (Fig. 8), which was part of a very warm-water tropical belt called the "Supertethys" (Kauffman and Johnson, 1988). In this setting some rudistids converged increasingly toward closely packed morphologies characterized by interlocking margins, resulting in appearances and functional attributes similar to those of colonial organisms like corals (see Chapter 9, this volume). As in corals, zooxanthellate photosymbiosis has been proposed for rudistids and it helps explain their extremely large size, occasionally exceeding 2 m in diameter.

Rudistids continued to diversify during the Late Cretaceous, taking over increasingly larger areas of reef habitat formerly occupied by corals. Since they produced impressive, calcified structures that rose above the seafloor, they are treated as bioconstructors (Fig. 7). Some workers such as Gili *et al.* (1995) have questioned the assumption that rudistids really were constructional reef builders. Kauffman and Johnson (1997) outlined three stages in the evolution and eventual domination of the reef habitat by rudistids, during an interval that spanned the Late Jurassic to Late Cretaceous time. Unlike other reef ecosystems whose reef guilds were occupied by taxonomically diverse calcifying organisms, those of the Cretaceous were dominated principally by rudistids. The group as a whole surprisingly, seemed unaffected by mass extinctions of the Cretaceous.

The takeover of the Cretaceous reef ecosystem by rudistids was unprecedented in the history of marine ecosystems because it did not follow a mass extinction. Much of the evolution of the Late Cretaceous rudistid-dominated ecosystem seems closely related to the extraordinary global warming and greenhouse interval postulated for that time. Colonial corals lived within otherwise rudistid-dominated reefs but corals never assumed constructional roles. Bivalves have not been considered competitors with corals for reef habitat today and in much of the geologic past. Both have different growth rates and often show different levels of tolerance to temperature, nutrients, and salinity. The question remains unresolved whether rudistids actively competed (in the ecological sense) with corals or whether these bivalves simply reclaimed ecospace vacated by corals? Rising temperature and salinity in the Supertethys during the Cretaceous have been cited as factors affecting the decline of reef corals and the rise of rudistids.

Near the end of the Cretaceous, in mid-Maastrichtian time, after prospering as a successful ecosystem for over 50 million years, rudistid reefs experienced global collapse (Johnson and Kaffman, Chapter 9, this volume). This coincided with extinctions among other groups of marine mollusks. While some workers have associated the rudistid demise with the disappearance of the dinosaurs and other terrestrial and marine life at the end of the Cretaceous (the K/T extinction), the rudistid ecosystem collapse actually began 1.5 to 3 million years before the end of the Cretaceous, not at the K/T boundary. The impact of comets or meteorites associated with the Alverez hypothesis at the K/T extinction appears to have come too late to cause the collapse of the rudist ecosystem. While no rudistid lived beyond the Cretaceous, scleractinian corals did survive and these survivors formed the seed stock for a subsequent revival of coral reefs in the Cenozoic Era.

3.9. Rise of Modern Coral Reefs

In a pattern similar to other reef collapses, the early part of the Tertiary or early Paleogene was a reef eclipse interval, one that may have lasted nearly 8 million years. The record of the early Tertiary corals that survived the K/T mass extinction is woefully incomplete and details of the reef recovery are poorly known, leading some workers to doubt whether a recovery actually occurred at this time. Some thickets and colonial coral associations are known to have survived in high-latitude and cool and deep water in Late Cretaceous and early Paleocene times. Following a Paleocene reduction in reef limestone and carbonate deposits, corals returned to the reef and by the Oligocene, coral communities were established. Why coral-dominated reef communities were delayed so long in reestablishing reef ecosystems possibly can be explained by the marine perturbations in effect after the K/T mass extinction. These conditions seem to have extended into part of Tertiary (early Paleogene) time. By Neogene time, a robust ecosystem dominated by scleractinian corals and coralline algae was established. The coral–coralline algal association was to become an important consortium on the Neogene reef.

Tertiary time witnessed biological marine revolutions and rapid adaptive radiations among reef corals. Although there were climatic cooling events near the end of the Paleogene, reefs survived, and during the Neogene, a robust coral-dominated reef ecosystem emerged in much of the world's oceans. Global temperature shifts and climatic changes of the Neogene were preludes to the ice ages. Cooling with reduction of the area of the tropics severely restricted the coral reef ecosystem. Coral diversity in the Mediterranean dropped between early and middle Miocene with extinctions near the end of Miocene time. The closing of the Tethys Seaway and drying of the Mediterranean had profound climatic effects on reef ecosystems. Emergence of the Isthmus of Panama in Pliocene time also had repercussions on reefs. It subdivided a previously broad ocean into today's Pacific and Atlantic basins and caused extinctions through changes in nutrients and ocean circulation patterns. It was the beginning of a great paleogeographic differentiation, still present today among marine faunas.

Glacial intervals of the Pleistocene ice ages brought more global cooling with significant drops in sea level, and corals responded to these changes. Despite severe sea-level drops and the more restricted area of the tropics, corals survived and evolved into numerous, rapidly growing reef-building species that, along with algae and other secondary constructors, constitute today's coral reef ecosystem.

3.10. Are There Recurring Patterns in Reef Ecosystem Evolution?

During the past 600 million years, a succession of reef ecosystems has emerged and disappeared in tropical oceans. All but today's reefs have experienced global collapse and it is a recognized fact that reefs, in comparison to other ecosystems, are more sensitive to instability of the environment. The importance of mass extinction in promoting the restructuring reefs was outlined by Newell (1971) and later Copper (1994b) emphasized the importance of global collapse on reef ecosystems. Collapse sometimes was followed by rebuilding and the emergence of entirely new ecosystems. The Permian to Triassic "lag, rebound, and expansion" scenario of Erwin (1993) fits this reef pattern. Players in the new ecosystems were recruited initially from survivors of the previous reef interval (as holdovers) and were later joined by newly evolving organisms that had emerged from the resulting adaptive radiations. In an overview of more than 600 million years of reef history and mass extinctions, Fagerstrom (1987) distinguished nine reef units coinciding closely with four first-order and either six or seven second-order mass extinctions. The "annihilation–collapse–rebuild" model described by Fagerstrom is similar to the collapse–rebuild hypothesis of Copper (1974, 1994a, 1996). A common pattern can be discerned.

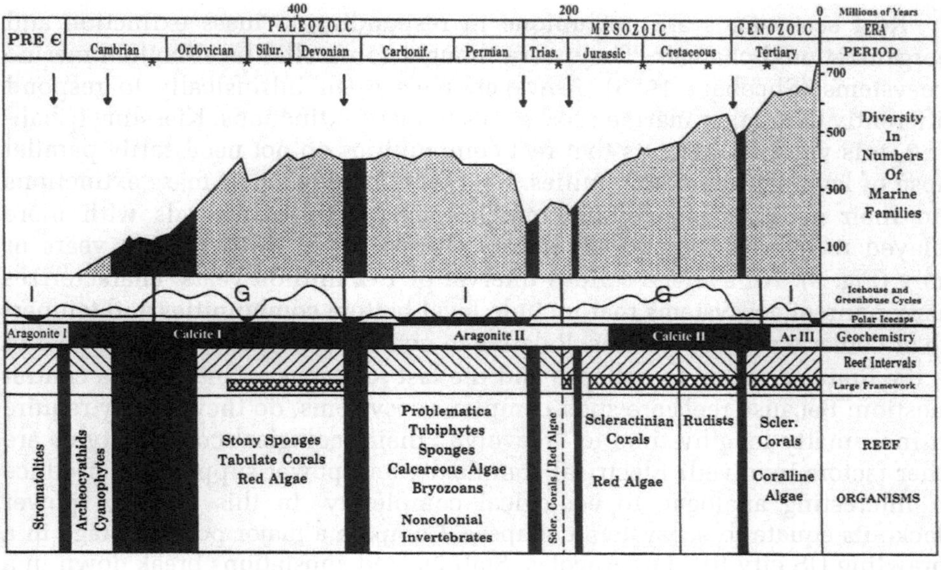

FIGURE 9. Some major characteristics of reef ecosystems through time. At top, arrows show the major mass extinctions while stars indicate second-order mass extinctions. Below is the diversity curve for marine families (Sepkoski, 1992) with black vertical bars representing the reef eclipse intervals which followed mass extinctions. Icehouse–greenhouse cycles are depicted and a relative temperature curve is sketched. Glaciation times and polar icecaps are indicated. Also included are changes in sea chemistry, intervals where either calcite-secreting or arganite-secreting organisms predominated (Stanley and Hardie, 1998). The principal reef intervals indicated are discussed in the text.

Eight relatively stable reef ecosystems are most easily recognized in the geologic record (Fig. 9). With the notable exception of the Cretaceous rudistid takeover, these ecosystems were preceded by mass extinctions. It appears that during specific times, global shocks were delivered to Earth's marine ecosystem and the ensuing ecosystem collapse and loss of diversity opened adaptive space into which new species could radiate. Following restructuring, new ecosystems emerged. There is no shortage of hypotheses to account for the ecosystem collapses (sea-level change, climatic deterioration, ocean anoxia, overnutrification, meteorite or comet impact, volcanic outgassing, etc.). In reviewing the changing character of reef ecosystems and the resulting diversity trends (Fig. 9), it is interesting to note how often ecosystem changes coincided with icehouse–greenhouse cycles and polar icecaps and to some extent changes from calcite-secreting to aragonite-secreting oceans (Stanley and Hardie, 1998). Is it possible to discover some common threads weaving all mass extinctions together? An examination of Fig. 9 can serve as a starting point. The final answer may not be explained by any single phenomenon and one must be reminded that a correspondence of reef episodes with geologic trends does not establish cause.

Reef ecosystems are not unique in responding to mass extinction and reorganizational changes. They were thought to parallel level-bottom marine ecosystems (Sheehan, 1985). However, reefs seem intrinsically to respond differently than other marine ecosystems to mass extinctions. Kiessling (Chapter 2, this volume) suggests that reef communities do not necessarily parallel those of level-bottom communities. Reef were harder hit by mass extinctions and their ecosystems exhibited much longer eclipse intervals with more delayed recoveries. Sometimes these eclipses lasted 2–10 million years or more (Fig. 9). An average eclipse interval of 1–2 million years, characterizes many nonreef ecosystems that include level-bottom communities and temperate or cooler water ecosystems. Relative to nonreef ecosystems, reefs appear to be the first ecosystem to collapse and the last to recover. This poses a central question: Because reefs are such complex ecosystems, do they merely require an inordinatly lengthy time to "reevolve" their ecological complexity or are other factors involved? Electrical complexity and power supply may produce an interesting analogue to ecological complexity. In this analogy, power blackouts equate to ecosystem collapses. Compare a major power outage in a sprawling US city like Los Angeles. Stations and substations break down in a cascading fashion and it takes days to resume electrical power. Compare this to power loss in a rural farming community which may require only a lineman to flip a switch to restart electrical power. An interesting question to pose is, following the principal "shock" and global collapse, could prolonged perturbations of the marine biosphere have held reef recovery at bay until conditions conducive to reef growth returned? The extraordinarily lengthy period of many reef eclipses suggests this idea may explain many Phanerozoic reef trends. Current research programs focus on recoveries from the various mass extinctions. Such endeavors are achieving the kind of biostratigraphic accuracy needed to resolve such issues (Kauffman and Erwin, 1995).

3.11. The Future of Reef Ecosystems

The fact that geologists can trace ancient reef ecosystems back at least a billion years or more is rather ironic. While such longevity suggests remarkable stability, we have noted stunning and frequent reef collapses in the geologic record. This may strike the reader as a real contradiction, implying the resilience of the coral reef ecosystem and at the same time its fragility. Since the devastating K/T mass extinction, scleractinian corals have withstood stresses of extraterrestrial impacts, anoxia, global warming, climatic cooling, sea-level change, hurricanes, and epidemics of voracious starfish, not to mention countless volcanic eruptions and other global earth forces.

Today this seemingly invulnerable ecosystem is being undermined not by natural agents but by the expansion of human populations. Human impact or the current degradation of the tropical ecosystems is undeniable. Tropical ecosystems have suffered the brunt of habitat destruction with concomitant loss of species diversity. Much reef degradation has been brought on by clear-cutting, overfishing, and gross mismanagement of resources. In addition, the release of agrochemicals, pesticides, sewage, and other pollutants into marine waters has had deleterious effects. Increasing numbers of humans live in island and coastal regions, and human exploitation is taking its toll. James W. Porter, a leading reef specialist, has stated (personal communication) that nearly 10% of coral cover worldwide has died, and if present trends continue, 20 to 30% of coral cover soon will be lost. Studies show that new species of pathogens—viruses, bacteria, and fungi—are killing corals at alarming rates. Many coral-damaging microbes can be traced to sewage discharge into the ocean.

Global warming, sea-level rise, and associated greenhouse effects—topics of concern directly or indirectly related to anthropogenic cause—also have caused further deletetrious effects to coral reefs (see Chapter 11, this volume). Global warming is associated with a rise in sea surface temperature that has been detected in many areas of the ocean. Kleypas *et al.* (1999) have discussed the past, present, and future rise in partial pressure of CO_2 and the detrimental effects this trend will have on carbonates and coral reefs. Considering these effects, how will our living reef ecosystem fare in the future? Unfortunately, the prognosis for reefs is not at all good. If predictions of greenhouse warming, rapid sea-level rise, and the surge in human populations in the next century are accurate, will reefs survive or will they be part of the next mass extinction?

Because of these concerns, a great deal of research is devoted to reef conservation. The year 1998 was designated the International Year of the Reef and publicity helped promote awareness of the issues and problems. There are international efforts now underway to survey and inventory all reefs, both flourishing and degraded ones. We need a reliable baseline to differentiate healthy reefs from declining ones. How do ancient reefs relate to our current problem? As we delve deeper into research on ancient reef ecosystems to gain more insight into how reef ecosystems responded to collapse, we produce valuable data with which to assess current problems of global change. As fragile entities, reefs are the first ecosystem to experience degradation and the last to recover. The public may fail to be concerned about the predicted reef decline, pointing to the fact that throughout their history, reef ecosystems have inevitably recovered. It is relevant, however, to be reminded of the magnitude of time. Reef eclipse intervals of the Phanerozoic spanned millions of years and millions of more years were needed before reef ecosystems recovered. Can we envision a place on earth for our present biological diversity, including humans, that does not include reefs and other tropical ecosystems? Placed in the time perspective for a short-lived, egocentric species like ourselves, the obvious implications are sobering. We have much to gain in understanding our current diversity crisis by studying ancient reefs and their tumultuous history of collapses and recoveries. We have a long road to travel in this worthy endeavor and paleontology has much to offer.

ACKNOWLEDGMENTS: This overview is derived from many sources and could not have been written without significant research and synthesis from many reef specialists. I thank my colleagues, Paul Copper and Dennis Hubbard, for reviews of this chapter, and I also thank Norman D. Newell for his insight and unwavering inspiration on the topic of reefs.

References

Beauvais, L., 1984, Evolution and diversification of Jurassic Scleractinia, *Palaeontogr. Am.* **54**:219–224.

Bowring, S. A., Erwin, D. H., Jin, Y. G., Martin, M. W., Davidek, K., and Wang, W., 1998, U/Pb zircon geochronology and tempo of the end-Permian mass extinction, *Science* **280**:1039–1045.

Brunton, F. R., Copper, P., and Dixon, O. A., 1997, Silurian reef-building episodes, in: *Proceedings of the 8th International Coral Reef Symposium* (H. A. Lessios and I. G. Macintyre, eds.), Smithsonian Tropical Research Institute, Balboa, Panama, pp. 1643–1650.

Chave, K. E., Smith, S. V., and Roy, J. K., 1972, Carbonate production by coral reefs, *Marine Geology* **12**:123–140.

Coney, P. J., Jones, D. L., and Monger, J. W. H., 1980, Cordilleran suspect terranes, *Nature* **288**:329–333.

Copper, P., 1974, Structure and development of Early Paleozoic reefs, in: *Proceedings of the Second International Coral Reef Symposium*, Vol. 1, Great Barrier Reef Committee, Brisbane, pp. 365–386.

Copper, P., 1994a, Reefs under stress: The fossil record, *Cour. Forsch. Inst. Senckenberg* **172**:87–94.

Copper, P., 1994b, Ancient reef ecosystems expansion and collapse, *Coral Reefs* **13**:3–11.

Copper, P., 1996, Reefs and carbonate productivity: Cambrian through Devonian, in: *Proceedings of the 8th International Coral Reef Symposium* (H. A. Lessios and I. G. Macintyre, eds.), Smithsonian Tropical Research Institute, Balboa, Panama, pp. 1623–1630.

Cowen, R., 1988, The rate of algal symbiosis in reefs through time, *Palaios* **3**:221–227.

Cummings, E. R., 1932, Reefs or bioherms? *Geol. Soc. Am. Bull.* **43**:331–352.

Cummings, E. R., and Schrock, R. R., 1928, Niagaran coral reefs in Indiana and adjacent states and their stratigraphic relations, *Geol. Soc. Am. Bull.* **39**:519–520.

Dunham, R. J., 1970, Stratigraphic reefs versus ecologic reefs, *Am. Assoc. Petroleum Geol. Bull.* **54**:1931–1932.

Embry, A. F., and Klovan, J. E., 1971, A Late Devonian reef tract on northeastern Banks Island, Northwest Territories, *Bull. Can. Petroleum Geol.* **33**:730–781.

Erwin, D. H., 1993, *The Great Paleozoic Crisis*, Columbia University Press, New York.

Ezaki, Y., 1997, The Permian coral *Numidiaphyllum*: New insights into Anthozoan phylogeny and Triassic Scleractinian origins, *Palaeontology* **40**(1):1–14.

Fagerstrom, J. A., 1985, Comparison of processes and guild structures in Holocene and ancient reef communities, *Fifth International Coral Reef Congress Proceedings* **2**:126.

Fagerstrom, J. A., 1987, *The Evolution of Reef Communities*, John Wiley and Sons, New York.

Fagerstrom, J. A., 1994, The history of Devonian–Carboniferous reef communities: Extinctions, effects, recovery, *Facies* **30**:177–192.

Fagerstrom, J. A., 1997, Reef building: A biological phenomenon, *Boll. Real Soc. Españ. Hist. Nat.* **92**:7–13.

Fagerstrom, J. A., and Weidlich, O., 1999, Origin of the upper Capitan–Massive limestone (Permian), Guadalupe Mountains, New Mexico-Texas: Is it a reef?, *Geol. Soc. Am. Bull.* **111**(2):159–176.

Flügel, E., 1982, Evolution of Triassic reefs: Current concepts and problems, *Facies* **6**:297–328.

Flügel, E., 1994, Pangean shelf carbonates: Controls and paleoclimatic significance of Permian and Triassic reefs, in: *Pangea: Paleoclimate, Tectonics, and Sedimentation during Accretion, Zenith, and Breakup of a Supercontinent* (G. D. Klein, ed.), Geological Society of America Special Paper, Boulder, CO, pp. 247–266.

Flügel, E., and Senowbari-Daryan, B., 1996, Evolution of Triassic reef biota: State of the art, in: *Global and Regional Controls on Biogenic Sedimentation* (J. Reitner, F. Neuwiler, and F. Gunkel, eds.), Gottinger Arbeiten Geologie Paläontologie, Gottingen, pp. 285–294.

Flügel, E., and Stanley, G. D., Jr., 1984, Reorganization, development and evolution of post-Permian reefs and reef organisms, *Palaeontogr. Am.* **54**:177–186.

Gautret, P., Cuif, J. P., and Freiwald, A., 1997, Composition of soluble mineralizing matrices in zooxanthellate and non-zooxanthellate scleractinian corals: Biochemical assessment of photosynthetic metabolism through the study of a skeletal feature, *Facies* **36**:189–194.

Gerhard, L. C., 1991, Reef modelling: Progress in simulation of carbonate environments, in: *Sedimentary Modelling* (E. K. Franseen, W. L. Watney, C. G. S. C. Kendall, and W. Ross, eds.), Kansas Geological Survey Bulletin 233, Lawrence, pp. 346–358.

Gili, E., Masse, J.-P., and Skelton, P. W., 1995, Rudists as gregarious sediment-dwellers, not reef-builders, on Cretaceous carbonate platforms, *Palaeogeogr. Palaeoclimatol. Palaeoecol.* **118**:245–267.

Glynn, P. W., 1997, Bioerosion and coral-reef growth: A dynamic balance, in: *Life and Death of Coral Reefs* (C. Birkeland, ed.), Chapman and Hall, New York, pp. 68–95.

Greenstein, B. J., and Moffat, H. A., 1996, Comparative taphonomy of Holocene and Pleistocene corals, San Salvador, Bahamas, *Palaios* **11**:57–63.

Greenstein, B. J., and Pandolfi, J. M., 1997, Preservation of community structure in modern reef coral life and death assemblages of the Florida Keys: Implications for the Quaternary fossil record of coral reefs, *Bull. Marine Sci.* **61**(2):431–452.

Grotzinger, J. P., and Knoll, A. H., 1995, Anomalous carbonate precipitates: Is the Precambrian the key to the Permian? *Palaios* **10**(6):578–596.

Hallam, A., 1990, The end-Triassic extinction event, *Geol. Soc. Am.* (Special Paper) **247**:577–583.

Hallam, A., 1996, Recovery of the marine fauna in Europe after the end-Triassic and early Toarcian mass extinctions, in: *Biotic Recovery from Mass Extinction Events* (M. B. Hart, ed.), Geological Society Special Publication, London, pp. 231–236.

Hallam, A., and Goodfellow, W. D., 1990, Facies and geochemical evidence bearing on the end-Triassic disappearance of the Alpine reef ecosystem, *Hist. Biol.* **4**:131–138.

Hallock, P., 1997, Reefs and reef limestones in earth history, in: *Life and Death of Coral Reefs* (C. Birkeland, ed.), Chapman and Hall, New York, pp. 13–42.

Hallock, P., and Schlager, W., 1986, Nutrient excess and the demise of coral reefs and carbonate platforms, *Palaios* **1**:389–398.

Harries, P. J., and Kauffman, E. G., 1996, The importance of crisis progenitors in recovery from mass extinction, in: *Biotic Recovery from Mass Events* (M. B. Hart, ed.), Geological Society, London, pp. 15–39.

Hatcher, B. G., 1997, Coral reef ecosystems: how much greater is the whole than the sum of the parts? *Coral Reefs* **16** (Suppl.):S77–S91.

Heckel, P. H., 1974, Carbonate buildups in the geological records: A review, *Soc. Econ. Paleontol. Mineral.* (Spec. Publ) **18**:90–154.

Hubbard, D. K., 1997, Reefs as dynamic systems, in: *Life and Death of Coral Reefs* (C. Birkeland, ed.), Chapman and Hall, New York, pp. 43–67.

Hutchinson, G. E., 1978, *An Introduction to Population Ecology*, Yale University Press, New Haven.

Insalaco, E., 1998, The descriptive nomenclature and classification of growth fabrics in fossil scleractinian reefs, *Sediment. Geol.* **118**:159–186.

Jablonski, D., 1993, The tropics as a source of evolutionary novelty through geological time, *Nature* **364**:142–144.

James, N. P., 1983, Reef environments, in: *Carbonate Depositional Environments* (P. A. Scholle, D. G. Bebout, and C. H. Moore, eds.), Memoir 33, American Association of Petroleum Geologists, Tusla, OK, pp. 346–462.

James, N. P., and Bourque, P.-A., 1992, Reefs and mounds, in: *Facies Models:* Response to Sea Level Change (R. G. Walker and N. P. James, eds.), Geological Association of Canada, St. John's, Newfoundland, pp. 323–347.

Kauffman, E. G., and Erwin, D. H., 1995, Surviving mass extinctions, *Geotimes* **40**:14–17.

Kauffman, E. G., and Johnson, C. C., 1988, The morphological and ecologic evolution of Middle and Upper Createaous reef-building rudists, *Palaios* **3**:194–216.

Kauffman, E. G., and Johnson, C. C., 1997, Ecological evolution of Jurassic–Cretaceous Caribbean reefs, in: *Proceedings of the 8th International Coral Reef Symposium* (H. A. Lessios and I. G. Macintyre, eds.), Smithsonian Tropical Research Institute, Balboa, Panama, pp. 1669–1676.

Kershaw, S., 1994, Classification and geologic significance of biostromes, *Facies* **31**:81–92.

Kleypas, J. A., Buddemeier, R. W., Archer, D., Gattuso, J. P., Langdon, C., and Opdyke, B. N., 1999, Geochemical consequences of increased atmospheric carbon dioxide on coral reefs, *Science* **284**(5411):118–120.

Knoll, A. H., Fairchild, I. J., and Swett, K., 1993, Calcified microbes in Neoproterozoic carbonates: Implications for our understanding of the Proterozoic/Cambrian transition, *Palaios* **8**:512–525.

Laporte, L. F. (ed.), 1974, *Reefs in Time and Space*, Society of Economic Paleontologists and Mineralogists, Tulsa, Special Publication 18.

Lehrmann, D. J., 1999, Early Triassic calcimicrobial mounds and biostromes of the Nanpanjiang Basin, south China, *Geology* **27**:359–362.

Lowenstam, H. A., 1950, Niagaran reefs of the Great Lakes area, *J. Geol.* **58**:430–487.

McGhee, G. R., Jr., 1996, *The Late Devonian Mass Extinction—The Frasnian/Famennian Crisis*, New York, Columbia University Press.

Muller-Parker, G., and D'Elia, C. F., 1997, Interactions between corals and their symbiotic algae, in: *Life and Death of Coral Reefs* (C. Birkeland, ed.), Chapman and Hall, New York, pp. 96–113.

Newell, N. D., 1971, An outline history of tropical organic reefs, *Am. Museum Novitiates* **2465**:1–37.

Newell, N. D., 1972, The evolution of reefs, *Sci. Am.* **226**:54–65.

Pandolfi, J. M., and Greenstein, B. J., 1997, Taphonomic alteration of reef corals: Effects of reef environment and coral growth form. I. The Great Barrier Reef, *Palaios* **12(1)**:27–42.

Parrish, J. T., 1993, Climate of the supercontinent Pangea, *J. Geol.* **101**:215–233.

Paulay, G., 1997, Diversity and distribution of reef organisms, in: *Life and Death of Coral Reefs* (C. Birkeland, ed.), Chapman and Hall, New York, pp. 298–353.

Playford, P. E., 1980, Devonian "Great Barrier Reef" of Canning Basin, Western Australia, *Am. Assoc. Petroleum Geol. Bull.* **64**:814–840.

Precht, W. F., 1994, The use of the term guild in coral reef ecology and paleoecology: A critical evaluation, *Coral Reefs* **13**(3):135–136.

Reid, R. P., 1985, The Facies and Evolution of an Upper Triassic Reef Complex in Northern Canada, PhD Thesis, University of Miami.

Riding, R., 1981, Composition, structure and environmental setting of Silurian bioherms and biostromes in northern Europe, *SEPM Spec. Pub.* **30**:41–83.

Riding, R., and Zhuravlev, A. Y., 1995, Structure and diversity of oldest sponge–microbe reefs: Lower Cambrian, Aldan River, Siberia, *Geology* **23**(7):649–652.

Romano, S. L., and Palumbi, S. R., 1996, Evolution of Scleractinian corals inferred from molecular systematics, *Science* **271**: 640–642.

Schubert, J. K., and Bottjer, D. J., 1995, Aftermath of the Permian–Triassic mass extinction event: Paleoecology of Lower Triassic carbonates in the western USA, *Palaeogeogr. Palaeoclimatol. Palaeoecol.* **116**:1–39.

Schumacher, H., and Zibrowius, H., 1985, What is hermatypic? A redefinition of ecological groups in corals and other organisms, *Coral Reefs* **4**:1–9.

Scotese, C. R., and Golonka, J., 1992, PALEOMAP Palaeogeographic Atlas, PALEOMAP Prog. Rep. **20**: 1–34.

Sepkoski, J. J., Jr., 1992, A compendium of fossil marine families, *Milwaukee Publ. Museums Contrib. Biol. Geol.* **83**:1–155.

Senowbari-Daryan, B., Zühlke, R., Beckstadt, T., and Flügel, E., 1993, Anisian (Middle Triassic) buildups of the northern Dolomites (Italy): The recovery of reef communities after the Permian/Triassic crisis, *Facies* **28**:186–256.

Sheehan, P. M., 1985, Reefs are not so different—they follow the evolutionary pattern of level-bottom communities, *Geology* **13**:46–49.

Soja, C. M., 1996, Island-arc carbonates: characterization and recognition in the ancient geologic record, *Earth-Sci. Rev.* **41:** 31–65.

Stanley, G. D., Jr., 1981, The early history of scleractinian corals and its geologic consequences, *Geology* **9:**507–511.

Stanley, G. D., Jr., 1988, The history of early Mesozoic reef communities: A three-step process, *Palaios* **3:**170–183.

Stanley, G. D., Jr., 1992, Tropical reef ecosystems and their evolution, in: *Encyclopedia of Earth System Science* (W. A. Nierenberg, ed.), Academic Press, New York, pp. 375–388.

Stanley, G. D., Jr., 1996, Confessions of a displaced reefer, *Palaios* **11**(1):1–2.

Stanley, G. D., Jr., and Beauvais, L., 1994, Corals from an Early Jurassic coral reef in British Columbia: refuge on an oceanic island reef, *Lethaia* **27:**35–47.

Stanley, G. D., Jr., and Cairns, S. D., 1988, Constructional azooxanthellate coral communities: An overview with implications for the fossil record, *Palaios* **3**(2):233–242.

Stanley, G. D., Jr., and Fagerstrom, J. A. (eds.), 1988, Ancient reef ecosystem, *Palaios* **3:**1–142.

Stanley, G. D., Jr., and Fautin, D. F., 2001, The origins of modern corals, *Science* **291**:1913–1914.

Stanley, S. M., and Hardie, L. A., 1998, Secular oscillations in the carbonate mineralogy of reef-building and sediment-producing organisms driven by tectonically forced shifts in seawater chemistry, *Palaeogeogr. Palaeoclimatol. Palaeoecol.* **144:**3–19.

Stanley, G. D., and McRoberts, C. A., 1993, Early Jurassic reef on an island arc in the Telkwa Range, Canadian Cordillera, BC: the first post-extinction coral reef, *Can. J. Earth Sci.* **30:**819–831.

Stanley, G. D., Jr., and Swart, P. K., 1995, Evolution of the coral–zooxanthellae symbiosis during the Triassic: A geochemical approach, *Paleobiology* **21:**179–199.

Sugiyama, T., and Nagai, K., 1994, Reef facies and paleoecology of reef-building corals in the lower part of the Akiyoshi Limestone Group (Carboniferous), Southwest Japan, *Courier Forsch.-Inst. Senckenberg* **172:**231–240.

Talent, J., 1988, Organic reef-building: Episodes of extinction and symbiosis? *Senckenbergiana Lethaea* **69:**315–368.

Teichert, C., 1958, Cold and deep-water coral banks, *Am. Assoc. Petrol. Geol. Bull.* **42:**1064–1082.

Webb, G. E., 1996, Was Phanerozoic reef history controlled by the distribution of non-enzymatically secreted reef carbonates (microbial carbonate and biologically induced cement)? *Sedimentology* **43(6):**947–971.

Wells, J. W., 1933, Corals of the Cretaceous of the Atlantic and Gulf coastal plains and interior of the United States, *Bull. Am. Paleontol.* **18**(67):83–292.

West, R. R., 1988, Temporal changes in Carboniferous reef mound ecosystems, *Palaios* **3:**152–169.

Wood, R., 1993, Nutrients, predation and the history of reef-building, *Palaios* **8:**526–543.

Wood, R., 1999, *Reef Evolution*, Oxford University Press, New York.

Woodley, J. D., Chornesky, E. A., Clifford, P. A., Jackson, J. B. C., Kaufman, L. S., Knowlton, N., Lang, J. C., Pearson, M. P., Porter, J. W., Rooney, M. C., Rylaarsdam, K. W., Tunnicliffe, V. J., Wahle, C. M., Wulff, J. L., Curtis, A. S. G., Dallmeyer, M. D., Jupp, B. P., Koehl, M. A. R., Nigel, J., and Sides, E. M., 1981, Hurricane Allen's impact on Jamaican reefs, *Science* **214:**749–755.

Yancey, T. E., and Stanley, G. D., Jr., 1999, Giant alatoform bivalves in the Upper Triassic of western North America, *Palaeontology* **42:**1–23.

Zhuravlev, A. Y., and Wood, R., 1995, Lower Cambrian reefal cryptic communities, *Palaeontology* **38**(2):443–490.

Chapter 2

Phanerozoic Reef Trends Based on the Paleoreef Database

WOLFGANG KIESSLING

1. Introduction

Although many review papers and books have discussed the Phanerozoic history of reefs in detail (Newell, 1971; Heckel, 1974; Wilson, 1975; James, 1983; Fagerstrom, 1987; Copper, 1988, 1989; Talent, 1988; Flügel and Flügel-Kahler, 1992; James and Bourque, 1992; Kauffman and Fagerstrom, 1993; Hallock, 1997; Wood, 1998, 1999), several open questions remain to be answered. The major limitations in current knowledge are due to the insufficient quantification of ancient reef attributes and consequently an often subjective evaluation. Reefs vary in terms of constructional types, dominant reef-building groups, environmental setting, and petrographic attributes. These differences have led to designations of an absence of reefs in particular

WOLFGANG KIESSLING • Department of Geophysical Sciences, University of Chicago, Chicago, Illinois, 60637.

The History and Sedimentology of Ancient Reef Systems, edited by George D. Stanley Jr., Kluwer Academic/Plenum Publishers, New York, 2001.

time intervals. For example, James (1983) and James and Bourque (1992) stated that metazoan reefs were present during the Middle Ordovician to Late Devonian, the Late Triassic, the Middle to Late Jurassic, the middle Cretaceous, and the younger Cenozoic, whereas the remainder of the Phanerozoic was exclusively characterized by mounds. Although it is correct to separate true reefs and mounds, this view limits our views of reefs as individual ecosystems. To allow a comparison of reefs through time; a broad definition of reefs has to be applied: In this chapter, reefs are regarded as laterally confined carbonate structures developing due to the growth or activity of aquatic sessile benthic organisms. Four basic reef types are defined: (1) true reefs with a rigid framework of skeletal reef builders; (2) reef mounds, where skeletal reef builders and matrix are about equally important; (3) mud mounds, where skeletal organisms are minor constituents; and (4) biostromes, where dense growth of skeletal organisms occurs but no significant depositional relief is evident. Although the broad definition of reef lumps many different carbonate bodies that are not commonly described as reefs, only this definition allows one to describe and compare the reef ecosystem through time and space.

The database for the evolutionary trends discussed in this chapter is a locality/paleolocality based collection of more than 3000 Phanerozoic reefs (Paleoreef database or simply Paleoreefs). Quaternary reefs are not included in Paleoreefs, in order to avoid bias by unequal data treatment (there are superior data on Quaternary reefs). Hence, only pre-Quaternary reef development is discussed in this chapter. Each reef in Paleoreefs is described numerically, as detailed as possible and as general as necessary, to allow a comparison of reef attributes through time and to account for the heterogeneous quality of reef data in published papers. The numerical description of each reef is done by assigning quantitative data to its measurable attributes. The most reliable measure for reef attributes are fairly rough interval classifications (2 to 4 intervals for any reef attribute). A detailed description of the database and initial interpretations were recently published (Kiessling *et al.*, 1999) and a book with detailed interpretations of the database by invited reef specialists is in preparation.

The numerical characterization of the reef ecosystem for a given time slice is possible by summing up all reef data for this time slice and calculating means or percentage values of particular reef attributes. Although Paleoreefs is principally designed for analysis of reef attributes on a supersequence level (time slices defined by second order sea-level fluctuations), it also can be used for finer or coarser stratigraphic resolutions. The evolutionary trends discussed in this chapter almost exclusively refer to stages. The high stratigraphic resolution goes at the expense of statistical confidence. Hence, the mean attributes of stages with few reefs have to be interpreted cautiously.

2. An Outline of Phanerozoic Reef Evolution

Phanerozoic reefs evolved in a complex way and are characterized by pronounced expansions and retreats, both in their abundance and their global extent. This chapter summarizes the mainstream of reef evolution taking into account only the prevailing reef types in time slices. The first sessile organisms capable of forming reef structures appeared as stromatolites roughly 3.5 Ga ago (Walter *et al.*, 1980). Cyanophyceans and other microbes started to form reefal structures as early as the Archean (Nisbet and Wilkins, 1989) and major reef complexes developed in the early Proterozoic (Hoffman, 1989; Grotzinger, 1989). The ecosystem already was moderately complex in the Proterozoic (Hoffmann and Grotzinger, 1985; Turner *et al.*, 1993), but biotically diverse, heterogeneous, and ecologically complex reef communities did not evolve before the Early Cambrian, with the rise of archaeocyath sponges. With the near extinction of archaeocyath sponges at the end of the Early Cambrian, the reef ecosystem suffered a significant deterioration and stromatolites and calcimicrobes were nearly the only reef builders until the Ordovician, with the notable exception of some demosponge reefs in Iran (Hamdi *et al.*, 1995).

Reef development in the Ordovician follows a general trend toward metazoan-dominated communities. Thrombolites and calcimicrobes dominated in the Tremadocian, occasionally accompanied by tabulate corals (Pratt and James, 1989), lithistid sponges, stromatoporoidlike pulchrilaminids, or receptaculitacean algae (Toomey and Nitecki, 1979). Some reefs already were dominated by sponges, algae, or bryozoans in the late Tremadocian (Rigby *et al.*, 1995), and those groups became increasingly important in the Arenigian. Nevertheless, large reef complexes were still dominated by microbes during the Early and early Middle Ordovician (Table 1). Bryozoans were the first colonial metazoans that dominated some reefs by volume in the Ordovician (Zhu *et al.*, 1995). The oldest reefs with a pronounced community succession were described from the Middle Ordovician (Alberstadt *et al.*, 1974). In the late Middle to Late Ordovician, tabulate and rugose corals as well as stromatoporoids diversified and dominated many reef structures. This was the start of a long-lasting period in reef building in which stromatoporoids and corals prevailed. One major mass extinction falls in this interval. During the end-Ordovician crisis, reef taxa were less affected than planktic and level-bottom communities and no significant change in the structure of the reef ecosystem was observed. However, few earliest Silurian (Rhuddanian) reefs are known. Another low in global reef abundance is noted in the earliest Devonian (Lochkovian). This time is characterized by the aftermath of the Caledonian Orogeny, which led to a global increase in siliciclastic deposits (Ronov *et al.*, 1980) and a simultaneous decrease of carbonate platform environments. This reef crisis was not accompanied by a mass extinction. Silurian and Devonian reefs exhibit a high degree of similarity (Copper, 1997) and formed major reef tracts (Table 1). Shallow-water reefs were dominated by

TABLE 1. Major Phanerozoic Reefs Tracts (>500 km extension)

Region	System	Stage/Epoch	Environment	Reef type	Lateral extent (km)	Reference
Yuktansk, Siberia	Cambrian	Tommotian	Platform	Microbial mound	700	Kuznetsov and Don (1984)
Lena River to Kotyou River, Siberia, Russia	Cambrian	Atdabanian	Platform, margin	Calcimicrobe–archaeocyath reefs	800	Rowland and Gangloff (1988)
Tunguska River, Siberia, Russia	Ordovician	Tremadocian	Intertidal platform	Stromatolitic mound	1500	Ioganson (1990), Copper, pers. comm. (1999)
Ellesmere Island to Melville Island, N.W.T., Canada	Ordovician	Tremadocian–Arenigian	Shelf and shelf margin	Microbial mound	1400	de Freitas and Mayr (1995)
Central Kazachstan	Ordovician	Caradocian–Ashgillian	Island arc	Microbial mound	1500	Zadoroshnaya and Nikitin (1990); Copper, pers. comm. (1999)
Northern Urals to Vaygach Island, Russia	Ordovician	Ashgillian	Shelf margin	Microbial reefs	1000	Antoshkina (1996, 1998), Bolshakova *et al.* (1994)
Northern Greenland	Silurian	Llandovery–(Wenlock)	Platform, platform margin, and slope	Stromatoporoid–coral mounds with Stromatactis	800	Hurst (1980), Sönderholm and Harland (1989)
Hudson Bay, Ontario, Canada	Silurian	Llandovery–Wenlock	Platform	Stromatoporoid–coral–microbe reefs	1400	Suchy and Steam (1992)
Great Lakes Area, US	Silurian	Llandovery–Wenlock	Platform and platform margin	Stromatoporoid–coral–microbe reefs	1600	Lowenstamm (1950)
Tyan-Shan, Siberia, Russia	Silurian	Wenlock	Shelf margin	—	1000	Dronov and Natalin (1990), Copper, pers. comm. (1999)
Ellesmere Island to Somerset Island, N.W.T., Canada	Silurian	(Wenlock)–Ludlow	Platform margin and slope	Microbe–lithistid sponge mounds	1100	de Freitas and Dixon (1995)
Baltic to Podolia, Ukraine	Silurian	Wenlock–Ludlow	Platform	Coral–stromatoporoid reefs	1300	Zadoroshnaya *et al.* (1982), Copper, pers. comm. (1999)
Mongolia to Inner Mongolia, China	Silurian	Ludlow	Shelf margin	Tabulate coral reefs	1200	Copper and Brunton (1991)
Urals, Russia	Devonian	Emsian–Eifelian	Back arc basin	—	2000	Zadoroshnaya *et al.* (1982). Copper, pers. comm. (1999)
Mongolia	Devonian	Emsian–Eifelian	Shelf margin	Tabulate coral reefs	1400	Sharkova (1986)
Keg River/Presqu'ile, British Columbia, Canada	Devonian	Eifelian–Givetian	Platform margin	Stromatoporoid–coral reefs	900	Moore (1989)
Okhotsk to Tas-Khaykhtakh Range, Russia	Devonian	Givetian	—	Tabulate coral reefs	2200	Bolshakova *et al.* (1994)
Kolyvan-Tomsk Trough to Minusinsk Basin, Russia	Devonian	Givetian	Platform	Stromatoporoid–coral reefs	1300	Bolshakova *et al.* (1994)
Hunnan to Guangxi, South China	Devonian	Givetian–Frasnian	Platform margin	Stromatoporoid–coral reefs	750	Tsien *et al.* (1988)
Timan-Pechora Basin to Novaya Zemlya, Russia	Devonian	Frasnian	Platform margin and slope	Stromatoporoid–coral reefs	1700	Heafford (1989)

Alberta, British Columbian	Devonian	Frasnian and slope	Platform margin	Stromatoporoid–coral reefs	900	Moore (1989)
Peri-Caspian Depression, Kazakhstan and Russia	Devonian	Frasnian–Famennian	Platform margin	Microbe–stromatoporoid–coral reefs	1800	Pol'ster *et al.* (1985)
Peri-Caspian Depression, Kazakhstan and Russia	Carboniferous	Visean–Baskirian	Platform margin	Tubiphytes–algal–microbe reefs and mounds	1500	Pol'ster *et al.* (1985)
Peri-Caspian Depression, Kazakhstan	Permian	Asselian–Artinskian	Shelf margin	*Tubiphytes*–bryozoan reef mounds	1000	Yaroshenko (1986)
Western and northern Urals, Russia	Permian	Asselian–Sakmarian	Shelf margin and slope	Algal mounds	1000	Chuvashov (1983), Heafford (1989)
Delaware Basin, US	Permian	Guadalupian	Platform margin and slope	Sponge–algal–cement reefs	500	Ward *et al.* (1986)
Zechstein Basin, Lithuania, Poland, Germany, Denmark, England	Permian	Guadalupian	Platform margin and slope	Stromatolite–bryozoan reefs	1500	Kuznetsov *et al.* (1984)
Guizhou, southern China	Permian	Late Permian	Shelf margin	Sponge–algal–cement–*Tubiphytes* reefs	500	Wang *et al.* (1994)
South China	Triassic	(Anisian)–Ladinian	Platform margin	Sponge–algal–cement–*Tubiphytes* reefs	650	Fan (1980)
Northern Alps, Carpathians	Triassic	Ladinian–(Carnian)	Shelf margin	Sponge–algal–cement–*Tubiphytes* reefs	850	Flügel (1981)
Northern Alps, Carpathians	Triassic	Norian	Shelf margin	Sponge–coral reefs	650	Flügel (1981)
Timor to Papua New Guinea	Triassic	Norian–(Rhaetian)	Shelf margin	Sponge–coral reefs	2600	Flügel, pers. comm. (1999)
NW Florida to South Texas	Jurassic	Oxfordian	Platform margin	Cyanophycean–*Tubiphytes*–coral reefs	1100	Baria *et al.* (1982), Montgomery *et al.* (1999)
Poland, Germany, France	Jurassic	Oxfordian	Epeiric sea	Siliceous sponge microbe mounds	1300	Database
East coast of North America	Jurassic	Oxfordian–Tithonian	Shelf margin to upper slope	Diverse	3300	Ryan and Miller (1981); Meyer (1989), Pratt and Jansa (1989)
Slovenia to Montenegro	Jurassic	Oxfordian–Kimmeridgian	Shelf margin	Coral–stromatoporoid–algal reefs	550	Turnšek (1968), Turnšek *et al.* (1981)
South Texas to Louisiana	Cretaceous	Aptian–Cenomanian	Shelf margin	Rudist–algal–stromatoporoid reefs	1100	Bebout and Loucks (1983)
Great Australian Bight, off Australia	Tertiary	Middle Miocene	Shelf margin	Coral reefs	500	Feary and James (1995)
Red Sea	Tertiary	Middle Miocene to Recent	Fringing	Coral reefs	2000	Sestini (1965), Purser *et al.* (1996)
Taiwan to Ryukyu Islands, Japan	Tertiary	Pliocene	Shelf margin	Coral reefs	1100	Yabe and Sugiyama (1935)
Great Barrier Reef, Australia	Tertiary-Quaternary	Pliocene to Recent	Shelf margin	Coral reefs	2000	Davies et al. (1989)

tabulate–rugosan corals, stromatoporoids, calcimicrobes, and calcareous algae. Neither the end-Ordovician extinction event nor the Lochkovian low in reef abundance had profound influences on the high-ranked taxonomic composition of reefs or ecosystem structure. However, the mass extinction terminating the Middle Ordovician to Late Devonian reef interval—the Frasnian–Famennian or Kellwasser event—can be seen as the Phanerozoic biotic event with the greatest impact on the reef ecosystem.

Reefs forming subsequent to this event were completely different in biotic composition and constructional reef type. This is mostly due to the strong decline of stromatoporoids and tabulate–rugosan corals. The majority of Famennian reefs were dominated by calcimicrobes and stromatolites. Starting in the Famennian (Cook *et al.*, 1994) and proliferating in the Tournaisian to early Visean, a distinct type of reef structure developed and was characterized by growth in deeper water, abundant micrite and marine cement, and few traces of macroscopic skeletal organisms. These so-called Waulsortian mounds or banks (Lees and Miller, 1995) dominated globally in Tournaisian deep-water ramp settings, but diverse shallow-water reef types developed in the Visean and Serpukhovian and evolved independently in isolated geographic areas (Webb, 1994). Microbial activity is generally thought to be the paramount control on Mississippian reef growth both for the shallow- and for the deep-water environment. Although rare, shallow reefs are known from the Tournaisran (Webb, 1998), widespread shallow water reef growth was not evident before the middle Viscan as exemplified by diverse reef communities in Great Britain (Mundy, 1994) and Japan (Nagai, 1978, 1985). However, the Pennsylvanian started with a major decline in global reef abundance and a subsequent takeover of various algae as main reef builders. Phylloid algae were the most important but problematic tubular algae and in higher latitudes the platy alga *Palaeoaplysina* also were common reef builders. In the Late Pennsylvanian and Early Permian, the enigmatic *Tubiphytes* (= *Shamovella*) became an important reef builder. Both algae and *Tubiphytes* were often associated with productive reefal hydrocarbon reservoirs (Kiessling *et al.*, 1999) and prolific amounts of marine cement (James *et al.*, 1988). Chaetetid sponges in the Pennsylvanian and inozoan and sphinctozoan sponges in the Permian played subordinate roles in reef building until the Middle Permian. During the Permian, sponges gradually became more important reef builders and dominated especially in the Late Permian of China (Rigby *et al.*, 1994).

The Permian–Triassic mass extinction set the clock back to microbial (mostly stromatolite) reefs, for some 10 million years. Considering the devastating effect of the Permian–Triassic event, recovery of the reef ecosystem was surprisingly rapid in the Triassic, and Middle Triassic to Carnian reefs have many aspects in common with Middle and Late Permian reefs (Flügel and Stanley, 1984), although there are some notable differences (Flügel, 1994). The prevailing microbe–sponge communities of the Middle Triassic were replaced by sponge–coral communities during the Late Triassic (Chapter 7, this volume). Late Triassic reefs experienced a major innovation with the rise of zooxanthellate corals (Stanley and Swart, 1995). This enhanced the growth potential of reefs and allowed them to thrive in nutrient-limited environments. Corals become increasingly dominant in Late Triassic reefs. This trend continued into the Jurassic and was only marginally affected by the Triassic–Jurassic mass extinction event. Although Early and Middle Jurassic reefs are much less abundant than Late Triassic or Late Jurassic reefs, coral reefs form the principal carbonate factory, assisted by widespread bivalve mounds in the Early Jurassic (Geyer, 1977) and siliceous sponge–microbe reefs during the entire Jurassic. Late Jurassic shallow-water reefs are abundant and mostly predominated by scleractinian corals. An increasing contribution to reef growth by stromatoporoid sponges also is evident.

The mechanisms of the major decline in reef abundance in the earliest Cretaceous time are poorly understood. It was not accompanied by a mass extinction event, nor did the ecosystem structure change significantly. During the Cretaceous a gradual shift from coral-dominated reefs to rudist-dominated reefs is observed. Older views attribute this change to competitive replacement, but the hypothesis of environmentally induced successions (Skelton *et al.*, 1997) is currently favored. Late Cretaceous rudist reefs are peculiar due to the scarcity of binders. Rudists were apparently rarely able to build a resistant framework and some authors claim that they did not form reefs at all (e.g., Gili *et al.*, 1995). Although rudists dominated the Late Cretaceous reef ecosystem, coral reefs continued to thrive, whereas other reef types were rare.

The only obvious effect of the Cretaceous–Tertiary mass extinction on the reef ecosystem is the total extinction of rudistid bivalves. Reef-building algae were hardly affected (Moussavian, 1992; Tragelehn, 1996; Roger *et al.*, 1998, but see Aguirre et al., 2000, for a contrary view) and although zooxanthellate scleractinian corals were reported to suffer significant extinction (Rosen and Turnšek, 1989; Rosen, 2000), the overall morphology and ecological role of corals changed little after the event. The recovery interval in tropical reefs was not significantly longer than in the calcareous plankton (Tragelehn, 1996), whereas the K/T boundary virtually cuts through cool-water bryozoan mounds (Surlyk, 1997). The rise of modern-type coral–algal reefs thus already was underway in the Paleocene and most of the major reef-building coral genera were present by the end Eocene (Frost, 1977). However, reefs were rare until the Oligocene. Accompanied by regional extinction events (Edinger and Risk, 1994), global reef abundance increased significantly by the late Oligocene and especially the early Miocene.

3. Reef Distribution Patterns

The global distribution of modern photosymbiotic coral reefs is strongly controlled by water temperature, carbonate saturation, nutrient level, and photosynthetically available light. Sea surface temperature is largely a function of latitude, but oceanic surface currents transport heat and major currents can substantially alter the zonal temperature distribution. The Gulf Stream and the Kuroshio are examples of warm currents that transport enough heat to permit reef growth in latitudes that are usually devoid of reefs. These currents are warm because they originate from water masses that traveled along the equator over a whole ocean (Atlantic and Pacific, respectively), thereby gaining heat. By approaching major land masses, the westward flowing currents are deflected and move along the continental margin toward higher latitudes where they are increasingly affected by the Coriolis force and cross the oceans in an eastward direction. Both the Gulf Stream and the Kuroshio are typical western boundary currents. On the other hand, cool water currents such as the Pacific Humboldt Current and the California Current, eastern boundary currents, originate from subpolar latitudes and travel along the eastern margin of the Pacific ocean. These currents considerably confine the latitudinal extent of zooxanthellate coral reefs. Besides their cool temperatures, eastern boundary currents also bring along high nutrient concentrations that are harmful to modern coral reefs (Hallock and Schlager, 1986; Hallock *et al.*, 1988; Chapter 11, this volume; Brasier, 1995). Nutrient levels are especially enhanced because offshore (trade) winds deflect eastern boundary currents away from the coast and allow nutrient-enriched cool intermediate water to enter the surface, a process known as coastal upwelling. Nutrients also are delivered by continental erosion and coastal areas are generally nutrient enriched in comparison with the open ocean. Therefore, the mouths of large rivers are usually devoid of reefs (but see De Moura *et al.*, 1999). In addition to nutrients, rivers deliver freshwater and often bring high suspension loads that reduce light penetration depth. As photosymbiosis in modern coral reefs requires light, fluvial input is harmful to reef growth, resulting in both reduced light and salinity. Clastic and dissolved material derived from the continents is increased if the sediment source is mountainous and humid, but modern mangrove systems may act as traps freeing reefs from siltation.

With the knowledge of modern latitudinal temperature distribution, global surface current patterns, and the location of upwelling zones, the global distribution pattern of photosymbiotic coral reefs can be predicted and vice versa some oceanographic parameters may be inferred from the reef distribution pattern. The latitudinal range of reefs is from nearly 34° N to nearly 32° S on western ocean margins but only 27° N to 2° S on their eastern side, except for the eastern Indian Ocean where the warm Leeuwin Current off Australia allows reef growth down to 31° S (all latitudinal ranges taken from ICLARM Reefbase, http://www.reefbase.org). The close match of physical parameters and reef distribution in the Recent has led many authors to use reefs as paleoclimatic indicators (e.g., Frakes *et al.*, 1992; Parrish, 1998). However, one has to take into account that reef ecology may have changed significantly through time. It is unlikely that reefs predominantly composed of non-photosymbiotic reef builders would present distribution patterns like modern reefs. Large-scale reef tracts are known that exhibit distribution patterns completely different from commonly depicted patterns. Mounds predominantly composed of the nonzooxanthellate coral *Lophelia* form a continuous belt along the Norwegian coast, in cold, deep, and nutrient-rich water (Teichert,

1958; Freiwald *et al.*, 1997). These mounds differ significantly from tropical zooxanthellate coral reefs in geometry, biodiversity, guild structure, and petrographic attributes, and thus are easily distinguished. In the geologic record, however, this is more problematic, and physical parameters and ecology cannot be reliably inferred from reef distribution patterns alone.

The global distribution of Phanerozoic reefs varied significantly, not only for the paleolatitudinal range but also for the longitudinal dispersal and tectonosedimentary settings. Global reef patterns are herein broadly assigned to actualistic and nonactualistic patterns. Reef settings that replicate a modern-type setting, that is, low latitude, western ocean margins, or open oceanic, subtropical gyre regions, are referred to as actualistic settings. Reef areas presumably influenced by eastern boundary currents, situated in high ($>30°$) paleolatitudes, far inside epeiric seas, or in siliciclastic environments, are termed nonactualistic settings. PaleoReefs is linked to 32 global paleogeographic maps permitting a detailed analysis of Phanerozoic reef distribution patterns. The evolution of reef patterns is exemplified by a selection of paleogeographic maps showing the major periods of reef development (Figs. 1–4).

The Early Cambrian reef expansion (Fig. 1a) exhibits a pattern that may be profoundly different from that of the Recent. Reefs were largely confined around 20° S at the northern margin of Laurentia and within continental blocks (Siberia). A considerable number of reef sites are located in high paleolatitudes and in eastern boundary settings (North Gondwana, South China). The Gondwana reefs in Spain, Morocco, and Sardinia (Italy) are additionally likely to have been affected by seasonal coastal upwelling (Golonka *et al.*, 1994). Only the Antarctic and South Australian reefs occur in a setting roughly comparable with the Recent western Pacific. Interpretation of the global pattern is restricted owing to still disputed plate tectonic reconstructions for the Cambrian.

The Middle Cambrian to Middle Ordovician is generally characterized by few reefs and the patterns are difficult to evaluate. The Middle and Late Cambrian pattern is largely a relict of the Early Cambrian and very few new reef areas are encountered. It is noteworthy that the only non-microbe-dominated reef (Hamdi *et al.*, 1995) is situated at the eastern margin of the Paleoasian Ocean. A reexpansion of reefs is evident in the Early Ordovician to Middle Ordovician (Webby, 1984). However, the great majority of reefs thrived at eastern boundary settings or along the southern margin of Laurentia. Reefs appear to be absent in the western boundary settings, and thus may be regarded as nutrient-opportunistic.

The Late Ordovician (Caradocian–Ashgillian) exhibits two pronounced and latitudinally restricted reef zones (Fig. 1b). The northern reef zone ranges longitudinally from the northwestern margin of Laurentia to eastern Australia and latitudinally from the equator to about 25° N. The southern hemisphere reef zone extends from southern Laurentia to South China and from 9° S to 33° S. No reefs developed in the equatorial southern hemisphere. Additional

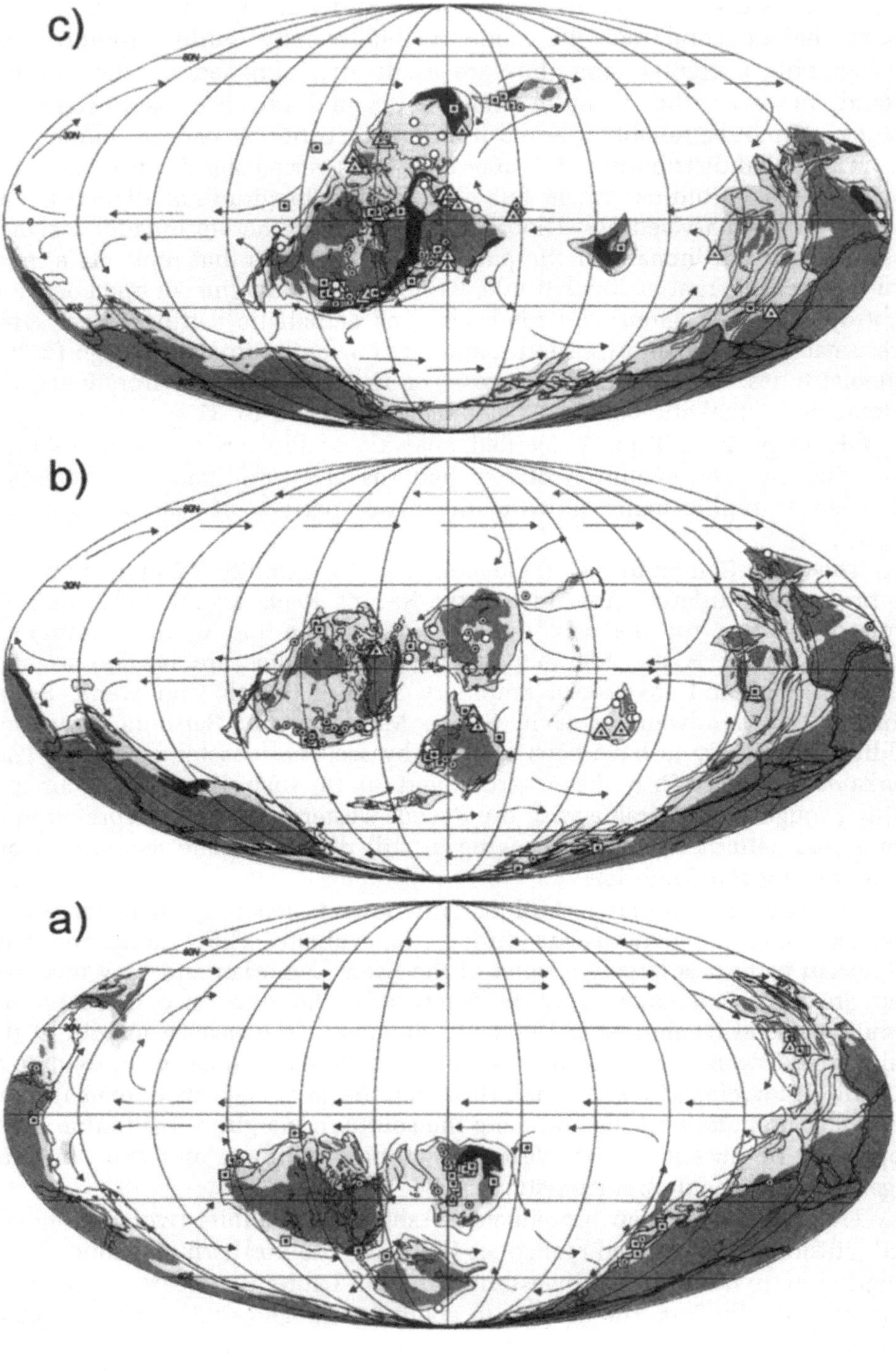

FIGURE 1. Early Paleozoic reef patterns: (a) Early Cambrian (520 Ma plate tectonic reconstruction); (b) Late Ordovician (452 Ma plate tectonic reconstruction); (c) Wenlockian–Ludlovian (425 Ma plate tectonic reconstruction). In this and Figs. 2–4, reef thickness is indicated: Small

reefs are known from extremely high southern latitudes (e.g., Vennin *et al.*, 1998), but those differ significantly in biotic composition from lower latitude sites. Reefs in the Timan–Pechora region and in northwestern Laurentia developed in modeled seasonal upwelling zones (Golonka *et al.*, 1994). Except for these occurrences, none of the Late Ordovician reef settings can be regarded as nonactualistic. All reefs were located in low latitudes and very rarely reefs were associated with eastern boundary currents. Additionally, although many reefs invaded epeiric seas, their distance from the shelf break was rarely more than 500 km. The largest reefs were constructed in low latitudes (20° S to 10° N). However, reefs associated with western boundary currents also are rare. Until now no reefs have been reported from east Greenland and Scotland, from Antarctica, or from the (former) eastern margin of the Siberian plate.

Early Llandoverian reefs were rare and occupy relict settings of the Ashgillian. During the course of the Llandoverian, reefs expanded rapidly and reef areas were occupied that previously had been lacking reefs in the Late Ordovician (e.g., Hudson Bay region and some places in Siberia). The Wenlockian–Ludlovian (Silurian) reefs are strongly concentrated on Laurussia ranging from 30° S to 45° N (Fig. 1c). The high latitude occurrences of reefs in the northern hemisphere are to be viewed with caution as tectonic reconstructions for Asian plates are still disputed. No latitudinal reef zones are observed. Major reef complexes existed from 30° S to 30° N with no significant concentration in western boundary settings. However, reefs also were rare in eastern boundary settings (Copper and Brunton, 1991). The overall reef distribution is similar to that of the Late Ordovician, but reefs now invaded deeply into epeiric seas. The Caledonian orogeny created a large mountain range along the southern and eastern margin of Laurentia, leaving behind a much more restricted epeiric sea. The fairly large reefs in southern Laurentia (northeastern United States to eastern Canada) were situated in subtropical latitudes where an arid climate can be assumed (evaporites). Therefore, the southern Laurentia setting is similar to the Miocene Mediterranean and with some restrictions to the Recent Red Sea. Reefs in areas of predicted upwelling (Moore *et al.*, 1993; Golonka *et al.*, 1994) are rare. The only occurrences are the reefs in the Himachal Himalayas (Bhargava and Bassi, 1986) that presumably are affected by continuous coastal upwelling. Although some Himalayan reefs additionally are situated in a siliciclastic sequence, none are markedly different from typical reefs in this time slice, being composed of coral–stromatoporoid–microbial framestones. This suggests that nutrients were not a major control in mid-Silurian reefs. Long reef tracts are especially common on the southern margin of Laurentia, but most reefs are small. The largest reefs

---

concentric hexagons—reefs less than 10 m thick; concentric squares—reefs 10–100 m thick; concentric triangles—reefs more than 100 m thick; circles—unknown thicknesses. Paleogeography based on Golonka (personal communication, 1996, modified from Scotese and Golonka, 1992). Surface ocean currents derived form continental configuration.

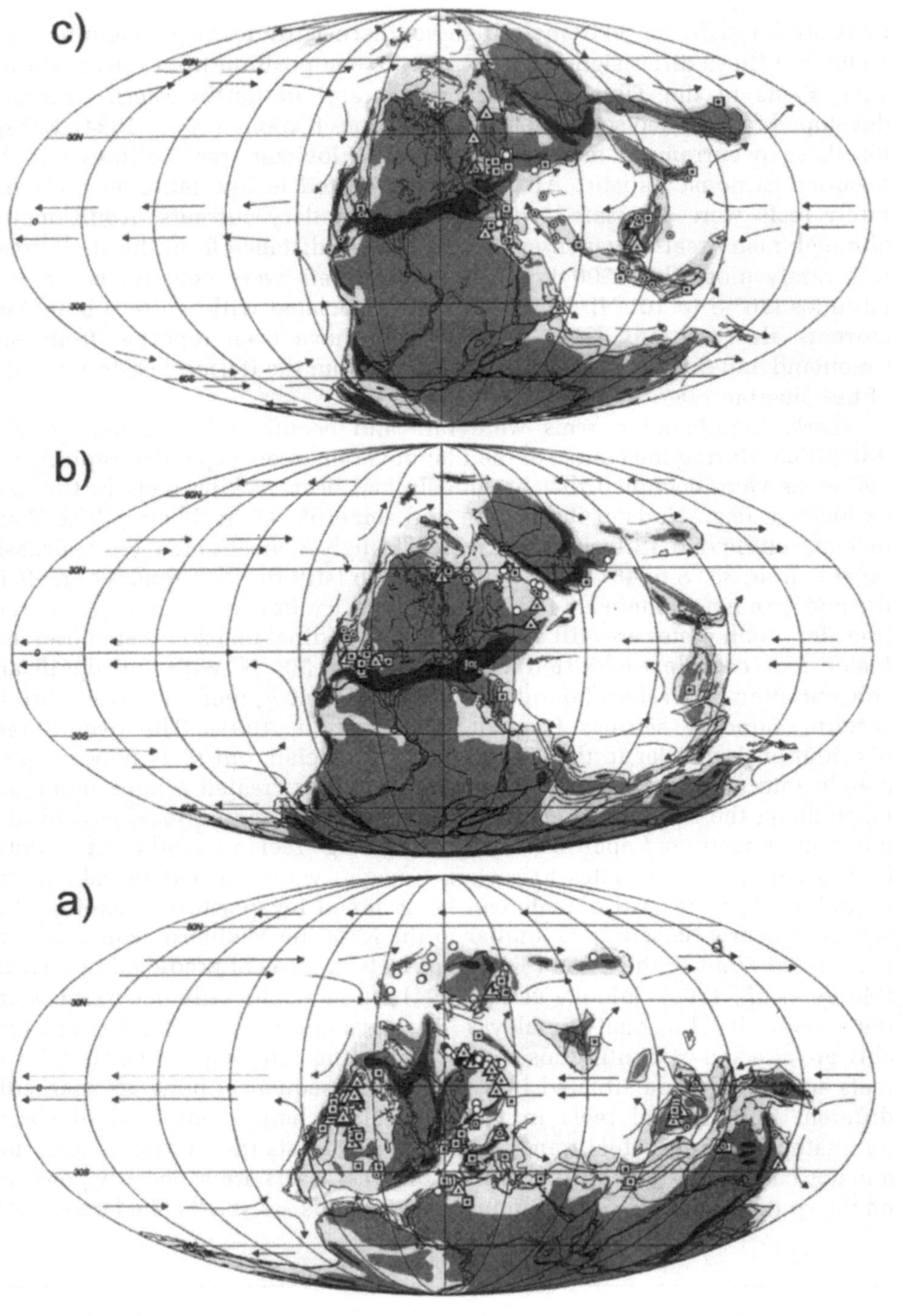

FIGURE 2. Devonian to Late Permian reef patterns: (a) Givetian–Frasnian (368 Ma plate tectonic reconstruction); (b) Moscovian–Kasimovian (302 Ma plate tectonic reconstruction); (c) Middle to Late Permian (255 Ma plate tectonic reconstruction).

attaining more than 500-m thickness are reported from the Timan–Pechora (Russia), and Gaspé Peninsula (Quebec, Canada).

In the Pridolian to Pragian, reef abundance declined, probably due to the widespread siliciclastic shedding of Caledonian mountain belts. However, reefs continued to grow close to mountain ranges such as in southern and northern Laurentia. Reef expansion in the Emsian–Eifelian was very significant with calculated paleopositions up to 65° N. The highest northern hemisphere occurrences refer to Asian continental blocks and data are not very reliable. Reefs directly bordering the eastern margins of oceans generally are rare.

Givetian–Frasnian reefs (Fig. 2a) were very widespread with a paleolatitudinal range from 42° S to more than 50° N. Even when considering the problematic position of Asian continental blocks, the reef zone obviously was extremely wide at this time. Major reef complexes of more than 100-m thickness range from 36° S to 35° N. Compared with older Devonian reefs, the number of reefs associated with eastern boundary currents is significantly higher. Additionally, Kiessling *et al.* (1999) have shown that many Givetian–Frasnian reefs occur in regions with a high probability of at least seasonal upwelling. Nearby upwelling on a global scale does not mean necessarily that nutrient-rich cool water affected the reefs, since marginal oceanographic barriers may have effectively protected the reefs from the adverse conditions typical of open eastern ocean basins (Whalen, 1995). Additionally, the oceans may have been nutrient-depleted on a global scale (Martin, 1996) and upwelling did not necessarily deliver cool and nutrient-enriched intermediate water to the shelves as it does today. Nevertheless, the overall distribution of reefs can be interpreted as strongly nonactualistic.

The Famennian reef pattern is essentially a relict of the Frasnian, although the biota were different. In many areas reefs continued to grow after the Kellwasser event but all are significantly smaller and nearly exclusively microbial. New reef areas were occupied during the Mississippian. Shallow-water reefs often are found in regions presumably influenced by western boundary currents (eastern Australia, Japan) or in landlocked settings such as the remnants of the Rheic ocean between Laurussia and Gondwana.

Pennsylvanian reefs (Fig. 2b) illustrate a latitudinally confined distribution pattern between 19° S and 32° N. Reefs are strongly concentrated on the western and northeastern margins of growing Pangea, that is, in southwestern North America and along the closing Ural ocean in Russia. These are the only sites where thick reef complexes formed. Framework reefs are rare but reef mounds continued to grow, especially in western boundary settings. The majority of shallow-water reefs occurred in areas where modern type reefs should be expected also.

The overall distribution pattern did not change in the latest Pennsylvanian and earliest Permian time, but reef numbers increased in the Urals region and Barents Sea, whereas they decreased in southwestern North America. This trend continued into the later Early Permian when reefs additionally encountered the transitional plates between the closing Paleotethys and the opening Neotethys.

In the Middle and Late Permian, many reefs occur in landlocked settings. This is the case for the North American Delaware Basin and for the Zechstein Basin. Equally abundant, however, are reefs along the margins of the Paleotethys and the Neotethys (Fig. 2c). Large reef complexes are concentrated in South China. The reef zone did extend beyond modern latitudes (43° S to 46° N) but reef occurrences outside the ±30° latitude zone are rare and poorly known. Eastern boundary settings are generally devoid of reefs. Where models indicate upwelling (Golonka *et al.*, 1994), reef occurrences are missing. The only exception is the South China reefs, which were separated, however, from the open ocean by land barriers. The overall pattern can be interpreted as moderately actualistic.

After the Permian–Triassic mass extinction, the reef zone narrowed considerably. Scythian mounds as well as Anisian to Ladinian reefs were largely confined between 21° N and 7° S. An expansion of the reef zone is observed in the Carnian and especially in the Norian–Rhaetian when reefs flourished from 38° S to more than 40° N (Fig. 3a). In spite of their wide latitudinal extent, the great majority of reefs thrived in what can be considered actualistic settings and thick reef complexes are mostly limited to western boundary settings. An important exception is the prolific reef growth at the western margin of Pangea (including North American terranes) and the northern margin of Australia, where reefs are associated with eastern boundary currents and are close to predicted upwelling sites (Kiessling *et al.*, 1999). Paradoxically, the reoccupation of eastern boundary settings appears at the same time that there is the first Mesozoic evidence of photosymbiosis in corals. Although the Norian reefs on east Panthallasa terranes rarely reach the size and complexity of Tethyan reefs, the biota exhibit a high degree of similarity (Stanley, 1988, 1994).

The Triassic–Jurassic extinction event led to a profound reduction of reef areas and reef carbonate production. The relict coral reef areas are situated in actualistic and nonactualistic settings (e.g., Stanley and Beauvais, 1994). Morocco is the only reef area newly occupied by coral reefs in the Early Jurassic. Bivalve reefs and banks are more widespread. During the Middle Jurassic coral reefs achieve a more global distribution but are strongly concentrated in the western Tethys. Reefs associated with eastern boundary settings are rare and small. The Late Jurassic witnessed a major global expansion of reefs (Fig. 3b). The overall pattern is clearly actualistic with nearly all reefs associated with the western boundary currents and very few reefs occur at the eastern margins of oceans. Three observations do not fit an actualistic pattern: (1) The latitudinal reef zone was strongly expanded; (2) many reefs grew far inside epeiric seas; and (3) some reef areas are associated with seasonal coastal upwelling as modeled by Golonka *et al.* (1994).

With few exceptions, the earliest Cretaceous reefs form only a relict pattern of the Late Jurassic. Starting with the Barremian, reefs became increasingly abundant in the open Pacific Ocean. Reef growth on Pacific atolls was especially prolific during the Aptian–Albian (Fig. 3c). The more or less continuous reef tract surrounding the North Atlantic region from the Late

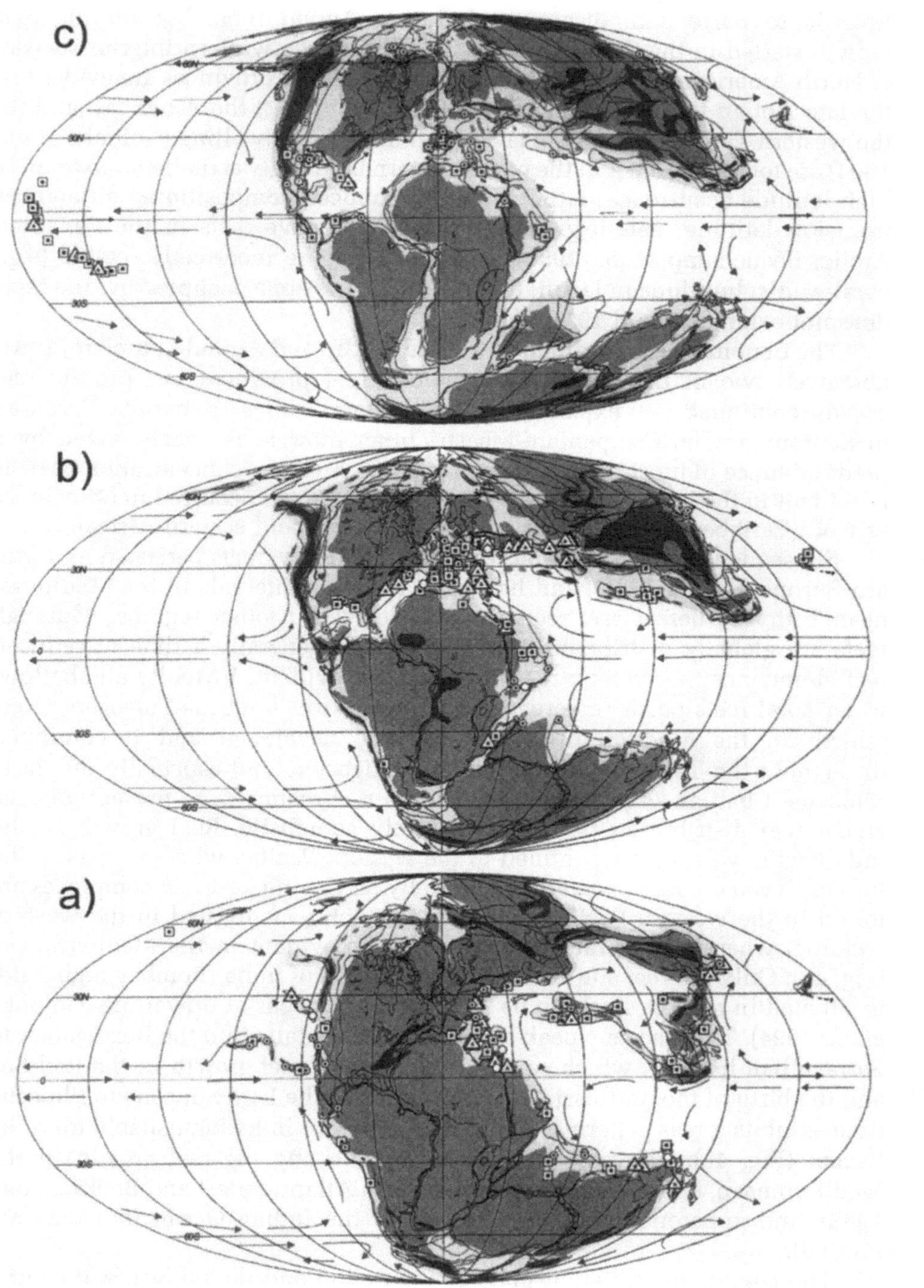

FIGURE 3. Mesozoic reef patterns: (a) Late Triassic (218 Ma plate tectonic reconstruction); (b) Late Jurassic (152 Ma plate tectonic reconstruction); (c) Albian-Cenomanian (105 Ma plate tectonic reconstruction).

Jurassic to Barremian, disappeared during Aptian time, but prolific reef growth started in the Gulf of Mexico area and in the western interior seaway of North America. Three longitudinally confined reef provinces are evident in the late Aptian to Cenomanian: (1) the mid-Pacific, (2) the Americas, and (3) the western Tethys. Reefs associated with modeled upwelling (Golonka *et al.*, 1994) are totally absent and the global pattern is strongly actualistic. Extremely high latitude occurrences either show pronounced compositional differences from low-latitude reefs [e.g., serpulid worm–bivalve reefs in the Canadian Arctic; Beauchamp *et al.*, 1989)] or are likely to be incorrectly rotated [e.g., coral–rudist buildups in Japan) (Sano, 1991)]. The zone occupied by true reefs does not extend beyond 19° S and 32° N.

The Cenomanian–Turonian reef crisis (Philip and Airaud-Crumière, 1991) destroyed two of the Albian–Cenomanian reef provinces but prolific reef growth continued and expanded in the western Tethys during the Turonian to Santonian. The Campanian–Maastrichtian interval is characterized by a predominance of biostromes rather than reefs. A strong concentration of reefs is evident in the Mediterranean Tethys but reef growth occurred in other areas as well (Caribbean, North America, Pacific, Oman, and eastern Africa).

Subsequent to the Cretaceous–Tertiary boundary, reefs persisted in northern Europe (Surlyk, 1997) and had short recovery intervals in the Mediterranean Tethys, whereas reef recovery was delayed in other regions. Although reefs are globally distributed, the Paleogene is generally a time of reduced reef abundance; no distinct reef provinces are evident. Virtually all shallow-water coral reefs occur in actualistic settings throughout the Paleogene. Reef growth in the Mediterranean was strongly developed and it continued throughout the Tertiary. During the Late Oligocene and especially the Early Miocene (Chattian–Aquitanian) all modern reef realms were present and the global reef distribution pattern was purely actualistic. Reef growth in the Indo-Pacific was mostly confined to the western Pacific, whereas reefs in the Indian Ocean are rare or rarely reported. By far the thickest reef complexes are found in the western Pacific, moderate to thick reefs existed in the western Atlantic, whereas moderate to small reefs are evident in the Mediterranean (Fig. 4a). Only the reef site at the western margin of India (Bombay high field) is situated in an area likely to be associated with coastal upwelling (Golonka *et al.*, 1994). The Neogene peak of reef abundance falls into the Burdigalian to Serravallian interval, which saw the initiation of reef growth in the Red Sea and the birth of the northern Great Barrier Reef. The Late Miocene to Pliocene time exhibits a reef pattern (Fig. 4b) that is almost indistinguishable from the Recent (Fig. 4c). The only exceptions are the ongoing reef growth in the Mediterranean, a reef occurrence in the North Atlantic (Best and Boekschoten, 1982), and apparently little reef growth in the Indian Ocean (but data are limited).

The completely different plate configuration and the widespread epeiric seas in older periods limit the direct comparison with Cenozoic reef distribution patterns. However, the above-described changes in reef patterns in relation to reconstructed ocean surface currents and latitude allow the

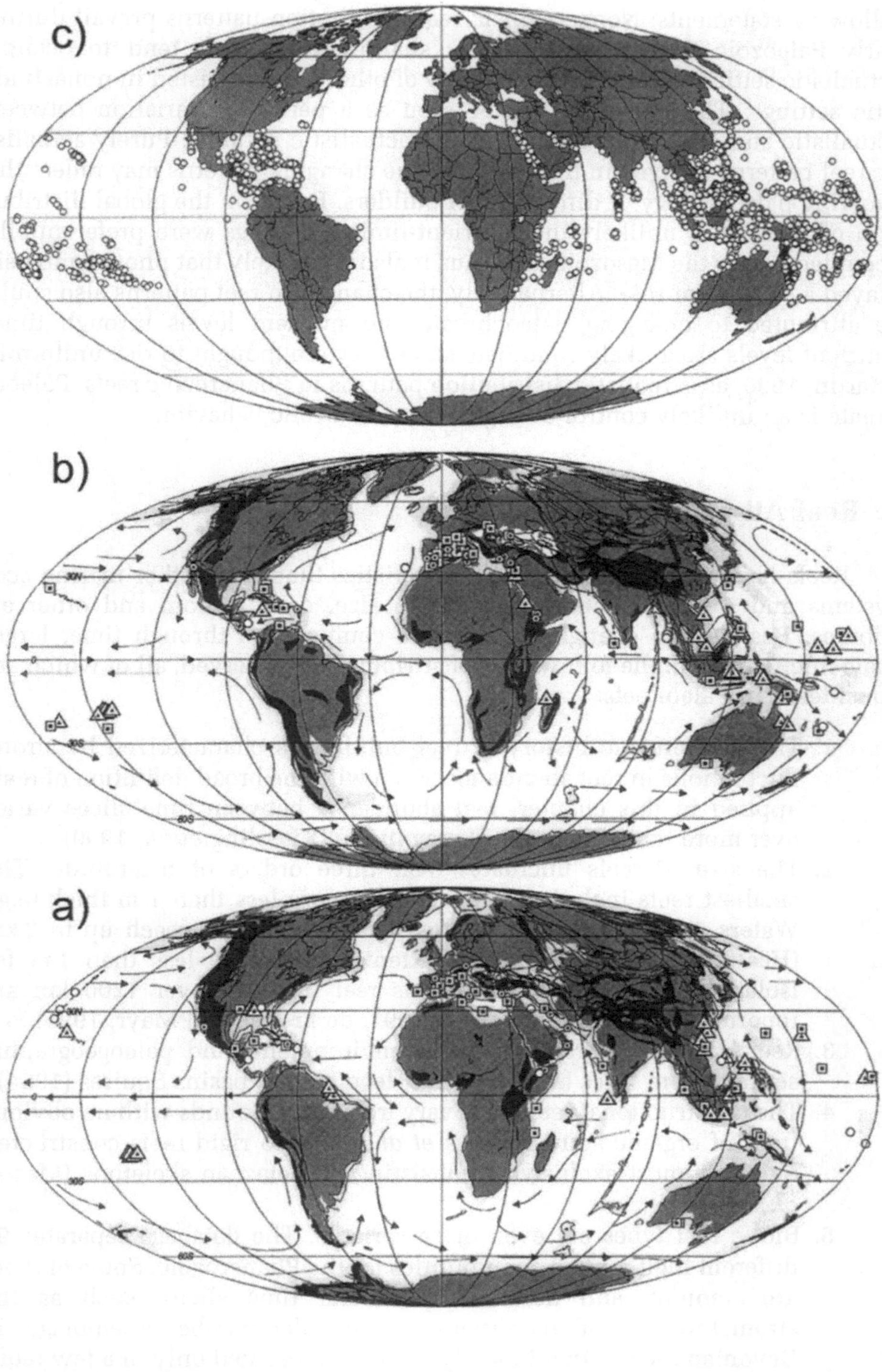

FIGURE 4. Cenozoic reef patterns: (a) Late Oligocene–Early Miocene (22 Ma plate tectonic reconstruction); (b) Late Miocene–Pliocene (6 Ma plate tectonic reconstruction); (c) Recent (data from Reefbase: http://www.reefbase.org/). See 1 for legend.

following statements: Nonactualistic reef distribution patterns prevail during Early Paleozoic time. Late Paleozoic shallow-water reefs tend to occupy actualistic settings, whereas the majority of other reefs persisted in nonactualistic settings. The Mesozoic can be seen as a period of variation between actualistic and mixed nonactualistic and actualistic patterns. Purely actualistic reef patterns evolved in the Tertiary. The changing patterns may reflect the evolving paleoecology of different reef builders. Based on the global distribution of reefs, it is unlikely that nutrient-limited settings were preferentially occupied before the Mesozoic, and thus it also is unlikely that photosymbiosis played a significant role. Alternatively, the changes in reef patterns also could be attributed to changing paleoclimate and nutrient levels through time. Nutrient levels are a likely candidate since they are thought to rise uniformly (Martin, 1996) as actualistic distribution patterns in Phanerozoic reefs. Paleoclimate is an unlikely control owing to its more cyclic behavior.

4. Reef Attributes through Time

Reefs have more measurable characteristics than most other marine ecosystems and exhibit greater variations in size, composition, and other attributes. Besides the changing taxonomic composition through time, large-scale fluctuations in the following reef attributes are observed, all of which are considered in PaleoReefs:

1. The Phanerozoic history of reef building is characterized by strong fluctuations in reef abundance. Even with the broad definition of reefs applied in this chapter, reef abundance between time slices varies over more than one order of magnitude (Kiessling *et al.*, 1999).
2. The size of reefs fluctuates over three orders of magnitude. The smallest reefs included in the database are less than 1-m thick (e.g., Waters, 1989), whereas the largest reef complexes reach up to 2 km (Heafford, 1989). The lateral extent also can be less than 1 m for isolated mounds, but continuous reef tracts of over 1200 km are reported (Copper and Brunton, 1991; de Freitas and Mayr, 1995).
3. Reefs form in a great variety of environments and paleogeographic settings from lakes (Arp, 1995) to deep marine basins Squires (1964).
4. The constructional reef types vary from mud mounds with no obvious trace of organic activity (Ross *et al.*, 1975) to rigid reefs constructed by and almost exclusively consisting of metazoan skeletons (Mas *et al.*, 1997).
5. Biotic reef types are even more variable. The database separates 90 different high-ranked communities for the Phanerozoic. Some of them are common and dominate particular time slices, such as the stromatoporoid–tabulate/rugose coral–calcimicrobe assemblage in Devonian reefs. Other biotic types are represented only in a few reefs. One example is the construction of reefs by encrusting foraminifers.

Biostromes of this type only are known from the Barremian of France (Wernli and Schulte, 1993), the early Eocene of southern France and northern Spain (Plaziat and Perrin, 1992), and the late Oligocene–early Miocene of Iran (unpublished data).

6. Reefs can be constructed by virtually monospecific associations but they may consist of more the 100 species of reef builders (Fürsich and Wendt, 1977), in addition to often very diverse reef dwellers.
7. Although all reefs are organic structures, they form by a variety of processes. The dominant process in modern reefs is construction of a dense framework by scleractinian corals and coralline algae. Binding of sediment by laminar biota is a common process in mud mounds is the precipitation of carbonate by microbial metabolic activity. The baffling of carbonate mud, formerly thought to be an important process in reefs, is now regarded only of minor significance.
8. Ecological succession in reefs can be very pronounced (Alberstadt *et al.*, 1974) but also can be totally absent. The same is true for the lateral zonation into fore-reef, reef crest, and back-reef. Reefs with a distinct development of facies zones are the exception rather than the rule.
9. Bioerosion in reefs can be very intense (Perry, 1996), but rarely is reported in pre-Late Triassic reefs.
10. The differentiation of carbonate between reefs and their environment also is subject to strong variations. Most modern and many ancient reefs export a high proportion of skeletal material to the fore-reef and back-reef environments (Hubbard *et al.*, 1990, 1998; Harris, 1994). This process leads to the interfingering of reefs and adjacent sediments, often obscuring the actual geometry of a reef construction. However, there also are many reefal structures that stand virtually isolated in a sedimentary sequence (e.g., Jeffery and Stanton, 1996) or even import (trap) sediment from the surrounding environment (Bosence *et al.*, 1985).
11. The petrographic features of reefs also vary considerably. Micrite is virtually ubiquitous in reefs, usually filling the voids between reef builders, but it is the dominant component in mud mounds. Spar usually is less abundant, but particular Paleozoic and Triassic reefs may contain extremely high proportions of marine cement (Webb, 1996; Chapter 5, this volume). Many ancient reefs form important hydrocarbon reservoirs. However, the reservoir quality of reefs varies extremely and rapidly through time (Kiessling *et al.*, 1999).

The Phanerozoic fluctuations of delineated attributes were defined by quantitative means using reef attributes in PaleoReefs. The variation of these means through time roughly may be divided into two categories: (1) attributes that vary through time without a significant overall trend, and (2) attributes that tend to either decrease or increase through time. This has particular value when searching for possible extrinsic controls on reefs.

4.1. Fluctuating Reef Attributes

The most widely discussed reef attributes are those that do not show a significant trend through time. Of particular importance are reef abundance (Fig. 5), reef dimensions (Fig. 5), ecological succession, and reef diversity (Fig. 6). All these attributes vary strongly through time, but none follows a distinct trend nor exhibit a regular cyclicity.

4.1.1. Reef Abundance

Pronounced peaks in reef abundance are separated by lows ranging in duration from 25 Ma to 90 Ma. The maximum number of reefs in a stage is noted in the Frasnian, followed by the Givetian, the Serravallian (Miocene), and the Oxfordian. However, as stages have different chronostratigraphic durations, the number of reefs constructed per million years gives a more meaningful measure. Here, the Phanerozoic maximum is in the Neogene, particularly in the Messinian, Burdigalian, and Serravallian, only then followed by the Frasnian and Oxfordian peaks (Fig. 1). Older reefs tend to have a progressively higher probability of being eroded or subducted/metamorphosed. Therefore, carbonate cycling has to be considered in order to calculate the original number of reefs. Sedimentary cycling models commonly assume an exponential decay curve (Gregor, 1985; Wilkinson and Walker, 1989; Wold and Hay, 1993), although the fit to the real-world data is quite poor. This approach allows the reconstruction of the original number of reefs for a given age (Fig. 1). With this reconstruction the Frasnian again represents the Phanerozoic maximum in reef construction, followed by the Wenlockian, Messinian, and Ludlovian.

4.1.2. Reef Dimensions

Similar fluctuations, although less pronounced, are evident in the mean dimensions of reefs. The mean thickness of reef structures shows no significant trend through time. Mean thickness is low to moderate in most Cambro-Ordovician reefs, moderate to high in Silurian to Mississippian time, low in most Pennsylvanian reefs, high in the Early Permian, moderate to low in the middle and Late Permian, low in the Early to early Middle Triassic, high in the late Middle and Late Triassic, moderate during the Jurassic (except for the Hettangian and Middle Jurassic lows), moderate in the Early Cretaceous, low in the Late Cretaceous, and moderate to high in the Cenozoic. Continuous trends toward increasing or decreasing reef thickness are also rare. The longest continuous rise in reef thickness is from the Bashkirian (Early Pennsylvanian) to the Artinskian (50 Ma), during the time of Gondwanan glaciation. Another continuous rise is evident in the Triassic, from the Scythian to the Ladinian, spanning about 20 Ma. Discontinuous trends toward increasing thickness are observed from the Late Ordovician to the Late Devonian and from the latest

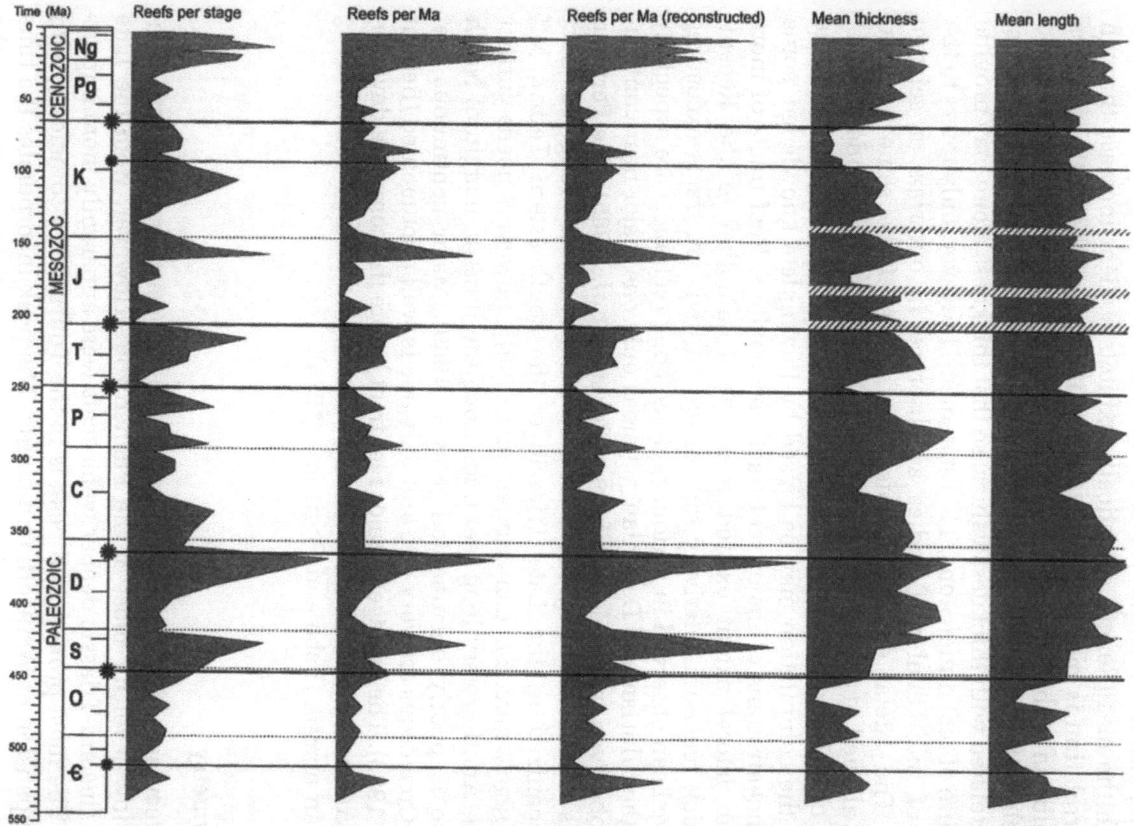

FIGURE 5. Variations in reef abundance and mean reef dimensions through the Phanerozoic; stage and epoch level (for simplicity termed stages in this and subsequent figure captions) stratigraphic resolution. The Phanerozoic maximum in reef numbers per stage lies in the Frasnian. "Reefs/Ma" refers to the normalized reef abundance balancing the different duration of stages. The Phanerozoic peak is now in the Neogene, particularly in the Messinian. "Reefs/Ma (reconstructed)" compensates the erosion of older reefs assuming an exponential decay curve. The thick lines marked with an asterisk denote important reef crises and mass extinctions. Diagonal-hatched bars indicated limited data.

Cretaceous to the Neogene. Decreases in reef thickness are even less continuous. Abrupt falls are followed by either thickness variations on lower levels or rapid increases. Significant declines in reef thickness are often but not always associated with reef crises, for example, the Frasnian–Famennian, the Permian–Triassic, and the Triassic–Jurassic boundaries.

Although the mean length statistically exhibits a slight increase through time, the correlation is mostly due to the significant but discontinuous rise from the Cambrian to the Devonian, whereas lateral extent subsequently varied on high levels without a pronounced trend (Fig. 5). Reef lengths are almost perfectly correlated with thickness values, so that thickness alone may provide a good measure of reef dimensions. This statement, however, only refers to the dimensions of individual reef bodies as observed in outcrops or seismic exploration. The lateral extent of reefs tracts (reefs of the same age and composition aligned in a near-continuous stripe) is almost independent of reef thickness. A compilation of major Phanerozoic reef tracts is provided in Table 1. Although the compilation may be biased by incomplete knowledge, some temporal concentrations of major reef tracts are obvious. Reef tracts of more than 500 km lateral extent existed from the Tommotian to the Recent. Although thick reefs are relatively rare in the Ordovician, tracts exceeding 1000 km have been observed in various regions. Long reef tracts are especially common in the Silurian and Devonian, whereas only one major near-continuous tract is observed in the Carboniferous (Pol'ster *et al.*, 1985). The Permian to Jurassic is characterized by fairly long reef tracts. A very long reef tract recently recognized in the Late Triassic of the southeastern Tethys [off Australia (Flügel, personal communication, 1999)], however, needs further confirmation. Another very long reef tract along the Jurassic margin of North America also is poorly documented (Meyer, 1989). Major continuous reef tracts in the Cretaceous are very rare and only one is well documented (Bebout and Loucks, 1983). The youngest major reef tract before the complete development of the Australian Great Barrier Reef has been noted in the Pliocene along the East Asian margin (Yabe and Sugiyama, 1935).

4.1.3. Bathymetry

The percentage of reefs growing below the fair-weather wave base fluctuated at a low level for most of the Phanerozoic. However, from the Late Devonian to the end-Triassic deeper water reefs were apparently more important (Fig. 6). The most pronounced peak is in the Tournaisian coinciding with the global expansion of Waulsortian mounds. The peak incorporates the time interval Frasnian to Visean, only interrupted by a slight decrease in the Famennian. Another major peak interval ranges from the Moscovian to Artinskian. Paradoxically, all these intervals are characterized by a dominance of presumably autotrophic organisms (microbes and algae) as the major reef builders. This observation also holds true for some minor Phanerozoic peaks in deeper water reefs such as the Ashgillian and the Oxfordian.

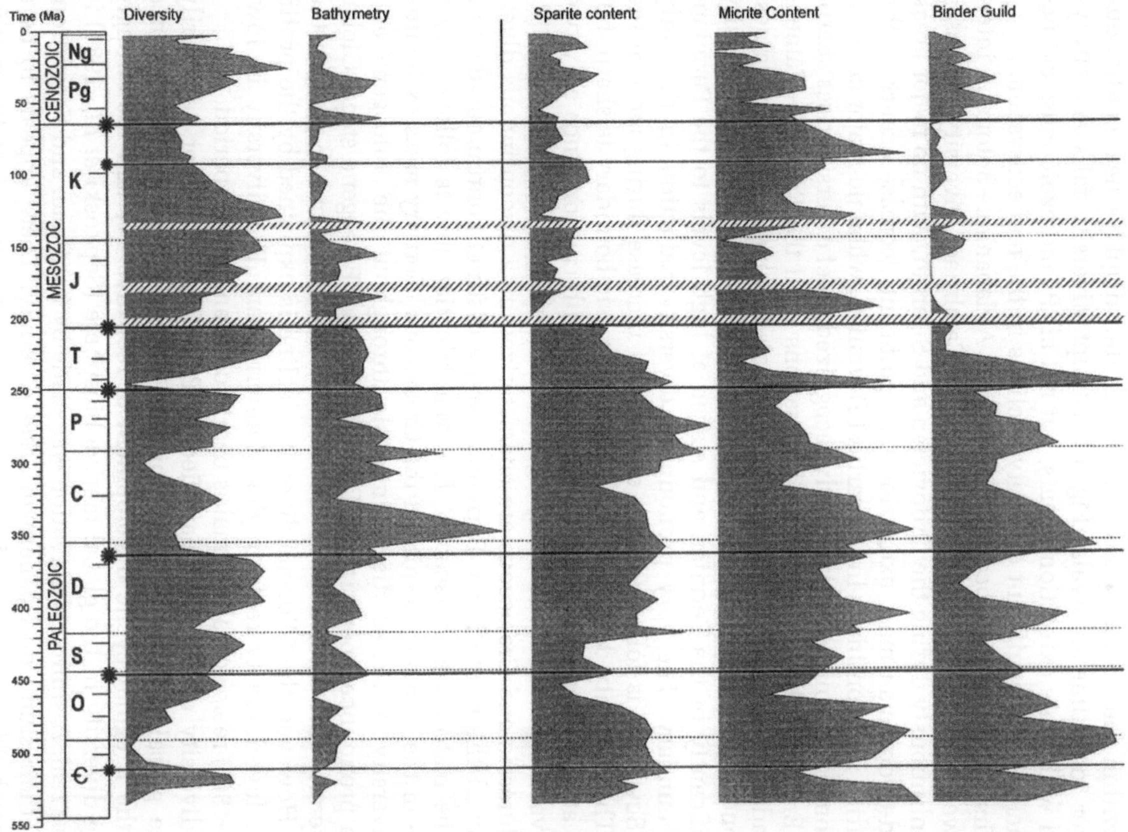

FIGURE 6. Variations in reef diversity, bathymetric setting, spar and micrite content, and the relative importance of the binder guild through the Phanerozoic; stage and epoch level stratigraphic resolution. Diversity and bathymetry exhibit no significant trend, whereas the binder guild and the spar and micrite content significantly decrease through time. Diagonal pattern indicates insufficient data to precisely define reef attributes.

4.1.4. Diversity

The diversity of reef builders was determined in PaleoReefs for each reef using three intervals: low (less than 5 species of reef-builders), moderate (5-25 species), and high (more than 25 species). The mean global diversity of reef builders within reefs fluctuates strongly but often displays near-continuous trends over long time intervals (Fig. 6). Abrupt diversity falls are always associated with mass extinction events, but not all Phanerozoic mass extinctions coincide with significant diversity declines in the reef ecosystem.

The first diversity peak occurred during the Atdabanian–Botomian interval, followed by a sharp drop in the late Early Cambrian. Diversity increases almost continuously during the Ordovician and Silurian, with no major break at the end-Ordovician mass extinction. A significant decrease occurs in the latest Silurian and continues in the earliest Devonian, when no major crisis in level-bottom communities is commonly recognized. Reef diversity rises sharply in the Emsian, followed by minor fluctuations until the Frasnian–Famennian boundary. The Carboniferous usually reveals low mean diversity values, but a second-order diversity peak occurs in the Visean to Bashkirian. Diversity rose significantly in the Permian and stayed at high levels to the Permian-Triassic boundary. The very pronounced diversity reduction in the Early Triassic (Scythian) is followed by a continuous increase during most of the Triassic. The diversity decline at the Triassic-Jurassic boundary is sharp, but the curve at this boundary has to be interpreted with caution since only two diversity values enumerate the Hettangian reef diversity.

The Jurassic and Cretaceous are characterized by a discontinuous diversity rise until the Hauterivian–Barremian and a distinct continuous decline through the rest of the Cretaceous. The Maastrichtian low is followed by a slight rise in the Danian, in spite of the Cretaceous–Tertiary mass extinction event. Diversity increases discontinuously throughout the Paleogene and exhibits a pronounced peak in the Chattian, the last Paleogene stage. Global mean diversity then decreases in the Miocene and has a minor peak in the Pliocene. However, the Miocene diversity decline is explained by the proliferation of low-diversity reefs in the Mediterranean. The diversity in low-latitude ($<30°$) reefs actually remains nearly constant in this period.

The diversity of reefs as quantified in PaleoReefs does not necessarily reflect the global diversity of reef builders. However, the great difference between the irregular cyclic development of reef diversity and the strongly increasing diversity in the global marine biosphere (Sepkoski *et al.*, 1981) is likely to be real. A similar observation was made by Kauffman and Fagerstrom (1993) in their study on Phanerozoic reef diversity. There appears to be a certain threshold for diversity in the reef ecosystem but not in the marine biosphere in general. Diversity as measured in PaleoReefs is not really a true measure of health of the global reef ecosystem. A decrease in mean reef diversity can be caused by the growth of many reefs in extreme habitats where only a low diversity of reef builders can be sustained. The just-mentioned example of the Mediterranean illustrates this bias. It may well be that a less healthy reef ecosystem would have been unable to cope with the events associated with the Messinian salinity crisis (Krijgsman *et al.*, 1999). Although the global diversity mean of Messinian reefs is reduced by the abundant low-diversity reefs in the Mediterranean region, diverse assemblages could grow in other areas (Budd and Johnson, 1999).

4.2. Evolving Reef Attributes

The majority of reef attributes stored in PaleoReefs follow a significant trend through time. None of these trends is continuous; they are masked by significant skips and reversals. Negative and positive trends can be subdivided.

4.2.1. Negative Trends

The most notable negative trends are seen in petrographic attributes and in features that are related to microbial activity. Both micrite and spar content decrease significantly through time. These attributes were quantified in PaleoReefs in their relative contribution to reef growth and in relation to the skeletal organisms. Their decrease through time consequently indicates that the relative contribution of skeletal reef builders significantly increased throughout the Phanerozoic. The amount of spar as quantified in PaleoReefs refers exclusively to synsedimentary or very early diagenetic sparitic cement. Sparite content fluctuates on high levels in the Paleozoic and the Triassic but decreases sharply at the Triassic–Jurassic boundary and fluctuates at low levels from the Jurassic to the Pliocene. There is no significant trend in spar content within the Paleozoic and the Mesozoic/Cenozoic, respectively; the negative trend is only significant for the whole Phanerozoic. This may indicate a different geochemical regime within the reefs in the two intervals. The abundance of spar in reefs as quantified by PaleoReefs agrees fairly well with the distribution of biocementstone reefs (Webb, 1996). However, mean spar content in Paleozoic reefs also is enhanced outside the biocementstones intervals of Webb (1996).

The relative amount of micrite fluctuates strongly and the decreasing trend is not evident at first glance. High values prevail in the Early Paleozoic and especially in the Mississippian. The Pennsylvanian and Permian exhibit moderate micrite contents. Except for the Scythian, micrite content in Triassic reefs is fairly low. In Jurassic reefs, micrite content is highest in Sinemurian and Pliensbachian and lowest in the Tithonian stages. Micrite content increases for most of the Cretaceous and discontinuously decreases from the Coniacian to the Neogene. The average amount of micrite in reefs of a particular stage is clearly linked to the prevailing reef type. It is highest if mud mounds or loosely packed reef mounds are abundant and low if framework reefs or densely packed biostromes predominate. Micrite content increases in the aftermath of most mass extinction events, especially after the Permian–Triassic event.

The guild concept as defined for reefs (Fagerstrom, 1987, 1991) has little in common with the original guild concept as defined for ecological studies (Precht, 1994). It nevertheless is useful to characterize the major constructional groups involved in reef building, with the exception of the ill-defined baffler guild (Fagerstrom and Weidlich, 1999). PaleoReefs separates three guilds: constructor, baffler, and binder. Binding of sediment as a predominant way of reef construction is usually done by microbes and algae but lamellar skeletal metazoans also can be binders. The precipitation of carbonate by microbial activity also is included in the binder guild. The abundance of reefs predominated by the binder guild decreases significantly through time, as does the proportion of mud mounds and microbial mounds. The binder guild prevails in most Paleozoic stages and in the Early to Middle Triassic. Very few reefs are dominated by the binder guild from the Late Triassic to the end-Cretaceous. The binder guild is more important again in Tertiary reefs but subordinate to the constructor guild.

4.2.2. Positive Trends

Significant increases are noted in the relative abundance of reef-derived debris (debris potential), bioerosion intensity, the relative importance of the constructor guild, the proportion of reefs growing at the shelf or platform margin, and the percentage of reefs growing in high latitudes. The debris potential of reefs is quantified in PaleoReefs as the relative amount of reef-derived debris produced by a reef. The absolute volume produced by a reef is termed "debris production." In modern reefs, more than 50% of the carbonate produced by reef builders can be transformed into sediment, mostly by bioerosion (Hubbard *et al.*, 1990). Roughly half the reefal debris remains in the reef body, whereas another 50% is exported into the adjoining environment, mostly the fore-reef area. However, ancient reefs can have a much higher debris potential: up to 90% of the carbonate produced by Triassic reefs were exported from the reef body (Harris, 1994). Even Holocene reefs may consist almost exclusively of loose sediment and rubble below the surface (Hubbard *et al.*, 1998). Debris potential in ancient reefs also can be very low: this is shown by surrounding sediments with very little or even without reefal material. Debris potential is increasing more continuously throughout the Phanerozoic than other reef attributes (Fig. 7). The peaks in debris potential are usually associated with peaks in skeletal framework reefs. However, reefs dominated by the constructor guild usually do not show as a high debris potential in the Paleozoic as they do in the Mesozoic and Cenozoic.

Bioerosion is a very important factor controlling reef growth in modern reefs (Hutchings, 1986; Glynn, 1997). However, bioerosion is commonly neglected in the description of ancient reefs. This is especially true for the intensity of bioerosion, which is indicated only by very few detailed studies (e.g., Perry, 1996). Consequently, PaleoReefs only contains data on the presence or absence

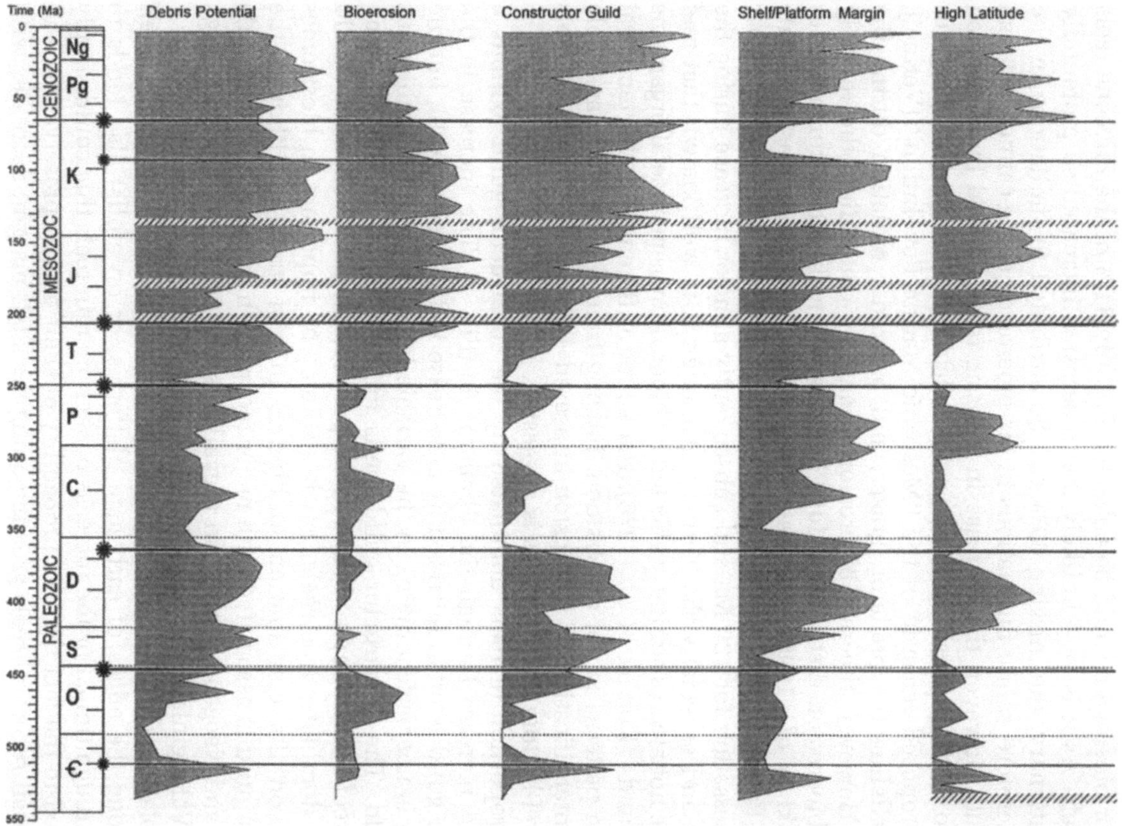

FIGURE 7. Reef attributes that tend to increase during the Phanerozoic. Debris potential (the relative amount of reefal debris produced by a reef), the number of reefs affected by bioerosion, and the dominance of the constructor guild increase most significantly. The increase in the relative amount of shelf–platform margin reefs is less pronounced but statistically significant. Paradoxically, also the percentage of high latitude reefs (>30°) increases through the Phanerozoic, although modern reefs are thought to be especially adapted to warm temperature. Diagonal pattern indicates insufficient data to precisely define reef attributes.

of bioerosion. The percentage of reefs evidently affected by bioerosion is commonly low in the Paleozoic, increases sharply in the Triassic, and fluctuates on high levels for the remainder of the Phanerozoic (Fig. 7). In comparison with most other curves depicted in this chapter, the confidence level of particular values is usually low. Virtually all modern reefs, especially in nutrient-rich areas, are affected by bioerosion to a variable degree, but evidence for bioerosion could be collected for only 31% of the Pliocene reefs in PaleoReefs. As it is very unlikely that bioerosion increased so profoundly in the Quaternary, the missing 69% can be attributed to limitations in the dataset. Therefore, minor fluctuations in the bioerosion curve cannot be taken at face value. However, the increase in Triassic reefs affected by bioerosion conforms to the hypothesis and appears to reflect the real situation. It agrees with the proposed scenario of the "Mesozoic Marine Revolution" (Vermeij, 1977), paralleling the predator, grazer, and bioeroder radiation during the Mesozoic. However, the reef bioerosion seemingly contradicts the recently described bioerosion distribution for Phanerozoic level-bottom communities (Kowalewski *et al.*, 1998). The data suggest that bioerosion was moderate during the Paleozoic and in the Early to Middle Triassic, followed by a near complete cessation for 120 Ma and subsequently a major increase during the Cretaceous and especially the Tertiary. Although it is well known that reef builders are not always similarly affected by bioerosion as perireefal organisms or hardgrounds (Kobluk *et al.*, 1978), the strong difference in the temporal distribution patterns is not understood. Judging from the currently available data, the intensification of bioerosion appeared significantly earlier in reefs than in level-bottom communities is suggested.

The proportion of reefs dominated by the constructor guild exhibits a pronounced increase through the Phanerozoic (Fig. 7). The increase in the constructor guild is often mirrored by a decrease in the binder guild, but the correlation is not perfect owing to the variable importance of the ambiguous baffler guild. The constructor guild tends to predominate in the Silurian–Devonian reef-building episode and in the post-Triassic. The constructor guild commonly dominates true reefs, but it also may be important in biostromes depending on the contribution of binders to reef growth. Although the constructor guild is associated with major reef growth in the Silurian, Middle Devonian, and Neogene, it also can prevail in stages with very few reefs (e.g., Aalenian, Valanginian).

The relative abundance of reefs bordering basins (growing at the shelf margin, along the edges of carbonate platforms, or on atolls) also tends to increase through time. For the increase of shelf margin reefs this could simply be attributed to the fact that the probability of shelf margin preservation, and thus the fossil record of reefs growing along the shelf margin, decreases with time. However, there is no reason to assume a similar relationship for the margins of carbonate platforms. Additionally, even if the platform margin is completely eroded it may be traced by its reefal debris in deeper water environments (e.g., Montañez, and Osleger, 1996). The observation of increasing platform or shelf margin reefs thus may be real and could be related to changing oceanographic conditions at the platform or shelf margin. The relative increase of margin reefs is very discontinuous (Fig. 7). Very few platform/shelf margin reefs are known in the Cambrian, Ordovician, and Early Silurian, whereas the proportion of platform margin reefs is fairly high in the Late Silurian and Devonian. Shelf/platform margin reefs usually are rare in the Carboniferous but common in the Permian. The Triassic exhibits a rapid increase of margin reefs with a peak from the Ladinian to the Norian and a decline in the Rhaetian. The proportion of shelf/platform margin reefs fluctuates extremely in the Jurassic and Cretaceous and shows a discontinuous rise in the Tertiary. Profound changes occur after mass extinction events. The latest Devonian Hangenberg event, the Permian–Triassic mass extinction, and the Triassic–Jurassic boundary events led to a substantial loss of shelf/platform margin reefs. However, no significant change is observed in the aftermath of the end-Ordovician or Frasnian–Famennian mass extinctions. A significant increase of shelf/platform margin reefs is noted after the Cretaceous–Tertiary boundary.

A paradox is obvious for the mean paleolatitude of reefs (absolute values) and for the percentage of high latitude (>30°) reefs (Fig. 7). As discussed above, reefs probably became progressively adapted to low-nutrient settings and the global reef distribution patterns became more comparable to the Recent during the Phanerozoic. Modern zooxanthellate coral reefs are highly adapted to warm water with an optimal range of 26–28°C (Hubbard, 1997). Since global paleoclimate did not "improve" through time, there is a problem explaining the reef trend towards higher latitudes. The paradox is partly resolved by the observation that cool-water high latitude reefs are more commonly observed in younger times. However, even if identified cool-water reefs are excluded from the analysis, the trend remains significant.

5. Reef Evolutionary Units

The categorization of the history of life into discrete units has attracted earth scientists ever since the beginning of modern geology. Major developments in animal evolution were traditionally used to define the three Phanerozoic eras. To interpret the marine fossil record Sepkoski (1981) proposed division into three great evolutionary faunas: Cambrian, Paleozoic, and modern fauna. This approach provided a more quantitative definition of metazoan evolutionary units than did previous views. However, the method was based purely on stratigraphic ranges of high-ranked taxa and did not take into account ecological innovations.

After realizing this shortcoming, the term ecological evolutionary units (EEUs) was introduced by Boucot (1983). Based on a qualitative comparison of communities characterized by long-term ecological stability, he defined 12 EEUs for the Phanerozoic. The ecological stability within EEUs is characterized by slow in-place evolution and a limited adaptation of clades to new habitats. The EEUs are separated by brief intervals of community reorganization and adaptive radiation. Most EEU boundaries coincide with mass extinction events. Sheehan (1996) revised the original EEUs by noting that three were related to recovery phases after mass extinction events (Fig. 8). The EEUs were defined by comparing marine level-bottom communities, that is, communities outside the reef ecosystem.

The next question is, can the EEU model be transferred to the reef ecosystem, and if so are both reef and nonreef units in phase with each other? There have been several attempts to subdivide the evolutionary history of reefs into discrete units or cycles. James (1983) distinguished two grand cycles, a Cambrian to Devonian cycle and a latest Devonian to Cenozoic cycle (Fig. 8). Each grand cycle was said to exhibit a similar evolution pattern in reef-building taxa, thus representing some kind of mega-succession. James (1983) also provided a subdivision on a finer scale based on the prevalence of large metazoans in reef-building communities. He recognized six intervals: (1) Middle and Late Ordovician; (2) Silurian and Devonian; (3) Late Triassic; (4) Jurassic; (5) middle Cretaceous; and (6) middle and Late Tertiary (Fig. 8). The periods in between these metazoan intervals are characterized by reef mounds and mounds without a major contribution of metazoans to reef construction. Copper (1988) used a similar approach to delimit major reef intervals. He applied the ecological succession model for reefs (e.g., Alberstadt *et al.*, 1974) to the history of reef building. In so doing, Copper (1988) defined six erathemic successions. Each erathemic succession starts after an extinction phase that is characterized by very poor reef development. The subsequent phase of reorganization exhibits so-called "arrested successions." The "climax phases" finally are periods when reefs were diverse, large, and widespread. Climax phases were defined by Copper (1988) in the (1) late Early Cambrian, (2) Silurian and Devonian, (3) Permian, (4) Late Triassic, (5) Late Cretaceous, and (6) post-Paleocene Cenozoic (Fig. 8). Both the demarcation of metazoan reef units (James, 1983) and the erathemic succession model (Copper, 1988) give the impression of a discontinuous reef evolution that may be difficult to reconcile with the paradigm utilizing ecological evolutionary units from level-bottom communities.

The same can be said for the major reef community complexes of Boucot (1983), the history of reef ecosystems of Stanley (1992) and the episodes of reef building as defined by Talent (1988). Talent (1988) produced the finest resolution of Phanerozoic reef-building episodes that is currently available. As much as 16 episodes of organic reef building were defined but many of these episodes are coincident with reef crises and their aftermath. Reef crises, however, demarcate boundaries between reef-building episodes rather than representing the episodes themselves. If we include the crisis and recovery episodes in the subsequent episodes, we acquire a new definition for 10 reef-building intervals (Fig. 8). James and Bourque (1992) assembled a more continuous subdivision of reef building, although they still claimed differences between times of major reef development and periods when only mounds occurred. For the Phanerozoic, six major reef intervals are again demarcated (Fig. 8). These are (1) Cambrian to Early Ordovician, (2) Middle Ordovician to

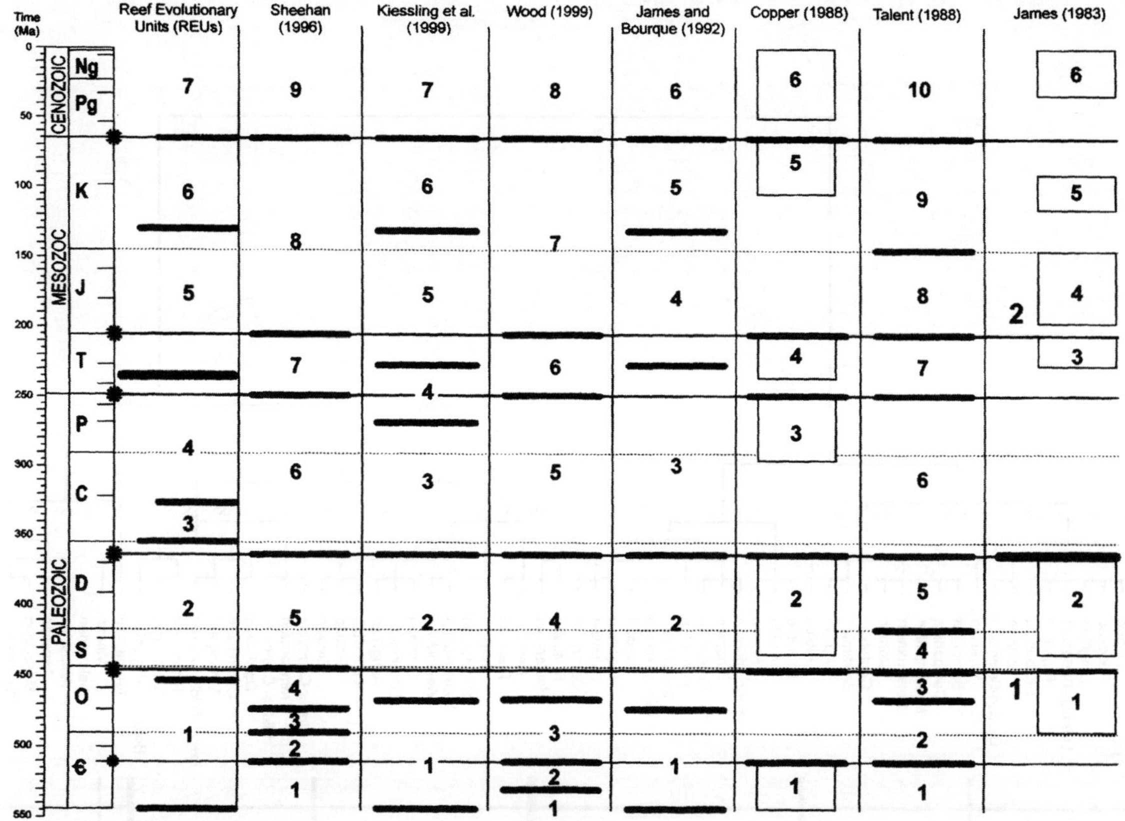

FIGURE 8. Comparison of some suggested subdivisions for the evolutionary history of reefs (Kiessling *et al.*, 1999; Wood, 1999; James and Bourque, 1992; Copper 1988; Talent, 1988; James, 1983), the currently used ecological evolutionary units (Sheehan, 1996), and the reef evolutionary units (REUs) proposed in this chapter (9). The length of bars for REUs indicates the magnitude of difference between adjacent REUs. Boxes for James (1983) and Copper (1988) indicate reef and "reef climax" intervals, in contrast to mound intervals and times of "arrested successions," respectively.

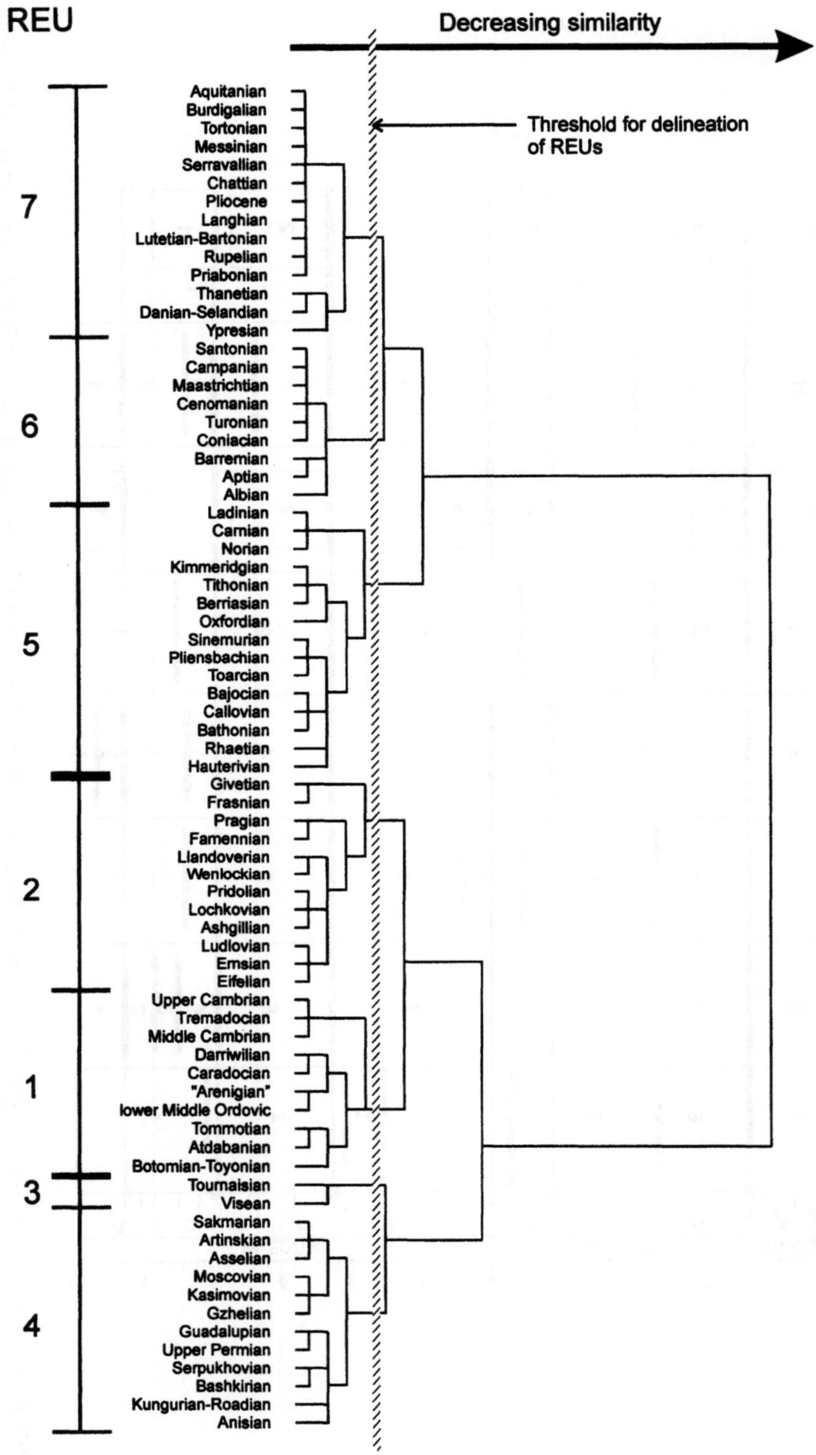
REU
Decreasing similarity
Threshold for delineation of REUs
7
6
5
2
1
3
4
Aquitanian
Burdigalian
Tortonian
Messinian
Serravallian
Chattian
Pliocene
Langhian
Lutetian-Bartonian
Rupelian
Priabonian
Thanetian
Danian-Selandian
Ypresian
Santonian
Campanian
Maastrichtian
Cenomanian
Turonian
Coniacian
Barremian
Aptian
Albian
Ladinian
Carnian
Norian
Kimmeridgian
Tithonian
Berriasian
Oxfordian
Sinemurian
Pliensbachian
Toarcian
Bajocian
Callovian
Bathonian
Rhaetian
Hauterivian
Givetian
Frasnian
Pragian
Famennian
Llandoverian
Wenlockian
Pridolian
Lochkovian
Ashgillian
Ludlovian
Emsian
Eifelian
Upper Cambrian
Tremadocian
Middle Cambrian
Darriwilian
Caradocian
"Arenigian"
lower Middle Ordovic
Tommotian
Atdabanian
Botomian-Toyonian
Tournaisian
Visean
Sakmarian
Artinskian
Asselian
Moscovian
Kasimovian
Gzhelian
Guadalupian
Upper Permian
Serpukhovian
Bashkirian
Kungurian-Roadian
Anisian

Late Devonian, (3) Mississippian to Middle Triassic, (4) Late Triassic earliest Cretaceous, (5) Cretaceous, and (6) Cenozoic. The most recent subdivision of reef history was done by Wood (1999), who distinguished eight periods of reef building: (1) Pre-Tommotian, (2) Early Cambrian, (3) Middle Cambrian to Middle Ordovician, (4) Middle Ordovician to Late Devonian, (5) Late Devonian to Permian, (6) Triassic, (7) Jurassic to Cretaceous, and (8) Cenozoic.

All the above subdivisions use a mixture of reef abundance and rough paleontological data to define reef periods. The subjective treatment of the available data resulted in several discrepancies (Fig. 8). The PaleoReef database offers a way to resolve evolutionary units in the ancient reef ecosystem based on a more objective evaluation of data. In addition, more reef attributes than just global abundance and dominant reef builders are utilized by the PaleoReefs database. A first evaluation of PaleoReefs led Kiessling *et al.* (1999) to define seven major Phanerozoic reef units (Fig. 8) based on the predominant reef-builders: (1) Cambrian to Early Ordovician: microbial mounds dominated, although archaeocyath sponges were additionally important for a limited time; (2) Middle Ordovician to Late Devonian: metazoan reef builders become important and reefs and reef mounds prevailed; (3) Latest Devonian to Early Permian: reef mounds and mounds with dominant microbes, algae, and bryozoans; *Tubiphytes* becomes additionally important later in this interval; (4) Middle Permian to early Late Triassic: reef mounds, mounds, and reefs dominated by coralline sponges, microbes, algae, or corals; (5) Late Triassic to earliest Cretaceous: shallow water reefs commonly dominated by corals; other reef types such as bivalve banks or siliceous sponge mounds were common; (6) Cretaceous: biostromes, reef mounds and reefs dominated by either rudists or scleractinian corals; and (7) Cenozoic: reefs predominated by corals and algae.

This subdivision, although more objective than previous approaches, also is biased by a limited utilization of data, namely, only prevailing reef builders and constructional reef types were considered. This approach, however, had the advantage of coinciding closely with the ecological evolutionary unit concept of Boucot (1983). If all reef attributes (abundance, size, environment, biota, diversity, constructional type, guilds, petrography) are considered and evaluated by techniques of cluster analysis, the emerging pattern is quite different. There is a problem, however, in placing the statistically defined similarities between stages into a stratigraphic context. For example, cluster

---

FIGURE 9. Cluster analysis of mean reef attributes in Phanerozoic stages and series. Quantified paleontological (dominant biota, diversity, dominant guild, bioerosion), geometrical (constructional reef type, size), environmental (paleogeographic setting, bathymetry), and petrographical (micrite and spar content) attributes were considered in the analysis. To define similarities in a stratigraphic context the age of the "stages" was included and given double weight in clustering. Note that stages with few data (Nemakit–Daldynian, Scythian, Hettangian, Valanginian) were excluded from the analysis. The clusters allow the definition of seven reef evolutionary units (REUs). Squared Euclidean distance and between groups average linkage method applied.

analysis reveals that Scythian reefs are similar to Middle and Late Cambrian reefs, but they hardly can be assigned to the same reef evolutionary unit. Therefore, the age was double-weighted in cluster analysis and out-of-sequence stages were assigned to the next appropriate cluster. The different cluster methods produce some controversial results in details but allow definition of high-ranked units (Fig. 9). Two major units are evident, equivalent to the grand cycles of James (1983), but with a different duration. The first unit ranges from the Early Cambrian to the Middle Triassic and the second unit ranges from the Middle Triassic to the Recent. Each unit consists of two major subunits. In the Paleozoic, one subunit ranges from the Early Cambrian to the Famennian and the other from the Tournaisian to the Anisian. In the Mesozoic, the two subunits include (1) the Ladinian to Hauterivian and (2) the Barremian to Pliocene. Below these clusters several others are evident, but here the analysis is limited to a cluster number that is comparable to the number of formally defined ecological units (Sheehan, 1996). In so doing, seven intervals can be defined (Fig. 9): (1) Earliest Cambrian to Caradocian, (2) Ashgillian to Famennian, (3) Tournaisian and Visean, (4) Serpukhovian to Anisian, (5) Ladinian to Hauterivian, (6) Barremian to Maastrichtian, and (7) Danian to Pliocene.

I propose the term reef evolutionary units (REUs) for the stratigraphic intervals with similar reef attributes. Due to the inclusion of many nonbiotic attributes, the definition of REUs differs considerably from the subdivisions of previous authors who recognized specific intervals of reef history. However, all these attributes are inherent in the reef ecosystem, and hence are measures of reef evolution, whereas other approaches mostly refer to the evolution of the reef macrofauna. The first important difference from previously recognized intervals is the extension of the Cambro-Ordovician cycle to the Caradocian. Most authors recognized major changes at the base or within the Middle Ordovician. The second major difference is the combination of the Famennian with the older Devonian and Silurian stages, whereas all previous subdivisions marked the Frasnian–Famennian boundary as the end of the Silurian–Devonian reef interval. The unusual composition of Tournaisian and Visean reefs has been noted previously but these stages usually were not assigned to an individual evolutionary unit. Tournaisian–Visean reefs, however, are as different from other Carboniferous and Permian reefs as are Tertiary reefs from Cretaceous reefs (Fig. 9). Admittedly, the differences would be less if only shallow-water reefs were considered, but it is the aim of this study to analyze the global reef ecosystem through time. The similarity of Middle/Late Permian reefs and Middle Triassic reefs is long known (Flügel and Stanley, 1984; Stanley, 1988) and some authors previously have demarcated a Late Paleozoic reef interval ranging into the Middle Triassic (Fig. 8). The only surprise is that Ladinian reefs are not included in the Paleozoic cluster, although their similarity with Middle Permian reefs explicitly has been mentioned (Flügel and Stanley, 1984). The next REU boundary in the earliest Cretaceous agrees with the divisions of

James and Bourque (1992) and Kiessling *et al.* (1999). It predates the actual "rudist takeover" by some 15 Ma, suggesting that the change in the reef ecosystem was not strictly speaking caused by the rudist radiation. The next REU boundary coincides with the Cretaceous/Tertiary (K/T) boundary agreeing with other models. Apparently, the ecosystem changes at the K/T boundary were more substantial than suggested above. Paleocene and Ypresian reefs form a separate subcluster, but the younger Tertiary reefs are very similar in overall composition.

The REUs are not directly comparable to the EEUs in the sense that they are not defined by coappearing, coexisting community groups. However, the data (dominant biota, diversity, guilds, bioerosion, paleoenvironment, geometry, petrography) are thought to define the reef ecosystem more accurately than the qualitative characterization of community types. Although many of the data considered in the delineation of REUs are not paleontological at first glance, most of them can be seen as geologic expressions of organic activity and synecological relationships. Thus their inclusion in the REU definition is justified.

Neither the reef units defined by the dominant biota (Kiessling *et al.*, 1999) nor the REUs defined herein agree well with EEUs. This already has been noted by Boucot (1983), although later Sheehan (1985) claimed that the match between reefs and level-bottom communities is better than proposed here. The dissimilarity could be explained by the slower rate of innovation and recruitment in the reef ecosystem. I conclude that reef ecosystem evolution versus evolution in level-bottom communities were largely decoupled throughout the Phanerozoic. To better understand the nature of reef evolutionary units, an understanding of the intrinsic and extrinsic controls on reef evolution is crucial.

6. Controls on Reef Evolution

The database summary presented above clearly indicates major temporal fluctuations of all attributes that define reefs. What is the driving force behind these variations? It is quite probable that no single factor is responsible for the variations observed, but what was the most likely combination of factors and which control predominates? Some general statements are realized. As a first approach, three end-member hypotheses can be formulated: (1) reef attributes are controlled by low-order variations in earth system parameters; (2) changes in reef attributes are controlled by dramatic short-term environmental changes (events); and (3) reef evolution is mostly driven intrinsically by biological parameters.

The first hypothesis can be tested by comparing the secular variations in reef attributes with variations in reconstructed or modeled physicochemical earth system variables. This has been done by Kiessling (in press). Trending reef attributes tend to correlate with trending earth system parameters. Amongst the strongest trending earth system parameters are supposedly the CO_2 concentration in the atmosphere (Berner, 1994) and the nutrient concentrations in the oceans (Martin, 1996). Most of the trending reef attributes correlate with these parameters and the developments in reef attributes.

The best way to test for actual links is to detrend the data using first differences and then check for significant correlations. This method eludes the effect of autocorrelation, a typical problem in time series analysis. With this method applied to the data, many correlations disappear and a great influence of pCO_2 on reef development appears unlikely. An example underlines the importance of this method: microbial reefs tend to be abundant during times of high inferred pCO_2 such as the early Paleozoic and are rare in times of relatively low pCO_2 like the Tertiary. However, as revealed by the detrending method, changes in pCO_2 rarely coincide with changes in microbial reef abundance. This implies that rather than being casually related, pCO_2 and microbial reef abundance coincidentally exhibit a parallel trend through time. The same applies to the inferred nutrient level. However, determining nutrient level in ancient seas is even more problematic than the modeling of CO_2 levels, which already has substantial uncertainties (Berner, 1994). Indirect measures of nutrient levels are available in the form of isotope measurements, especially carbon isotopes, strontium isotopes, and sulfur isotopes. Common correlations between long-term changes of those measures and changes in reef attributes suggest a casual link. However, although the isotopic measurements are precise and often reflect global change, they do not directly indicate nutrient levels in ancient seas, but are masked by additional geological factors so that the actual influence of nutrient level on reef development cannot be exactly determined.

Eustatic sea level change, plate tectonic evolution, and paleoclimatic change are additional potential controls on reef development. These parameters, as far as they have been quantitatively determined or modeled, exhibit significant correlations with at least one reef attribute in PaleoReefs, although the correlation coefficients are usually low. The only parameter that will be discussed here is paleoclimate, in particular paleotemperature. An enormous amount of literature provides paleoclimatic data (see Parrish, 1998, for a recent review), but only few publications summarize the paleoclimate of the Phanerozoic. Frakes *et al.* (1992) provided the most detailed summary. A comparison of their paleotemperature curve with the averaged reef attributes in PaleoReefs produces disappointingly few significant correlations. The correlations with other, more idealized temperature curves (Veevers, 1990; Berner, 1994; Worsley *et al.*, 1994) gave no better correlations. Only two very significant ($P < 0.01$) correlations with a regression slope of more than 0.6 are evident. One is with the percentage of reefal reservoirs (Kiessling *et al.*, 1999), indicating that reefs form better reservoirs during cool climatic periods. The other is the percentage of calcareous algal-dominated reefs in a particular time slice. The latter correlation is especially interesting. Calcareous algae are well known to replace corals during cooling periods in the Tertiary (Bourrouilh-Le

Jan and Hottinger, 1988) and also to be the major reef builders during the Permo-Carboniferous Gondwanan glaciation (Kiessling *et al.*, 1999). In addition, coralline algae bioherms occur well outside the tropical coral reef belt today (Freiwald, 1993). Therefore, the predominance of algal bioherms appears to have some paleoclimatic significance. The fact that tropical reefs change in composition during icehouse periods may be explained by oceanographic changes related to the climatic change (Bourrouilh-Le Jan and Hottinger, 1988).

There is no apparent correlation between global paleotemperatures and the relative number of high latitude reef occurrences or the mean paleolatitude of reefs. Some cold periods in earth history appear to even show an especially high percentage of high-latitude reefs. Examples are the latest Pennsylvanian to Early Permian and the Miocene (Fig. 7). Ziegler *et al.* (1984) made a similar observation and indicated that tropical carbonates do not significantly extend poleward during warm intervals. Nelson (1988) noted that cool-water carbonates appear to be particularly common during glacial episodes, that is, the Late Ordovician, the Permo-Carboniferous, and the younger Cenozoic. While Nelson (1988) claimed that these observations are due to a biased database, PaleoReefs supports the observations with some limitations. The lack of correlation between climate and reef occurrences cannot be attributed to problematic paleogeographic reconstructions, since high-latitude reef occurrences are especially common in younger stages (Fig. 7) where plate tectonic reconstructions are well constrained. High-latitude reefs are common in greenhouse (Middle Devonian, Late Jurassic) as well as icehouse periods, but only in icehouse periods are the high-latitude reefs significantly different in comparison to the tropical reefs. For the Permo-Carboniferous this is exemplified by the distribution of *Palaeoaplysina*-rich reefs. This enigmatic alga is limited to reefs occurring either in high latitudes or close to eastern boundary currents (Kiessling *et al.*, 1999). In the Miocene, reefs at the highest latitudes are usually different from lower-latitude reefs. Bryozoan mounds (Boreen and James, 1995; Pisera, 1996), serpulid–microbe biostromes (Friebe, 1994), vermetid-algal reefs (Pisera, 1985), and coralline algal frameworks (Baluk and Radwanski, 1977) have been reported, whereas coral reefs prevailed in lower latitudes, although in higher latitudes than modern coral reefs occur. Pending further studies, high-latitude reefs appear to differ from tropical reefs in being mostly nutrient opportunistic (Copper, 1994) and limited to cool periods. Only then were the high latitudes significantly enriched in nutrients.

In summary, fluctuations in earth system parameters are well reflected by the variations in many reef attributes. However, the often abrupt changes in reef attributes at certain stage boundaries suggest that geologic events also may play a substantial role. Mass extinction events have been claimed by Kauffman and Fagerstrom (1993) to be a major trigger of reef evolution, in particular to be the dominant control on diversity. Indeed, many mass extinction events are associated with rapid changes in reef attributes. The most profound short-term changes in the reef ecosystem are evident at the Frasnian–Famennian boundary, the Permian–Triassic boundary, and the Triassic–Jurassic boundary. At all these boundaries the number of reefs decline significantly (72 to 92%) in the subsequent stage and diversity is strongly reduced. Significant changes also are noted in nearly all other reef attributes associated with these mass extinction events. Without a doubt, mass extinction events can have profound consequences for reefs. It is obvious from Figs. 5–7, however, that not all major mass extinction events resulted in equally profound changes. Although the small reduction in global reef abundance at the Ordovician–Silurian boundary is likely due to the poor stratigraphic resolution of PaleoReefs, there certainly is no significant decline at the Cretaceous–Tertiary boundary (comparison at stage level). Danian/Selandian reefs are even more diverse than average Maastrichtian reefs. There also are significant changes at stage boundaries that are not related to any widely recognized mass extinction. Examples are the reef reductions at the Silurian–Devonian boundary and the Mississippian–Pennsylvanian boundary. The fact that rapid changes in reef attributes are not always associated with mass extinction events does not exclude the possibility that geologic events are the general agents of change in reef evolution. Sea-level and oceanographic changes and modifications of sedimentation patterns can occur very rapidly in relation to plate tectonic events (Hay, 1996) and can have a profound impact on the reef ecosystem without necessarily leading to a mass extinction event. In summary, short-term (stage to substage) changes in the earth system appear to influence virtually all reef attributes. The crucial question is do such events also control the long-term (supersequence to period) evolution of reefs? The answer is negative for reef attributes exhibiting significant trends through time. Negative and positive trends are commonly interrupted by geologic events but continue after a recovery time of variable duration. For example, debris potential (Fig. 7) decreases profoundly at the Frasnian–Famennian, Permian–Triassic, and Triassic–Jurassic boundaries, but the overall increasing trend is only interrupted but not reversed at these boundaries. The patterns of fluctuational reef attributes (Figs. 5–6), do not allow a conclusive statement, but it appears that an abrupt decline in reef abundance was often followed by a long-lasting depression in the number of reefs, whereas reef dimensions, diversity, and bathymetry usually recovered more rapidly after short-term events in the reef ecosystem. The only reef attribute that exhibits an irreversible modification by an event is the spar content after the Triassic–Jurassic mass extinction.

The third hypothesis mentioned above predicts that reef evolution is mostly intrinsically driven, that is, mostly controlled by biotic coevolution and escalation processes (Vermeij, 1987). This hypothesis is unlikely considering the common correlations of reef attributes with long- and short-term changes in the earth system. On the other hand, correlations within the reef ecosystem are much more common and more significant than those with extrinsic, physicochemical parameters. For example, the strong correlation between bioerosion and debris potential favors a predominantly intrinsic control of debris potential. However, although the evolution of bioerosion can be largely explained by evolutionary innovations in the biosphere (Vermeij, 1987), the regional intensity of bioerosion is likely to be driven by physical or biological factors influencing coral mortality (Glynn, 1997), and thus predominantly extrinsically controlled. Nevertheless, some reef attributes are not correlated or only weakly correlated with any known earth system parameter. These are reef abundance, reef diversity, succession in reefs, and reef carbonate production. Although Kiessling and Flügel (1999) noted a weak correlation of diversity with the Exxon sea-level curve and the strontium isotope curve of Veizer *et al.* (1997), the correlation of diversity with intrinsic factors such as debris potential and environmental setting is much stronger. Hence, there is obviously an important intrinsic component in the evolution of reefs.

In conclusion reef attributes tend to change in parallel with several measured or modeled earth system parameters. The correlations suggest that long term changes in nutrient levels, eustatic sea level, plate tectonic configuration, and paleoclimate had a significant influence on particular reef attributes. However, reef abundance and diversity do not strongly correlate with earth system parameters and are likely to be largely driven by a combination of geologic events and poorly understood biotic interactions and recruitment patterns within the reef ecosystem.

7. Conclusions

The Phanerozoic history of reefs is so multifaceted that some authors claimed: "... there has never been a single global-evolving reef ecosystem" (Wood, 1999, p. 33). However, a wide definition of reefs and a large database on Phanerozoic reefs (PaleoReefs) allow treatment of all reefs as one ecosystem and quantitative analysis of reef attributes through time. PaleoReefs also permits the detection of evolutionary trends and cycles in the Phanerozoic history of reefs and the database helps unravel possible controls on reef evolution.

The global geographic distribution of reefs varies significantly throughout the Phanerozoic. Global reef distribution can be roughly assigned to actualistic patterns that agree with the distribution of modern zooxanthellate coral reefs (low latitude, western margin of oceans, little clastic sedimentation) and nonactualistic patterns. A set of paleogeographic reef distribution maps have been compiled to document the gradual change from prevailing nonactualistic patterns in the Early Paleozoic to actualistic patterns in the younger Mesozoic and Cenozoic.

Measurable reef attributes usually vary strongly through time but can be assigned to two categories: (1) attributes that show a fluctuational behavior, and (2) attributes that exhibit a significant trend through time. All evolutionary trends, whether fluctuating or sloping, are prone to sudden changes that sometimes agree with mass extinction events but often occur without an obvious connection to geological events.

Based on the statistical analysis of all reef attributes stored in PaleoReefs, seven reef evolutionary units can be defined. The reef evolutionary units are different in duration and boundaries from ecological evolutionary units as defined by Boucot (1983) and Sheehan (1996). This suggests that reef evolution was largely decoupled from the evolution of level-bottom communities in the Phanerozoic.

The paleoclimatic significance of ancient reefs is limited. Latitudinal expansions or contractions of reefs are rarely correlated with global climatic change, and reef occurrences cannot be used for paleogeographic reconstructions in any straightforward way. However, reefs growing during icehouse periods exhibit common characteristics that are significantly different from those growing during greenhouse periods. The development of cool-water reefs with particular biotic compositions is observed only during icehouse periods, such as the Ashgillian, the Late Carboniferous–Early Permian, and the later Cenozoic. Greenhouse reefs tend to have a uniform composition throughout their latitudinal range which can be quite high such as in the Devonian or low as in Middle Triassic time.

Many trends and fluctuations in the Phanerozoic reef ecosystems are comprehensively outlined by the PaleoReef database. This underscores the great potential of large databases in paleobiology and paleoecology (Benton, 1999). While PaleoReefs cannot provide firm solutions to all questions asked by reef paleontologists and sedimentologists, it certainly can be utilized to test current hypotheses and to frame appropriate new questions. Further refinements and the predictions made by this database are expected to guide the field geologist in the study and understanding of ancient reefs.

ACKNOWLEDGMENTS: Many thanks to George D. Stanley, Jr., for giving me the opportunity to contribute to this volume. I am grateful to Paul Copper, Erik Flügel, and George Stanley for their critical and insightful remarks on the manuscript.

References

Aguirre, J., Riding, R. and Braga, J. C. 2000, Late Cretaceous incident light reduction: evidence from benthic algae. *Lethaia* **33**:206–214.

Alberstadt, L. P., Walker, K. R., and Zurawski, R. P., 1974, Patch reefs in the Carters Limestone (Middle Ordovician) in Tennessee, and vertical zonation in Ordovician reefs, *Geol. Soc. Am. Bull.* **85**:1171–1182.

Antoshkina, A. I., 1996, Ordovician reefs of the Ural Mountains, Russia: A review, *Facies* **35**:1–8.

Antoshkina, A. I., 1998, Organic buildups and reefs on the Paleozoic carbonate platform margin, Pechora Urals, Russia, *Sediment. Geol.* **118**:187–211.

Arp, G., 1995, Lacustrine bioherms, spring mounds, and marginal carbonates of the Ries-Impact-Crater (Miocene, Southern Germany), *Facies* **33**:35–90.

Baluk, W., and Radwanski, A., 1977, Organic communities and facies development of the Korytnica basin (Middle Miocene; Holy Cross Mountains, central Poland), *Acta Geol. Polonica* **27**:85–123.

Baria, L. R., Stoudt, D. L., Harris, P.M., and Crevello, P. D., 1982, Upper Jurassic reefs of Smackover Formation, United Stated Gulf Coast, *AAPG Bull.* **66**:1449–1482.

Beauchamp, B., Harrison, C., Nassichuk, W. W., and Eliuk, L. S., 1989, Lower Cretaceous (Albian) serpulid-bivalve carbonate "mounds" related to hydrocarbon seeps, Canadian Arctic Archipelago, in: *Reefs—Canada and adjacent areas*, Vol. 13 (H. H. J. Geldsetzer, N. P. James, and G. E. Tebbutt, eds.), Canadian Society of Petroleum Geologists Memoir, Calgary, pp. 706–712.

Bebout, D. G., and Loucks, R. G., 1983, Lower Cretaceous reefs, south Texas, in: *Carbonate Depositional Environments*, Vol. 33 (P. A. Scholle, D. G. Bebout, and C. H. Moore, eds.), AAPG Memoir, Tulsa, OK, pp. 441–444.

Benton, M. J., 1999, The history of life: Large databases in palaeontology, in: *Numerical Palaeobiology* (A. T. Harper, eds.), Wiley, New York, pp. 249–283.

Berner, R. A., 1994, Geocarb II: A revised model of atmospheric CO_2 over Phanerozoic time, *Am. J. Sci.* **294**:56–91.

Best, M. W., and Boekschoten, G. J., 1982, On the coral fauna in the Miocene reef at Baixo Porto-Santo eastern Atlantic, *Netherland J. Zool.* **32**:412–418.

Bhargava, O. N., and Bassi, U. K., 1986, Silurian reefal buildups: Spiti-Kinnaur, Himachal Himalaya, India, *Facies* **15**:35–52.

Bolshakova, L. N., Geller, M. P., Goryunova, R. V., Dubatolov, V. N., Ivanovskiy, A. B., Kosmynin, V. N., Luchinina, V. A., Sokolov, B. S., and Tesako, Y. I., 1994, Paleozoic coral reefs in Russia, *Stratigrafiya Geologicheskaya Korrelyatsiya* **2**:46–54 (in Russian).

Boreen, T. D., and James, N. P., 1995, Stratigraphic sedimentology of Tertiary cool-water limestones, SE Australia, *J. Sediment. Res.* **B65**:142–159.

Bosence, D. W. J., Rowlands, R. L., and Quine, M. L., 1985, Sedimentology and budget of a Recent carbonate mound, Florida Keys, *Sedimentology* **32**:317–343.

Boucot, A. J., 1983, Does evolution take place in an ecological vacuum? II, *J. Paleontol.* **57**:1–30.

Bourrouilh-Le Jan, F. G., and Hottinger, L. C., 1988, Occurrence of rhodolites in the tropical Pacific—A consequence of Mid-Miocene paleo-oceanographic change, *Sediment. Geol.* **60**:355–367.

Brasier, M. D., 1995, Fossil indicators of nutrient levels. 2: Evolution and extinction in relation to oligotrophy, in: *Marine Palaeoenvironmental Analysis from Fossils*, Vol. 83 (D. W. J. Bosence and P. A. Allison, eds.), Geological Society Special Publication, London, pp. 133–150.

Budd, A. F., and Johnson, K. G., 1999, Origination preceding extinction during late Cenozoic turnover of Caribbean reefs, *Paleobiology* **25**:188–200.

Chuvashov, B. I., 1983, Permian reefs of the Urals, *Facies* **8**:191–212.

Cook, H. E., Zhemchuzhnikov, V. G., Buvtyshkin, V. M., Golub, L. Y., Gatovsky, Y. A., and Zorin, A. Y., 1994, Devonian and Carboniferous passive-margin carbonate platform of southern Kazakhstan: summary of depositional and stratigraphical models to assist in the exploration and production of coeval giant carbonate platform oil and gas fields in the North Caspian Basin, Western Kazakhstan, in: *Pangea: Global Environments and Resources*, Vol. 17 (A. F. Embry, B. Beauchamp, and D. J. Glass, eds.), Canadian Society of Petroleum Geologists Memoir, Calgary, pp. 363–381.

Copper, P., 1988, Ecological succession in Phanerozoic reef ecosystems: is it real? *Palaios* **3**: 136–152.

Copper, P., 1989, Enigmas in Phanerozoic reef development, *Mem. Assoc. Australasian Palaeontol.* **8**:371–385.

Copper, P., 1994, Reefs under stress: The fossil record, *Courier Forschungsinstitut Senckenberg* **172**:87–94.

Copper, P., 1997, Reefs and carbonate productivity: Cambrian through Devonian, in: *Proceedings 8th International Coral Reef Symposium*, Vol. 2, Balboa, Panama by Smithsonian Tropical Research Institute, pp. 1623–1630.

Copper, P., and Brunton, F., 1991, A global review of Silurian reefs, *Special Papers Palaeontol.* **44**:225–259.

Davies, P. J., Symonds, P. A., Feary, D. A., and Pigram, C. J., 1989, The evolution of the carbonate platforms of northeast Australia, in: *Controls on carbonate platform and basin development*, Vol. 44 (P. D. Crevello, J. L. Wilson, F. Sarg, and J. F. Read, eds.) SEPM Special Publication, Tulsa, Oklahoma, pp. 233–258.

de Freitas, T. A., and Dixon, O. A., 1995, Silurian microbial buildups of the Canadian Arctic, in: *Carbonate Mud-Mounds—Their Origin and Evolution*, Vol. 23 (C. L. V. Monty, D. W. J.

Bosence, P. H. Bridges, and B. R. Pratt, eds.), Special Publications of the International Association of Sedimentologists, Oxford, pp. 151–169.

de Freitas, T., and Mayr, U., 1995, Kilometre-scale microbial buildups in a rimmed carbonate platform succession, Arctic Canada: New insight on Lower Ordovician reef facies, *Bull. Can. Petroleum Geol.* **43**:407–432.

De Moura, R. L., Rodrigues, M. C. M., Francini-Filho, R. B., and Sazima, I., 1999, Unexpected richness of reef corals near the southern Amazon River mouth, *Coral Reefs* **18**:170.

Dronov, A. V., and Natalin, B.A., 1990, Southern Tyan-Shan, in: *Phanerozoic Reefal and Sulfate Bearing Formations of the SSSR* (G. A. Belenitskaya and N. M. Zadoroshnaya, eds.), Ministerstvo Geologii SSSR, Nedra, Moscow (in Russian), pp. 56–65.

Edinger, E. N., and Risk, M., 1994, Oligocene–Miocene extinction and geographic restriction of Caribbean corals: Roles of turbidity, temperature, and nutrients, *Palaios* **9**:576–598.

Fagerstrom, J. A., 1987, *The Evolution of Reef Communities*, Wiley, New York.

Fagerstrom, J. A., 1991, Reef-building guilds and a checklist for determining guild membership, *Coral Reefs* **10**:47–52.

Fagerstrom, J. A., and Weidlich, O., 1999, Strengths and weaknesses of the reef guild concept and quantitative data: Application to the Upper Capitan-massive community (Permian), Guadalupe Mountains, New Mexico-Texas, *Facies* **40**:131–156.

Fan, J., 1980, The main features of marine Triassic sedimentary facies in Southern China, *Rivista Italiana di Paleontologia e Stratigrafia* **85**:1125–1146.

Feary, D. A., and James, N. P., 1995, Cenozoic biogenic mounds and buried Miocene(?) barrier reef on a predominantly cool-water carbonate continental margin—Eucla basin, western Great Australian Bight, *Geology* **23**:427–430.

Flügel, E., 1981, Paleoecology and facies of Upper Triassic reefs in the Northern Calcareous Alps, in: *European Fossil Reef Models*, Vol. 30 (D. E. Toomey, ed.), SEPM Special Publication, Tulsa, OK, pp. 291–359.

Flügel, E., 1994, Pangean shelf carbonates: Controls and paleoclimatic significance of Permian and Triassic reefs, in: Pangea: *Paleoclimate Tectonics and Sedimentation during Accretion Zenith and Breakup of a Supercontinent*, Vol. 288 (G. D. Klein, ed.), Geological Society of America Special Paper, Boulder, pp. 247–266.

Flügel, E., and Flügel-Kahler, 1992, Phanerozoic reef evolution: Basic questions and data base, *Facies* **26**:167–278.

Flügel, E., and Stanley, G. D., 1984, Reorganization, development and evolution of post-Permian reefs and reef organisms, *Palaeontogr. Am.* **54**:177–186.

Frakes, L. A., Francis, J. E., and Syktus, J. I., 1992, *Climate Modes of the Phanerozoic: The History of the Earth's Climate over the Past 600 Million Years*, Cambridge University Press, Cambridge.

Friebe, J. G., 1994, Serpulid–bryozoan–foraminiferal biostromes controlled by temperate climate and reduced salinity: Middle Miocene of the Styrian Basin, Austria, *Facies* **30**:51–62.

Freiwald, A., 1993, Coralline algal maerl frameworks—Islands within the phaeophytic kelp belt, *Facies* **29**:133–148.

Freiwald, A., Henrich, R., and Pätzold, J., 1997, Anatomy of a deep-water coral reef mound from Stjernsund, west Finnmark, northern Norway, in: *Cool-Water Carbonates*, Vol. 56 (N. P. James and J. A. D. Clarke, eds.), SEPM Special Publication, Tulsa, OK, pp. 141–162.

Frost, S. H., 1977, Cenozoic reef systems of Caribbean—Prospects for paleoecologic synthesis, in: *Reefs and Related Carbonates—Ecology and Sedimentology*, Vol. 4 (S. H. Frost, M. P. Weiss, and J. B. Saunders, eds.), AAPG Studies in Geology, Tulsa, OK, pp. 93–110.

Fürsich, F. T., and Wendt, J., 1977, Biostratinomy and palaeoecology of the Cassian Formation (Triassic) of the Southern Alps, *Palaeogeogr. Palaeoclimatol. Palaeoecol.* **22**:257–323.

Geyer, O. F., 1977, The "*Lithiotis*-Limestones" in the Early Jurassic Tethys region, *Neues Jahrbuch Geol. Paläontol. Abhandlungen* **153**:304–340 (in German).

Gili, E., Masse, J. P., and Skelton, P. W., 1995, Rudists as gregarious sediment dwellers, not reef builders, on Cretaceous carbonate platforms, *Palaeogeogr. Palaeoclimatol. Palaeoecol.* **118**:245–267.

Glynn, P. W., 1997, Bioerosion and coral-reef growth: Adynamic balance, in: *Life and Death of Coral Reefs* (C. Birkeland, ed.), Chapman and Hall, New York, pp. 68–95.

Golonka, J., Ross, M. I., and Scotese, C. R., 1994, Phanerozoic paleogeographic and paleoclimatic modeling maps, in: *Pangea: Global Environments and Resources*, Vol. 17 (A. F. Embry, B. Beauchamp, and D. J. Glass, eds.), Canadian Society of Petroleum Geologists Memoir, Calgary, pp. 1–47.

Gregor, C. B., 1985, The mass-age distribution of Phanerozoic sediments, in: *The Chronology of the Geologic Record*, Vol. 10 (N. J. Snelling, ed.), Geological Society of America Memoir, Boulder, pp. 284–289.

Grotzinger, J. P., 1989, Construction of Early Proterozoic (1.9 Ga) barrier reef complex, Rocknest platform, Northwest Territories, in: Reefs—*Canada and Adjacent Areas*, Vol. 13 (H. H. J. Geldsetzer, N. P. James, and G. E. Tebbutt, eds.), Canadian Society of Petroleum Geologists Memoir, Calgary, pp. 30–37.

Hallock, P., 1997, Reefs and reef limestones in earth history, in: *Life and Death of Coral Reefs* (C. Birkeland, eds.), Chapman and Hall, New York, pp. 13–42.

Hallock, P., and Schlager, W., 1986, Nutrient excess and the demise of coral reefs and carbonate platforms, *Palaios* **1**:389–398.

Hallock, P. A., Hine, A. C., Vargo, G. A., Elrod, J. A., and Jaap, W. C., 1988, Platforms of the Nicaraguan rise: Examples of the sensitivity of carbonate sedimentation to excess trophic resources, *Geology* **16**:1104–1107.

Hamdi, B., Rozanov, A. Y., and Zhuravlev, A. Y., 1995, Latest Middle Cambrian metazoan reef from northern Iran, *Geol. Mag.* **132**:367–373.

Harris, M. T., 1994, A volume-based approach to reef productivity and submarine erosion rates—A case-study of a Middle Triassic reef margin (Latemar buildup, northern Italy), *J. Geol.* **102**:603–610.

Hay, W. W., 1996, Tectonics and climate, *Geol. Rundschau* **85**:409–437.

Heafford, A. P., 1989, Reef plays in the Timan–Pechora Province, Cambridge Arctic Shelf Programme (CASP) Report No. 461.

Heckel, P. H., 1974, Carbonate buildups in the geologic record: a review, in: *Reefs in Time and Space*, Vol. 18 (L. F. Laporte, ed.), SEPM Special Publication, Tulsa, OK, pp. 90–154.

Hoffman, P. F., 1989, Pethei reef complex (1.9 Ga), Great Slave Lake, N.W.T., in: *Reefs—Canada and Adjacent Areas*, Vol. 13 (H. H. J. Geldsetzer, N. P. James, and G. E. Tebbutt, eds.), Canadian Society of Petroleum Geologists Memoir, Calgary, pp. 38–48.

Hoffmann, H. J., and Grotzinger, J. P., 1985, Shelf-facies microbiotas from the Odjick and Rocknest formations (Epworth Group: 1.89 Ga), northwest Canada, *Can. J. Earth Sci.* **22**:1781–1792.

Hubbard, D. K., 1997, Reefs as dynamic systems, in: *Life and Death of Coral Reefs* (C. Birkeland, ed.), Chapman and Hall, New York, pp. 43–67.

Hubbard, D. K., Miller, A. I., and Scaturo, D., 1990, Production and cycling of calcium carbonate in a shelf-edge reef system (U.S. Virgin Islands): Applications to the nature of reef systems in the fossil record, *J. Sediment. Petrol.* **60**:335–360.

Hubbard, D. K., Burke, R. B., and Gill, I. P., 1998, Where's the reef: the role of framework in the Holocene, *Carbonates Evaporites* **13**:3–9.

Hurst, J. M., 1980, Paleogeographic and stratigraphic differentiation of Silurian carbonate buildups and biostromes of north Greenland, *AAPG Bull.* **64**:527–548.

Hutchings, P. A., 1986, Biological destruction of coral reefs, *Coral Reefs* **4**:239–252.

Ioganson, A. K., 1990, Siberian Platform, in: *Phanerozoic Reefal and Sulfate Bearing Formations of the SSSR* (G. A. Belenitskaya and N. M. Zadoroshnaya, eds.), Ministerstvo Geologii SSSR, Nedra, Moscow (in Russian), pp. 131–161.

James, N. P., 1983, Reef environment, in: *Carbonate Depositional Environments*, Vol. 33 (P. A. Scholle, D. G. Bebout, and C. H. Moore, eds.), AAPG Memoir, Tulsa, OK, pp.345–440.

James, N. P., and Bourque, P.-A., 1992, Reefs and mounds, in: *Facies Models: Response to Sea Level Change* (R. G. Walker and N. P. James, eds.), Geological Association of Canada, St. John's, pp. 323–347.

James, N. P., Wray, J. L., and Ginsburg, R. N., 1988, Calcification of encrusting aragonitic algae Peyssonneliaceae: Implications for the origin of Late Paleozoic reefs and cements, *J. Sediment. Petrol.* **58**:291–303.

Jeffery, D. L., and Stanton, R. J., 1996, Growth history of Lower Mississippian Waulsortian mounds: Distribution, stratal patterns, and geometries, New Mexico, *Facies* **35**:29–58.

Kauffman, E. G., and Fagerstrom, J. A., 1993, The Phanerozoic evolution of reef diversity, in: *Species Diversity in Ecological Communities* (R. E. Ricklefs and D. Schluter, eds.), University of Chicago Press, Chicago, pp. 315–329.

Kiessling, W., and Flügel, E., 1999, The Phanerozoic diversity of reef builders, *Geol. Soc. Am. Abstr. Progr.* **31**:A-336.

Kiessling, W., in press, Secular variations in the Phanerozoic reef ecosystem, in: *Phanerozoic Reef Patterns* (W. Kiessling, E. Flügel, and J. Golonka, eds.) SEPM Special Publications, Tulsa, Oklahoma.

Kiessling, W., Flügel, E., and Golonka, J., 1999, Paleo reef maps: Evaluation of a comprehensive database on Phanerozoic reefs, *AAPG Bull.* **83**:1552–1587.

Kobluk, D. R., James, N. P., and Pemberton, S. G., 1978, Initial diversification of macroboring ichnofossils and exploitation of the macroboring niche in the Lower Paleozoic, *Paleobiology* **4**:163–170.

Kowalewski, M., Dulai, A., and Fürsich, F. T., 1998, A fossil record full of holes: The Phanerozoic history of drilling predation, *Geology* **26**:1091–1094.

Krijgsman, W., Hilgen, F. J., Raf, I., Sierro, F. J., and Wilson, D. S., 1999, Chronology, causes and progression of the Messinian salinity crisis, *Nature* **400**:652–655.

Kuznetsov, V. G., and Don, O. V., 1984, Organic buildups in the Cambrian of the Nepsko-Botuobinskoi anteclise (Siberian Platform), *Bull. Moscow Soc. Natural. Section Geol.* **59**:63–71 (in Russian).

Kuznetsov, V. G., Kurze, M., Smilgis, I. I., Suveizis, P. I., and Schmidt, G., 1984, Reefs in the Lower Zechstein of eastern Europe, *Akad. nauk SSSR Izvest. Ser. Geol. Otdel. ottisk* **1**: 51–65 (in Russian).

Lees, A., and Miller, J., 1995, Waulsortian banks, in: *Carbonate Mud-Mounds—Their Origin and Evolution*, Vol. 23 (C. L. V. Monty, D. W. J. Bosence, P. H. Bridges, and B. R. Pratt, eds.), Special Publications of the International Association of Sedimentologists, Blackwell, Oxford, pp. 191–271.

Lowenstamm, H. A., 1950, Niagaran reefs of the Great Lakes area, *J. Geol.* **58**:430–487.

Martin, R. E., 1996, Secular increase in nutrient levels through the Phanerozoic: Implications for productivity, biomass, and diversity of the marine biosphere, *Palaios* **11**:209–219.

Mas, J. R., Alonso, A., and Benito, M. I., 1997, Depositional and diagenetic evolution of Late Jurassic coral reefs in northern Iberian Ranges (North Spain), *Bol. Real Soc. Español. Hist. Natur. Sección Geol.* **92**:143–160.

Meyer, F. O., 1989, Siliciclastic influence on Mesozoic platform development: Baltimore Canyon Trough, western Atlantic, in: *Controls on Carbonate Platform and Basin Development*, Vol. 44 (P. D. Crevello, J. L. Wilson, F. Sarg, and J. F. Read, eds.), SEPM Special Publication, Tulsa, OK, pp. 213–232.

Montañez, I. P., and Osleger, D. A., 1996, Contrasting sequence boundary zones developed within cyclic carbonates of the Bonanza King Formation, Middle to Late Cambrian, southern Great Basin, *Geol. Soc. Am. Spec. Paper* **306**:7–21.

Montgomery, S. L., Walker, T. H., Wahlman, G. P., Tobin, R. C., and Ziegler, D., 1999, Upper Jurassic "Reef" play, East Texas basin: An updated overview: Part 1—Background and outboard trends, *AAPG Bull.* **83**:707–726.

Moore, G. T., Hayashida, D. N., and Ross, C. A., 1993, Late Silurian (Wenlockian) general circulation model-generated upwelling, graptolotic black shales, and organic-rich source rocks—An accident of plate tectonics? *Geology* **21**:17–20.

Moore, P. F., 1989, Devonian reefs in Canada and some adjacent areas, in: *Reefs—Canada and Adjacent Areas*, Vol. 13 (H. H. J. Geldsetzer, N. P. James, and G. E. Tebbutt, eds.), Canadian Society of Petroleum Geologists Memoir, Calgary, pp. 367–390.

Moussavian, E., 1992, On Cretaceous bioconstructions: composition and evolutionary trends of crust-building associations, *Facies* **26**:117–144.

Mundy, D. J. C., 1994, Microbialite–sponge–bryozoan–coral framestones in Lower Carboniferous (Late Visean) buildups of northern England (UK), in: *Pangea: Global Environments and Resources*, Vol. 17 (A. F. Embry, B. Beauchamp, and D. J. Glass, eds.), Canadian Society of Petroleum Geologists Memoir, Calgary, pp. 713–729.

Nagai, K., 1978, Litho- and bio-facies of reef limestones in the Ryugoho area of the Akiyoshi Limestone Plateau, *Bull. Akiyoshi-Dai Museum Nat. Hist.* **13**:15–34.

Nagai, K., 1985, Reef-forming algal chaetetid boundstone found in the Akiyoshi Limestone Group, Southwest Japan, *Bull. Akiyoshi-Dai Museum Nat. Hist.* **20**:1–16.

Nelson, C. S., 1988, An introductory perspective on non-tropical shelf carbonates, *Sediment. Geol.* **60**:3–12.

Newell, N. D., 1971, An outline history of tropical organic reefs, *Am. Museum Novitates* **2465**:1–37.

Nisbet, E. G., and Wilkins, M. E., 1989, Archaean stromatolite reef at Steep Lake, Atikokan, northwestern Ontario, in: *Reefs—Canada and Adjacent Areas*, Vol. 13 (H. H. J. Geldsetzer, N. P. James, and G. E. Tebbutt, eds.), Canadian Society of Petroleum Geologists Memoir, Calgary, pp. 89–92.

Parrish, J. T., 1998, *Interpreting Pre-Quarternary Climate from the Geologic Record*, Columbia University Press, New York.

Perry, C. T., 1996, Distribution and abundance of macroborers in an Upper Miocene reef system, Mallorca, Spain: Implications for reef development and framework destruction, *Palaios* **11**:40–56.

Philip, J. M., and Airaud-Crumière, C., 1991, The demise of the rudist-bearing carbonate platforms at the Cenomanian/Turonian boundary: Alglobal control, *Coral Reefs* **10**:115–125.

Pisera, A., 1985, Paleoecology and lithogenesis of the Middle Miocene (Badenian) algal–vermetid reefs from the Roztocze Hills, south-eastern Poland, *Acta Geol. Polonica* **35**:89–155.

Pisera, A., 1996, Miocene reefs of the Paratethys: A review. *SEPM Concepts Sedimentol. Paleontol.* **5**:97–104.

Plaziat, J.-C., and Perrin, C., 1992, Multikilometer-sized reefs built by foraminifera (Solenomeris) from the early Eocene of the Pyrenean domain (S. France, N. Spain): Palaeoecologic relations with coral reefs, *Palaeogeogr. Palaeoclimatol. Palaeoecol.* **96**:195–231.

Pol'ster, L. A., Cherpelyugin, A. B., Sheremet'yev, Y. F., and Shermet'yeva, G. A., 1985, The formation of major zones and areas of oil and gas accumulation in the Paleozoic reefs of the Peri-Caspian syneclise, in: *Conditions of Formation of Major Oil–Gas Accumulation Zones*, Moscow (in Russian), pp. 163–168.

Pratt, B. R., and James, N. P., 1989, Coral-*Renalcis*-thrombolite reef complex of Early Ordovician age, St. George Group, western Newfoundland, in: *Reefs—Canada and Adjacent Areas*, Vol. 13 (H. H. J. Geldsetzer, N. P. James, and G. E. Tebbutt, eds.), Canadian Society of Petroleum Geologists Memoir, Calgary, pp. 224–230.

Pratt, B. R., and Jansa, L. F., 1989, Upper Jurassic shallow water reefs of offshore Nova Scotia, in: *Reefs—Canada and Adjacent Areas*, Vol. 13 (H. H. J. Geldsetzer, N. P. James, and G. E. Tebbutt, eds.), Canadian Society of Petroleum Geologists Memoir, Calgary, pp. 741–747.

Precht, W. F., 1994, The use of the term guild in coral reef ecology and paleoecology: A critical evaluation, *Coral Reefs* **13**:135–136.

Purser, B. H., Plaziat, J.-C., and Rosen, B. R., 1996, Miocene reefs of the northwest Red Sea, *SEPM Concepts Sedimentol.* **5**:347–365.

Rigby, J. K., Fan, J., Zhang, W., Wang, S., and Zhang, X., 1994, Sphinctozoan and inozoan sponges from the Permian reefs of South China, *Brigham Young University Geol. Stud.* **40**:43–109.

Rigby, J. K., Nitecki, M. H., Zhu, Z., Liu, B., and Jinag, Y., 1995, Lower Ordovician reefs of Hubei, China, and the western United States, in: *Ordovician Odyssey: Short Papers for the 7th International Symposium on the Ordovician System*, Vol. 77 (J. D. Cooper, M. L. Droser, and S. C. Finney, eds.), SEPM Pacific Section, Fullerton, CA, pp. 423–426.

Roger, J., Bourdillon, C., Razin, P., Le Callonnec, L., Renard, M., Aubry, M.-P., Philip, J., Platel, J.-P., Wyns, R., and Bonnemaison, M., 1998, Palaeoenvironmental and biotic changes across the Cretaceous/Tertiary boundary in the Oman Mountains, *Bull. Soc. Géol. France* **169**:255–270 (in French).

Ronov, A. B., Khain, V. E., Balukhovsky, A. N., and Seslavinsky, K. B., 1980, Quantitative analysis of Phanerozoic sedimentation, *Sediment. Geol.* **23**:311–325.

Rosen, B. R., 2000, Algal symbiosis, and the collapse and recovery of reef communities: Lazarus corals across the k-T boundary, in: *Biotic Response to Global Change: The Last 145 Million*

Years (S. J. Culver and P. F. Rwason, eds.) Cambridge University Press, Cambridge, pp. 164–180.

Rosen, B. R., and Turnšek, D., 1989, Extinction patterns and biogeography of scleractinian corals across the Cretaceous/Tertiary boundary, *Memoirs Assoc. Australasian Palaeontol.* **8**:355–370.

Ross, R. J., Jaanuson, V., and Friedman, I., 1975, Lithology and origin of Middle Ordovician calcareous mud mound at Meiklejohn Peak, *US Geol. Survey Prof. Papers* **871**:1–45.

Rowland, S. M., and Gangloff, R. A., 1988, Structure and paleoecology of Lower Cambrian reefs, *Palaios* **3**:111–135.

Ryan, W. B. F., and Miller, E. L., 1981, Evidence of a carbonate platform beneath Georges Bank, *Marine Geol.* **44**:213–228.

Sano, S.-I., 1991, Discovery of a coral–rudist builup in the Miyako Group, northeast Japan, *Trans Proc. Palaeontol. Soc. Japan, New Series* **162**:794–800.

Scotese, C. R., and Golonka, J., 1992, PALEOMAP Paleogeographic Atlas, *PALEOMAP Prog. Rep.* **20**:1–34.

Sepkoski, J. J., Jr., 1981, A factor analytic description of the Phanerozoic marine fossil record, *Paleobiology* **7**:36–53.

Sepkoski, J. J., Jr., Bambach, R. K., Raup, D. M., and Valentine, J. W., 1981, Phanerozoic marine diversity and the fossil record, *Nature* **293**:435–437.

Sestini, J., 1965, Cenozoic stratigraphy and depositional history, Red Sea coast, Sudan, *AAPG Bull.* **49**:1453–1472.

Sharkova, T. T., 1986, Coral associations in Devonian reefs of South Mongolia, in: *Phanerozoic Reefs and Corals of the SSSR*, Akad. Nauk SSSR, Moscow (in Russian), pp. 188–190.

Sheehan, P. M., 1985, Reefs are not so different—They follow the evolutionary pattern of level-bottom communities, *Geology* **13**:46–49.

Sheehan, P. M., 1996, A new look at ecologic evolutionary units (EEUs), *Palaeogeogr. Palaeoclimatol. Palaeoecol.* **127**:21–31.

Skelton, P. W., Gili, E., Rosen, B. R., and Valldeperas, F. X., 1997, Corals and rudists in the late Cretaceous: A critique of the hypothesis of competitive displacement, *Bol. Real Soc. Española Hist. Natural (Sección Geol.)* **92**:225–239.

Sönderholm, M., and Harland, T. L., 1989, Franklinian reef belt, Silurian, North Greenland, in: *Reefs—Canada and Adjacent Areas*, Vol. 13 (H. H. J. Geldsetzer, N. P. James, and G. E. Tebbutt, eds.), Canadian Society of Petroleum Geologists Memoir, Calgary, pp. 356–366.

Squires, D. F., 1964, Fossil coral thickets in Wairarapa, New Zealand, *J. Paleontol.* **38**:904–915.

Stanley, G. D., Jr., 1988, The history of Early Mesozoic reef communities: A three-step process, *Palaios* **3**:170–183.

Stanley, G. D., Jr., 1992, Tropical reef ecosystems and their evolution, in: *Encyclopedia of Earth System Science*, Vol. 4, Academic Press, London, pp. 375–388.

Stanley, G. D., Jr., 1994, Late Paleozoic and early Mesozoic reef-building organisms and paleogeography: The Tethyan–North American connection, *Courier Forschungs-Institut Senckenberg* **172**:69–75.

Stanley, G. D., Jr. and Beauvais, L., 1994, Corals from an Early Jurassic coral reef in British Columbia on an oceanic island reef, *Lethaia* **27**:35–47.

Stanley, G. D., Jr., and Swart, P. K., 1995, Evolution of the coral–zooxanthellae symbiosis during the Triassic: A geochemical approach, *Paleobiology* **21**:179–199.

Suchy, D. R., and Stearn, C. W., 1992, Lower Silurian sequence stratigraphy and sea-level history of the Hudson Bay Platform, *Bull. Can. Petrol. Geol.* **40**:335–355.

Surlyk, F., 1997, A cool-water carbonate ramp with bryozoan mounds: Late Cretaceous- Danian of the Danish Basin, in: *Cool-Water Carbonates*, Vol. 56 (N. P. James and J. A. D. Clarke, eds.), SEPM Special Publication, Tulsa, OK, pp. 293–307.

Talent, J. A., 1988, Organic reef-building: Episodes of extinction and symbiosis? *Senckenbergiana Lethaea* **69**:315–368.

Teichert, C., 1958, Cold- and deep-water coral banks, *AAPG Bull.* **42**:1064–1082.

Toomey, D. F., and Nitecki, M. H., 1979, Organic buildups in the Lower Ordovician (Canadian) of Texas and Oklahoma, *Fieldiana Geol. New Series* **2**:1–181.

Tragelehn, H., 1996, Maastrichtian and Paleocene at the southern margin of the Northern Calcareous Alps (Lower Austria, Styria)—Facies, stratigraphy, and fossil contents of the "Kambühel Limestone" and associated sediments, unpublished PhD, Erlangen (in German).

Tsien, H. H., Hou, H. F., Zhou, W. L., Wu, Y., Yin, D. W., Dai, Q. Y., and Liu, W. J., 1988, Devonian reef development and paleogeographic evolution in south China, in: *Devonian of the World, Vol. I: Regional Syntheses*, Vol. 14 (N. J. McMillan, A. F. Embry, and D. J. Glass, eds.), Canadian Society of Petroleum Geologists Memoir, Calgary, pp. 619–633.

Turner, E. C., Narbonne, G. M., and James, N. P., 1993, Neoproterozoic reef microstructures from the Little Dal Group, northwestern Canada, *Geology* **21**:259–262.

Turnšek, D., 1968, Some hydrozoans and corals from Jurassic and Cretaceous strata of southwestern Jugoslavia, *Slov. Akad. Znan. Umet. Razprave Classis IV* **11**:351–376.

Turnšek, D., Buser, S., and Ogorelec, B., 1981, An Upper Jurassic reef complex from Slovenian, Yugoslavia, in: *European Fossil Reef Models*, Vol. 30 (D. F. Toomey, ed.), SEPM Spec. Publ., Tulsa, OK, pp. 361–369.

Veevers, J. J., 1990, Tectonic-climatic supercyle in the billion-year plate-tectonic eon; Permian Pangean icehouse alternates with Cretaceous dispersed-continents greenhouse, *Sediment. Geol.* **68**:1–16.

Veizer, J., Buhl, D., Diener, A., Ebneth, S., Podlaha, O. G., Bruckschen, P., Jasper, T., Korte, C., Schaaf, M., Ala, D., and Azmy, K., 1997, Strontium isotope stratigraphy: Potential resolution and event correlation, *Palaeogeogr. Palaeoclimatol. Palaeoecol.* **132**:65–77.

Vennin, E., Alvaro, J. J., and Villas, E., 1998, High-latitude pelmatozoan-bryozoan mud-mounds from the late Ordovician northern Gondwana platform, *Geol. J.* **33**:121–140.

Vermeij, G. J., 1977, The Mesozoic marine revolution: Evidence from snails, predators and grazers, *Paleobiology* **3**:245–258.

Vermeij, G. J., 1987, *Evolution and Escalation—An Ecological History of Life*, Princeton University Press, Princeton.

Walter, M. R., Buick, R., and Dunlop, J. S. R., 1980, Stromatolites 3.000–3500 Myr old from the North Pole are, Western Australia, *Nature* **284**:443–445.

Wang, S., Fan, J., and Rigby, J. K., 1994, The Permian reefs in Ziyun County, southern Guizhou, China, *Brigham Young University Geol. Stud.* **40**:155–183.

Ward, R. F., Kendall, C. G. S. C., and Harris, P. M., 1986, Upper Permian (Guadalupian) facies and their association with hydrocarbons—Permian Basin, West Texas and New Mexico, *AAPG Bull.* **70**:239–262.

Waters, B. B., 1989, Upper Cambrian *Renalcis–Girvanella* framestone mounds, Alberta, in: *Reefs—Canada and Adjacent Areas*, Vol. 13 (H. H. J. Geldsetzer, N. P. James, and G. E. Tebbutt, eds.), Canadian Society of Petroleum Geologists Memoir, Calgary, pp. 165–170.

Webb, G. E., 1994, Non-Waulsortian Mississippian bioherms: a comparative analysis, in: *Pangea: Global Environments and Resources*, Vol. 17 (A. F. Embry, B. Beauchamp and D. J. Glass, eds.), Canadian Society of Petroleum Geologists Memoir, Calgary, pp. 701–712.

Webb, G. E., 1996, Was Phanerozoic reef history controlled by the distribution of nonenzymatically secreted reef carbonates (microbial carbonate and biologically induced cement), *Sedimentology* **43**:947–971.

Webb, G., 1998, Earliest known Carboniferous shallow-water reefs, Gudman Formation (Tn 16), Queensland, Australia: Implications for Late Devonian reef collapse and recovery. *Geol.* **26**:951–954.

Webby, B. D., 1984, Ordovician reefs and climate: a review, in: *Aspects of the Ordovician System*, Vol. 295 (D. L. Bruton, ed.), Palaeontological Contributions from the University of Oslo, Oslo, pp. 89–100.

Wernli, R., and Schulte, S., 1993, "*Bdelloidina*" *urgonensis* n. sp., a bioherm–building foraminifer in the Urgonian of Haute-Savoie (France), *Ecolgae Geol. Helvetiae* **86**:529–541 (in French).

Whalen, M. T., 1995, Barred basins: A model for eastern ocean basin carbonate platforms, *Geology* **23**:625–628.

Wilkinson, B. H., and Walker, J. C. G., 1989, Phanerozoic cycling of sedimentary carbonate, *Am. J. Sci.* **289**:525–548.

Wilson, J. L., 1975, *Carbonate Facies in Geologic History*, Springer, Berlin.

Wood, R., 1998, The ecological evolution of reefs, *Annu. Rev. Ecol. Systemat.* **29**:179–206.

Wood, R., 1999, *Reef Evolution*, Oxford University Press, Oxford.

Wold, C. N., and Hay, W. W., 1993, Reconstructing the age and lithology of eroded sediment, *Geoinformatics* **4**:137–144.

Worsely, T. R., Moore, T. L., Fraticelli, T. M., and Scotese, C. R., 1994, Phanerozoic CO_2 levels and global temperatures inferred from changing paleogeography, in: *Pangea: Paleoclimate Tectonics and Sedimentation during Accretion Zenith and Breakup of a Supercontinent*, Vol. 228 (G. D. Klein, ed.), Geological Society of America Special Paper, Boulder, CO, pp. 57–73.

Yabe, H., and Sugiyama, T., 1935, Geological and geographical distribution of reef–corals in Japan, *J. Paleontol.* **9**:183–217.

Yaroshenko, A. V., 1986, Role of the structural–tectonic factor in formation of Late Paleozoic organic buildups in the North Caspian Depression, *Petrol. Geol.* **21**:472–475.

Zadoroshnaya, N. M., and Nikitin, I. F., 1990, Kazachstan region, in: *Phanerozoic Reefal and Sulfate Bearing Formations of the SSSR* (G. A. Belenitskaya and N. M. Zadoroshnaya, eds.), Ministerstvo Geologii SSSR, Nedra, Moscow, pp. 41–52 (in Russian).

Zadoroshnaya, N. M., Osadchaya, D. V., and Novoselova, L. N., 1982, *Growth and Regional Development of Strata with Organogenic Buildups*, Nedra, Leningrad (in Russian).

Zhu, Z., Jinag, Y., and Liu, B., 1995, Paleoecology of late Tremadocian reef-bearing strata in western Hubei Province of China, in: *Ordovician Odyssey: Short Papers for the 7th International Symposium on the Ordovician System*, Vol. 77 (J. D. Cooper, M. L. Droser, and S. C. Finney, eds.) SEPM Pacific Section, Fullerton, CA, pp. 427–428.

Ziegler, A. M., Hulver, M. L., Lottes, A. L., and Schmachtenberg, W. F., 1984, Uniformitarianism and palaeoclimates: Inferences from the distribution of carbonate rocks, in: *Fossils and Climate*, Vol. 11 (P. J. Brenchley, ed.) Geological Journal Special Issues, Wiley, Chichester, England, pp. 3–25.

Chapter 3

Evolution, Radiations, and Extinctions in Proterozoic to Mid-Paleozoic Reefs

PAUL COPPER

1. Introduction

Reefs have an immensely long fossil record: those built by prokaryotes (Eubacteria and Archaea) have been with us for over 3 billion years, at a time when the atmosphere was enormously enriched in CO_2 (possibly as much as the 95–98% CO_2 atmospheres of the nearest planets Mars and Venus) and virtually devoid of oxygen. Skeletal reefs built by single-celled or multicellular $CaCO_3$ prokaryotes and simple eukaryotes (i.e., calcimicrobes) have been here since the Late Proterozoic (ca 1 billion–700 million years ago), and metazoan reefs, built by multicellular animals often in conjunction with calcimicrobes, have been here for the last half billion (ca 530 million years ago: Fig. 1). In the 180 million year time interval spanning the appearance of the first metazoan–calcimicrobe reef consortium of the Early Cambrian to the close of the Middle Paleozoic (Devonian) tropical reef ecosystem, with abundant corals and other

PAUL COPPER • Department of Earth Sciences, Laurentian University, Sudbury, Ontario, Canada P3E 2C6.

The History and Sedimentology of Ancient Reef Systems, edited by George D. Stanley Jr., Kluwer Academic/Plenum Publishers, New York, 2001.

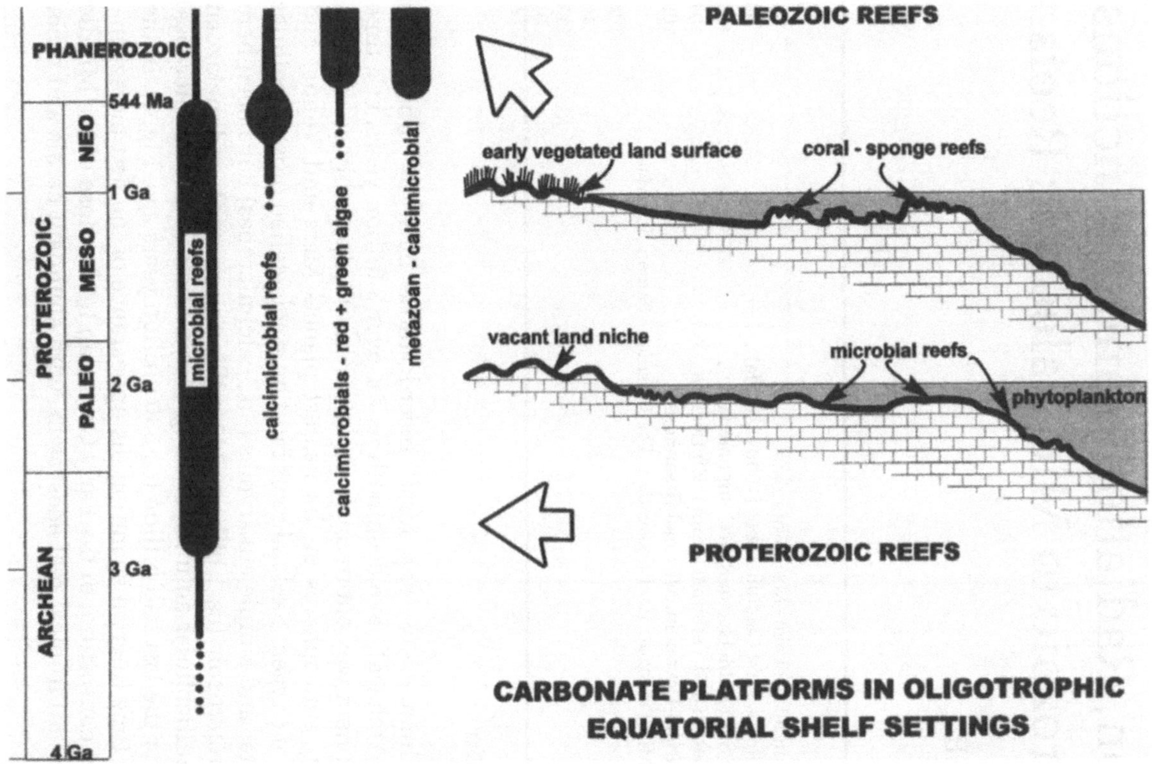

FIGURE 1. Contrast between Late Precambrian (Neoproterozoic) and the mid-Paleozoic, oligotrophic, tropical shelf, and slope regimes and their biotas during three major steps in their development from microbial (I—Archean-Mesoproterozoic), to calcimicrobial (II—Neoproterozoic), to metazoan-dominated (III—Phanerozoic). Note that the evolution of calcified microbes and rigid structures in the Neoproterozoic assisted in the production of reef cavity space, resulting in large-scale, microbial–calcimicrobial reefs in the early Phanerozoic. By Tommotian (Early Cambrian) time the metazoan (archeocyath)–calcimicrobial reef ecosystem was installed as part of the "Cambrian" community. By late Middle Ordovician time, the basic tabulate and rugose coral–stromatoporoid reef consortium was in place, displacing the Cambro–Ordovician reef community: This reached full swing in the Siluro-Devonian, and by the end of the Devonian tropical rainforests on land began to have an impact on coastal reefs. (Modified from Copper, 1994a.)

biota, metazoan reefs survived three global mass extinctions, each of which at least temporarily reset their ecological and in part evolutionary "clocks." Moreover, by the close of the Devonian, land areas adjacent to reefs were occupied by tropical pteridophyte rainforests, pumping up atmospheric levels of oxygen to near-modern levels, adding another dimension to coastal erosion, sediment supply, and fluctuating greenhouse to icehouse climates, thus providing new controls and new settings for reef growth.

The early evolution and developmental stages of skeletal, metazoan-built reef ecosystems around the Precambrian–Phanerozoic (P/P) transition are still not fully understood. The physicochemical setting for these early reefs also is controversial: What were atmospheric levels of O_2 some 544 million years ago at the P/P boundary? Did periodic episodes of hypoxia kill off reefs? What were nutrient levels and pH in tropical shelf habitats and how were organisms adapted to these at the time? How complex or simple were early food chains and trophic webs? Considering much higher levels of P_{CO_2} in the mid-Paleozoic oceans, were there even greater numbers of zooxanthellae or other photosymbionts assisting in Paleozoic metazoan $CaCO_3$ fixation than today? Were the skeletons of reef formers a significant sink for carbon or did living tissue shed a net surplus of CO_2, and what role did reefs play in the carbon cycle at this early stage? What were the links between biological processes in reefs (photosynthesis, respiration, assimilation of phytoplankton and zooplankton, recycling of wastes) and calcification or dissolution of skeletons?

The definition of "reefs" commonly hinges on the individual interpretation of the significance of the organisms and processes that construct them, their overall morphology, and sedimentary context. Thus reefs have been variably named and classified by different specialists: by biologists, paleontologists, and sedimentologists alike. This terminology also is hampered by such terms as ecological or organic reef (all reefs are ecological/organic structures), stratigraphic reef (a mound-shaped body, usually in the subsurface, formed by sedimentologic or structural processes, therefore not a true reef), bioherm and biostrome: some of these terms are now abandoned or redundant, or some have reduced the terminology to the word "buildup" as a substitute for reef. Fagerstrom (1987) defined the two most important characters of reefs as being rigid and possessing a framework; he added that they contained densely packed, rapidly growing, mostly fixosessile colonial or gregarious organisms. Nevertheless geologists often do not realize that many modern and fossil reefs either lack a framework or that taphonomic and storm effects may eliminate evidence for a framework, thus leaving a reef essentially as a living veneer on a pile of skeletal debris (Hubbard *et al.*, 1990; Blanchon and Jones, 1997). Reefs also need not be rigid, or be located in high-energy, shallow-shelf conditions. James and Geldsetzer (1989) used the term "reef" for massive or layered, laterally restricted carbonate buildups formed *in situ*, possessing topographic relief and stabilized syndepositionally by organic growth and/or submarine cementation: this includes buildups where organisms play only minor roles and is the definition followed here. James and Bourque (1992) later confined the term "reef" to those structures produced by clonal biota (greater than 5 cm in size), that thrived in energetic environments, and thus are located primarily in shallow waters above wave base and in the photic zone. However, there is no reason why such structures should not be deeper and well below the wave base, i.e., slope and basinal settings, as complete gradations exist between shallow and deeper water complexes, and small (less than 1 mm in size) and large frame builders.

Webb (1996) stressed that many Phanerozoic reefs have a rigidity that resulted from not only frame builders, but also from early lithification of microbial carbonates, and biocements; a significant and frequently major volumetric component of ancient reefs is the presence of cavities and their infill. Indeed, even reefs that are indisputably created by large frame builders, such as sponges and corals or large-scale calcimicrobial frames, may be 70% or more infilled by syn- and postdepositional muds (automicrites) and cements. It also is very difficult to tell what component of a fossil reef was soft, firm, or hard what the original shape and nature of biofilms once was (as in spongiostromate structures common to many fossil reefs), and whether these were formed by microbes or sponges (Reitner and Neuweiler, 1995; Flügel, 1996). This, and the gradation between mounds that contain larger or smaller skeletalized metazoans and mounds that contain mostly or nearly entirely carbonate muds (produced or mediated by a range of bacteria and archaeans) suggest that there is a very wide spectrum of reef types and reef processes. Only a broad definition of reefs as biogenic structures raised above the prevailing seafloor seems usable.

Mudmounds, reefs that contain in the order of 50% or more muds (an arbitrary amount), also played a significant role in the Phanerozoic, particularly at specific phases in earth history. The term "mudmound" evolved originally from the Florida Bay area, where mud is trapped by seagrasses and/or green algae such as *Penicillus*, *Halimeda*, and so forth, in waters less than 3 m deep, and thus involves structures raised above the surrounding substrate by a mix of photosynthetic, biogenic, and physicochemical processes (Ginsburg and Lowenstam, 1958). Green algal mudmounds also may occupy deeper waters at the limit of the photic zone in very clear waters, e.g., 100–150 m, as the *Halimeda* reefs of the Java Sea near Kalimantan (Roberts *et al.*, 1987; Roberts and Sydow, 1997). Mudmounds later came to have other connotations, particularly deeper water ones, and most mudmounds now are considered to be largely formed by microbial or organomineralization processes (Reitner and Neuweiler, 1995; Monty *et al.*, 1995; Neuweiler *et al.*, 1999). Early Carboniferous Waulsortian mounds are called mudmounds, and such reefs are now recognized throughout the Phanerozoic, though Fagerstrom (1987, p. 12) stated that mudmounds were not reefs. In contrast, structures formerly called reefs have sometimes been renamed mudmounds (e.g., Monty, 1995; Pratt, 1995). To James and Bourque (1992), mounds or reef mounds were built by small, delicate, or solitary elements in tranquil waters, and both reefs and mounds were said to be biologically constructed. They subdivided mounds (i.e., mudmounds in the broad sense) into microbial, skeletal, and mud types, all being dominated by mud content. Bosence and Bridges (1995) defined mudmounds as carbonate buildups having depositional relief and being composed dominantly of carbonate mud, peloidal mud or micrite, and recognized only microbial and biodetrital mudmounds. Brunton and Dixon (1994) showed that sponge–microbial reefs, largely mud-dominated, occupied shallow shelf and deep waters in the Cambrian through Jurassic at specific cyclical phases in the Phanerozoic. Reitner and Neuweiler (1995) focused on mudmound genesis, specifically (1) interaction of microbial processes, seawater, and macrobiota, and (2) microbially influenced diagenesis, thus using a broad mudmound definition and stating that mudmounds represent an ultraconservative life strategy for buildups defining the onset of metazoan involvement in the reef ecosystem, an interpretation favored herein.

Some sponge–microbial mudmounds appear to have grown well below the photic zone and thus favored heterotrophic metazoans such as sponges, bryozoans, tube-building worms, shelly organisms, and some cool water corals, which do not requite photosynthetic symbionts. Those adapted to this cryptic niche could live in the dark and probably grew best under higher nutrient input, cooler waters, and lower energies. However, mudmounds also form in shallow subtidal settings and even peritidal zones (e.g., Florida Bay), so that depth is not a primary factor in mudmound formation. During globally warm climatic episodes, mudmounds appear to have retreated to deeper water, off-shelf, or deep ramp conditions following the Caradoc–Ashgill radiation of the mid-Paleozoic reef fauna (Copper, 1997). During periods of mass extinction, when the shallow shelf tropical reef community was most severely disturbed, generally because of lowered sea-level and lowered temperatures reducing $CaCO_3$ precipitation rates, mudmounds moved up into shallower shelf conditions (Fig. 2). Deep-water mudmounds (50–100 m) were defined by Pratt (1995) as ecologic reefs that lack a dominant metazoan component (but this need not necessarily be the case as some of these may have a significant sponge, bryozoan or crinozoan veneer); he also remarked that they have been around since the Paleoproterozoic and that they were restricted to deep waters after the Early Ordovician. Lithoherms, steep-sided deep-water mounds like those of the modern day Florida Straits (Neumann *et al.*, 1977; Messing *et al.*, 1990), produced around cold methane or oil seeps by organomineralization processes, may be different from pseudoreefs that cut across strata and are the result of inorganic mineralization processes via hydrothermal venting.

Most mid-Paleozoic metazoan reefs grew ideally in warm, equatorial, marine, subtidal settings within the photic zone at carbonate supersaturation levels with optimal $CaCO_3$ precipitation rates under normal marine salinities, alkaline pH, oligotrophic conditions, and minimal siliciclastic sediment input. Skeletal production appears to be the key factor, commonly tied to photosymbionts living within soft tissues of reef dwellers; reef components tended to grow vertically, against the law of gravity, whereas sediment tends to be distributed horizontally. Metazoans adapted to those conditions have been called "photozoans" by James (1997). Yet a number of reefs were capable of growth under stressed or well below optimal conditions and a range of

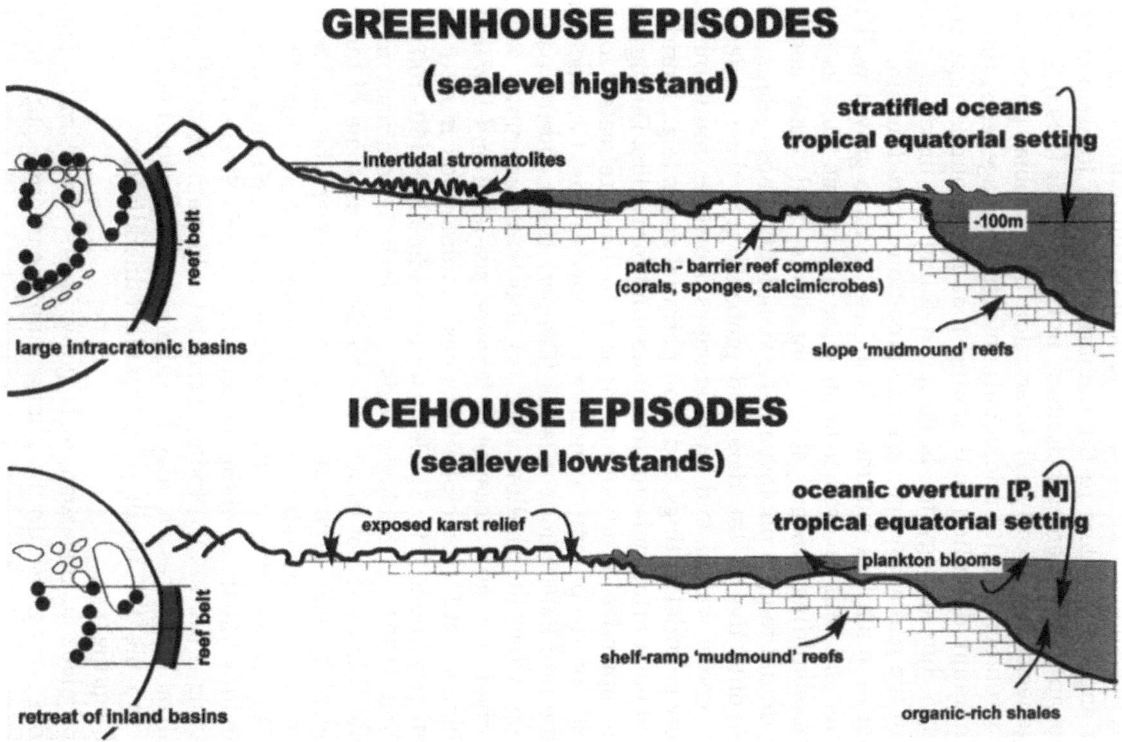

FIGURE 2. The response of reef evolution in the Paleozoic to periods of global warming or cooling, sea-level high- and low-stand, and latitudinal ranges of the tropical reef ecosystem: Sponge–microbial reefs were largely confined to deeper off-shelf, slope facies. During mass extinction episodes, as the shallow shelf was depopulated, sponge–microbial mounds moved from deeper-water settings into the narrowed shelf regime (e.g., Hirnantian, Frasnian–Famennian extinction episodes, and Carboniferous Waulsortian mounds). (Modified from Copper, 1997.)

organisms have adapted to such stressed conditions, either along a facies gradient, during succession or initiation of reef growth, or during regional or global perturbations (Copper, 1994b). Ancient reefs, for example, are known from below the photic zone in deep or turbid waters (slope to basin settings), yet because these are so inaccessible in the modern ocean, they are relatively little known or barely explored today. Such deeper and/or cooler water biota thus are heterotrophic (they live on nutrients derived from the plankton above) and light independent; James (1997), in referring primarily to cool-water carbonates, called these biota "heterozoans." Other stress factors come from life in (1) warm temperate, even to cool waters at high latitudes, or on the upwelling west sides of continents in low latitudes, (2) under relatively higher rates of sediment accumulation, (3) under abnormally high or low salinities, (4) under higher nutrient (phosphorus, nitrogen) content, or (5) also from stress in the deeper intertidal zone (e.g., worm reefs, oyster reefs; consult also Fig. 6). The biota in such reefs differs substantially, and a reef gradient, or reef "reaction stress series," is detectable, ranging from those adapted to optimal reef growth to those under conditions where reefs are absent (see Fig. 6). The most adaptable to stress are microbial or calcimicrobial reefs dominated by prokaryotic bacteria. At the opposite end of the spectrum are larger skeletal metazoans such as calcifying, highly integrated colonial corals (the scleractinians today, the rugosans and tabulates in the Paleozoic). Organisms therefore are useful in providing an "ecothermometer" (or ecobathymeter, etc.) for optimal to stressed habitats in the past.

2. Precambrian Prelude: Archean–Mesoproterozoic

What was the environmental setting and architecture of the earliest reefs? This is not a topic generally discussed in reviews of Precambrian life (e.g., Bengtson, 1992, 1994). Reefs most likely evolved in tropical to warm temperate, oligotrophic oceans, on the east sides of continents, as these climatic–oceanic conditions set the stage for optimum carbonate precipitation and fixation. "Modern" Mesozoic–Cenozoic tropical reef organisms tend to favor aragonite as a skeletal medium; for much of the mid-Paleozoic the prevailing mineral was low to higher magnesium calcite. Reefs were probably initially constructed in simple fashion as soft to firm microbial carpets, raised above the surrounding seafloor, probably with a thin, living veil of photosynthetic cyanobacteria or chloroxybacteria in the upper few millimeters and a decomposing layer of other bacteria and archaeans (including methanogens) recycling organic matter in the muddy sediments underneath. A significant, indirect, cumulative component of this ecosystem may have been the input of "marine snow" by falling calcareous phytoplankton in the upper water mass; modern marine plankton contain similar significant cyanobacteria such as *Trichodesmium* (Capone *et al.*, 1997) and archaeans (Fuhrman *et al.*, 1992), which fix nitrogen and release oxygen. Such life survived initially under a greenhouse

CO_2-dominated atmosphere devoid of free oxygen and such microbial reefs were circumequatorial oases of life, with cyanobacteria absorbing CO_2 required for photosynthesis, mediating the fixation of $CaCO_3$ while pumping out oxygen in return. It is relatively undisputed that the Precambrian atmosphere was effectively anoxic or hypoxic until surplus oxygen no longer captured by free iron or other oxidizing metals became available. Such an atmosphere–ocean system is thought to have prevailed from the first appearance of life around 4 billion years ago, through the Paleoproterozoic (until about 2.0–1.8 Ga), as seen in the seafloor precipitation of banded iron formations, which ceased at that time. Such ancient microbial reefs, and carbonate platforms as a whole, were scarce in Archean time; reef mounds a few meters high and tens of meters in diameter across are known from a few late Archean localities of Canada and Australia (Walter, 1983; Schopf, 1992a), and carbonate platforms and ramps became established during the late Archean–early Proterozoic, ca 2.5 Ga ago (Grotzinger, 1994; Walter, 1994).

The evolution of the first Eucarya, somewhere between 2.1 and 1.9 Ga or possibly even at 2.7 Ga (Doolittle *et al.*, 1996), marked a new stage: rimmed carbonate platforms surrounded continents and microbial reefs became widespread in the tropics, forming in slope settings and as barrier, patch, and shoreline fringing reefs (Copper, 1974). The arrival of eukaryotes, most of which require minimal amounts of oxygen for life processes, and the abundance of carbonate as muds, storing CO_2 but also requiring some O_2 for precipitation, suggest that at least small amounts of free biogenic oxygen circulated in the atmosphere. Eukaryotes may have made a start with archaeans as hosts and cyanobacteria as symbionts (Martin and Müller, 1998; Doolittle, 1998) or by union of prokaryotic bacteria (Margulis, 1981). Stromatolite (microbial) reefs appear to have reached their greatest diversity and largest sizes in the Mesoproterozoic, with prokaryotes and eukaryotes coexisting and forming integrated reef structures (Fig. 2). The arrival of eukaryotes may have facilitated early cementation of reef fabrics and sedimentary structures, accelerating the production of carbonate ooids and pebbly oncoids, some of the first "bioclasts." There is no direct fossil evidence for the presence of multicellular animal fossils at this time.

3. Neoproterozoic Reefs: First Calcimicrobes (1000–544 Ma)

By the late Precambrian (Neoproterozoic), stromatolite diversity and abundance had apparently declined (Awramik, 1971; Grotzinger, 1990), though calcimicrobial reefs with skeletalizing calcimicrobes became large and prominent, accumulating as huge mounds several hundreds of meters thick and up to a kilometer in diameter (Aitken, 1989; Turner *et al.*, 1993). The reasons for decline of microbial diversity and rise of calcimicrobes are still obscure, and some argue it is just an artifact of estimating diversity based on shape and size. In contrast, ideas for the expansion of (calcimicrobial) reefs have hinged around one or a combination of the following driving factors: (1) ocean chemistry, e.g., decreased carbonate saturation and decreased supply for carbonate-trapping microbial communities [Grotzinger (1990, 1994); note that Buddemeier and Fautin (1996) argued the opposite—supersaturation—for the rise of skeletogenic metazoans]; (2) oceanic pH flux, toward levels of 7–8, as a driving factor stimulated by the "sulfate pump," with increased precipitation of calcium sulfates such as gypsum and anhydrite (Kempe and Kazmierczak, 1994); (3) construction–destruction of the first supercontinent Rodinia in the Neoproterozoic and rapid reorganization of the early Phanerozoic continents with concomitant climate effects (Kirschvink *et al.*, 1997); (4) higher ocean temperatures, accelerating $CaCO_3$ precipitation rates (Riding, 1993, 1996); (5) competitive exclusion on the seafloor from arriving metazoans, pushing out the nonskeletal stromatolites, and establishing metazoans (Reid and Macintyre, 1992); (6) the arrival of the first grazing, bulldozing, and burrowing metazoans disturbing substrate microbial mats (Garrett, 1970); (7) the rapid evolution of large phytoplankton in the upper ocean water mass, as seen in the fossil record of acritarchs, dinoflagellates, and other enigmatic plankton (Schopf, 1992b; Butterfield and Rainbird, 1998), large-scale production and precipitation of organic matter, and the explosive increase in biotic phosphate precipitation as oceans reached saturation levels for phosphorus [this would enhance $CaCO_3$ precipitation if inhibiting phosphorus was removed; see also Bengtson (1998)]; (8) the evolution of plankton or nekton with a linear gut, capable of consuming large amounts of surface plankton and transferring this into fecal pellets that could sink rapidly to the seafloor, thus freeing up oxygen in surface waters (Logan *et al.*, 1995); (9) evolution of a complex food chain, with a cascading effect through marine ecosystems (Bambach, 1993; Debrenne and Zhuravlev, 1997); and (10) a multicausal factor, with rapid cumulative rise in surplus oxygen content of the ocean–atmosphere system by the end of the Precambrian, reaching threshold conditions for precipitation of skeletons, accentuated by high-latitude glaciation, sinking of deep cold oxygen-rich waters, and evolution of macrophytoplankton and the linear gut in animals.

A particularly striking aspect of the Neoproterozoic equatorial reef and perireefal subtidal community was the arrival of calcified microbes or skeletal stromatolites (porostromatolites), mostly the cyanobacteria (Riding, 1992; Monty, 1995). These were in the form of *Girvanella*-like calcified filamentous sheaths and *Renalcis*-like calcified coccoids. Prokaryotes may calcify by at least three processes: (1) internal calcification of cells within the cell membrane during life; (2) development of $CaCO_3$ sheaths around filaments and coccoids; and (3) the production of microenvironments within mats or reef niches where calcification is enhanced either during life or via very early, postmortem processes. Calcification of cyanobacteria (and possibly chloroxybacteria and archaeans) may be the result of large-scale environmental factors during global periods of intense marine carbonate precipitation, with cyclic development favoring bacterial calcification episodes (the carbonate calcification events of Riding, 1992, 1996). Alternatively, calcification may be an evolutionary factor, genetically inherent though not necessarily obligate, in some prokaryote taxa and not others (Monty, 1995). What might or would trigger environmental factors such as global warming, changes in CO_2–O_2 concentration, or concomitant flux of the oceanic pH is still undecided (Walter *et al.*, 1992), and the effect of such processes on calcification is also unclear. Preservation of minute calcified prokaryote sheaths or coccoids is undoubtedly another problem (Pratt, 1995).

Several biotic factors seem to coincide in the Neoproterozoic tropical carbonate shelf, though almost none of the research on Precambrian life has specifically focused on reef settings; indeed, reefs are almost never mentioned in the evolutionary spectrum (Bengtson, 1992, 1994). First is the decline of the soft or firm, nonskeletal stromatolites (Schopf, 1992b), though this may have started slightly earlier with the explosion of eukaryotes. Second, multicellular algae and large-sized macrophytoplankton increased dramatically (Schopf, 1992c; Vidal and Moczydlowska-Vidal, 1997). Third, the soft-bodied metazoans began their radiation (including burrowers, grazers, and forms with a linear gut) and the first skeletal metazoans arrived around this transition time, with more than 38 lineages that independently began mineralization in the latest Precambrian–earliest Cambrian (Lipps *et al.*, 1992a; Bengtson and Runnegar, 1992; Bengtson, 1998). Fourth, skeletal prokaryotes and protists began to form calcimicrobial crusts, carpets, and reefs (Riding, 1996). Fifth, in view of the arrival of complex multicellular benthic and planktic life in the Neoproterozoic, it seems almost inevitable that the evolution of trophic strategies such as herbivory, suspension feeding, detritus feeding, and predation resulted in the construction of food webs (Lipps *et al.*, 1992b), and that these in turn stimulated mineralization of skeletons. Sixth, the confirmation of fossil dinoflagellates in the Neoproterozoic (Moldowan *et al.*, 1996; Vidal and Moczydlowska-Vidal, 1997; Butterfield and Rainbird, 1998) suggests that the symbionts so prominent in Recent scleractinian calcification were in place as early 800–900 Ma and that they were likely to have participated in the construction of earliest metazoan and calcimicrobial reefs. Diversity is described as highest in the tropical latitudes in the Early Cambrian (Lipps *et al.*, 1992a) and possibly also in the Neoproterozoic. Rapid calcification of reef biotas forming basal, walled, and internal skeletons, allowing reef structures to grow and expand above the surrounding seafloor and permitting new niches to be created within reef cavities, were the hallmarks of the latest Neoproterozoic and Early Cambrian reefs. Cyanobacteria were accompanied in the late Neoproterozoic by multicellular green and red algae (Zhang, 1989; Butterfield and Rainbird, 1998), which occurred in nutrient-rich settings alongside phosphatized stromatolites but which are not clearly known to play a role in reefs at this time.

Possibly as early as 800 Ma ago, the age of the oldest spicules, sponges formed biofilm aggregates with microbial communities, though no whole sponges or sponge reefs are presently known for strata of that age (Reitner and Mehl, 1995). Complete sponges are known in Neoproterozoic strata at least 570 Ma old in both Australia and China (Gehling and Rigby, 1996; Li *et al.*, 1998).

The inability of the early siliceous Precambrian sponges to form reefs, as far as known, may have been due to a lack of sediment-baffling capacity (too low or too soft to have formed rigid or semirigid structures that could trap sediment); their inability to cement, encrust, and secrete carbonate (i.e., taphonomy); or possibly the earliest sponges inhabited oceans too cool or too deep in which to precipitate carbonates. In the 1990s, a series of striking discoveries by paleontologists in China and elsewhere illuminated a treasure trove of skeletalized, multicellular animal fossils showing wall tissues, whole skeletons, and embryos of Late Precambrian age (Bengtson and Zhao, 1997; Li *et al.*, 1998; Xiao *et al.*, 1998; Bengtson, 1998). At about the same time biologists have used gene sequence differences to push the molecular biological clock for the origin of major animal phyla back to between 670–1300 Ma (Wray *et al.*, 1997), splitting the branch between protostomes (worms, arthropods, mollusks) and deuterostomes (echinoderms and chordates) to at least 670 Ma. All these may have had an indirect impact on the establishment of the first calcimicrobial reefs.

The oldest protists (single-celled animals) have a protein clock date of about 1230 Ma (Doolittle *et al.*, 1996), but again these have not yet produced any direct evidence in the form of skeletal fossils. Protistan forams are known to have been important calcite secreters and reef builders in much younger rocks, first becoming especially abundant in the Late Paleozoic [Late Devonian (Famennian)-Carboniferous]. The late Neoproterozoic is best known for its "soft-bodied," nonskeletal Ediacaran fauna, a short-lived distinctive group of benthic organisms with a "mattresslike" structure that lasted for about 20 million years (Jenkyns, 1992; Narbonne, 1998). These Ediacaran organisms, also known as vendozoans (Seilacher, 1989, 1992), are nearly always found in siliciclastic sediments such as sandstones, mudstones, and tuffs, reflecting rapid event burial; none are known to have formed reefs or sediment piles. Fedonkin (1992) proposed that Ediacaran organisms lived in shallow, cool oceans, a factor that would have prevented them from forming limestone reefs. The last of the soft-bodied Ediacaran fauna stretched just to the Precambrian–Cambrian boundary ca 544 Ma (Grotzinger *et al.*, 1995), which indicates that this group suffered extinction during a relatively short time interval. Although cnidarians appear to have constituted part of the Ediacara fauna, as imprints of planktic jellyfishlike structures or possibly benthic seapenlike structures (Jenkyns, 1992), the affinity of these with cnidarians is disputed and some suggest these as the first large-scale "failed experiment" in the evolution of animal life (Seilacher, 1989). Cnidarians appear to have lacked the ability to secrete a $CaCO_3$ skeleton in the Precambrian and earliest Cambrian (Nemakit-Daldynian through Tommotian) and lagged in this skeleton-secreting capacity nearly 10–30 million years after the sponges; this is still a very puzzling feature for which no adequate explanation has been found. The first skeletal cnidarians (tabulate corals) do not appear until the later part of the Early Cambrian (Scrutton, 1997, 1998), and corals did not become common until the end of the Middle Ordovician, ca 460 Ma ago.

Were there other possible reef builders in the Neoproterozoic? Soft-bodied metazoans of an erect size large enough to baffle sediments appeared about 800–1000 Ma (Conway Morris, 1993). The first shelly animals date back to the latest Precambrian and include a range of small tube-building animals such as thicket-building *Cloudina* and *Cambrotubulus*, *Curvolithus*, and *Anabarites*, which built walls of aragonite, phosphates (e.g., apatite), or opaline silica, with some forms agglutinating walls of sand grains (Bengtson and Conway Morris, 1992). However, whether these built reefs raised above the seafloor is debated, though small cuplike and tube-building skeletal metazoans (sipunculids or polychaetes?) are abundantly attached to stromatolite reefs or form independent small patch reefs in the late Neoproterozoic of Namibia (Germs, 1972; Grotzinger *et al.*, 1995; Reitner, personal communication). *Skolithos*, a widespread agglutinating tube builder of the early Paleozoic and especially abundant in the Early Cambrian, also is present in the late Precambrian (it is often erroneously called a burrow structure: Crimes, 1992); strata packed with such agglutinating, tube-forming and -building organisms are similar to modern sabellarid worms that build "sand reefs" in temperate to tropical settings by using an organic mucilage (Ekdale and Lewis, 1993; Gruet and Brodeur, 1995). Modern sabellarid sand reefs form fringing, patch reef, and barrier reef structures with spur-and-groove morphology and even play a role in the intertidal surf zone around tropical coral reefs (Pandolfi and Kirtley, 1998). It is not difficult to visualize such structures as the earliest metazoan reefs in the late Precambrian and Phanerozoic. Polychaete worms, therefore, may have been among the oldest metazoan reef constructors, making their first appearance in the late Neoproterozoic, ca 600 Ma, with greatest abundance in the Early Cambrian. In the Phanerozoic, polychaetes generally occupy a marginal reef role, becoming abundant when and where environments were under stress, e.g., during episodes of mass extinction or in coastal marginal settings with high sediment input, high rates of fresh water runoff, or intertidal zones (Carey, 1987; Copper, 1994b). In the Late Ordovician, for example, the tube-building polychaete *Tymboochos* formed small patch reefs and mounds in equatorial, shoreline siliciclastic settings flanking the Precambrian Shield (Steele-Petrovich and Bolton, 1998) and the spiral polychaete tube builder *Spirapora* formed small mounds during the end Rawtheyan (Late Ordovician) mass extinctions [here this was mistakenly called a tabulate coral (Copper, 1981)].

The late Neoproterozoic features a series of spectacular glaciations, in some areas associated with either temperate or subtropical carbonates, suggesting rapid climate shifts with a periodic "snowball earth" pushing ice cover into the near tropics (Hoffman *et al.*, 1998a,b) (others suggest that paleomagnetic data also can be interpreted so as to place these tillites in polar positions). These were accompanied by rapid, large-scale excursions of $\delta 13C$ at a scale unknown in the Phanerozoic, except during mass extinction episodes (Kauffman and Knoll, 1995), suggesting cycles of large-scale phytoplankton expansion and collapse associated with global climate change even more severe than the Pleistocene ice ages. This first may have stimulated the rise of the cool water Ediacara fauna (Fedonkin, 1992; Jenkyns, 1992) and then delayed the arrival of metazoan reefs until the Tommotian, favored by warmer equatorial ocean temperatures. Buddemeier and Fautin (1996) and Buddemeier (1997) point out that increased growth and calcification of coralline algae and reef-building corals respond to increased supersaturation of aragonite in the modern ocean in the tropical belts (as well factors such as temperature and light) and that the evolution of nonskeletal scleractinians may be linked to reduced mineral saturation in the oceans, with less supersaturated high latitudes favoring solitary or small skeletogenic corals or anemones lacking a skeleton. This might explain the delay in the onset of reef-building metazoans as a result of Neoproterozoic glaciation, as well as the response of reefs to cooling mass extinctions. The paradox would be to explain the massive size of shallow water carbonate platforms and coral–sponge reefs in the Middle Paleozoic, at a time of much higher CO_2 concentrations (14 to 16 times that of the modern and 3–4 times that of the Cretaceous) (Berner, 1994). One such explanation might be in highly efficient photosymbiosis in Paleozoic corals and sponges, removing CO_2 from surface waters, and increasing supersaturation of carbonate ions.

4. Cambrian Reefs: Start of Metazoan Reef Components

Calcimicrobial reefs of types similar to the late Precambrian continued into the Nemakit–Daldynian, the first stage of the earliest Cambrian (Zhuravlev, 1986; Zhuravlev and Wood, 1995). This meant that for reefs, the so-called "Cambrian explosion" of skeletalized animals, as portrayed in most textbooks, correlated not with any instantaneous introduction of a complex metazoan reef ecosystem but was represented by a phased-in reef development period lasting about 10 to 15 million years (the Nemakit–Daldynian) and which continued gradually into the Tommotian stage, when archeocyath sponges still played only a small role. The same appears to be generally true of perireefal and nonreefal skeletal organisms; brachiopods, mollusks, sponges, and cnidarians were not introduced en masse at the Precambrian–Phanerozoic boundary but in steps. Many enigmatic small shelly fossils of the first two Cambrian stages, often difficult to assign to specific phyla because they may consist of disjointed skeletal parts, and many secreting minerals such as phosphates, aragonite, or calcite (Bengtson and Conway Morris, 1992) preceded the classic Cambrian fauna. Why should organisms that built a basal skeleton (like corals) or internal walled skeleton (like sponges) be favored in the rapid growth of reefs? What is the significance of the introduction of skeletal animals into reef ecosystems? Was it the evolution, expansion, and rise of organisms with skeletons that reorganized reefs, or were there global changes in the atmosphere–ocean system that permitted groups with a long soft-bodied record to build hard skeletons?

The first animal pioneers in Early Cambrian reefs were the archeocyath sponges, usually cup shaped to cylindrical sponges with a double or single, thin or thick, finely granular calcite wall containing numerous pores or canals. *Archeocyathus*, which gave its name to the whole group, is a branching colonial form consisting of cylindrical, finger-sized digits, each of which has a relatively dense, solid wall penetrated by upwardly directed canals. Many archeocyaths appear to consist of a solitary conical cup with a double wall separated by porous partitions, but repent sheets that grew flat or gently arched on the substrate are also known (e.g., *Retilamina* from the late Early Cambrian of Labrador). The first true tabulate corals as well as coralomorphs (resembling corals but that may or may not be corals, since their skeleton is poorly preserved) also are present, though very rare and difficult to find at most localities (Savarese *et al.*, 1993) and probably only playing a small role in the ecosystem. Earliest Cambrian archeocyath reefs and their communities have been well documented and interpreted, e.g., from Canada, Siberia, Mongolia, and Australia, where they are especially well preserved in a wide range of carbonate facies (see James and Kobluk, 1978; James and Gravestock, 1990; Debrenne, 1992; Debrenne and Courjault-Radé, 1994; Kruse, 1991; Wood *et al.*, 1993; Kruse *et al.*, 1995; Zhuravlev and Wood, 1995). By the late Early Cambrian (Toyonian), in Labrador a whole range of archeocyath patch reef types adapted to various settings had evolved, from pouffe-sized mounds that were largely calcimicrobial (with *Angusticellularia, Girvanella*), to those with significant cup-shaped or branching archeocyaths, to biostromal, spongiostromate structures with dish-shaped archeocyaths, to haystack-sized mounds with flat *Retilamina* archeocyaths and *Girvanella*, and to archeocyath thickets lacking calcimicrobes. Reitner and Mehl (1995) described a very different sponge mound assemblage from the Early Cambrian (Tommotian) of China consisting of mud bound by spicule mats derived from hexactinellids; these may be prototype sponge mounds for deeper-water facies. They also noted that such sponges were very different from Calcarea and demosponges associated with archeocyath reefs, suggesting that more than one sponge reef or mound complex was evolving at the time.

Savarese *et al.* (1993) suggested a third type of reef assemblage in the Early Cambrian from Australia: these were coral-bearing archeocyath reefs, which flourished in nearshore, higher energy, mixed siliciclastic–carbonate facies. Corals also are known from other early Cambrian reef faunas, though the true coral affinity of these is questioned by some (Zhuravlev *et al.*, 1993). The first tabulate corals determined from archeocyath reefs were only rare representatives. Sorauf and Savarese (1995) identified forms such as *Flindersipora* and *Moorowipora* as tabulate corals, with which Scrutton (1997) concurred. Cambrian rugose corals have not been found; the first of these occur in the late Llanvirn (Ordovician), some 60 million years later.

The archeocyath–calcimicrobial reef system collapsed by the end of the Early Cambrian (end Toyonian), followed by a long period with very little or no metazoan reef development worldwide, though subtidal calcimicrobial reefs and intertidal to shallow subtidal stromatolite–thrombolite reefs were prominent and omnipresent in shallow shelf and infracratonic seas. Zhuravlev (1996), however, claimed (for the loss of the archeocyath–calcimicrobe consortium) that this was not the case, and said the biggest Cambrian reef frameworks were in the Middle Cambrian, adding, though without metazoans. Zhuravlev (1996) proposed that there was a double Early Cambrian mass extinction, one called the "Sinsk event" at the end of the mid-Botomian marked by anoxia and phytoplankton blooms and the second, in the early Toyonian, the "Hawke Bay event," which featured a worldwide regression (neither of these removed the archeocyath sponges, either from reefs or perireefal settings). Zhuravlev and Wood (1996) suggested anoxia as the cause of the demise of the archeocyath community. However, in Labrador, which has some of the last archeocyath reefs known, black shale horizons are unknown at the top of the succession and the postreefal succession is not marked by sea level drawdown sufficient to produce an unconformity, karst, intertidal sediments, or other sedimentary signatures common to subaerial exposure. Zhuravlev (1996) further proposed a recovery scenario for the mid-Cambrian reef ecosystem, which entailed reduced grazing pressures, unhealthy metazoan–calcimicrobial interactions, and Elvis taxa that intruded into the thrombolite–stromatolite community.

Post-Early Cambrian carbonate platforms essentially repeated calcimicrobial reef morphologies of the Neoproterozoic; regionally and locally there is little to distinguish stromatolitic mounds of Proterozoic age superficially from those of the earliest Cambrian and Early Ordovician. Microbial and calcimicrobial reefs dominated, with thrombolite–stromatolite communities in nearshore and intertidal settings, and calcimicrobial reefs, with generally rare metazoan components, in offshore mid- to distal shelf and slope settings (Hamdi *et al.*, 1995). The Late Cambrian was a time of extensive cratonic regression–transgression cycles (at least four major episodes), displacing biomeres carrying trilobites in North America and elsewhere, and marked by a continued phase of globally warm climates. Excluding Cambrian archeocyaths and the mid-Paleozoic stromatoporoids, Rigby (1971) identified relatively few localities where sponges were known to be associated with stromatolitic reefs in the Cambrian, but this number has increased more recently, though not significantly (Hamdi *et al.*, 1995). Zhuravlev (1996) stated that the biggest Middle to Late Cambrian reefs were known from Laurentia (citing Lohmann, 1976; McIlreath, 1977) and the Siberian Platform. However, the North American localities cited have disputed interpretations and many may represent intertidal, restricted hypersalinity or exposed, karst-eroded rimmed shelf sequences, instead of being normal marine subtidal, shelf reefs. Impressively thick, well-documented Mid- to Upper Cambrian carbonate successions occur in northeastern Canada and northern Greenland, ranging from slope to shelf to onshore mixed siliciclastic–carbonate facies stretching out over 1500 km and several kilometers thick (Trettin, 1991; Ineson *et al.*, 1994; Ineson and Peel, 1997). These show a ramp or high-energy platform rim, a shallow platform interior, and carbonate slope aprons with debris flows, dated by trilobites and other shelly benthos. Almost the only reefs were small microbial mounds on the tropical platform interior and rim, with thrombolitic and laminated microbial fabrics; metazoans are not yet known to be significant participants in these Middle to Late Cambrian reefs. This unusual aspect, the almost total loss or absence of calcifying metazoans in the postarcheocyath Cambrian (post-Toyonian) over continent-sized areas with huge carbonate platforms and ramps, such as North America, requires explanation. This could be partly explained by Buddemeier and Fautin's (1996) hypothesis that, similar to the late Cretaceous and Eocene at a time of high atmospheric CO_2 and warm temperatures, with a possibly shallower carbonate compensation depth, saturation of $CaCO_3$ was lowered, reducing possibilities for skeleton accretion and stimulating the evolution of nonskeletal scleractinians such as sea anemones from skeletal types. However, the counterpart would be the paradox of explaining the vast Mid-Paleozoic coral–sponge reefs and tropical carbonate platforms at a time of much higher pCO_2 (Berner, 1994), greatly exceeding that of the Cretaceous.

5. Ordovician Radiation and Terminal Ordovician Decline

One of the most significant marine metazoan radiations and expansions of life during the whole Phanerozoic, especially at the family to order taxonomic level, took place during Ordovician time, spanning a period of about 60 million years (500–440 Ma). This saw the demise of the "Cambrian" fauna (Sepkoski and Sheehan, 1983), followed by the rise of the mid-Paleozoic fauna (Droser *et al.*, 1996; Droser and Sheehan, 1997). This scenario applies equally as well to overall marine biodiversity as it did to reefs. During the Ordovician, the lithistid–calcimicrobe reef association, which prevailed initially, was gradually replaced by the rise and radiation of the first reefal Mid-Paleozoic coral–stromatoporoid–calcareous algae consortium in shallow-shelf settings (Webby, 1984; Copper, 1997; Webby *et al.*, 1997). By the end of the Middle to early Late Ordovician (late Llanvirn–early Caradoc), this shallow-water, "mid-Paleozoic" coral, stromatoporoid, and skeletal green–red algal consortium had displaced the lithistid microbial reefal mudmound elements into deeper-water slope and basinal settings (Fig. 4), a facies association that continued through much of the Silurian and Devonian (Copper, 1997).

Webby (1984, 1992) first documented the rise of Ordovician reefs on a global basis, drawing on data from around the world, especially in regard to the distribution of stromatoporoids and early corals. Since then a number of other reef occurrences have cropped up, particularly those from South America, which contain Early and Middle Ordovician reefs and faunas remarkably similar to those of the United States and arctic Canada (Cañas and Keller, 1993; Lehnert and Keller, 1993; Cañas, 1995; Keller and Bordonaro, 1993; Keller *et al.*, 1994, 1995; De Freitas and Mayer, 1995); this has redrawn the latitudinal map for the North and South American plates, since both reef belts occurred in the paleotropics, with the South American reefs extended westward into the United States. The North and South American Tremadoc-Llanvirn reefs show the presence of the simplest, oldest stromatoporoids, the

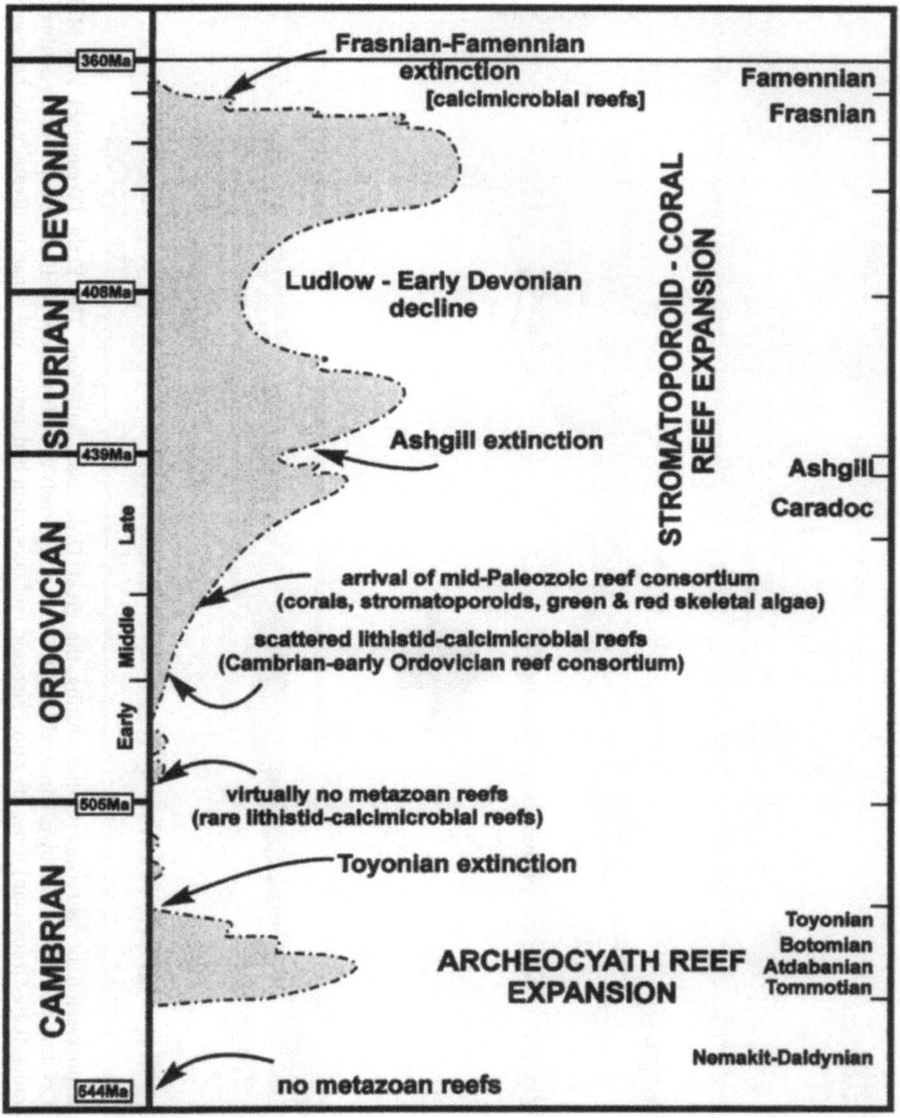

FIGURE 3. Generalized range chart of Cambrian through Devonian states of reef expansion and collapse. Corals played no significant role in reefs in the Cambrian through Middle Ordovician, though isolated occurrences are reported. The mid-Paleozoic reef community took over by late Llanvirn–early Caradoc time in the Ordovician. (Modified from Copper, 1994a.) Reef biodiversity suffered relatively limited losses at the genus and family level during the Late Ordovician (Hirnantian) glacial episodes: This was comparable to losses in the Pleistocene. (Based on unpublished data: Copper.)

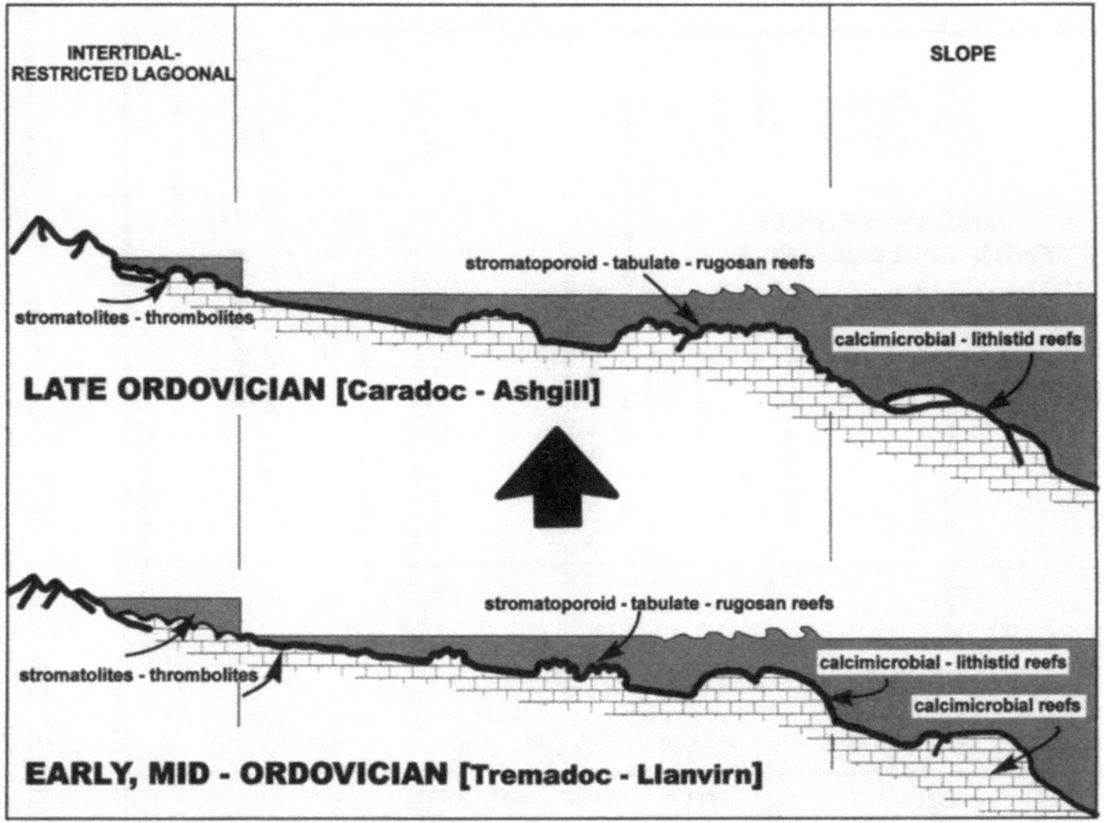

FIGURE 4. Replacement and dispossession of reef communities during the great Ordovician radiation. The Early (Tremadoc) and Middle (Arenig-Llanvirn) Ordovician reef system was dominated by calcimicrobial–lithistid sponge reefs in both shallow shelf and deeper-water settings. By the end of Llanvirn and early Caradoc (Late Ordovician) time, the mid-Paleozoic reef ecosystem had arrived and began to diversify in shallow shelf waters, pushing the lithistid–calcimicrobial reefs into off-shelf, slope, and deeper habitats. (Modified from Copper, 1997.)

encrusting laminar, aragonitic sheets of *Pulchrilamina* (Webby, 1994). Pratt and James (1989) demonstrated the presence of a very simple, early encrusting tabulate coral, *Lichenaria*, in some Tremadoc reefs of Newfoundland, Canada. From Hubei (China) have come new records of Early Ordovician lithistid–microbial reefs (Zhu *et al.*, 1993, 1995); still undescribed reefs remain from Gansu and Shaanxi in the north and Jiangsu to the east (Wu and Yuan, personal communication). The data from such new localities supplement and do not contradict older reef records in the Early and Middle Ordovician (through most of the Llanvirn), e.g., the dominant elements in reefs at this time were calcimicrobes and simple early "lithistid" sponges. Corals and stromatoporoids beyond the simple *Pulchrilamina*-grade were absent or very rare.

The Late Ordovician (end Ashgill, Hirnantian) was marked by multiple ice ages, global oceanic cooling, sea level drawdowns estimated at 30–60 m, and mass extinctions (Berry and Boucot, 1973; Brenchley *et al.*, 1995). This terminated a long Cambro-Ordovician episode of global greenhouse climates, but the sharply demarcated glacial events appear to have lasted no more than about 700,000 to 1 million years. Climatic rewarming of the Early Silurian, as indicated by stable δ13C isotopes and pCO_2 (Berner, 1994), was rapid. The extinction took place at least in two phases, the first wiping out much of the classic Rawtheyan (Richmondian) faunas and the second pushing many of the remaining Ordovician relicts out at the Ordovician Silurian (O/S) boundary by the end of the Hirnantian (Copper and Jin, 1996). The Hirnantian mass extinction interval began and ended with cooling episodes, but much of the <1-Ma period of global perturbation appears to have been an interglacial climate mode, possibly similar to that of the warm cycles during the ca 2-Ma-long Pleistocene icehouse. This meant that reefs of Hirnantian age and shelly nearshore or cool water Hirnantian faunas themselves were relatively widespread and cosmopolitan. Diversity during the Hirnantian was higher than the period immediately preceding and following the glaciation events (Copper, 1999). Hirnantian age reefs are known from eastern Canada (Copper, 1989), and the Canadian northeast arctic, Greenland (Higgins *et al.*, 1991), northern Ireland (Chair of Kildare limestone), Girvan (Scotland), the Boda Limestone in Sweden (Thorslund, 1935; Beutinger, 1996), the Baltic (Nestor, 1995), and further east in Kazakhstan (Nikitin *et al.*, 1996). Except for the up to 8- to 10-m-thick diverse, coral–stromatoporoid reefs from the Ellis Bay Formation of Anticosti and some of the coral–stromatoporoid reefs from the early Hirnatian Porkuni Formation of the Baltic, most of the Hirnantian reefs are microbial mudmounds with a sparse fauna of calcimicrobes, green algae, lithistid sponges, bryozoans, and brachiopods. Yet all seem to occur in relatively shallow onshore to distal shelf settings, in waters from ca 10–100 m deep. This corroborates the relatively stressed, marginal environmental settings of Hirnantian carbonate shelf areas, even in the tropical low latitudes. The situation was exacerbated by sea-level drawdown exposing most of the Late Ordovician shelf epicontinental sea and area formerly available for reef development in North America and elsewhere; very few tropical Hirnantian strata and almost no continuous sections across the O/S boundary are known in shelf settings. This strongly parallels what is seen in the Late Devonian (Famennian) and end-Permian, in both sea-level drawdown and cooling.

6. Reefs in the Silurian–Devonian: Maximal Greenhouse

During the Silurian and Devonian, reef construction proceeded on a spectacular and large scale in many parts of the world, accompanied by global warming episodes and major marine transgressions, producing extensive epicontinental and infracratonic seas, as reviewed exhaustively elsewhere (Copper and Brunton, 1991; Brunton *et al.*, 1998; Copper, 1994a, 1997, in press). The initial start-up to reef growth was slow in the Early Silurian, with few reefs and reefs of generally small sizes known from the early Llandovery (Rhuddanian: Copper and Brunton, 1991). Reef growth did not accelerate until mid-Aeronian time, 4 to 6 million years later, which was paralleled by increasing biodiversity of the marine benthos in the Llandovery. Reefs in shallow tropical waters were dominated by a mix of complex colonial tabulate and rugose corals, including a wide range of constructional types from highly integrated (losing their walls: thamnasterioid–aphroid), to meandroid (tollinaform, cateniform), to phaceloid, and to very large solitary forms with individual calices up to 20 cm diameter (Scrutton, 1997, 1998). Corallite morphological diversity and growth rates were comparable to those of modern scleractinians (Scrutton, 1997, 1998), even including acroporid morphologies as seen in the thamnoporid tabulates. Many calcimicrobes participated in reef growth in shallow water settings (Soja and Antoshkina, 1997). Some Silurian microbial reefs also favored deeper water slope to basinal settings, many as pinnacle structures and some over 1100 m thick and 25 km wide, as found in the Canadian arctic (Brunton and Dixon, 1994; De Freitas *et al.*, 1995, 1999; De Freitas and Dixon, 1995). Silurian shelf reefs reached an acme in size and biodiversity during the Wenlockian and episodic reef expansion seems to have been achieved during eight global cycles, probably in response to cosmopolitan warmer, transgressive oceans at higher latitudes, with retreats during periodic regressive cycles and bottom water turnovers (Brunton *et al.*, 1998). Episodic gains and losses of benthic taxa broadly appear to have correlated in planktic and nektic biota (though conodont–graptolite extinctions may at times have been decoupled, e.g., in the early Ludlow) and δ87Sr/δ86 Sr ratios (Ruppel *et al.*, 1998). By the end of the Silurian, progressive introduction of Devonian-type taxa and disappearance of earlier Silurian forms, with increasing diversity losses, marked a changeover in faunas from more cosmopolitan to endemic marine reef faunas typical of the Early Devonian.

7. Collapse of the Mid-Paleozoic Reef Ecosystem: The Frasnian–Famennian Mass Extinctions

Toward the end of the Middle Devonian (Givetian, Fig. 5), the reef ecosystem suffered a major series of setbacks, with dramatic biodiversity

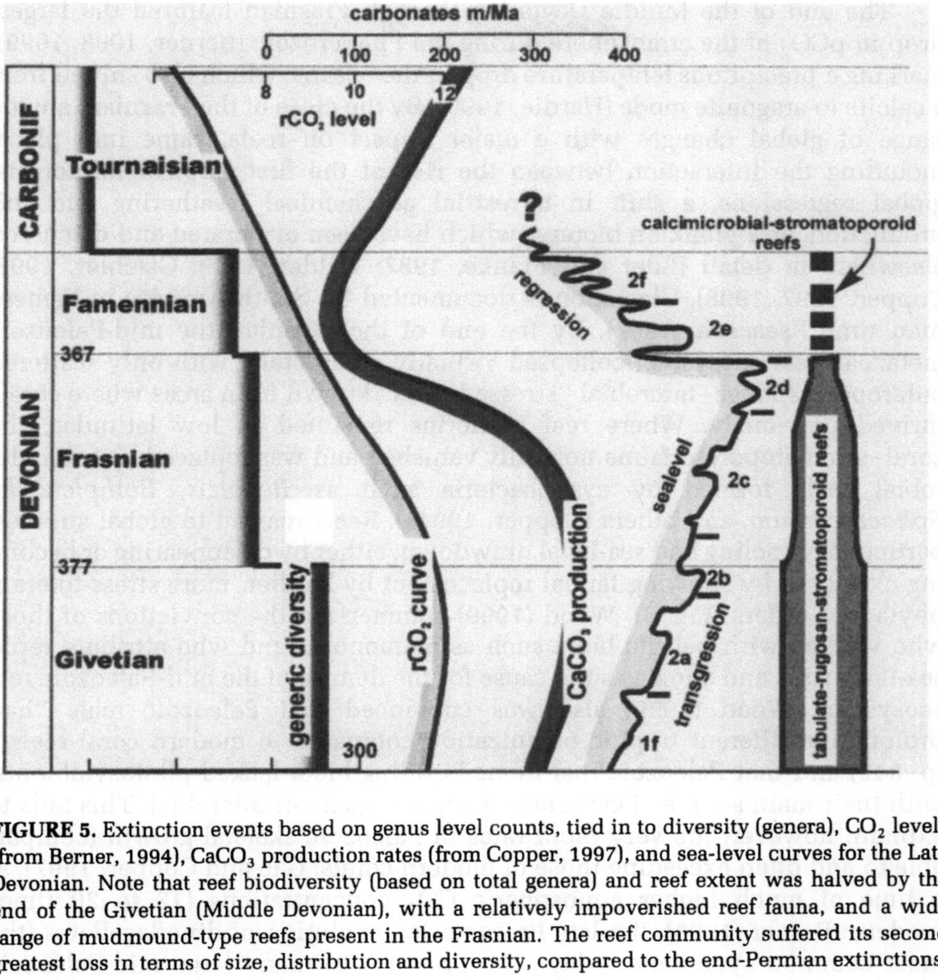

FIGURE 5. Extinction events based on genus level counts, tied in to diversity (genera), CO_2 levels (from Berner, 1994), $CaCO_3$ production rates (from Copper, 1997), and sea-level curves for the Late Devonian. Note that reef biodiversity (based on total genera) and reef extent was halved by the end of the Givetian (Middle Devonian), with a relatively impoverished reef fauna, and a wide range of mudmound-type reefs present in the Frasnian. The reef community suffered its second greatest loss in terms of size, distribution and diversity, compared to the end-Permian extinctions.

losses, disappearance of reef tracts, and downgrading of many reef types to mudmounds through the Frasnian (Copper, 1994b, 1997). Coral diversity (calcitic tabulates, rugosans) continued to drop by >50% during the Frasnian, following maximal Middle Devonian diversity (Scrutton, 1998). This was paralleled by other biota, especially reef and perireefal tropical calcitic brachiopods, with more than 70% losses (Copper, 1998). The aragonitic stromatoporoids declined from 39 Frasnian genera to 11 Famennian genera (a 72% loss), coinciding with the introduction of 5 new genera during the extinction crises and total loss thereafter (Stearn *et al.*, 1999). Partial Famennian reef "recovery" was dominated by cyanobacterial calcimicrobes such as *Renalcis* and *Sphaerocodium*, which inhabited the carcasses or karst-eroded platforms of the coral–stromatoporoid metazoan reefs before them or produced entirely new reefs (Yu and Shen, 1998).

The end of the Middle Devonian through Frasnian featured the largest drop in pCO_2 of the atmosphere during the Phanerozoic (Berner, 1998, 1999), marking a precipitous temperature drop in the oceans, which also shifted from a calcite to aragonite mode (Hardie, 1996). By the close of the Frasnian, a wide range of global changes with a major impact on reefs, came into place, including the interaction between the rise of the first tropical rainforests, global regressions, a shift in terrestrial geochemical weathering and soil production, and plankton blooms, which have been elaborated and discussed elsewhere in detail (Eder and Franke, 1982; Wilder, 1989; Gischler, 1992; Copper, 1997, 1998). Glaciation is documented for South America in Famennian time (Isaacson, 1997). By the end of the Frasnian the mid-Paleozoic metazoan reef ecosystem collapsed virtually completely, with only scattered outcrops of sponge–microbial "stressed reefs" known from areas where corals thrived previously. Where reef platforms remained in low latitudes, the coral–stromatoporoid fauna normally vanished and was replaced by calcimicrobial reefs formed by cyanobacteria such as *Renalcis*, *Rothpletzella*, *Sphaerocodium*, and others (Copper, 1994a). Reefs reacted to global stresses, particularly cooling and sea-level drawdown, either by disappearing or becoming extinct or by showing faunal replacement by hardier, more stress-tolerant phyla and orders (Fig. 6). Wood (1999) summarized the convictions of those who worked with pelagic biota such as ammonoids and who attribute rapid sea-level rises and anoxia as the cause for the demise of the mid-Paleozoic reef ecosystem. Wood (1999) also was convinced that Paleozoic reefs "had profoundly different trophic organization compared to modern coral reefs," (p. 342) and that Paleozoic reef frame-building biota lacked photosymbionts, with their main source of carbonate being inorganic or microbial. This fails to explain, however, the very rapid rates of Paleozoic skeletal growth (comparable to and often exceeding those of modern corals: Gao and Copper, 1997), at a time of much higher atmospheric CO_2 concentrations (12 to 20 times modern: Berner, 1998), the late Precambrian evolution of dinoflagellates (the most common symbionts today), the presence of many large mid-Paleozoic coral and sponge reefs without significant microbial content, and very high rates of Paleozoic bioerosion.

8. Summary

1. The precursors to Phanerozoic reefs were Neoproterozoic calcimicrobial mounds, with the first skeletal prokaryotes from tropical carbonate shelf and intracratonic basin settings.
2. Metazoan reefs (the archeocyath–calcimicrobe consortium) trailed behind the evolution of other skeletal organisms by 10–15 million years in the earliest Cambrian tropical shelves.
3. The first prominent, widespread archeocyath sponge–calcimicrobial reefs became extinct at the end of the Toyonian, via a downstep sequence of biodiversity losses.

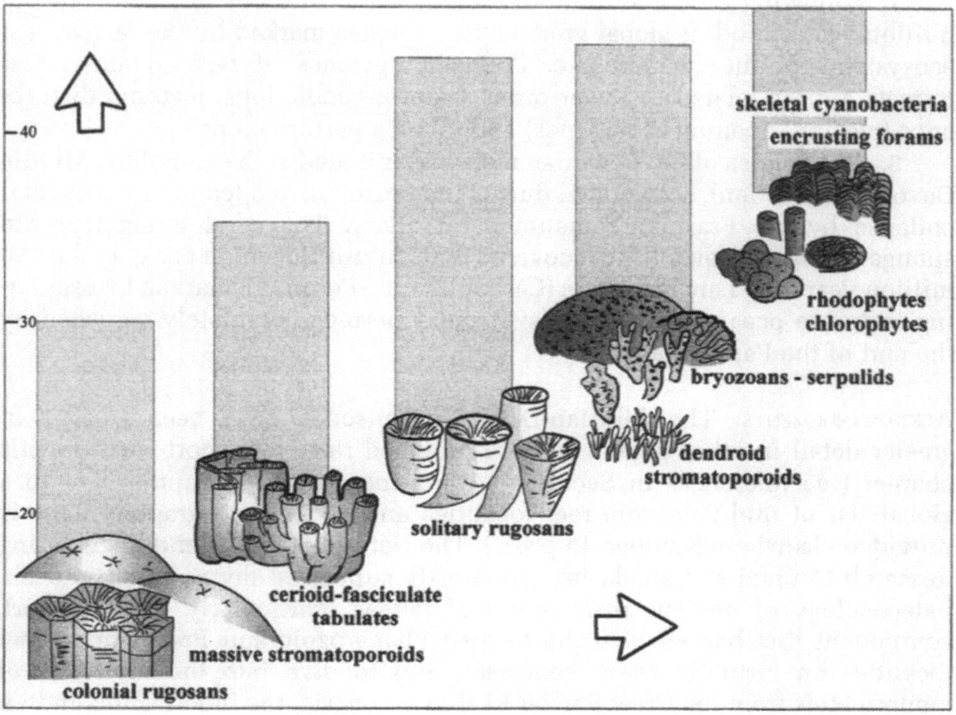

FIGURE 6. Response of reef organisms (via changing dominance of orders and phyla, and constructional modes, e.g., solitary versus colonial, and general morphology) to stress in the reef ecosystem, especially temperature- and latitude-related factors. Paleozoic reefs were made of a wide spectrum of organisms with a different response to stresses from global change: Those most sensitive are seen on the left side of the diagram, those least sensitive on the right. The order of biotic response represents a gradient that may be used as a measure of global environmental stress; thus, solitary colonial rugosans were able to survive many mass extinctions, but complex colonial forms with highly integrated skeletons were not. The solid bars show ranges to the higher latitudes: all of these can be extended to the equator. (Modified from Copper, 1994b.)

4. A long lag time, ca 50–70 million years from mid-Cambrian through Llanvirn (mid-Ordovician), was required before sponges fully reoccupied the reef niche in large numbers over wide areas, though isolated records of reefs are known.

5. The Caradoc (Late Ordovician) saw the rapid spread of complex skeletal metazoans (tabulate and rugose corals, stromatoporoids) and dasycladacean chlorophytes and solenoporid rhodophytes, marking the rise of the mid-Paleozoic reef ecosystem.

6. The Hirnantian, beginning and ending with at least two sharp sealevel drawdowns forced by Saharan glaciation, shifted the mudmound community back onto the shallow shelf, which lost much of its coral–stromatoporoid reef faunas, except locally. This interval introduced Silurian coral and tropical shelf-dwelling brachiopod elements in large numbers.

7. Following a 3–4 million year delay the Siluro-Devonian saw an 80-million-year period of global greenhouse climates marked by the largest reef ecosystems of the Phanerozoic. Frequent episodes of reef advances and retreats appear related to lower order sealevel oscillations, tectonic disturbance (closing–opening of sealanes) and climate perturbations.

8. The demise of the Devonian reefs was initiated at the end of the Middle Devonian (Givetian), accelerated during the Frasnian, and led to virtually total collapse by the Frasnian-Famennian boundary. The coral–stromatoporoid sponge reef fauna never fully recovered to its Siluro-Devonian peaks in the 120 million year long Late Paleozoic (Carboniferous–Permian), marked by oscillating icehouse oceans and climates; stromatoporoids completely vanished by the end of the Famennian.

ACKNOWLEDGMENTS: The Silurian-Devonian episodes have been treated in greater detail in other papers and are provided relatively short shrift in this chapter (see references in Sections 6–7); a more detailed approach, with a global list of mid-Paleozoic reef localities and 12 paleogeographic maps is provided elsewhere (Copper, in press). The Natural Sciences and Engineering Research Council of Canada has generously supported my research into the paleoecology of ancient reefs and reef faunas, particularly the fieldwork component that has enabled me to visit Phanerozoic and Precambrian reef localities on virtually every continent, and to dive into the mysteries of modern reefs from the Great Barrier Reef to Indonesia, the Indo-Pacific, Indian Ocean, and Red Sea, as well as the Caribbean. These have taken me not only to almost all major representative Phanerozoic reef sites throughout North America from the high arctic and Hudson Bay, to the Gulf of Mexico, South America, and from the Rockies and the Great Basin east to Newfoundland, but also to many classic Eurasian reefs sites from Gotland south to southern Spain, Devon (England) through Germany, Poland, Ukraine, Russia (Russian Platform, Urals, Siberia), Uzbekistan, Georgia, Armenia; in North Africa (Morocco, Algeria, Tunisia, Libya, Mauritania); east Asia, i.e., China (Shaanxi, Sichuan, Guizhou, Guangxi, Yunnan, Jiangsu, Hebei, Xinjiang), Japan, and Burma, Thailand, Indonesia, and Australia. I thank also Shell Oil Canada (who trained me for 5 months in the Cambrian–Carboniferous of the Rockies), J. C. Sproule Engineering (Calgary), for the privilege of my first taste of arctic geology (Melville, Bathurst, Devon, Cornwallis islands), and the Japan Society for the Promotion of Science (for a look at Paleozoic carbonates in a tectonically active belt). Seeing fossil reefs firsthand under the expert eye of regional geologists too numerous to mention and observing living reefs have given me barely a peek into the ancient reef spectrum.

References

Aitken, J. D, 1989, Giant "algal" reefs, middle-upper Proterozoic Little Dal group (770–120 Ma), Mackenzie Mountains, N.W.T., Canada, *Can. Soc. Petrol. Geol. Memoir* **13**:13–23.

Awramik, S. M. 1971, Precambrian columnar stromatolite diversity, *Science* **174**:825–827.

Bambach, R. K, 1993, Seafood through time; changes in biomass, energetics, and productivity in the marine ecosystem, *Paleobiology* **19**:372–397.

Bengtson, S., 1992, Proterozoic and earliest Cambrian skeletal metazoans, in: *The Proterozoic Biosphere* (J. W. Schopf and C. Klein, eds.), Cambridge University Press, Cambridge, England, pp. 397–411.

Bengtson, S., 1994, The advent of animal skeletons, in: *Early Life on Earth*, Nobel Symposium No. 84 (S. Bengtson, ed.), Columbia University Press, New York, pp. 412–425.

Bengtson, S. 1998, Animal embryos in deep time, *Nature* **391**:529–530.

Bengtson, S., and Conway Morris, S. 1992, Early radiation of biomineralizing phyla, in: *Origin and Early Evolution of the Metazoa,* Topics in Geobiology, Vol. 10 (J. H. Lipps and P. W. Signor, eds), Plenum Press, New York, pp. 448–482.

Bengtson, S., and Runnegar, B. N., 1992, Origins of biomineralization in metaphytes and metazoans, in: *The Proterozoic Biosphere* (J. W. Schopf and C. Klein, eds.), Cambridge University Press, Cambridge, England, pp. 447–452.

Bengtson, S., and Zhao, Y., 1997, Fossilized metazoan embryos from the earliest Cambrian, *Science* **277**:1645–1648.

Berner, R. A., 1994, Geocarb II, a revised model of atmospheric CO_2 over Phanerozoic time, *Am. J. Sci.* **294**:56–91.

Berner, R. A., 1998, The carbon cycle and CO_2 over Phanerozoic time. The role of land plants, *Royal Soc. Philosoph. Trans.* **B353**:75–82.

Berner, R. A., 1999, Atmospheric oxygen over Phanerozoic time, *Proc. Nat. Acad. Sci.* **96**:10955–10957.

Berry, W. B. N., and Boucot, A. J., 1973, Glacio-eustatic control of Late Ordovician—Early Silurian platform sedimentation and faunal changes, *Bull. Geol. Soc. Am.* **84**:275–284.

Beutinger, B. W., 1996, *Mikro-fazielle und paläökologische Untersuchung des ober-ordovizischen Boda-Kalkes, Siljan Gebiet* (Sweden), Diplomarbeit Universität Stuttgart, Stuttgart.

Blanchon, P., and Jones, B., 1997, Hurricane control on shelf-edge reef architecture around Grand Cayman, *Sedimentology* **44**:479–506.

Bosence, D. W. J., and Bridges, P. H., 1995, A review of the origin and evolution of carbonate mud-mounds, in: *Carbonate Mud-mounds: Their origin and evolution* (C. L. V. Monty, D.W. J. Bosence, P. H. Bridges, and B. R. Pratt, eds.), Special Publications International Association Sedimentologists, Vol. 23, Blackwell, Oxford, United Kingdom, pp. 3–9.

Brenchley, P. J., Carden, G. A. F., and Marshall, J. D., 1995, Environmental changes associated with the "first strike" of the Late Ordovician mass extinction, *Mod. Geol.* **20**:69–82.

Brunton, F. R., and Dixon, O. A., 1994, Siliceous sponge-microbe biotic associations and their recurrence through the Phanerozoic as reef mound constructors, *Palaios* **9**: 370–387.

Brunton, F. R., Smith, L., Dixon, O. A., Copper, P., Nestor, H., and Kershaw, S., 1998, Silurian reef episodes, changing seascapes and paleobiogeography, *NY State Museum Bull.* **491**:265–282.

Buddemeier, R. W., 1997, Making light work of adaptation, *Nature* **388**:229–230.

Buddemeier, R. W., and Fautin, D. G., 1996, Global CO_2 and evolution among the Scleractinia, *Bull. Instit. Oceanogr. Monaco* (Numéro Spéc.) **14**:33–38.

Butterfield, N. J., and Rainbird, R. H., 1998, Diverse organic-walled fossils including "possible dinoflagellates" from the Early Neoproterozoic of arctic Canada, *Geology* **26**:963–966.

Cañas, F., 1995, Early Ordovician carbonate platform facies of the Argentine Precordillera: Restricted shelf to open platform evolution, in: *Ordovician Odyssey* (J. D. Cooper, M. L. Droser, and S. C. Finney, eds.), SEPM, Riverside, CA, pp. 221–224.

Cañas, F., and Keller, M., 1993, Arrecifes y monticulos arrecifales en la Formación San Juan (Precordillera Sanjuanina, Argentina): Los arrecifes más antiguas de Sudámerica. *Bol. Real Soc. Española Hist. Nat.* **88**:127–136.

Capone, D. G., Zehr, J. P., Paerl, H. W., Bergman, B., and Carpenter, E. J., 1997, *Trichodesmium*, a globally significant marine cyanobacterium, *Science* **276**: 1221–1227.

Carey, D. A., 1987, Sedimentological effects and palaeoecological implications of the tube-building polychaete *Lanice conchilega* Pallas, *Sedimentology* **34**:49–66.

Conway Morris, S., 1993,The fossil record and the early evolution of the Metazoa, *Nature* **361**:219–225.
Copper, P., 1974, Structure and development of Early Paleozoic reefs, *Proc. Second Int. Symp. Coral Reefs* **1**:365–386.
Copper, P., 1981, *Spirapora*, a new Late Ordovician tabulate coral from Manitoulin Island, Ontario, Canada, *J. Paleontol.* **55**:1071–1075.
Copper, P., 1989, Upper Ordovician and Lower Silurian reefs from Anticosti Island, Quebec, *Can. Soc. Petrol. Geol. Memoir* **13**: 271–276.
Copper, P., 1994a, Ancient reef ecosystem expansion and collapse, *Coral Reefs* **13**:3–11.
Copper, P., 1994b, Reefs under stress: The fossil record, *Courier Forschungsinstitut Senckenberg* **172**:87–94.
Copper, P., 1997, Reefs and carbonate productivity: Cambrian through Devonian, *Proc. 8th Int. Coral Reef Symp., Panama* **2**:1623–1630.
Copper, P., 1998, Evaluating the Frasnian–Famennian mass extinction: Comparing brachiopod faunas, *Palaeontolog. Polonica* **43**:137–154.
Copper, P., 1999, Brachiopods during and after the Late Ordovician mass extinctions on Anticosti Island, E Canada, *Acta Universitatis Carolinae—Geologica* **43**:207–209.
Copper, P. (in press), *Silurian and Devonian Reefs:* 80 My of Global Greenhouse between Two Ice Ages, SEPM Special Paper, Riverside, CA.
Copper, P., and Brunton, F. R., 1991, A global review of Silurian reefs, *Spec. Papers Palaeontol.* **44**:225–259.
Copper, P., and Jin, J. S. (eds.), 1996, *Brachiopods*, Balkema Press, Rotterdam.
Crimes, T. P., 1992, The record of trace fossils across the Proterozoic-Cambrian boundary, in: *Origin and Early Evolution of the Metazoa* (J. H. Lipps, and P. W. Signor, eds.), Plenum Press, New York, pp. 177–202.
Debrenne, F., 1992, Diversification of Archeocyatha, in: *Origin and Early Evolution of the Metazoa* (J. H. Lipps and P. W. Signor, eds.), Plenum Press, New York, pp. 425–443.
Debrenne, F., and Courjault-Radé, P., 1994, Répartition paléogeographique des archéocyathes et délimitation des zones intertropicales au Cambrien inférieur, *Bull. Soc. Géol. France* **165**:459–467.
Debrenne, F., and Zhuravlev, A., 1997, Cambrian food web: A brief review, *Geobios Monogr. Spéc.* **20**:181–188.
De Freitas, T. A., and Dixon, O. A., 1995, Silurian microbial buildups of the Canadian Arctic, *Spec. Publ. Int. Sedimentol. Assoc.* **23**:151–169.
De Freitas, T. A., and Mayer, U., 1995, Kilometre-scale buildups in a rimmed carbonate platform succession, arctic Canada: New insight on Lower Ordovician reef facies, *Bull. Can. Petrol. Geol.* **43**:407–432.
De Freitas, T. A., Dixon, O. A., and Mayer, U., 1993, Silurian pinnacle reefs of the Canadian arctic, *Palaios* **8**:172–182.
De Freitas, T. A., Trettin, H. P., Dixan, O. A., and Mallamo, M., 1999, Silurian system of the Arctic Archipelago, *Bull. Can. Petrol. Geol.* **47**:136–193.
Doolittle, W. F., 1998, A paradigm gets shifty, *Nature* **392**:15–16.
Doolittle, W. F., Feng, D. F., Tsang, S., Cho, G., and Little, E., 1996, Determining divergence times of the major kingdoms of living organisms with a protein clock, *Science* **271**:470–477.
Droser, M. L., and Sheehan, P. M., 1997, Paleoecology of the Ordovician radiation: Resolution of large scale patterns with individual case histories, paleogeography and environments, *Geobios Mém. Spéc.* **20**:221–229.
Droser, M. L., Fortey, R. A., and Li, X., 1996, The Ordovician radiation, *Am. Sci.* **84**:122–131.
Eder, W., and Franke, W., 1982, Death of Devonian reefs, *Neues Jahrbuch Geol. Paläontol. Abhandlungen* **163**:241–243.
Ekdale, A. A., and Lewis, D. W., 1993, Sabellariid reefs in Ruby Bay, New Zealand, *Palaios* **8**:614–620.
Fagerstrom, J. A., 1987, *The Evolution of Reef Communities*, Wiley Interscience, New York.
Fedonkin, M. A., 1992, Vendian faunas and the early evolution of Metazoa, in: *Origin and Early Evolution of the Metazoa* (J. H. Lipps and P. W. Signor, eds.), Plenum Press, New York, pp. 87–129.

Flügel, E. (ed.), 1996, Global and regional controls on biogenic sedimentation, 1, Reef evolution, *Göttinger Arbeiten Geol. Paläontol.* **Sb2:**1–428.

Fuhrman, J. A., McCallum, K., and Davis, A. A., 1992, Novel marine archaebacterial group from marine plankton, *Nature* **356:**148–150.

Gao, J. G., and Copper, P., 1997, Growth rates of Middle Paleozoic corals and sponges: early Silurian of eastern Canada, *Proc. 8th Int. Coral Reef Symp.* **2:**1651–1656.

Garrett, P., 1970, Phanerozoic stromatolites: noncompetitive ecologic restriction by grazing and burrowing animals, *Science* **169:**171–173.

Gehling, J. G., and Rigby, J. K., 1996, Long expected sponges from the Neoproterozoic Ediacara fauna of South Australia, *J. Paleontol.* **70:**185–195.

Germs, G. J. B., 1972, New shelly fossils from the Nama group, Southwest Africa, *Am. J. Sci.* **272:**752–761.

Ginsburg, R. N., and Lowenstam, H. A., 1958, The influence of marine bottom communities on the depositional environment of sediments, *J. Geol.* **66:**310–318.

Gischler, E., 1992, Das devonische Atoll von Iberg und Winterberg im Harz nach Ende des Riffwachstums, *Geol. Jahrbuch* **A129:**5–193.

Grotzinger, J. P., 1990, Geochemical model for stromatolite decline, *Am. J. Sci.* **290A:**80–103.

Grotzinger, J. P., 1994, Trends in Precambrian carbonate sediments and their implication for the understanding of the evolution of life, in: *Early Life on Earth* (S. Bengtson, ed.), Columbia University Press, New York, pp. 245–258.

Grotzinger, J. P., Bowring, S. A., Saylor, B. Z., and Kauffman, A. J., 1995, Biostratigraphic and geochronologic constraints on early animal evolution, *Science* **270:**598–604.

Gruet, Y., and Brodeur, Y., 1995, Ecological conditions of modern sabellaran reefs development: Geological Implications. *Publ. Service Géol. Luxembourg* **29:**73–80.

Hamdi, B., Rozanov, A. Y., and Zhuravlev, A. Y., 1995, Latest Middle Cambrian metazoan reef from northern Iran, *Geol. Mag.* **132:**367–373.

Hardie, L. A., 1996, Secular variation in seawater chemistry: an explanation for coupled secular variation in the mineralogies of marine limestones and potash evaporites over the past 600 m.y., *Geology* **24:**279–283.

Higgins, A. K., Ineson, J. R., Peel, S. J., Surlyk, F., and Sønderholm, M., 1991, Lower Palaeozoic Franklinian Basin of north Greenland, *Bull. Grønlands Geol. Undersøgelse* **160:**71–140.

Hoffman, P. F., Kauffman, A. J., and Halverson, G. P., 1998a, Comings and goings of global glaciations on a Neoproterozoic tropical platform in Namibia, *GSA Today* **8:** 1–9.

Hoffman, P. F., Kauffman, A. J., and Halverson, G. P., 1998b, A snowball earth, *Science* **281:**1342–1346.

Hubbard D. K., Miller, A. I., and Scaturo, D., 1990, Production and cycling of calcium carbonate in a shelf-edge reef system (St. Croix, U.S. Virgin Islands): Applications to the nature of reef systems in the fossil record, *J. Sediment. Petrol.* **60:**335–360.

Ineson, J. R., and Peel, J. S., 1997, Cambrian shelf stratigraphy of north Greenland, *Geol. Greenland Survey Bull.* **173:**1–120.

Ineson, J. R., Surlyk, F., Higgins, A. K., and Peel, J. S., 1994, Slope, apron and deep shelf sediments of the Brønlund Fjord and Tavskens Iskappe groups (Lower Cambrian-Lower Ordovician), *Bull. Grønlands Geol. Undersøgelse* **169:**7–24.

Isaacson, P. E., 1997, Late Devonian (Famennian) glaciation in Gondwana and forced marine regression in North America, in: *Abstracts Geological Society America Annual Meeting*, Salt Lake City, USA, p. 41.

James, N. P., 1997, The cool-water carbonate depositional realm, *SEPM Spec. Publ.* **56:**1–20.

James, N. P., and Bourque, P. A., 1992, Reefs and mounds, in: *Facies Models* (R. G. Walker and N. P. James, eds.), Geological Association of Canada, St. Johns, pp. 323–347.

James, N. P., and Geldsetzer, H. H. J., 1989, Reefs—Canada and adjacent areas, introduction, *Can. Soc. Petrol. Geol. Memoir* **13:**13–23.

James, N. P., and Gravestock, D. I., 1990, Lower Cambrian shelf and shelf margin buildups, Flinders Ranges, South Australia, *Sedimentology* **37:**455–480.

James, N. P., and Kobluk, D. R., 1978, Lower Cambrian patch reefs and associated sediments: Southern Labrador, Canada, *Sedimentology* **25:**1–35.

Jenkyns, R. J. F., 1992, Functional and ecological aspects of Ediacara assemblages, in: *Origin and Early Evolution of the Metazoa* (J. H. Lipps and P. W. Signor, eds.), Plenum Press, New York, pp. 131–176.

Kauffman, A. J., and Knoll, A. H., 1995, Neoproterozoic variations in the C-isotopic composition of seawater: Stratigraphic and geochemical implications, *Precambrian Res.* **73**:27–49.

Keller, M., and Bordonaro, O., 1993, Arrecifes de estromatopóridos en el Ordovicico inferior del oeste Argentino y sus implicaciones paleogeográficas, *Rev. Española Paleontol.* **8**:165–169.

Keller, M., Bordonaro, O., and Cañas, F., 1995, Some plate tectonic aspects of Early Ordovician reefs, western Argentine Precordillera, *XII Congr. Geol. Argentino y II Congr. Exp. Hidrocarb.* **1**:235–240.

Keller, M., Cañas, F., Lehnert, O., and Vaccari, N. E., 1994, The Upper Cambrian and Lower Ordovician of the Precordillera (western Argentina): Some stratigraphic reconsiderations, *Newslett. Stratigr.* **31**:115–132.

Kempe, S., and Kazmierczak, J., 1994, The role of alkalinity in the evolution of ocean chemistry, organization of living systems, and biocalcification processes, *Bull. Inst. Océanogr. Monaco* **133**:61–116.

Kirschvink, J. L., Ripperdan, R. L., and Evans, D. A., 1997, Evidence for a large scale reorganization of Early Cambrian continental masses by inertial interchange true polar wanders, *Science* **277**:541–545.

Lehnert, O., and Keller, M., 1993, Posicion estratigrafiya de los arrecifes Arenigianos en la Precordillera Argentina, *Documents Lab. Géol. Lyon* **125**:263–275.

Li, C. W., Chen, J. Y., and Hua, T. E., 1998, Precambrian sponges with cellular structures, *Science* **279**:879–882.

Lipps, J. H., Bengtson, S, and Farmer, J. D., 1992a, The Precambrian–Cambrian evolutionary transition, in: *The Proterozoic Biosphere* (J. W. Schopf and C. Klein eds.), Cambridge University Press, Cambridge, England, pp. 453–457.

Lipps, J. H., Bengtson, S., and Fedonkin, M. A., 1992b, Ecology and biogeography, in: *The Proterozoic Biosphere* (J. W. Schopf and C. Klein, eds.), Cambridge University Press, Cambridge, England, pp. 437–441.

Logan, G. A., Hayes, J. M., Hieshima, G. B., and Summons, R. E., 1995, Terminal Proterozoic reorganization of biogeochemical cycles, *Nature* **376**: 53–56.

Lohmann, K. C., 1976, Lower Dresbachian (Upper Cambrian) platform to deep shelf transitions in eastern Nevada and western Utah, *Brigham Young Univ. Stud.* **23**:111–132.

Margulis, L., 1981, *Symbiosis in Cell Evolution*, W. H. Freeman, San Francisco.

Martin, W., and Müller, M., 1998, The hydrogen hypothesis for the first eukaryote, *Nature* **392**:37–41.

McIlreath, I., 1977, Association of a Middle Cambrian deep water limestone debris apron adjacent to a vertical submarine carbonate escarpment, *Soc. Econ. Paleontol. Mineral. Spec. Publ.* **25**:113–124.

Messing, C. G., Neumann, A. C., and Lang, J. C., 1990, Biozonation of deep-water lithoherms and associated hardgrounds in the northeastern straits of Florida, *Palaios* **5**:15–33.

Moldowan, J. M., Dahl, J., Jacobson, S. R., Huizinga, B. J., Fago, F. J., Shetty, R., Watt, D. S., and Peters, K. E., 1996, Chemostratigraphic reconstruction of biofacies, *Geology* **24**:159–162.

Monty, C. L. V., 1995, The rise and nature of carbonate mudmounds: An introductory actualistic approach, *Spec. Publ. Int. Assoc. Sedimentol.* **23**:11–48.

Monty, C. L. V., Bosence, D. W. J., Bridges, P. H., and Pratt, B. R. (eds.), 1995, Carbonate mud-mounds: Their origin and evolution, *Spec. Publ. Int. Assoc. Sedimentol.* **23**: 538.

Narbonne, G. M., 1998,The Ediacara biota: A terminal Neoproterozoic experiment in the evolution of life, *GSA Today* **8**:11–16.

Nestor, H., 1995, Ordovician and Silurian reefs in the Baltic area, *Publ. Service Géol. Luxembourg* **29**:39–47.

Neumann, A. C., Kofoed, J. W., and Keller, G., 1977, Lithoherms in the Straits of Florida, *Geology* **5**:4–10.

Neuweiler, F., Gautret, P., Thiel, V., Lange, R., Michaelis, W., and Reitner, J., 1999, Petrology of Lower Cretaceous carbonate mudmounds (Albian, N. Spain): Insights into organomineralic deposits of the geological record, *Sedimentology* **46**:837–859.

Nikitin, I. F., Popov, L. E., and Holmer, L. E., 1996, Late Ordovician brachiopod assemblage of Hiberno-Salairan type from central Kazakhstan, *Geol. Föreningens Förhandlingar* **118**:83–96.

Pandolfi, J. M., and Kirtley, D. W., 1998, Roles for worms in reef building, *Coral Reefs* **17**:120.

Pratt, B. R., 1995, The origin, biota and evolution of deep-water mud-mounds, *Spec. Publ. Int. Assoc. Sedimentol.* **23**:49–123.

Pratt, B. R., and James, N. P., 1989, Coral-*Renalcis* reef complex of Early Ordovician age. St George Group, western Newfoundland, *Can. Soc. Petrol. Geol. Mem.* **13**:224–230.

Reid, R. P., and Macintyre, I. G., 1992, Stromatolites: Newly discovered members of the Holocene reef community, *Proc. 7th Int. Coral Reef Symp., Guam,* **2**:1227–1228.

Reitner, J., and Mehl, D., 1995, Early Paleozoic diversification of sponges: New data and evidence, *Geol. Paläontol Mitteilungen Innsbruck* **20**:335–347.

Reitner, J., and Neuweiler, F., 1995, Mud mounds: Recognizing a polygenetic spectrum of fine-grained carbonate buildups, *Facies* **32**:2–4.

Riding, R., 1992, Temporal variation in calcification in marine cyanobacteria, *J. Geol. Soc. London* **149**:979–989.

Riding, R., 1993, Phanerozoic patterns of marine $CaCO_3$ precipitation, *Naturwissenschaften* **80**:513–516.

Riding, R., 1996, Long term changes in marine $CaCO_3$ precipitation, *Mem. Soc. Géol. France* **169**:157–166.

Rigby, J. K., 1971, Sponges and reef and related facies through time, *Proc. North Am. Paleontol. Convention* **J**:1374–1388.

Roberts, H. H., and Sydow, J., 1997, Siliciclastic-carbonate interactions in a tropical deltaic setting: Mahakam Delta of east Kalimantan (Indonesia), *Proc. 8th Int. Coral Reef Symp.* **2**:1773–1777.

Roberts, H. H., Phipps, C. V., and Effendi, L., 1987, *Halimeda* bioherms of the eastern Java Sea, Indonesia, *Geology* **15**:371–374.

Ruppel, S. C., James, E. W., Barrick, J. E., Nowlan, G., and Uyeno, T. T., 1998, High resolution Silurian δ87Sr/86Sr record: Evidence of eustatic control of seawater chemistry? *NY State Museum Bull.* **491**:285–295.

Savarese, M., Mount, J. F., Sorauf, J. E., and Bucklin, L., 1993, Paleobiologic and paleoenvironmental context of coral-bearing Lower Cambrian reefs, *Geology* **21**:917–920.

Schopf, W. J., 1992a, Paleobiology of the Archean, in: *The Proterozoic Biosphere* (J. W. Schopf and C. Klein, eds.), Cambridge University Press, Cambridge, England, pp. 25–40.

Schopf, W. J., 1992b, Patterns of Proterozoic microfossil diversity: An initial tentative analysis, in: *The Proterozoic Biosphere* (J. W. Schopf and C. Klein, eds.), Cambridge University Press, Cambridge, England, pp. 529–552.

Schopf, W. J., 1992c, Evolution of the Proterozoic biosphere: benchmarks, tempo, mode, in: *The Proterozoic Biosphere* (J. W. Schopf and C. Klein, eds.), Cambridge University Press, Cambridge, England, pp. 581–600.

Scrutton, C. T., 1997, The Palaeozoic corals, I: Origins and relationships, *Proc. Yorkshire Geol. Soc.* **51**:177–208.

Scrutton, C. T., 1998, The Palaeozoic corals, II: Structure, variation and palaeoecology, *Proc. Yorkshire Geol. Soc.* **52**:1–57.

Seilacher, A., 1989, Vendozoa: organismic construction in the Proterozoic biosphere, *Lethaia* **22**:229–239.

Seilacher, A., 1992, Vendobionta and Psammocorallia: lost constructions of Precambrian evolution, *J. Geol. Soc. London* **149**:607–613.

Sepkoski, J. J., and Sheehan, P. M., 1983, Diversification, faunal change and community replacement during the Ordovician radiations, in: *Recent and Fossil Benthic Communities* (M. J. C. Tevesz and P. L. McCall, eds.), Plenum Press, New York, pp. 673–718.

Soja, C., and Antoshkina, A. I., 1997, Coeval development of Silurian stromatolite reefs in Alaska and the Ural Mountains: implications for geography of the Alexander terrane, *Geology* **25**:539–542.

Sorauf, J. E., and Savarese, M., 1995, A Lower Cambrian coral from South Australia, *Palaeontology* **38**:757–770.

Stearn, C. W., Webby, B. D., Nestor, H., and Stock, C. W., 1999, Revised classification and terminology of Palaeozoic stromatoporoids, *Acta Palaeontol. Polonica* **44:**1–70.

Steele-Petrovich, M., and Bolton, T. E., 1998, Morphology and palaeoecology of a primitive mound-forming tubiculous polychaete from the Ordovician of the Ottawa Valley, Canada, *Palaeontology* **41:**125–145.

Thorslund, P., 1935, Über den Brachiopodenschiefer und den jüngeren Riffkalk in Dalarne, *Nova Acta Regiae Soc. Sci. Upsaliensis* **9:**1–50.

Trettin, H. P., 1991, Geology of the Innuitian Orogen and arctic platform of Canada and Greenland, *Geol. North Am.* **E:**1`63–238.

Turner, E. C., Narbonne, G. M., and James, N. P., 1993, Neoproterozoic reef microstructures from the Little Dal Group, northwest Canada, *Geology* **21:**259–262.

Vidal, G., and Moczydlowska-Vidal, M., 1997, Biodiversity, speciation, and extinction trends of Proterozoic and Cambrian phytoplankton, *Paleobiology* **23:**230–246.

Walter, M. R., 1983, Archean stromatolites: evidence of the earth's earliest benthos, in: *Earth's Earliest Biosphere* (J. W. Schopf, ed.), Princeton University Press, Princeton, NJ, pp. 187–213.

Walter, M. J., 1994, Stromatolites: The main geological source of information on the evolution of early benthos, in: *Early Life on Earth*, Nobel Symposium No. 84 (S. Bengtson, ed.), Columbia University Press, New York, pp. 270–286.

Walter, M. J., Grotzinger, J. P., and Schopf, J. W., 1992, Proterozoic stromatolites, in: *The Proterozoic Biosphere* (J. W. Schopf and C. Klein, eds.), Cambridge University Press, Cambridge, England, pp. 253–260

Webb, G. E., 1996, Was Phanerozoic reef history controlled by the distribution of nonenzymatically secreted reef carbonates (microbial carbonate and biologically induced cement)? *Sedimentology* **43:**947–971.

Webby, B. D., 1984, Ordovician reefs and climate: a review, in: *Aspects of the Ordovician System* (D. L. Bruton, ed.), Universitetsforlaget, Oslo, pp. 89–100.

Webby, B. D., 1992, Global biogeography of Ordovician corals and stromatoporoids, in: *Global Perspectives on Ordovician Geology* (B. Webby and J. R. Laurie, eds.), Balkema, Rotterdam, pp. 261–276.

Webby, B. D., 1994, Evolutionary trends in Ordovician stromatoporoids, *Courier Forschungsinstitut Senckenberg* **172:**373–380.

Webby, B. D., Zhen Y. Y., and Percival, I. G., 1997, Ordovician coral- and sponge-bearing associations: Distribution and significance in volcanic island shelf to slope habitats, eastern Australia, *Bol. Real Soc. Española Hist. Nat.* **92:**163–175.

Wilder. H., 1989, Neue Ergebnisse zum oberdevonischen Riffsterben am Nordrand des mitteleuropäischen Variscikums, *Fortschritte Geol. Nordrhein-Westfalens* **35:**57–74.

Wood, R. A., 1999, *Reef Evolution*, Oxford University Press, Oxford.

Wood, R. A., Zhuravlev, A. Yu., and Anaaz, C. T., 1993, The ecology of Lower Cambrian buildups from Zuune Arts, Mongolia: implications for early metazoan reef evolution, *Sedimentology* **40:**829–858.

Wray, G. A., Levinton, J. S., and Shapiro, L. H., 1997, Molecular evidence for Precambrian divergences among metazoan phyla, *Science* **274:**568–573.

Xiao, S., Zhang, Y. and Knoll, A. H., 1998, Three-dimensional preservation of algae and animal embryos in a Neoproterozoic phosphorite, *Nature* **391:**553–558.

Yu, C., and Shen, J., 1998, *Devonian Reefs and Reef Complexes in Guilin, Guangxi, China*, Jiangsu Science Technology Publishing House, Beijing.

Zhang, Y., 1989, Multicellular thallophytes with differentiated tissues from late Proterozoic phosphatic rocks of South China, *Lethaia* **22:**113–132.

Zhu, Z. D., Guo, C. X., Liu, B. L., Hu, M. Y., Hu, A. M., Xiao, C. T., Meng, X. F., and Li, X. M., 1993, Lower Ordovician reefs at Huanghuachang, Yichang, east of the Yangtze Gorge, *Sci. Geol. Sinica* **2:**79–90.

Zhu, Z. D., Jiang, Y. W., and Liu, B. L., 1995, Paleoecology of late Tremadocian reef-bearing strata in western Hubei province of China, in: *Ordovician Odyssey: Short Papers for the Seventh International Symposium on the Ordovician System* (J. D. Cooper, M. L. Droser, and S. C. Finney, eds.), SEPM, Riverside, CA, pp. 427–428.

Zhuravlev, A. Yu., 1986, Evolution of archeocyaths and palaeobiogeography of the Early Cambrian, *Geol. Mag.* **123:**377–385.

Zhuravlev, A. Yu., 1996, Reef ecosystem recovery after the Early Cambrian extinction, in: *Biotic Recovery from Mass Extinction Events* (M. B. Hart, ed.), Geological Society London Special Publications 2, pp. 79–96.

Zhuravlev, A. Yu., and Wood, R. A., 1995, Lower Cambrian reefal cryptic communities, *Palaeontology* **38:**443–470.

Zhuravlev A. Yu., and Wood, R. A., 1996, Anoxia as the cause of the mid-Early Cambrian (Botomian) extinction event, *Geology* **24:**311–314.

Zhuravlev, A. Yu., Debrenne, F., and Lafuste, J., 1993, Early Cambrian microstructural diversification of Cnidaria, *Courier Forschungsinstitut Senckenberg* **164:**365–372.

Chapter 4

Paleoecology of Cambrian Reef Ecosystems

ANDREY YU. ZHURAVLEV

1. Introduction

Modern tropical reefs, together with tropical rain forests, comprise the principal centers of biodiversity. The tropical rain forest, however, is a relatively fresh phenomenon in a geological time sense. Therefore, to estimate ancient biodiversity, we first of all look to fossil reefs. Mutual intergrowths as well as processes of marine lithification and minor diagenetic alternation allow us to study many aspects of interaction between organisms in long-ago vanished ecosystems, which are poorly preserved in other settings. All these turn fossil reefs into highly attractive areas for scientific exploration. Cambrian reefs, which formed between 545 Ma and 490 Ma, are especially interesting among fossil reefs because they retain their history. They recall the blooming and are a testament to the failure of the earliest metazoan reef ecosystem.

ANDREY YU. ZHURAVLEV • Paleontological Institute, Academy of Russian Sciences, Moscow 117337, Russia.

The History and Sedimentology of Ancient Reef Systems, edited by George D. Stanley Jr., Kluwer Academic/Plenum Publishers, New York, 2001.

Cambrian reefs appeared as a subject of importance in the scientific literature for more than a century. At first, archaeocyathan buildups were compared with the Great Barrier Reef (Hyatt, 1885). Much later, when joint sedimentological and paleontological work on these Cambrian reefs began, their rather small size and muddy nature were noted (Zhuravleva and Zelenov, 1955; Debrenne, 1959). Some workers even denied the presence of true reefs in Cambrian strata, calling them instead "taphoherms" due to the prevalence of toppled archaeocyath cups (Jazmir, 1960). Detailed lithological and bio-sedimentological investigation has revealed a number of features, namely *in situ* organism–organism intergrowth, abundant marine cements, stromatactis structures, micro- and macroborings, and primary cavities containing diverse cryptobionts, all of which are typical of recent reefs (Copper, 1974; James and Kobluk, 1978; Kobluk and James, 1979; Kobluk, 1981).

Rowland and Gangloff (1988), in their review of Early Cambrian reefs, proposed that there were framework reefs in the Cambrian comparable both ecologically and constructionally to modern-type reefs with a photosymbiont trophic web. Afterward, several detailed studies including precise systematic study of reef-inhabiting species, mostly on Early Cambrian reefs, suggested that they were very different ecologically from the Recent reefs even in sedimentologically and climatically comparable environments (e.g., Rees *et al.*, 1989; James and Gravestock, 1990; Wood *et al.*, 1993; Kruse *et al.*, 1995; Pratt, 1995). This difference is seen first in the composition of reef builders which consisted primarily of archaeocyathan calcified sponges and calcimicrobes (such as renalcids) (Table 1). The interpretation of ecology and affinities of these organisms strongly influenced understanding of the nature of Cambrian reefs. Ecological parallels between archaeocyaths and renalcids and modern scleractinian corals and calcified algae, respectively, provoke comparisons of Cambrian reef ecosystems with living tropical coral reefs in terms of nutrient availability, energy flow, and trophic nucleus. Another approach using recent analogies would assign Cambrian reefs to either deep-water or cold-water sponge mudmound facies.

In general, most Cambrian reefs are loaf- or pillow-shaped mounds termed "kalyptrae" (1–2 m wide and 1 m high). They occur singly or collectively, in some cases, stacked on top of one another, forming complexes up to 20 m high and 200 m wide. Rarely, individual bioherms achieved a thickness of 200 m and length over 600 m. The total relief, however, scarcely exceeded one meter (Fig. 2B, D). Such bioherms, nonetheless, may be termed "ecological" reefs (*sensu* Lowenstam, 1950; Heckel, 1974) because they (1) possessed a three-dimensional framework with topographic relief above the surrounding sea-floor, (2) were built by *in vivo* calcified modular organisms mutually *in situ* encrusting each other, (3) were wave-resistant lithified structures exhibiting abundant synsedimentary carbonate cements strengthening the reef builders and providing a further rigid substrate for them, and (4) at least some of them grew from deeper, quiet-water settings upward into the shallow zone of continual wave agitation (Figs. 1A–E and 2A–D).

TABLE 1. Vendian, Cambrian, Early and Middle Ordovician Reef-Building Organisms and Associations[a]

Age	Setting stage	Deep marine (aphotic zone)	Shallow marine (photic zone, 30–40 ppm): Cold	Shallow marine (photic zone, 30–40 ppm): Warm (tropical–subtropical)	Restricted marine (brackish or hypersaline)
Middle Ordovician	Llandeilo, Llanvirn	Stromatactis-bearing clotted mud	Bryozoans	Renalcids, spicular demosponges, stromatoporoids, tabulates, bryozoans, red and green calcified algae, soanitids, microbial stromatolites	Microbial stromatolites
Early Ordovician	Arenig, Tremadoc	Stromatactis-bearing clotted mud, thrombolites, renalcids, spicular demosponges	Clotted mud (spicular sponges?)	Renalcids, spicular demosponges, soanitids, *Pulchrilamina*, tabulates bryozoans, thrombolites, microbial stromatolites	Microbial stromatolites
Late Cambrian	Datsonian, Sunwaptan, Steptoean	Stromatactis-bearing clotted mud, thrombolites, renalcids		Microbial stromatolites, thrombolites, renalcids, spicular demosponges, eocrinoids	Microbial stromatolites, thrombolites
Middle Cambrian	Marjumian, Amgan	Stromatactis-bearing clotted mud, thrombolites, renalcids		Renalcids, spicular demosponges, microbial stromatolites, thrombolites, calcified algae	Microbial stromatolites, renalcids
Early Cambrian	Toyonian, Botoman, Atdabanian, Tommotian	Stromatactis-bearing clotted mud, thrombolites, renalcids, spicular sponges, solitary ajacicyathids	Microbial stromatolites	Renalcids, archaeocyaths, radiocyaths, coralomorphs, cribricyaths, pharetronid sponges, thrombolites, microbial stromatolites	Microbial stromatolites, renalcids
Early Cambrian/Vendian	Nemakit–Daldynian	Thrombolites	Microbial stromatolites	Renalcids, thrombolites, microbial stromatolites	Microbial stromatolites
Vendian	Kotlinian, Redkinoan, Laplandian	Microbial stromatolites	Microbial stromatolites	Microbial stromatolites, thrombolites	Microbial stromatolites

[a]Data sources: Ordovician reefs: Nikitin *et al.*, 1974; Dronov and Fedorov, 1994; Pratt, 1995; Webby, 1999; Cambrian reefs: this chapter; Vendian reefs: Semikhatov *et al.*, 1970; Aitken and Narbonne, 1989; Geldsetzer *et al.*, 1989.

FIGURE 1. Field view of Cambrian reefs: (A) stromatolite–demosponge (anthaspidellid) mounds on echinoderm packstone, Mila Formation, middle Marjumian, Middle Cambrian, eastern Elburs, Iran; (B) stromatactis-bearing mudmound, Sierra Gorda Member, Alconera Formation, early Botoman, Early Cambrian, Ossa-Morena Zone, Spain (Elena Moreno-Eiris in the middle); (C) paleorelief of *Tubomorphophyton* dendrolite patch reefs (kalyptra), Pestrotsvet Formation, early Atdabanian, Early Cambrian, middle Lena River, Siberian Platform, Russia; (D) microbial thrombolites, Ust'mil' Formation, middle Marjumian, Middle Cambrian, Aldan River, Siberian Platform (thrombolite head is 20 cm across); (E) stromatactis-bearing mudmound, Pestrotsvet Formation, late Atdabanian, Early Cambrian, Yudoma River, Siberian Platform, Russia.

The Cambrian system still lacks an official stage subdivision. Most commonly used is the Lower and lower Middle Cambrian stage scale (Nemakit–Daldynian through Amgan) which is based on the Siberian Platform sections. The upper Middle and Upper Cambrian stages (Marjumian to Sunwaptan) based on North American strata are included in this scale. Finally, the Australian Datsonian stage is used to fill the gap created after the official acceptance of the Cambrian/Ordovician boundary defined at the base of *Cordylodus lindstromi* conodont zone in China. This level lies slightly above the traditionally recognized base of the Tremadoc series which begins the Ordovician system. Thus, this chapter follows stage subdivision of the Cambrian system in ascending order: Nemakit–Daldynian (beginning about 550 Ma); Tommotian (540 Ma); Atdabanian, Botoman, and Toyonian (520 Ma) (Lower Cambrian); Amgan and Marjumian (500 Ma) (Middle Cambrian); and Steptoean, Sunwaptan, and Datsonian (490 Ma) (Upper Cambrian) (Table 1 and Fig. 11). [Radiometric age data in brackets are after Grotzinger *et al.* (1995) and Tucker and McKerrow (1995) and corrected biostratigraphically by Zhuravlev (1995)].

2. Builders, Destroyers, and Dwellers

Two principal groups of Cambrian reef builders were filter/suspension-feeding metazoans and microbial autotrophs. Calcified aspiculate sponges—archaeocyaths—were probably close relatives of demosponges (Debrenne and Zhuravlev, 1994). Two orders—ajacicyathids and archaeocyathids—composed the core of this group and played quite different roles in the Cambrian reef ecosystem (Debrenne and Zhuravlev, 1992; Wood *et al.*, 1992, 1993). Both orders had primary and secondary calcareous skeletons. The primary calcareous skeleton, developed *in vivo*, was part of a complex aquiferous system, and thus has taxonomic value. A secondary calcareous skeleton was formed facultatively either as a response of organism to external disturbances or to the firm attachment of the organism to the substrate. Archaeocyathids (and the small order, Kazachstanicyathida) included a number of modular forms. These possessed a mobile aquiferous system, which was not fixed in the skeleton and which produced an abundant secondary calcareous skeleton. The skeleton served as a firm base and also may have suppressed competitors and encrusters. As reef builders they were able to form a framework, which especially in archaeocyathids was characterized by branching and domal modular growth (Fig. 3b–d). On the other hand, ajacicyathids were mostly solitary forms indicating antagonistic rejection of conspecific individuals, with an aquiferous system intimately connected to the complicated primary skeleton. With rare exceptions, most ajacicyathids were not framebuilders, but nevertheless were significant components in Early Cambrian reefs. Ajacicyathids were capable of inhabiting soft muddy substrates and formed thickets that prepared the substrate for further colonization and framework. Often frameworks initiated

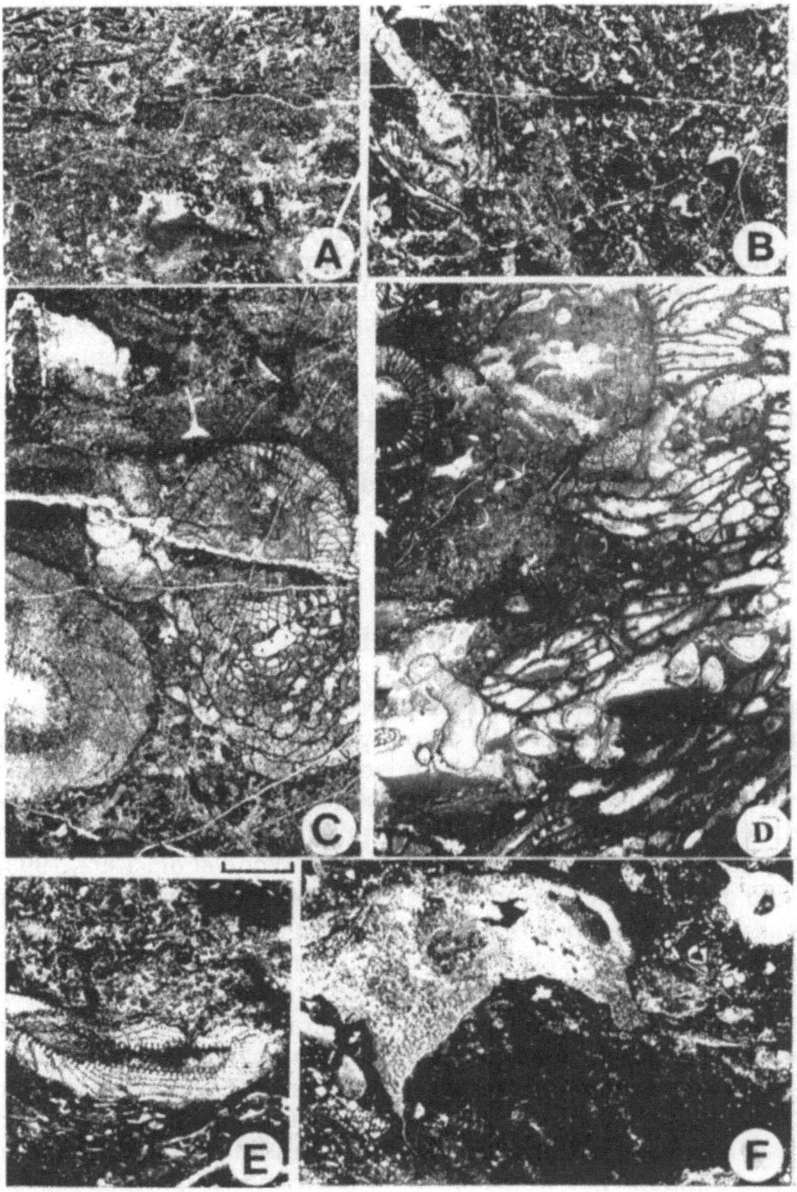

FIGURE 2. Thin sections of Cambrian reefs at different stages of succession: (A) climax stage (*Razumovskia* crust) on *Epiphyton–Tarthinia* skeletal thrombolite overlain by *Razumovskia* grainstone, Nokhoroy Member, early Atdabanian, Early Cambrian, middle Lena River, Siberian Platform, Russia; (B) climax stage (*Razumovskia* crust) on *Gordonophyton* dendrolite embraced by skeletal packstone, Burgasutay Formation, Atdabanian, Early Cambrian, Lake Province, Mongolia; (C) diversity stage of archaeocyathan–*Renalcis–Tarthinia* framework reef containing pharetronid sponge *Bottonaecyathus condensus* (Vologdin) (left), solitary coscinocyathid *Clathricoscinus vassilievi* (Vologdin) (middle) encrusted by massive modular kazachstanicyathid

on skeletal packstones and grainstones consisted of ajacicyathids. Second, under calm-water conditions such a settlement was a precondition for the appearance of a rigid reef formed by the binding of ajacicyathid skeletons with renalcids (Fig. 2E). Third, ajacicyathids possibly formed a baffle around microbial communities, preventing them from being buried by suspended sediment. Only a minority of ajacicyathids (about 1%), which were modular, formed reefs (Fig. 2D).

Archaeocyaths often constitute the majority of metazoan in reef communities of Early Cambrian. Interpretation of their nature is important in understanding the Early Cambrian reef ecosystem. Attempts to look at reef history through the prism of the modern reef ecosystem, which relies on photosymbiont-bearing animals, suggest archaeocyaths also may have been photosymbiontic (Rowland and Gangloff, 1988; Talent, 1988). Superficially, Cowen's (1988) criteria fit archaeocyaths: They display a photoautotropic morphology including platelike forms, possess heavily calcified skeletons, and are restricted to low latitudes. Further, they are suspected to have prefered oligotrophic waters (Rowland and Savarese, 1988).

However, the overwhelming majority of archaeocyaths possessed either narrow conical to subcylindrical or branching morphology with a reverse surface to volume ratio and with soft tissue hidden in the depths of the skeleton. Besides, many archaeocyaths (e.g., the entire order Capsulocyathida) were obligate cryptobionts, while others did not display a phototropic behavior being members of both cryptic and noncryptic communities (Wood *et al.*, 1992; Zhuravlev and Wood, 1995). Archaeocyaths (especially modular branching ones) often inhabited areas with increased terrigenous supply and were able to initiate reefs in siliciclastic-dominated settings (Zadorozhnaya *et al.*, 1973; Gangloff, 1976; Morgan, 1976; Read, 1980; Courjault-Radé, 1988; Savarese *et al.*, 1993; Bechstdät *et al.*, 1994) (Table 2, Fig. 3C–D). The first archaeocyathan reefs were restricted to the strata characterized by increased phosphate content (Rozanov, 1979). Most sponges contained diverse symbionts including photosymbiontic bacteria and algae, but a few sponges entirely depended on their metabolites. Such sponges preferred well-illuminated, calm environments with little suspended sediment (Wilkinson and

---

Altaicyathus notabilis Vologdin, and remains of branching coralomorph *Yakovlevites granulosus* (Vologdin) and chancelloriid *Allonnia* sp., Burgasutay Formation, Botoman, Early Cambrian, Lake Province, Mongolia; (D) diversity stage of archaeocyathan framework reef built by massive modular ajacicyathid *Sajanocyathus ussovi* Vologdin and kazachstanicyathid *Korovinella sajanica* (Yaworsky) with cryptic bivalved animals, toppled solitary coscinocyathid *Clathricoscinus spatiosus* (Vologdin) and ajacicyathid *Carinacyathus bagenovi* Vologdin are in surrounding mudstone, Verkhnemonok Formation, Botoman, Early Cambrian, Western Sayan, Russia; (E) stabilization stage, initiation of *Tubomorphophyton* dendrolite on toppled solitary ajacicyathid *Squamosocyathus taumatus* Zhuravleva, Perekhod Formation, late Atdabanian, Early Cambrian, middle Lena River, Siberian Platform, Russia; (F) stabilization stage, initiation of stromatolite–demosponge (anthaspidellid)–eocrinoid mound on echinoderm–trilobite grainstone, Mila Formation, middle Marjumian, Middle Cambrian, eastern Elburs, Iran. Scale bar 3 mm.

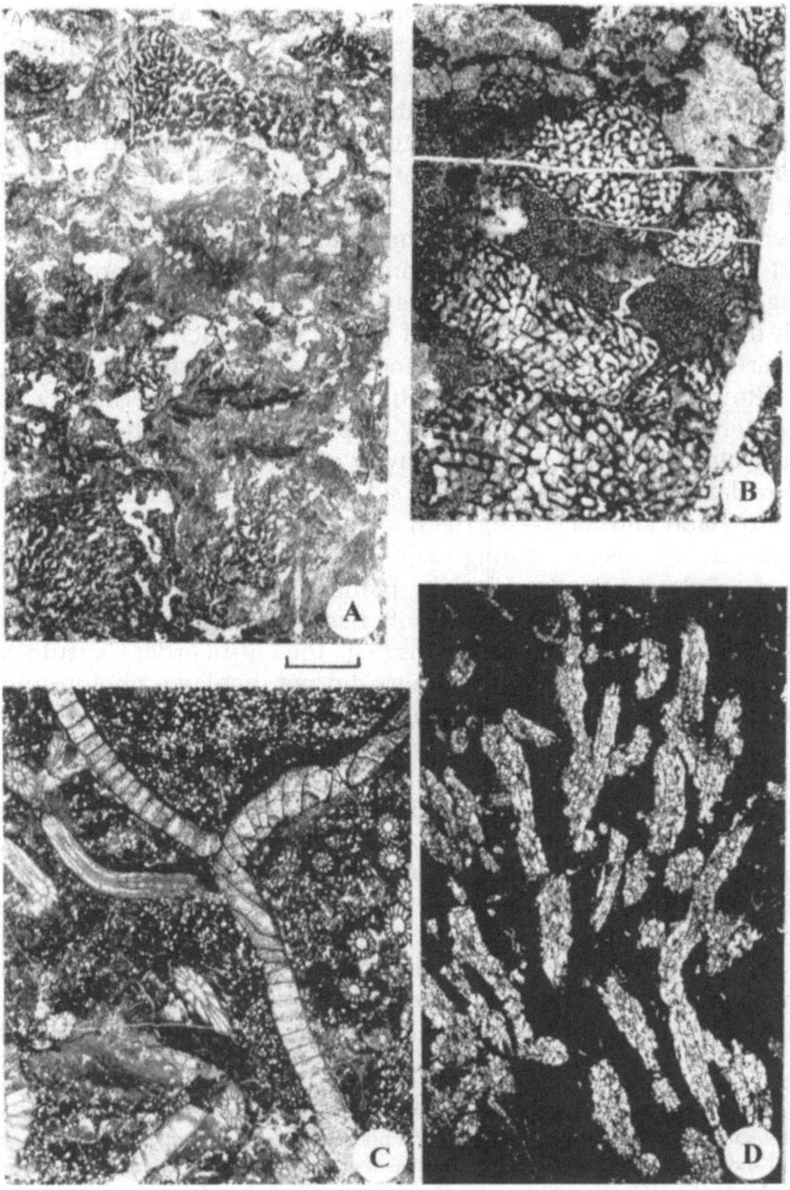

FIGURE 3. Thin sections of Early Cambrian reefs from Mongolia: (A) *Razumovskia–Kordephyton* skeletal stromatolite, Burgasutay Formation, Botoman, Lake Province; (B) modular archaeocyathid *Archaeocyathus* sp. framework reef with cryptic solitary coscinocyathids *Clathricoscinus vassilievi* (Vologdin), Burgasutay Formation, Botoman, Lake Province; (C) modular archaeocyathid *Mikhnocyathus zolaensis* Maslov–*Cambrocyathellus foraminosus* (Fonin) bafflestone in siliciclastics, Sortantuin Formation, early Atdabanian, Ider Province; (D) modular archaeocyathid *Cambrocyathellus foraminosus* (Fonin) bafflestone in volcanic ash, Sortantuin Formation, early Atdabanian, Ider Province. Scale bar 3 mm.

TABLE 2. Influence of Siliciclastic and Volcanic Ash Supply on the Composition of Early Atdabanian (Early Cambrian) Reef Communities in Western Mongolia[a]

Locality	Salaany Gol	Zuune Arts	Telmen
Formation	Salaany Gol	Salaany Gol	Sortantuin
Diversity of sessile reef-dweller species	50	16	14
Modular Archaeocyath species diversity	12	6	7
Modular/solitary Archaeocyath species ratio	34%	38%	50%
Modular/solitary Archaeocyath abundance ratio	1:1	20:1	100:1
Renalcid generic diversity	13	8	5
Environmental features	Moderate siliciclastic input	Increased siliciclastic input	High siliciclastic and volcanic ash input

[a]Modified from Zhuravlev (1999).

Cheshire, 1989). Many archaeocyaths inhabited a range of extremely shallow, turbid, turbulent onshore sites directly influenced by UV irradiation, disruptive for pigments, as well as deep, dark conditions (James and Gravestock, 1990; Bechstädt *et al.*, 1994; Pratt, 1995; Debrenne and Zhuravlev, 1996). Both extremely shallow-water and deep-water habitats do not favor the behavior of photosymbiont-bearing sponges. Judging from analyses of morphogenetic trends in archaeocyathan evolution and from the distribution and direction of currents in skeleton models, archaeocyathan skeleton morphology fills a passive filter-feeding function, dependent on currents bearing plankton (Zhuravlev, 1993; Savarese, 1995). These observations correspond with actual distribution of archaeocyaths and different skeletal morphologies in the reef environment: The simplest morphologies are restricted to facies formed under active hydrodynamics, while the most elaborate morphologies provide essential pumping in calm-water settings (Debrenne and Zhuravlev, 1996). In comparison with photosymbiont-bearing calcified animals, archaeocyaths grew relatively slowly and as a result were easily buried by mud or overgrown by renalcids and even by stromatolites (James and Kobluk, 1978; Wood *et al.*, 1993; Zhuravlev, 1996). In addition, a geochemical approach reveals that archaeocyaths precipitated skeletal carbonate in equilibrium with ambient seawater, and such findings did not support the contention that they possessed photosymbionts (Surge *et al.*, 1997). These features allow us to surmise that archaeocyaths did not need oligotrophic conditions. It was hydrodynamic conditions rather than light that governed archaeocyathan communities (Wood *et al.*, 1992, 1993; Debrenne and Zhuravlev, 1996).

Pharetronid calcarean sponges participated in Early Cambrian reef building either as cryptic encrusters, e.g., small *Gravestockia* (Zhuravlev and Wood,

1995) or as framebuilders such as relatively large branching *Bottonaecyathus* (Fig. 2C). Rigid spiculate demosponges (mainly Anthaspidellidae and Axinellidae) nearly formed a monopoly among metazoan reef-forming group during the Middle and Late Cambrian and Early to Middle Ordovician (Hamdi *et al.*, 1995; Spincer, 1996; Wood, 1999) (Fig. 2F).

Radiocyaths, with calcareous skeletons composed of bipolar starlike calcareous elements, were some of the largest Early Cambrian reef-builders (Kruse, 1991; Wood *et al.*, 1993; Kruse *et al.*, 1995) (Fig. 5C). Their branching skeletons produced highly cavernous reefs inhabited by numerous cryptobionts. In the Early Ordovician time, this group revived as soanitids (*Calathium*), which possessed skeletons of similar structure, locally became major reef builders until other modular metazoans appeared (Toomey and Nitecki, 1979; Church, 1991). Both groups may be related to receptaculitids, which were initially identified as sponges and later as a type of calcified algae. The latest data on the skeletal microstructure again hint at metazoan affinities (Nitecki and Mutvei, 1996). By their inferred ecology, both radiocyaths and soanitids resembled filter-feeding animals rather then autotrophic algae (Church, 1991; Wood *et al.*, 1993).

Coralomorphs are corallike Early Cambrian problematic calcified organisms some of which may be corals (Scrutton, 1997). Massive modular coralomorphs (*Flindersipora*, tabulaconids, and tannuolaiids) built frameworks, and some relatively large solitary and branching forms (*Cysticyathus*, *Rackovskia*, hydroconozoans) participated in archaeocyath–renalcid buildups (Savarese *et al.*, 1993; Kruse *et al.*, 1995, 1996) (Figs. 2C, 4A, D). Inferred ecology, together with the small size of individual zooids, imply that the majority of coralomorphs were filter feeders rather than microcarnivores (Wood *et al.*, 1993; Kruse *et al.*, 1995). Cribricyaths were tiny (<1 cm long) but locally abundant calcareous problematica that occasionally took part in reef formation (Wood *et al.*, 1993) (Fig. 5A).

In addition, a number of calcified organisms contributed to the accumulation of reef sediments, especially chancelloriids, rhynchonelliform brachiopods, orthothecimorph hyoliths, echinoderms, and macluritid gastropods (Kobluk and James, 1979; Yochelson and Stinchcomb, 1987; Wood *et al.*, 1993; Kruse *et al.*, 1995; Pratt, 1995; Riding and Zhuravlev, 1995; Spincer, 1996; Zhuravlev *et al.*, 1996; Wood, 1999) (Figs. 2C, F, and 4C). They were mostly sessile filter and suspension feeders (Debrenne and Zhuravlev, 1997).

The presence of diverse filter- and suspension-feeding animals resulted in complicated tiering within Early Cambrian reef communities (Fig. 5). This tiering, as well as community differentiation into distinctive open surface and cryptic (with obligate cryptobionts) habitats as well as ecological zonation, resulted in relatively high diversity for Early Cambrian reef communities. Species diversity reached over 50–60 species per several dozens square meters (Fig. 11b). Thus, it is not surprising that the later extinction of Early Cambrian metazoan reefs led to a most severe drop in overall diversity (Fig. 11A).

Among the Cambrian autotrophic reef builders, renalcids played an important role. Renalcids were tiny dendritic, chambered, fanlike, or tubular

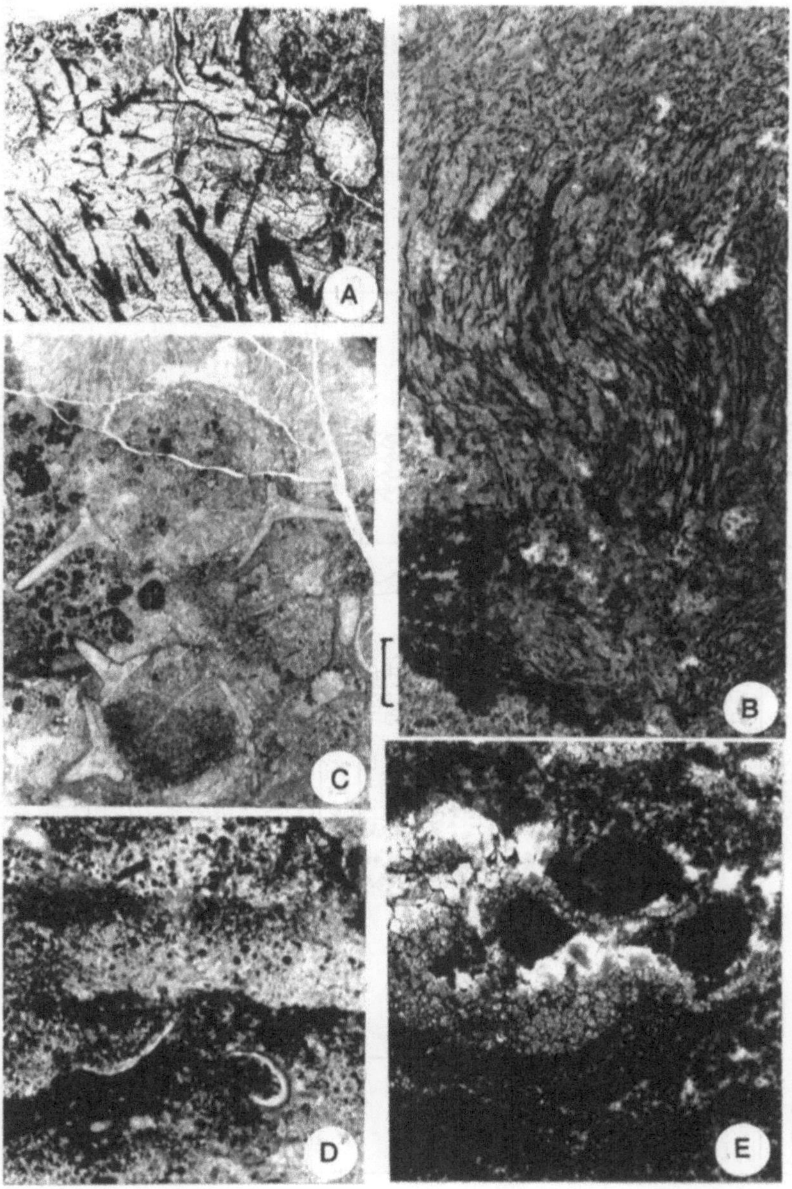

FIGURE 4. Thin sections of Cambrian reef builders and dwellers: (A) massive modular tabulaconid coral *Yaworipora khalfinae* A. Zhuravlev with encrusting coscinocyathid *Tubericyathus clathratus* Vologdin and renalcid *Gordonophyton*, Usa Formation, Botoman, Early Cambrian, Kuznetsky Alatau, Russia, scale bar 3 mm; (B) tubular renalcid *Proaulopora glabra* (Krasnopeeva) Luchinina, Amga Formation, Amgan, Middle Cambrian, Amga River, Siberian Platform, Russia, scale bar 0.5 mm; (C) two scleritomes of chancelloriid *Allonnia* sp., Verkhnemonok Formation, Botoman, Early Cambrian, Western Sayan, Russia, scale bar 10 mm; (D) clotted fabric of radiocyath bafflestone with hydroconozoans *Hydroconus tenuis* (Vologdin), Salaany Gol Formation, early Atdabanian, Early Cambrian, Zavkhan Province, Mongolia, scale bar 0.5 mm; (E) microbial thrombolite with cryptic renalcids *Gordonophyton* and *Renalcis*, Tangha Formation, middle Marjumian, Middle Cambrian, Amga River, Siberian Platform, Russia, scale bar 0.5 mm.

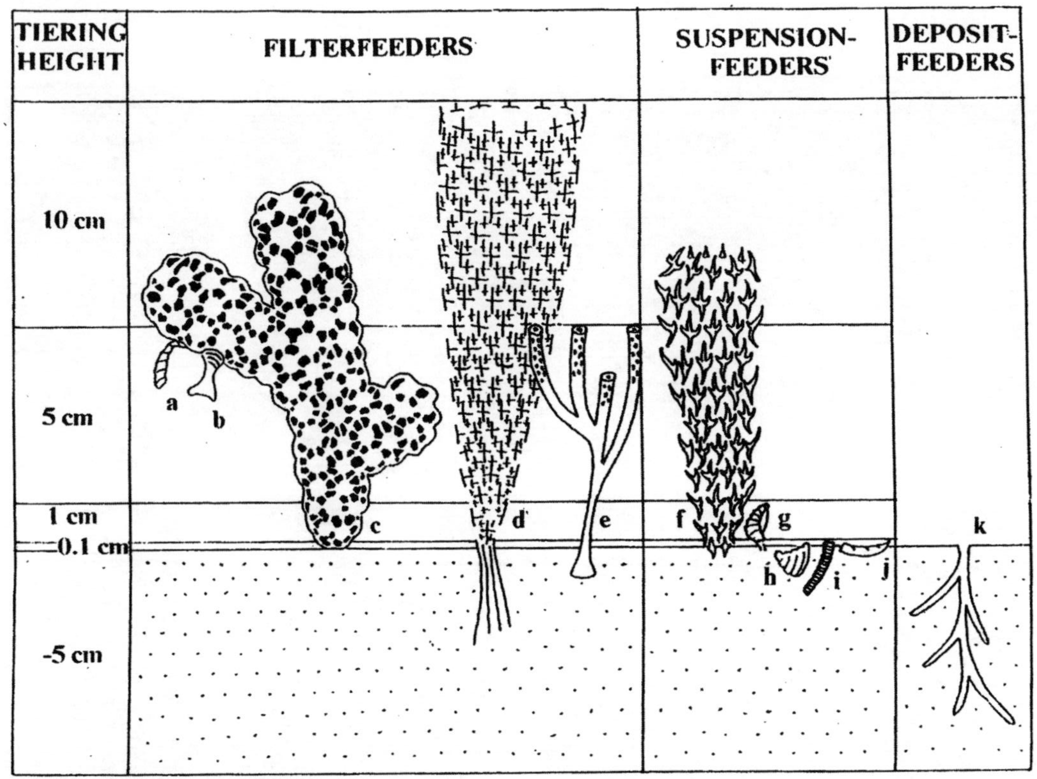

FIGURE 5. Tiering in the early Atdabanian, Early Cambrian reef community of the Zavkhan Basin, western Mongolia: (A) cribricyath; (B) hydroconozoan coralomorph *Hydroconus tenuis* (Vologdin); (C) branching modular radiocyath *Girphanovella georgensis* (Rozanov); (D) solitary hexactinellid sponge of Protospongiidae family; (E) branching modular archaeocyathid sponge *Cambrocyathellus tuberculatus* (Vologdin); (F) chancelloriid *Allonnia* sp.; (G) lingulate brachiopod *Dzunartsina elenae* Ushatinskaya; (H) helcionelloid mollusk *Ilsanella* sp., (I) tube-dwelling worm *Koksuja* sp.; (J) orthothecimorph hyolith; (K) burrower.

calcareous microfossils (*Renalcis*, *Epiphyton*, *Girvanella*, and their relatives). Their simple morphology and common facultative occurrence in cryptic cavities imply that they were calcified bacteria, probably cyanobacteria, rather than algae (Riding, 1991). However, the distinct microgranular microstructure, typical of eukaryotes (Rozanov and Sayutina, 1982; Barskov, 1984), the appearance of obligate cryptobionts (Zhuravlev and Wood, 1995), and the absence of vital effects on carbon isotope values (Surge *et al.*, 1997) do not square with their interpretation as cyanobacteria. Thus, I prefer to call the group renalcids after the most typical Paleozoic genus and to restrict this name to microfossils possessing microgranular microstructure to avoid confusion with calcified cyanobacteria and different enigmatic microfossils of similar morphology. Observed interactions with skeletal metazoans indicate that renalcids were initially calcified or became cemented rapidly during reef growth (Kruse *et al.*, 1995; Zhuravlev and Wood, 1995; Zhuravlev, 1996). The most common Cambrian reef builders among renalcids were dendritic *Gordonophyton* and *Tubomorphophyton*, and threadlike *Razumovskia* (Figs. 2A, B, 3A, and 4B). Dendritic forms are often described by sedimentologists as *Epiphyton*, and *Razumovskia* is ascribed to *Girvanella* (e.g., Ahr, 1971; James, 1981; Read and Pfeil, 1983; Rees *et al.*, 1989). These were renalcids that built the largest individual Cambrian reefs (Debrenne, 1975; Mel'nikov *et al.*, 1991). They produced microframework reefs called "dendrolites" by Riding (1991). Dendrolite is a useful term to separate such reefs from stromatolites, thrombolites, and micritic mudmounds.

Stromatolites and thrombolites comprised a significant part of the Cambrian reef ecosystem, especially Middle and Late Cambrian reefs (Kennard and James, 1986; Zhuravlev, 1996). The term "thrombolite" was introduced by Aitken (1967) to refer to "non-laminated cryptalgal bodies characterized by a clotted fabric". Mesoclots producing such a fabric may consist of renalcids (e.g., *Renalcis*, *Tarthinia*), or cryptomicrobial bodies (Kennard and James, 1986; Riding, 1991). Thus, in analogy with skeletal and agglutinated stromatolites (Riding, 1991) (Fig. 3A), it would be useful to discriminate between skeletal thrombolites formed by renalcids and other calcified bacteria and algae (Fig. 2A) and cryptomicrobial thrombolites (Figs. 1D and 4E). Skeletal stromatolites may be built by a variety of tubular renalcids (*Girvanella*, *Razumovskia*, and *Batinevia*), while noncalcified bacteria and algae produce by the trapping/binding of particulate sediment agglutinated stromatolites. The algal stromatolites, however, are largely a modern phenomenon (Riding *et al.*, 1991).

Micritic mudmounds are another group of common Cambrian reeflike features that are typical of relatively deeper marine settings (James and Gravestock, 1990; Pratt, 1995; Debrenne and Zhuravlev, 1996). A comprehensive compilation by Reitner and Neuweiler (1995) provides a plausible explanation of the nature of micritic mudmounds. The primary organic matrix mediated *in situ* precipitation of micrite and initiated mound development. A sponge–microbial consortium is commonly said to be responsible for autochthonous micrite production. As a result, mounds consist of abundant sponge spicules and peloidal micrite (Reitner and Neuweiler, 1995). This is exactly the fabric that is observed in many Cambrian mudmounds (e.g., James and Gravestock, 1990; Wood *et al.*, 1993; Debrenne and Zhuravlev, 1996) but is also locally present in reefs from shallow settings (Fig. 4D).

In addition to reef builders, destroyers play a significant role in the shaping of the modern reef ecosystem (Hallock, 1988; Wood, 1999). Because many Early Cambrian metazoan reefs formed on muddy substrates, they may have suffered from a biological bulldozing effect from burrowers (Thayer, 1983) more than from borers. Indeed, macroborers (sponges and *Trypanites*-like forms), although already present in the Early Cambrian, were uncommon in reefs (James and Kobluk, 1978; Kobluk, 1981; Gandin and Debrenne, 1984; Waters, 1989; James and Gravestock, 1990; Kennard, 1991). Diverse microborers were ubiquitous members of reef communities, which included cyanobacteria and fungi (Kobluk, 1985; Zhuravlev and Wood, 1995). A few examples of inferred rasping grazer traces are recorded in Cambrian reefs (Kobluk, 1985). Only by the very end of the Cambrian did probable polyplacophoran and monoplacophoran mollusks, interpreted as reefal browsers, appear (Stinchcomb, 1975; Runnegar *et al.*, 1979). Even in later cases we cannot be sure of the diet due to an absence of data on their radular structures.

Available information on the diet of Cambrian animals based on the analyses of gut content, appendage morphology, and general body features (Debrenne and Zhuravlev, 1997), does not allow us to consider how prevalent grazing was in the Cambrian. Indeed, the most powerful grazers, such as teleost fishes, chitons, limpets, insects, crustaceans, and sea urchins did not appear until late Mesozoic–Cenozoic time (e.g., Shunula and Ndibalema, 1986; Stenek, 1986; Bruggemann, 1995) and their rise paralleled Vermeij's (1987) "great escalation" and the diversification of major predators. Both processes facilitated the development of modern reef ecosystems. Modern zooxanthellate–coral reefs thrive under moderate grazing pressure that controls the settlement of algae that normally compete with corals and in turn are controlled by predators who reduce grazing pressure on the corals themselves (Van Treeck *et al.*, 1996). Unlike light-limited coral reefs (Hallock, 1988; Wood, 1993), the proliferation of Cambrian reef communities may not have depended on grazing pressure.

Among the vagrant reef dwellers, trilobites were some of the most diverse Cambrian animals. Already by the middle Early Cambrian, trilobite communities specialized for reef habitats had appeared (Repina and Zharkova, 1974; Repina, 1977; Debrenne *et al.*, 1989a). During the Middle and Late Cambrian, trilobites became prominent reef dwellers and some trilobite species, even genera, were possible obligate reef animals (Pegel, 1982; Sukhov and Pegel, 1986; Westrop, 1989, 1996; Loch and Taylor, 1997). The redlichioid *Giordanella* in the Early Cambrian and plethopeltids in the Middle and Late Cambrian were adapted to agitated reef settings. Such trilobites are characterized by an inflated and smooth cephalon with small eyes, angular articulation of the cephalon and thorax, and postcephalic segments with wide axes, which suggest a suspension-feeding semi-infaunal life habit (Stitt, 1976; Pillola,

1996). Such trilobites likely deployed a sessile suspension-feeder niche vacated after the extinction of other Early Cambrian animals.

In summary, the Early Cambrian tropical reef ecosystem differ from modern coral reefs in the following ways: (1) Its trophic nucleus consisted of filter and suspension feeders, whereas in modern reefs, photosymbiont-containing scleractinian corals compose the trophic nucleus. As a result the modern tropical reef ecosystem prefers oligotrophic conditions excluding phytoplankton proliferation (Hallock, 1988), while Cambrian reef ecosystem did not require such conditions. The appearance of Early Cambrian metazoan reefs in siliciclastic-dominated coastal setting and even in areas of increased phosphate supply indicates that they were adapted to mesotrophic to mildly eutrophic conditions. (2) The roles of large-scale destroyers and grazers were negligible in Cambrian reefs in contrast to modern coral reefs.

3. Spatial Distribution and Temporal Evolution of Cambrian Reefs and Reef Communities

Most, if not all Cambrian reefs, formed at low latitudes (Courjault-Radé *et al.*, 1992) (Fig. 6). The Cambrian continent of Avalonia was situated at southern high latitudes and its earliest Cambrian strata contain planar iron oxide and silica-impregnated stromatolites that grew in peritidal conditions

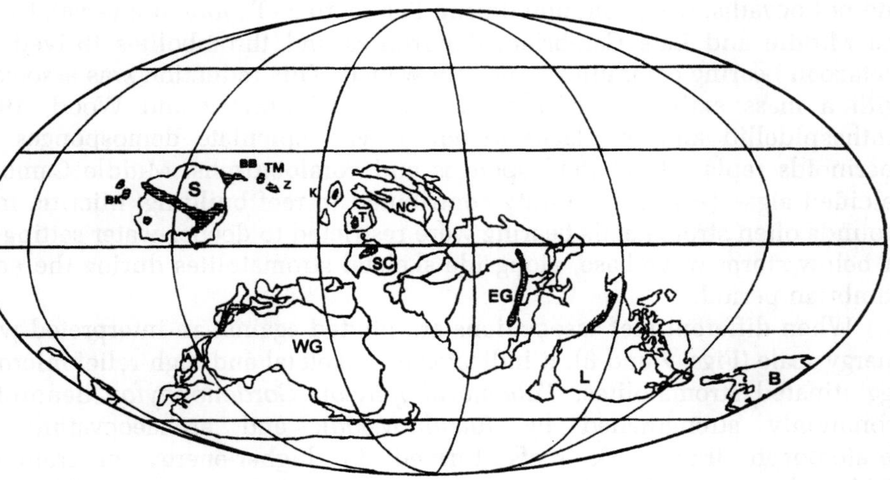

FIGURE 6. Early Cambrian paleogeography for the early Botoman time (modified after Zhuravlev and Maidanskaya, 1998). Reef areas are stippled. Letters mark the continents and microcontinents: A, Avalonia; B, Baltia; BB, Bateni–Baratal microcontinent (Altay–Sayan Foldbelt in part); BK, Bureya–Khanka microcontinent (Russian Far East in part); EG, East Gondwana; K, Kazakhstania; L, Laurentia; NC, North China; SC, South China; S, Siberia; T, Tarim; TM and Z, Tuva–Mongolian and Zavkhan microcontinents of Mongolia, respectively.

and carbonate stromatactis mudmounds formed by sediment baffling and/or microbial precipitation in subtidle conditions (Landing, 1996). According to Copper (1974), *Skolithos* tubes, which are dense assemblages of simple agglutinated vertical structures, developed in nearshore shifting sandy substrates. This could result in worm mounds resembling modern sabellariid reefs. Such fabrics were indeed quite common during the Cambrian time but, according to Droser (1991), were deep bioturbation traces rather than hard, rigid structures.

Depth-dependent distribution of different Cambrian associations of reef-building organisms is plotted on Fig. 7. Only microbial agglutinated stromatolites cover a wide spectrum of environments during the Vendian (Table 1). Tube-dwelling taxa such as *Cloudina* and *Anabarites* formed thickets coalescing into reef bodies during the latest Vendian (Luchinina, 1985; Mattes and Conway Morris, 1990). These thickets, however, were scarce among renalcid patches. During the Cambrian, this community was generally restricted to inter- and supratidal settings, with the exception of a short interval in the Late Cambrian (Sunwaptan) when subtidal settings became common once again (Zhuravlev, 1996). In the Cambrian skeletal stromatolites and in the Early Cambrian archaeocyathan framework reefs sometimes occupied peritidal settings (Debrenne and Zhuravlev, 1996). Shallow subtidal settings above storm wave base contain a wide spectrum of reefs. The most common subtidal buildups were thrombolites, which have appeared in the Neoproterozoic (Aitken and Narbonne, 1989). Dendrolites and skeletal stromatolites, as well as framework reefs and bafflestones constructed by modular archaeocyathids, ajacicyathids, kazachstanicyathids, coralomorphs, and radiocyaths, were common during Tommotian–Toyonian interval. During the Middle and Late Cambrian, dendrolites and thrombolites thrived and metazoan-bearing communities greatly waned. This reduction was associated with a mass extinction of reef metazoans (Zhuravlev and Wood, 1996). Anthaspidellid and in places axinellid rigid spiculate demosponges and eocrinoids replaced calcified sponges and coralomorphs. Middle Cambrian calcified algae (*Amgaella*) rarely contributed to reef building. Micritic mudmounds often stromatactis bearing were restricted to deeper-water settings, at or below storm wave base, alongside skeletal stromatolites during the entire Cambrian period.

When different reef associations are plotted against an interpreted water energy scale (Figs. 7 and 8), it indicates that skeletal and high-relief microbial agglutinated stromatolites, *Tubomorphophyton–Gordonophyton* dendrolites (commonly strengthened by *Razumovskia*), and archaeocyathid and coralomorph framework reefs thrived in higher-energy environments (Zadorozhnaya, 1974; Lohmann, 1976; Cook and Taylor, 1977; Astashkin, 1979; Markello and Read, 1981; Rees and Robison, 1989; Waters, 1989; James and Gravestock, 1990; Debrenne *et al.*, 1989b, 1990, 1993; Bao *et al.*, 1991; Geyer *et al.*, 1995). *Epiphyton–Renalcis* dendrolites, skeletal and microbial thrombolites, ajacicyathid bafflestones, and micritic stromatactis-bearing mudmounds preferred calmer water settings (Morgan, 1976; Schmitt and Monniger, 1977; Zamarreño, 1977; Read, 1980; Gandin and Debrenne, 1984; Stepanova, 1986; Sychev, 1986; Courjault-Radé, 1988; Pratt, 1989; Alexander and Gravestock, 1990; Debrenne *et al.*, 1990; James and Gravestock, 1990; Kennard, 1991; Kruse, 1991).

Communities changed as reefs developed across a water depth gradient. Ajacicyathid–*Renalcis*–*Epiphyton* associations commonly turned into a pure *Razumovskia* crust when a reef mound top reached the fair-weather wave base (Zhuravlev and Gravestock, 1994) (Fig. 2A, B). Shoaling could cause microbial stromatolite development over archaeocyathid–radiocyathan bafflestone (Zhuravleva, 1966). The appearance of domelike archaeocyathid *Retilamina*, which firmly anchored to a substrate, capping archaeocyathid–renalcid mounds, could mark a similar response (Copper, 1974). Both sea-level change and reef shallowing upward changed complex, alternated, diverse communities and consequently reef fabrics (Chafetz, 1973; Read and Pfeil, 1983; Rowland, 1984; Rees and Robison, 1989; Wood *et al.*, 1993; Kruse *et al.*, 1995, 1996).

4. Metazoans versus Nonmetazoans

Some Cambrian reef organisms managed to form edifices in various marine environments, but when other groups of reef builders prevailed, they were largely restricted to marginal conditions. For instance, microbial agglutinated stromatolites formed complicated buildups with archaeocyaths in normal marine subtidal areas, although their role was usually minor (Zhuravleva, 1966; Kruse, 1991; Savarese *et al.*, 1993; Wood *et al.*, 1993). In the absence of other reef types (pre-Nemakit–Daldynian Vendian and late Sunwaptan), microbial stromatolites spread out over the entire spectrum of marine environments. A number of Cambrian reef communities had a bimodal distribution, where other reef communities were present: Skeletal stromatolites built by *Girvanella* occurred either in extremely shallow or in relatively deep-water settings (Fig. 7). On the other hand, dendrolite-producing renalcids and modular archaeocyathids had a strong preference for shallow subtidal conditions, from approximately above fair-weather wave base to storm wave base. An inferred ecology of renalcids and archaeocyaths suggests that such conditions are near optimal. Renalcids had to avoid both extremely shallow marine settings because of UV radiation and suspended sediment and extremely deep settings because of decreased illumination; archaeocyaths, being passive filtrators, relied on constant currents bringing plankton and dissolved organic matter. The same conditions possibly existed for radiocyaths and coralomorphs as well.

On a small scale, between renalcids and archaeocyaths and within these groups, a competition for space is evident, which is expressed in suppression of the weakest individuals (Debrenne and Zhuravlev, 1992, 1994; Wood *et al.*, 1992, 1993; Kruse *et al.*, 1995; Zhuravlev and Wood, 1995; Zhuravlev, 1996).

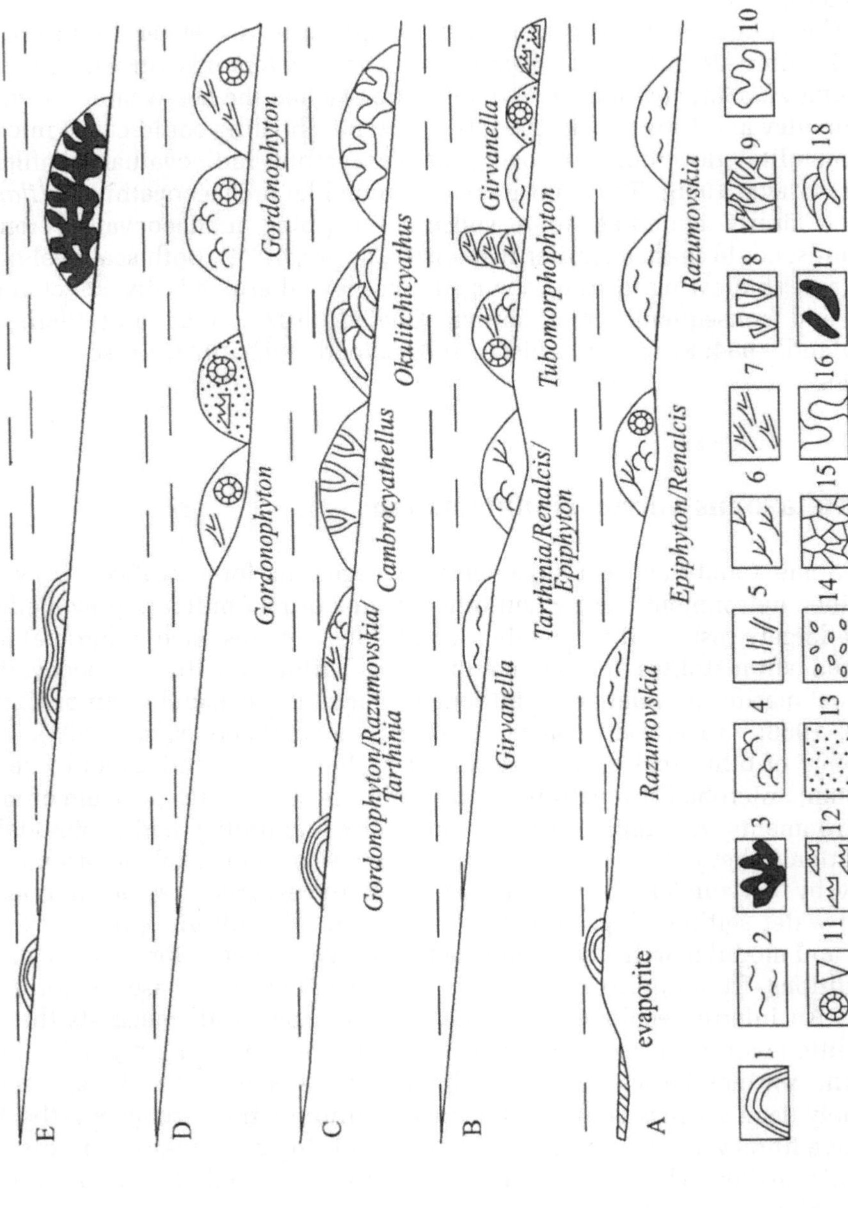
E
D
Gordonophyton
Gordonophyton
C
Gordonophyton/Razumovskia/
Tarthinia
Cambrocyathellus
Okulitchicyathus
B
Girvanella
Tarthinia/Renalcis/
Epiphyton
Tubomorphophyton
Girvanella
A
evaporite
Razumovskia
Epiphyton/Renalcis
Razumovskia
1 2 3 4 5 6 7 8 9 10
11 12 13 14 15 16 17 18

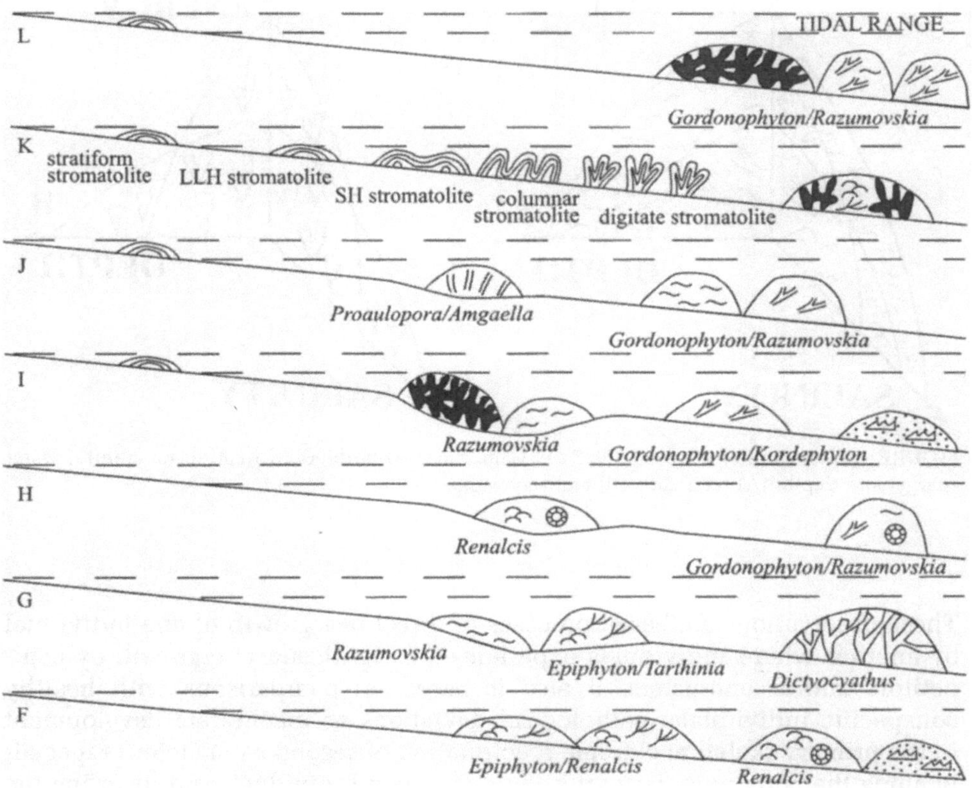

FIGURE 7. Distribution of Cambrian reef communities along seafloor paleoprofiles. (A) Osa Horizon, early Atdabanian, Early Cambrian, Turukhansk–Irkutst–Olekma Basin, Siberian Platform (after Usychenko, 1988); (B) Churan Member/Nochoroy Member/Pestrotsvet Formation, early Atdabanian, Early Cambrian, Anabar–Sinsk and Yudoma–Olenek basins, Siberian Platform (after Stepanova, 1986; Debrenne and Zhuravlev, 1996; pers. observations); (C) Salaany Gol Formation, early Atdabanian, Early Cambrian, Zavkhan Basin, Mongolia (after Wood *et al.*, 1993); (D) lower Wilkawillina Limestone, late Atdabanian, Early Cambrian, Arrowie Basin, Australia (after James and Gravestock, 1990); (E) Sierra Gorda Member, Alconera Formation, early Botoman, Early Cambrian, Ossa–Morena Province, Spain (after Moreno-Eiris, 1987); (F) La Hoya Member, Alconera Formation, early Botoman, Early Cambrian, Ossa–Morena Province, Spain (after Moreno-Eiris, 1987); (G) Matoppa Formation, early Botoman, Early Cambrian, Sulcis Basin, Sardinia (after Debrenne *et al.*, 1993); (H) Shackleton Limestone, late Botoman, Early Cambrian, Antarctica (after Rees *et al.*, 1989); (I) Upper Shady Dolomite, late Botoman, Early Cambrian, Appalachians (after Read and Pfeil, 1983; Barnaby and Read, 1990); (J) Diringde Reef Massif, late Marjumian, Middle Cambrian, Siberian Platform (after Shishkin *et al.*, 1978); (K) Maynardville Limestone, late Marjumian, Middle Cambrian, Appalachians (after Glumac and Walker, 1997); (L) Cow Head Group, Late Cambrian, western Newfoundland (after Coniglio and James, 1985). 1, agglutinated microbial stromatolite; 2, skeletal (*Razumovskia/Girvanella*) stromatolite and *Razumovskia* crust; 3, microbial thrombolite; 4, skeletal (*Renalcis/Tarthinia/Epiphyton*) thrombolite and binding *Renalcis*; 5, skeletal (*Proaulopora/Amgaina*) dendrolite; 6, *Epiphyton* dendrolite; 7, *Gordonophyton/Tubomorphophyton* dendrolite; 8, archaeocyathid bafflestone and modular archaeocyathids; 9, archaeocyathid framestone; 10, radiocyath bafflestone; 11, solitary ajacicyathids; 12, stromatactis-bearing mudmound; 13, mud to mudstone; 14, clastics to packstone/grainstone; 15, synsedimentary cement; 16, encrusting monocyathids; 17, borings; 18, domelike archaeocyathids.

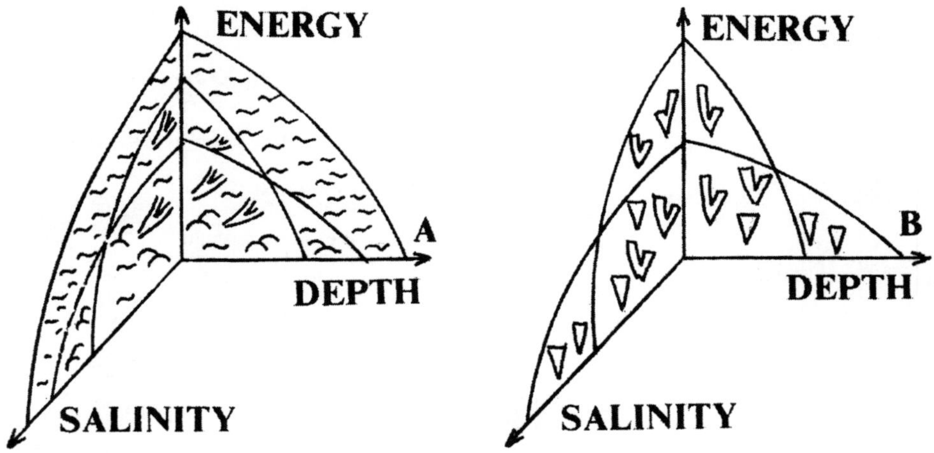

FIGURE 8. Distribution of principal Early Cambrian reef-builders in relation to salinity, water energy, and depth: (A) renalcids; (B) archaeocyaths.

These observations are based on cases of direct overgrowth of one individual by another where individuals experienced a significant overgrowth by competitors and/or encrusters. It also is based on comparisons with healthy conspecific individuals: pathological deviations or incomplete development in the primary skeleton; extreme precipitation of secondary skeleton, especially along the body boundary with the overgrown individual; and dwarfing the ontogeny of the other. This is the reason why any study of reef community is in great need of precise taxonomy. Judging by such observations, hierarchy of competitive interactions may be inferred (Fig. 9): Both modular archaeocyathids and dendritic renalcids outcompeted solitary ajacicyathids, but even modular archaeocyathids were subdued, in places by renalcids. In turn, ajacicyathids as well as their superiors were able to outcompete chambered and tubular renalcids and stromatolite-producing microbes. Large radiocyaths outcompeted even modular archaeocyaths (Debrenne and Zhuravlev, 1992: Pl. 38, Fig. 3). These observations on paleontological interactions do not support an assumption about an overall temporal replacement of smaller renalcid reefs by larger archaeocyathan reefs (e.g., Fagerstrom, 1987; Rowland and Gangloff, 1988; Talent, 1988).

Besides, by analogy with Recent sponges and bacteria and algae, other possibilities for competitive interaction can be suggested, such as production of various toxins and inhibition of larval settlement. Common monospecific settlements of archaeocyaths, which show a very narrow size range, suggest that larval spat falls were restricted to specific, short intervals. Modern sponges use such a strategy when their larvae are inhibited by faster-growing algae, and sponges have to restrict their reproduction to a season when conditions unfavorable for algae dominate (Ilan and Loya, 1988).

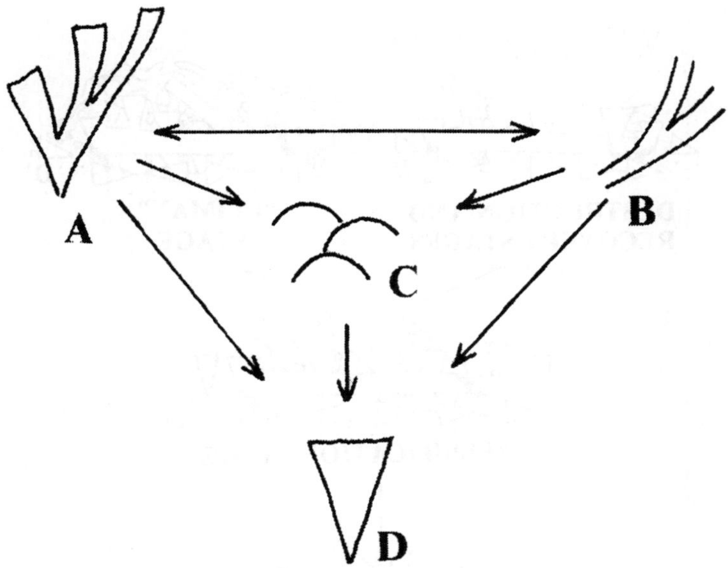

FIGURE 9. Hierarchy of interactions between principal Early Cambrian reef builders. Arrows are directed to subdued organisms: (A) modular archaeocyathid; (B) branching renalcid; (C) chambered renalcid; (D) solitary ajacicyathid.

Competitive interactions resulted in the displacing of solitary ajacicyathids and some renalcids to marginal environments (Figs. 2D, E and 10) and in the establishing of three principal associations of organisms approaching the optimum of their ecological requirement: (1) modular archaeocyathids and *Gordonophyton–Tubomorphophyton* renalcids occupied a position closest to the ecological optimum; (2) microbial and skeletal stromatolite-producing microbes were marginal; and (3) solitary ajacicyathids and *Renalcis–Tarthinia* occurred in intermediate positions (Table 1, Figs. 7 and 8).

Three groups of species, which differ according to their ecological strategies, are well-known to botanists (Ramenskiy, 1935; Grime, 1979). Among them a group of species is distinguished that is dominant under favorable conditions and optimally using available resources. Usually they are fast-growing and large "competitors" (in terms of Grime, 1979). Another group of species is subordinated to competitors and subsequently dominant under permanently or discontinuously unfavorable conditions due to their ability to resist severe conditions. These are fast-growing and small "ruderals." The third group of species is inferior to both aforementioned groups. As a result, they dominated under pioneer conditions only when competition was low. They are slow-growing, small "stress tolerators" following Grime (1979). Rosen (1981) demonstrated that such concepts may be applicable to sessile reef animals that specialize as space occupiers. Figures 7 and 8 show the distribution of three principal groups of renalcids (*Girvanella–Razumovskia*,

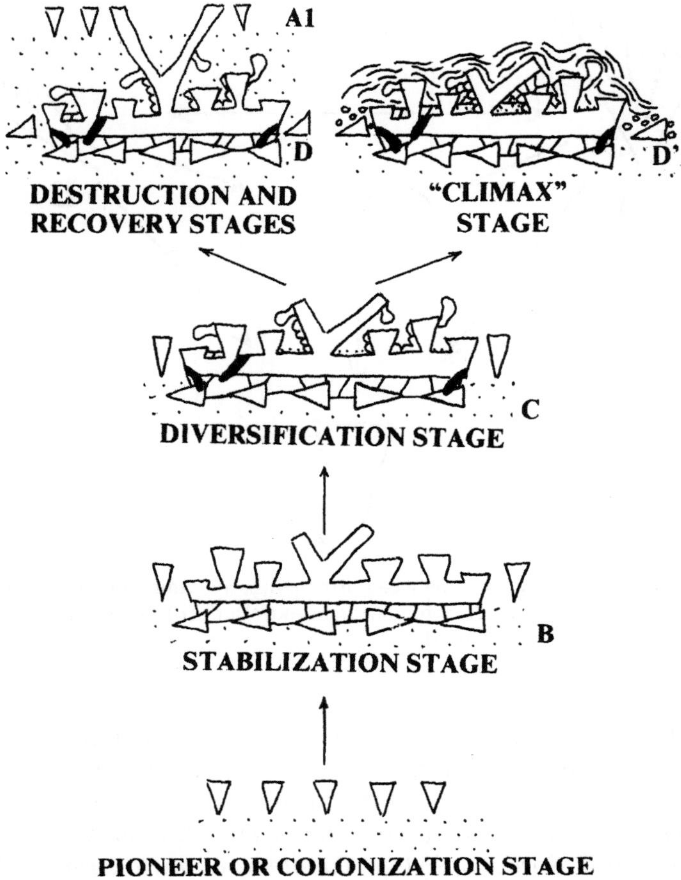

FIGURE 10. Successions on Early Cambrian reefs: (A) pioneer or colonization stage: settlement of solitary ajacicyathids on soft substrate; (B) stabilization stage: lithification of substrate providing by solitary ajacicyathids and settlement of modular archaeocyathids forming a framework; (C) diversification stage: appearance of encrusters, borers, and cryptobionts; (D) destruction stage: burial of framework with loose sediment due to sea-level drop; (D′) climax stage: appearance of monospecific *Razumovskia* (filamentous renalcid) crust due to reef growth into shallow agitated conditions; (A1) recovery stage: a new settlement of solitary ajacicyathids and recovery of some modular archaeocyathids.

Renalcis–Tarthinia, and *Gordonophyton–Tubomorphophyton*) and two major groups of archaeocyaths (solitary ajacicyathids and modular archaeocyathids) according to three environmental parameters (salinity, depth, energy). The entire area, which can be occupied by each group, approximates the conditions satisfying their ecological requirements (Fig. 8). The competition, however, restricted subdued groups to the peripheral areas where they escaped direct conflict with more competitive groups. This phenomenon is expressed in the observed spatial and temporal distribution of Cambrian reef communities (Fig. 7).

Data plotted on Figs. 8 and 9 coupled together indicate that Cambrian reef builders can be conditionally subdivided into competitors (*Gordonophyton*, *Tubomorphophyton*, modular archaeocyathids, radiocyaths), ruderals (*Renalcis*, *Tarthinia*, *Epiphyton*, solitary ajacicyathids), and stress-tolerants (*Girvanella*, agglutinated stromatolite-producing microbes). For instance, communities associated with mudmounds consisted of ruderals and stress-tolerants (K-strategists in Reitner and Neuweiler, 1995) that were characterized by reduced reproductive rates, slow growth rates, and specialized feeding behavior. The latter feature is especially well expressed in mudmound-inhabiting solitary ajacicyathids that possess very complicated skeletal structures.

During the Tommotian epoch, superior competitors like renalcids and archaeocyaths displaced stromatolite-producing microbes in marginal environments (Table 1 and Fig. 7). At the same time, the differentiation of archaeocyaths into strong competitors (modular archaeocyathids) and ruderals (solitary ajacicyathids) as well as the introduction of competitive dendritic forms among renalcids resulted in further moving of renalcid (*Renalcis–Tarthinia–Epiphyton*)–ajacicyathid reef community from the optimum and displaced skeletal and agglutinated microbial stromatolites to the severest settings. The extinction of the majority of reef-building metazoans by the beginning of the Toyonian epoch led to the domination by dendritic renalcids, which, as powerful competitors, could outcompete the surviving reef-building metazoans (Zhuravlev, 1996). Still existing during Toyonian time, archaeocyaths and radiocyaths played only a minor role (Debrenne *et al.*, 1991; Kruse, 1991). As a result, Middle to early Late Cambrian reefs occurring in environmentally optimal conditions were dendrolites (Astashkin *et al.*, 1984; Shabanov *et al.*, 1987; Pratt, 1989; Bao *et al.*, 1991). Newly evolved reef-building metazoans (spiculate demosponges and eocrinoids) coexisted with stromatolites and thrombolites but avoided dendrolites (Hamdi *et al.*, 1995; Wood, 1999). Further collapse of the reef community may have been caused by the lost of spatial heterogeneity (Zhuravlev, 1996). The collapse liberated the environmental optimum for the stress-tolerating stromatolite- and thrombolite-producing microbes during the Sunwaptan epoch. With the revival of modular metazoans (soanitid receptaculitids, tabulates, *Pulchrilamina*, and later on, stromatoporoids, chaetetids, and bryozoans) in the Ordovician, a new wave of differentiation of reef builders into competitors and subordinated species began (Table 1). For instance, bryozoans evolved different strategies for competitive interaction with microbial stromatolites (Hillmer and Scholz, 1996).

5. Biotic Factors versus Abiotic Factors

Microbial stromatolites spread over wide environmental spectrum eventually including normal marine, level-bottom conditions during the Proterozoic, Late Cambrian (Sunwaptan), and Early Triassic (Spathian) and some other postextinction intervals (Copper, 1988, 1997; Awramik, 1992; Schubert and Bottjer, 1992; Zhuravlev, 1996). Schubert and Bottjer (1992) even discussed stromatolites as disaster forms, the definition of which is very close to the definition of stress-tolerants. Similarly, renalcids were principal reef builders in different environments of the Middle Cambrian, but also after the Frasnian–Famennian mass extinction (Wray and Playford, 1970; Playford and Cockbain, 1989; Copper, 1994) and the Middle and Late Triassic (Flügel, 1991; Riding, 1992).

Commonly, abiotic factors explain the decline–proliferation of microbial and algal reefs. The decline of stromatolites in normal marine conditions meanwhile is explained by the decrease in the saturation of seawater (e.g., Grotzinger and Knoll, 1995). However, this factor hardly was responsible for the repetitive invasions of stromatolite communities into normal marine subtidal settings. The relative success of renalcids has been ascribed to elevated temperatures, enhancing rates of carbonate precipitation (Riding, 1992). In a different frame of reference, such a suggestion is to a some extent controversial. For instance, Berner (1994) calculated a considerable lowering of temperature starting near the Frasnian–Famennian boundary. Thus, the oscillating pattern in the development of microbial and renalcid reefs does not coincide with a precise abiotic agent.

However, some extrinsic factors should not be completely neglected. Cambrian and Triassic periods coincided with Sandberg's (1983) shifts from an aragonite-facilitating sea state to an aragonite-inhibiting sea. Indeed, if Early Cambrian reefs were rich in aragonite marine cements and in organisms whose skeletons were built either of aragonite or high-magnesium calcite (James and Klappa, 1983; Read and Pfeil, 1983; Pratt, 1991; Wood *et al.*, 1993) (aragonite botryoids were commonly described as calcified alga *Zaganolomia* in older papers: Drosdova, 1980; Stepanova, 1986), late Middle and Late Cambrian reefs lack such fabrics. From the Early to Middle Cambrian, a transition from aragonite to calcite micrites occurred (Lasemi and Sandberg, 1996) and first hardground communities bearing a rich fauna appeared (Zhuravlev *et al.*, 1996). The development of hardground communities needs aragonite-inhibiting conditions (Wilson *et al.*, 1992; Myrow, 1995). In turn, the change of aragonite sea state to calcite sea state is probably related to the commencement of greenhouse conditions. Middle Cambrian strata show low-amplitude sea-level fluctuations which are typical of greenhouse conditions (Adams and Grotzinger, 1996). Such conditions could favor the expansion of renalcid reefs due to enhanced carbonate precipitation (cf. Riding, 1992). At the same time, the progressive development of carbonate hardgrounds may have stimulated the proliferation of multiserial modular metazoan encrusters (Wood, 1993) which started during the Ordovician. This included the displacement of renalcids and low modular metazoans (spiculate sponges, soanitids) to the marginal settings. The same sedimentological and ecological pattern is observed across the Triassic–Jurassic boundary. Reefs of aragonite-facilitating intervals show voluminous marine cements, microbial crusts, renalcids, low modular and solitary calcified sponges, and very similar calcified problematica such as the coralomorph *Rackovskia* and the "foraminifer" *Shamovella*, the coralomorph *Edelsteinia* and problematic

Ladinella. For Triassic examples, see Flügel and Senowbari-Daryan (1996). The establishment of aragonite-inhibiting greenhouse conditions may have led in part to the displacement and partial extinction of former reef communities.

The modern aragonite-facilitating icehouse interval is very different from the previous intervals in the proliferation of giant reefs created by highly modular metazoans. This phenomenon is related to development of the coral–zooxanthellate symbiosis, which is an immanent biotic factor, allowing reefs to occupy oligotrophic environments in spite of the coming of an icehouse epoch. This may have started in the Late Triassic (Stanley and Swart, 1995).

On the other hand, the decline of stromatolites was ascribed to grazing and burrowing marine animals (e.g., Garrett, 1970), or to competitive exclusion by encrusting eukaryotes (e.g., Monty, 1973). In the Mesozoic–Cenozoic interval, principal grazers appeared and the limited effect of modern grazers on the proliferation of stromatolite-building communities has been demonstrated (Wickstrom and Castenholz, 1985; Stenek, 1986; Farmer, 1992; Bruggemann, 1995). But during the Proterozoic and Paleozoic time, grazing pressure may not have been exclusively responsible for the regulation of stromatolite. The presence of calcified algae and microbes including renalcids in the Neoproterozoic reefs (Grant *et al.*, 1991; Knoll *et al.*, 1993; Turner *et al.*, 1997) hints that the appearance of such advanced forms was reason enough for the decline of microbial stromatolites in the normal marine settings. The developmental history of reef communities after the Permian–Triassic crisis very much resembles the pattern observed during the Late Cambrian to Middle Ordovician interval (Flügel and Senowbari-Daryan, 1996).

6. Ecological Succession in Cambrian Reef Ecosystems

Since the pioneer work by Walker and Alberstadt (1975), reef succession in the geological record is commonly ascribed to a regular sequence of pioneer (colonization), stabilization, diversification, and climax stages. Blanchon *et al.* (1997), following earlier works, suggest a scenario of reef community succession including colonization, destruction, and recovery stages and describe reefs as storm-adapted structures. Connell (1978) emphasized that the amazing species diversity of modern coral reefs may be sustained by a nonequilibrium state maintained by disturbances. Ecological succession includes pioneer or colonization, stabilization, diversification, and climax stages (e.g., Walker and Alberstadt, 1975; Copper, 1988; Kosmynin and Hecker, 1997).

The earliest Cambrian reefs show that the reef rock included frequent alternations of grainstones/packstones, consisting mostly of solitary ajacicyathids and framestones built by a modular archaeocyathid further strengthened by *Renalcis* (Kruse *et al.*, 1995; Riding and Zhuravlev, 1995). The following ecological succession may be suggested for these Cambrian reefs: (1) settlement of solitary ajacicyathids, tolerant in turbid conditions, and able to occupy a muddy substrate (pioneer stage; Fig. 10A); (2) encrusting of resulting grainstone/packstone by modular archaeocyathids, creating a framework (stabilization stage; Fig. 10B); and (3) strengthening of the framework by binding *Renalcis* and the occupation of the resulting reef cavities by diverse organisms such as monocyathids, hydroconozoans, and borers, which together constitute the diversification stage (Fig. 10C). This example demonstrates clearly that the first three stages of reef development were biologically driven. Some species (solitary ajacicyathids) created new environments for further occupation by other species such as modular archaeocyathids. In turn, the subsequent organisms further modified the environment by the creation of a three-dimensional framework densely inhabited by diverse species. The entire succession developed according to Clements (1916) by biotic factors and the initial pioneering community.

The associations of these stages developed in similar environmental conditions, and thus represent an ecological succession. The fourth stage depends on either extrinsic factors, such as a sea-level drop, or on an intrinsic factor, namely the reef growth into the low tide or surf zone. In the first case, resulting shallowing and increased agitation brought about the development of *Razumovskia* crusts (or *Retilamina* settlements) (Fig. 10D). In the second case, a storm action led to destruction and recovery stages (Fig. 10D, and A1). The latter was provided by surviving modular archaeocyathids. In both cases, the new asssociation was species impoverished and almost always monospecific. Such an association is comparable with the climax stage community (Walker and Alberstadt, 1975; Copper, 1988) but it always developed as a community response to an environmental change, thus satisfying Hoffman and Narkiewicz's (1977) criteria. Later researchers actually present a modern paradigm of this complicated problem. Useful here is Zherikhin's (1997) distinction between the "climax" stage as an intrinsically stable final stage, achieved due to autogenic development of the reef ecosystem, and the "subclimax" stage, extrinsically stabilized by an outer physical factor. Rather than a climax stage capping the succession, the subclimax stage is a transitional state reverting to a replay of the entire succession, beginning with the pioneer stage. It is totally identical, in terms of species composition and biovolume, to that of the previous succession. Many Cambrian reefs, despite their small sizes, frequently represent such a replay of entire successions.

7. Mass Extinction in Cambrian Reefs

Another topic of complex interplay between biotic and abiotic factors is mass extinction. The elimination of the Early Cambrian reef ecosystem was a part of the first Phanerozoic mass extinction. It was comparable in severity to that of the end-Permian and end-Cretaceous time, which reduced metazoan generic diversity by 50% (Newell, 1972; Zhuravlev and Wood, 1996) (Fig. 11A). Even more striking was the effect of this extinction on

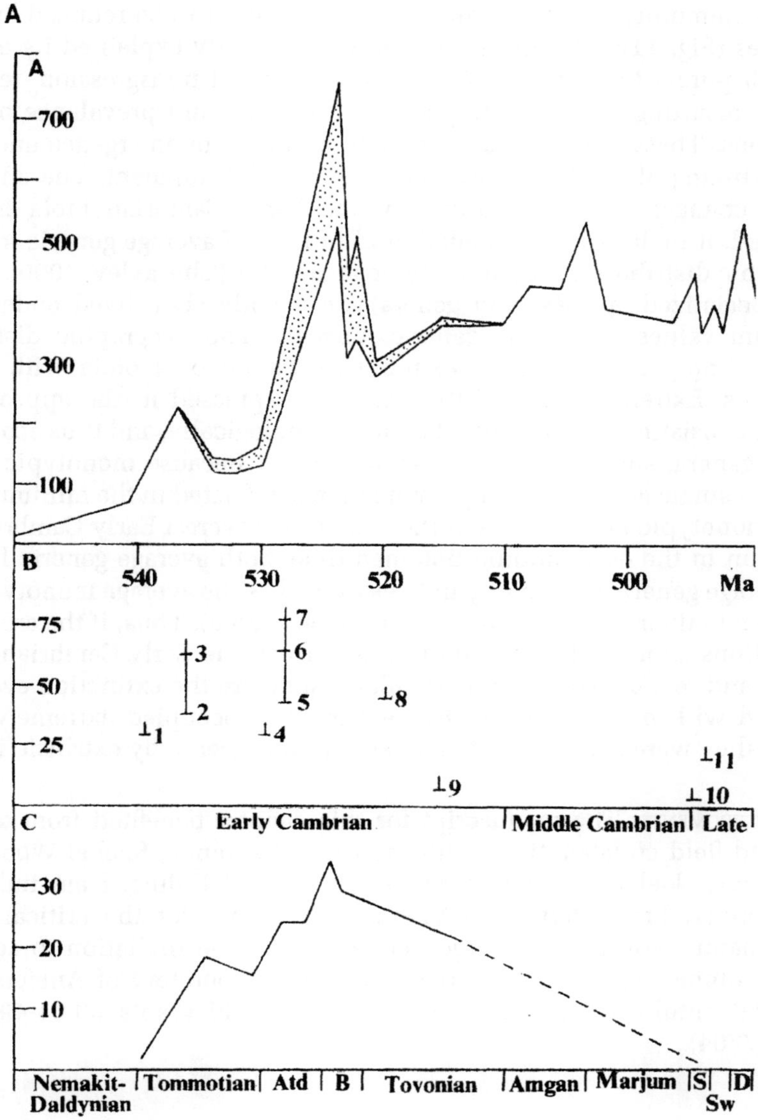

FIGURE 11. Metazoan diversity in Cambrian reefs: (A) generic diversity of reef builders (stippled) plotted against overall generic diversity of skeletal metazoans (modified after Zhuravlev, 2000); (B) species diversity in selected reef communities [1, Pestrotsvet Formation, Siberian Platform (Riding and Zhuravlev, 1995); 2–4, Pestrotsvet Formation, Siberian Platform (Kruse *et al.*, 1995); 5, 7, Salaany Gol Formation, western Mongolia (Wood *et al.*, 1993; Kruse *et al.*, 1996); 6, Oymuran Reef Massif, Siberian Platform (pers. obs.); 8, Forteau Formation, western Newfoundland (James and Kobluk, 1978; Kobluk and James, 1979; Debrenne and James, 1981; Spencer, 1981); 9, Wirrealpa Limestone, South Australia (Kruse, 1991; Brock and Cooper, 1993); 10, Mila Formation, northern Iran (Hamdi *et al.*, 1995; pers. obs.); 11, Diringde Reef Massif, Siberian Platform (Shishkin *et al.*, 1978; Pegel, 1982; pers. obs.)]; (C) species diversity of cryptobionts in reef communities (modified after Zhuravlev and Wood, 1995). Stages: Atd, Atdabanian; B, Botoman; Marjum, Marjumian; S, Steptoean; Sw, Sunwaptan; D, Datsonian.

within-community species richness which caused it to be reduced by almost ten times (Fig. 11B, C). This extinction can be easily explained by extrinsic, although purely terrestrial, factors such as a global transgression–regression couplet, resulting in carbonate platform drowning and prevalence of anoxic conditions. These were replaced quickly by platform emergence and significant shrinking of shelf areas available for reef development. The analysis of secular changes of some features of the Early Cambrian biota has been presented. It included the calculation of indices of average generic longevity, geographic distribution, and monotypic taxa index (Zhuravlev, 2000). Because most specialized species (and genera) are usually short-lived endemic, the minimum values of average generic longevity and geographic distribution indices simply indicate the maximum saturation of a biota with extreme specialists. Extreme niche splitting may be expressed in the appearance of families consisting of a large number of morphologically, and thus ecologically similar genera, shoehorned into narrow niches. Because monotypic families include a single genera, such a phenomenon is reflected in the minimum value of the monotypic taxa index. On the eve of the severest Early Cambrian mass extinction in the early middle Botoman time, both average generic longevity and average generic endemicity indices as well as the average monotypy index fell down to their minimum values (Zhuravlev, 2000). Thus, if the assumption of functions expressed by the indices is correct, the Early Cambrian biota in general and reef ecosystem in particular just before the extinction event were saturated with narrow specialists. Because they occupied extremely narrow niches, they were susceptible to an extinction triggered by extrinsic factors.

ACKNOWLEDGMENTS: The manuscript for this chapter benefited from conversations and field collaborations with Françoise Debrenne, Rachel Wood, David Gravestock, Noel James, Pierre Kruse, and Robert Riding. I am indebted to Paul Copper, Brian Pratt, and Vladimir Zherikhin for the critical reviews of the manuscript and to George Stanley, Jr. for the invitation to contribute to this volume. The work is supported by the Laboratory of Ancient Organisms (Paleontological Institute, Moscow 2nd RFFI grants 00-04-49182 and 00-15-97764).

References

Adams, R. D., and Grotzinger, J. P., 1996, Lateral continuity of facies and parasequences in Middle Cambrian platform carbonates, Carrara Formation, southeastern California, USA, *J. Sediment. Res.* **66**:1079–1090.

Ahr, W. M., 1971, Paleoenvironment, algal structures, and fossil algae in the Upper Cambrian of central Texas, *J. Sediment. Petrol.* **41**:205–216.

Aitken, J. D., 1967, Classification and environmental significance of cryptalgal limestones and dolomites with illustrations from the Cambrian and Ordovician of southwestern Alberta, *J. Sediment. Petrol.* **37**:1163–1178.

Aitken, J. D., and Narbonne, G. M., 1989, Two occurrences of Precambrian thrombolites from the Mackenzie Mountains, northwestern Canada, *Palaios* **4**:384–388.

Alexander, E. M., and Gravestock, D. I., 1990, Sedimentary facies in the Sellick Hill Formation, Fleurieu Peninsula, South Australia, in: *The Evolution of a Late Precambrian–Early Palaeozoic Rift Complex: The Adelaide Geosyncline* (J. B. Jago and P. S. Moore, eds.), *Geol. Soc. Austral. Spec. Publ.* **16**:269–289.

Astashkin, V. A., 1979, Osnovnye tipy organogennykh postroek v rivovykh sistemakh kembriya severnogo sklona Aldanskogo shchita [Principal types of organogeneous buildups in the Cambrian on the northern slope of the Aldan Shield], in: *Geologiya Rifovykh Sistem Kembriya Zapadnoy Yakutii* [Geology of the Cambrian Reef Systems in Western Yakutia] (V. E. Savitskiy, ed.), *Tr. Sibirsk. Nauch.-Issled. Inst. Geol. Geofiz. Miner. Syr'ya* **270**:19–30 [in Russian].

Astashkin, V. A., Varlamov, A. I., Gubina, N. K, Ekhanin, A. E., Pereladov, V. S., Romenko, V. I., Sukhov, S. S., Umperovich, N. V., Fedorov, A. B., Fedyanin, A. P., Shishkin, B. B., and Khobnya, E. I., 1984, *Geologiya i Perspektivy Neftegazonosnosti Rifovykh Sistem Kembriya Sibirskoy Platformy* [Geology and Prospects of Oil-Gas-Bearing of the Cambrian Reef Systems of the Siberian Platform], Nedra, Moscow [in Russian].

Awramik, S. M., 1992, The history and significance of stromatolites, in: *Early Organic Evolution: Implications for Mineral and Energy Resources* (M. Schidlowski, ed.), Springer-Verlag, Berlin, Heidelberg, pp. 435–449.

Bao, H., Mu, X., and Riding, R., 1991, Middle Cambrian dendrolite biostromes, Jinan, East China, in: *5th International Symposium Fossil Algae, April 1991, Abstracts*, Capri, Italy, pp. 7–8.

Barnaby, R. J., and Read, J. F., 1990, Carbonate ramp to rimmed shelf evolution: Lower to Middle Cambrian continental margin, Virginian Appalachians, *Geol. Soc. Am. Bull.* **102**:391–404.

Barskov, I. S., 1984, Paleontological aspects of biomineralization, in: *Proceedings 27th International Geology Congress, Moscow* (4–14 August 1984), Vol. 2, *Palaeontology*, VNU Sciences Press, Utrecht, pp. 63–69.

Bechstädt, T., Boni, M., and Fröhler, M., 1994, Facies development in early and middle Cambrian time, in: *Sedimentological, Stratigraphical and Ore Deposits Field Guide of the Autochthonous Cambro-Ordovician of Southwestern Sardinia*, Pt I (T. Bechstädt and M. Boni, eds.), *Mem. Descr. Carta Geol. d'It.* **48**:47–106.

Berner, R. A., 1994, GEOCARB II: A revised model of atmospheric CO_2 over Phanerozoic time, *Am. J. Sci.* **294**:56–91.

Blanchon, P., Jones, B., and Kalbfleisch, W., 1997, Anatomy of a fringing reef around Grand Cayman: Storm rubble, not coral framework, *J. Sediment. Res.* **67**:1–16.

Brock, G. A., and Cooper, B. J., 1993, Shelly fossils from the Early Cambrian (Toyonian) Wirrealpa, Aroona Creek, and Ramsay limestones of South Australia, *J. Paleont.* **67**:758–787.

Bruggemann, J. S., 1995, *Parrotfish Grazing on Coral Reefs: A Trophic Novelty*, Rijksuniversitet Groningen, Utrecht.

Chafetz, H. S., 1973, Morphological evolution of Cambrian algal mounds in response to a change in depositional environment, *J. Sediment. Petrol.* **43**:435–446.

Church, S. B., 1991, A new lower Ordovician species of *Calathium*, and skeletal structure of western Utah calathids, *J. Paleont.* **65**:602–610.

Clements, F. E., 1916, Plant succession: An analysis of the development of vegetation. *Carnegie Inst. Publ.* **242**:1–512.

Coniglio, M., and James, N. P., 1985, Calcified algae as sediment contributors to early Paleozoic limestones: Evidence from deep-water sediments of the Cow Head Group, western Newfoundland, *J. Sediment. Petrol.* **55**:746–754.

Connell, J. H., 1978, Diversity in tropical rain forests and coral reefs, *Science* **199**:1302–1310.

Cook, H. E., and Taylor, M. E., 1977, Comparison of continental slope and shelf environments in the Upper Cambrian and lowest Ordovician of Nevada, in: *Deep-Water Carbonate Environments* (H. E. Cook and P. Enas, eds.), *Soc. Econ. Paleontol. Mineral. Spec. Publ.* **25**:51–81.

Copper, P., 1974, Structure and development of early Palaeozoic reefs, in: *Proceedings Second International Coral Reef Symposium*, Great Barrier Reef Committee, Brisbane, pp. 365–386.

Copper, P., 1988, Ecological succession in Phanerozoic reef ecosystems: Is it real? *Palaios* **3**:136–152.

Copper, P., 1994, Ancient reef ecosystem expansion and collapse, *Coral Reefs* **13**:3–11.

Copper, P., 1997, Reefs and carbonate productivity: Cambrian through Devonian, *Proc. 8th Int. Coral Reef Symp.* **2**:1623–1630.

Courjault-Radé, P., 1988, Analyse sédimentologique de la formation de l'Orbiel ("alternances gréso-calcaires" auct., Cambrien inférieur). Evolution tectono-sédimentaire et climatique (versant sud de la montagne Noire, Massif central, France), *Bull. Soc. géol. France 8e Sér.* **4**:1003–1013.

Courjault-Radé, P., Debrenne, F., and Gandin, A., 1992, Paleogeographic and geodynamic evolution of the Gondwana continental margins during the Cambrian, *Terra Nova* **4**:657–667.

Cowen, R., 1988, The role of algal symbiosis in reefs through time, *Palaios* **3**:221–227.

Debrenne, F., 1959, Récifs, biohermes ou bancs fossilifères d'Archaeocyatha, *Bull. Soc. géol. France 7e Sér.* **1**:393–395.

Debrenne, F., 1975, Formations organogènes du Cambrien inférieur du Maroc, in: *Iskopaemye Cnidaria*, T. 2 [Fossil Cnidaria, Vol. 2] (B. S. Sokolov, ed.), *Tr. Inst. Geol. Geofiz. Sibirsk. otd. Akad. Nauk SSSR* **202**:19–24.

Debrenne, F., and James, N. P., 1981, Reef-associated archaeocyathans from the Lower Cambrian of Labrador and Newfoundland, *Palaeontology* **24**:343–378.

Debrenne, F., and Zhuravlev, A., 1992, *Irregular Archaeocyaths*, Cahiers de Paléontologie, CNRS Editions, Paris.

Debrenne, F., and Zhuravlev, A., 1994, Archaeocyathan affinity: How deep can we go into the systematic affiliation of an extinct group? in: *Sponges in Time and Space: Biology, Chemistry, Paleontology* (R. W. M. Van Soest, T. M. G. Van Kempen, and J. C. Braekman, eds.), Balkema, Rotterdam, pp. 3–12.

Debrenne, F., and Zhuravlev, A. Yu., 1996, Archaeocyatha, palaeoecology: A Cambrian sessile fauna, *Bol. Soc. Paleont. Ital. Spec. Vol.* **3**:77–85.

Debrenne, F., and Zhuravlev, A. Yu., 1997, Cambrian food web: A brief review, *Géobios, Mém. Spéc.* **20**:181–188.

Debrenne, F., Gandin, A., and Pillola, G. L., 1989a, Biostratigraphy and depositional setting of Punta Manna Member type-section (Nebida Formation, Lower Cambrian, SW Sardinia, Italy), *Riv. It. Paleont. Strat.* **94**:483–514.

Debrenne, F., Gandin, A., and Rowland, S. M., 1989b, Lower Cambrian bioconstructions in northwestern Mexico (Sonora). Depositional setting, paleoecology and systematics of archaeocyaths, *Géobios* **22**:137–195.

Debrenne, F., Gandin, A., and Gangloff, R. A., 1990, Analyse sédimentologique et paléontologie de calcaires organogènes du Cambrien inférieur de Battle Mountain (Nevada, U.S.A.), *Ann. Paléont. (Vert.-Invert.)* **76**:73–119.

Debrenne, F., Gandin, A., and Zhuravlev, A., 1991, Palaeoecological and sedimentological remarks on some Lower Cambrian sediments of the Yangtze platform (China), *Bull. Soc. géol. Fr.* **162**:575–584.

Debrenne, F., Gandin, A., and Debrenne, M., 1993, Calcaires à archéocyathes du Membre de la Vallée de Matoppa (Formation de Nebida), Cambrien inférieur du sud-ouest de la Sardaigne (Italie), *Ann. Paléont. (Vert.-Invert.)* **79**:77–118.

Dronov, A. V., and Fedorov, P. V., 1994, Novye dannye o stroenii i rasprostranenii gekkerovykh gorbov v nizhneordovikskikh karbonatnykh otlozheniyak okrestnostey Sankt-Petersburga [New data on the structure and distribution of Hecker's humps in the Lower Ordovician carbonate strata of St. Petersburg vicinity], *Vestn. St. Petersburg. Univ. Ser. 7: Geol. Geogr.* **2**:89–93 [in Russian].

Droser, M. L., 1991, Ichnofabric of the Paleozoic *Skolithos* ichnofacies and the nature and distribution of *Skolithos* piperock, *Palaios* **6**:316–325.

Drosdova, N. A., 1980, Vodorosli v organogennykh postroykakh nizhnego kembriya Zapadnoy Mongolii [Algae in the Lower Cambrian Organogenous Buildups of Western Mongolia], *Tr. Sovm. Sovet.-Mongol. Paleont. Eksp.* **10**:1–140 [in Russian].

Fagerstrom, J. A., 1987, *The Evolution of Reef Communities*, John Wiley and Sons, New York.

Farmer, J. D., 1992, Grazing and bioturbation in modern microbial mats, in: *The Proterozoic Biosphere: A Multidisciplinary Study* (W. J. Schopf and C. Klein, eds.), Cambridge University Press, Cambridge, England, pp. 295–297.

Flügel, S. M., 1991, Triassic and Jurassic marine calcareous algae: A critical review, in: *Calcareous Algae and Stromatolites* (R. Riding, ed.), Springer-Verlag, Berlin, pp. 481–503.

Flügel, S. M., and Senowbari-Daryan, B., 1996, Evolution of Triassic reef biota: State of the art, in: *Globale und Regionale Steuerungsfaktoren Biogener Sedimentation. I. Riff-Evolution. DFG-Schwerpunktprogr* (J. Reitner, F. Neuweiler, and F. Gunkel, eds.), *Göttinger Arbeiten Geol. Palont. Sonderband* **2**:285–294.

Gandin, A., and Debrenne, F., 1984, Lower Cambrian bioconstructions in southwestern Sardinia (Italy), *Géobios, Mém. Spéc.* **8**:231–240.

Gangloff, R. A., 1976, Archaeocyatha of eastern California and western Nevada, in: *Depositional Environments of Lower Paleozoic Rocks in the White-Inyo Mountains, Inyo County, California, Pacific Coast Paleogeography Field Guide* 1 (J. N. Moore and A. E. Fritsche, eds.), Pacific Sec., SEPM, Los Angeles, CA, pp. 19–30.

Garrett, P., 1970, Phanerozoic stromatolites: Noncompetitive ecologic restriction by grazing and burrowing animals, *Science* **169**:171–173.

Geldsetzer, H. H. J., James, N. P., and Tebbutt, E., 1989, Reefs, Canada and adjacent area, *Can. Soc. Petrol. Geologists Mem.* **13**.

Geyer, G., Landing, E., and Heldmaier, W., 1995, Faunas and depositional environments of the Cambrian of the Moroccan Atlas regions, *Beringeria Spec. Issue* **2**:47–119.

Glumac, B., and Walker, K. R., 1997, Selective dolomitization of Cambrian microbial carbonate deposits: A key to mechanisms and environments of origin, *Palaios* **12**:98–110.

Grant, S. W. F., Knoll, A. H., and Germs, G. J. B., 1991, Probable calcified metaphytes in the latest Proterozoic Nama Group, Namibia: Origin, diagenesis and implications, *J. Paleont.* **65**:1–18.

Grime, J. P., 1979, *Plant Strategies and Vegetation Processes*, John Wiley and Sons, Chichester.

Grotzinger, J. P., and Knoll, A. H., 1995, Anomalous carbonate precipitates: Is the Precambrian the key to the Permian? *Palaios* **10**:578–596.

Grotzinger, J. P., Bowring, S. A., Saylor, B. Z., and Kaufman, A. J., 1995, Biostratigraphic and geochronologic constraints on early animal evolution, *Science* **270**:598–604.

Hallock, P., 1988, The role of nutrient availability in bioerosion: Consequences to carbonate buildups, *Palaeogeogr. Palaeoclimatol., Palaeoecol.* **66**:275–291.

Hamdi, B., Rozanov, A. Yu., and Zhuravlev, A. Yu., 1995, Latest Middle Cambrian metazoan reef from northern Iran, *Geol. Mag.* **132**:367–373.

Heckel, P. H., 1974, Carbonate buildups in the geologic record: a review, in: *Reefs in Time and Space* (L. F. Laporte, ed.), *Soc. Econ. Paleontol. Mineral. Spec. Publ.* **18**:90–154.

Hillmer, G., and Scholz, J., 1996, Structure and dynamics of bryozoan communities and microbial mats, in: *Globale und Regionale Steuerungsfaktoren Biogener Sedimentation. I. Riff-Evolution. DFG-Schwerpunktprogr.* (J. Reitner, F. Neuweiler, and F. Gunkel, eds.), *Göttinger Arbeiten Geol. Paläont. Sonderband* **2**:53–57.

Hoffman, A., and Narkiewicz, M., 1977, Developmental pattern of Lower to Middle Paleozoic banks and reefs, *N. Jb. Geol. Paläont. Mh.* **5**:272–283.

Hyatt, A., 1885, Cruise of Arethusa (Letters to the editor), *Science* **7**:384–386.

Ilan, M., and Loya, Y., 1988, Reproduction and settlement of the coral reef sponge *Niphates* sp. (Red Sea), *Proc. 6th Intern. Coral Reef Symp., Australia* **2**:745–749.

James, N. P., 1981, Megablocks of calcified algae in the Cow Head Breccia, western Newfoundland: Vestiges of a Cambro-Ordovician platform margin, *Geol. Soc. Am. Bull. Pt I* **92**:799–811.

James, N. P., and Gravestock, D. I., 1990, Lower Cambrian shelf and shelf margin buildups, Flinders Ranges, South Australia, *Sedimentology* **37**:455–480.

James, N. P., and Klappa, C. F., 1983, Petrogenesis of Early Cambrian reef limestone, Labrador, Canada, *J. Sediment. Petrol.* **53**:1051–1096.

James, N. P., and Kobluk, D. R., 1978, Lower Cambrian patch reefs and associated sediments: Southern Labrador, Canada, *Sedimentology* **25**:1–35.

Jazmir, M. M., 1960, O prirode nizhnekembriyskikh biogermov poberezh'ya srednego techeniya r. Aldana [On the nature of Lower Cambrian bioherms from the banks of the Aldan River middle courses], *Uchenye Zapiski Saratov. Gos. Univ. Vyp. Geol.* **74**:157–166.

Kennard, J. M., 1991, Lower Cambrian archaeocyathan buildups, Todd River Dolomite, northeast Amadeus Basin, central Australia: Sedimentology and diagenesis, in: *Geological and Geo-*

physical Studies in the Amadeus Basin, Central Australia (R. J. Korsch, and J. M. Kennard, eds.), *Bull. Bur. Miner. Res.* **236**:195–225.

Kennard, J. M., and James, N. P., 1986, Thrombolites and stromatolites: Two distinct types of microbial structures, *Palaios* **1**:492–503.

Knoll, A. H., Fairchild, I. J., and Swett, K., 1993, Calcified microbes in Neoproterozoic carbonates: Implications for our understanding of the Proterozoic/Cambrian transition, *Palaios* **8**:512–525.

Kobluk, D. R., 1981, Lower Cambrian cavity-dwelling endolithic (boring) sponges, *Can. J. Earth Sci.* **18**:972–980.

Kobluk, D. R., 1985, Biota preserved within cavities in Cambrian *Epiphyton* mounds, Upper Shady Dolomite, southwestern Virginia, *J. Paleont.* **59**:1158–1172.

Kobluk, D. R., and James, N. P., 1979, Cavity-dwelling organisms in Lower Cambrian patch reefs from southern Labrador, *Lethaia* **12**:193–218.

Kosmynin, V. N., and Hecker, M. R., 1997, Suktsessii na rifakh [Successions on reefs], in: *Rifogennye Postroyki v Paleozoie Rossii* [Reefogenous Buildups in the Paleozoic of Russia] (B. S. Sokolov and A. B. Iwanowsky, eds.), Nauka, Moscow, pp. 117–136 [in Russian].

Kruse, P. D., 1991, Cyanobacterial-archaeocyathan-radiocyathan bioherms in the Wirrealpa Limestone of South Australia, *Can. J. Earth Sci.* **28**:601–615.

Kruse, P. D., Zhuravlev, A. Yu., and James, N. P., 1995, Primordial metazoan-calcimicrobial reefs: Tommotian (Early Cambrian) of the Siberian Platform, *Palaios* **10**:291–321.

Kruse, P. D., Gandin, A., Debrenne, F., and Wood, R., 1996, Early Cambrian bioconstructions in the Zavkhan Basin of western Mongolia, *Geol. Mag.* **133**:429–444.

Landing, E., 1996, Avalon: Insular continent by the latest Precambrian, *Geol. Soc. Am. Spec. Pap.* **304**:29–63.

Lasemi, Z., and Sandberg, P. A., 1996, Micrites as potential clues to changes in Phanerozoic ocean-atmosphere composition, in: *Carbonates and Global Change: An Interdisciplinary Approach*, Wildhaus, Switzerland, pp. 90–91.

Loch, J. D., and Taylor, J. F., 1997, Trilobite biofacies associated with microbialitic bioherms from the Upper Cambrian Ore Hill Member of the Gatesburg Formation in south-central Pennsylvania, in: *2nd International Trilobite Conference*, Brock University, St. Catharines, Ontario, pp. 34–35.

Lohmann, K. C., 1976, Lower Dresbachian (Upper Cambrian) platform to deep-shelf transition in eastern Nevada and western Utah: An evaluation through lithologic cycle correlation, *Brigham Young Univ. Geol. Stud.* **23**:111–122.

Lowenstam, H. A., 1950, Niagaran reefs of the Great Lakes area, *J. Geol.* **58**:430–487.

Luchinina, V. A., 1985, Vodoroslevye postroyki rannego paleozoya Sibirskoy platformy [Algal buildups of the early Paleozoic of the Siberian Platform], in: *Sreda i zhizn' v Geologicheskom Proshlom: Paleobasseyny i ikh Obitateli* [Environment and Life in the Geological Past: Paleobasins and Their Inhabitants] (O. A. Betekhtina, and I. T. Zhuravleva, eds.), *Tr. Inst. Geol. Geofiz. Sibirsk. Otd. Akad. Nauk SSSR* **628**: 45–49 [in Russian].

Markello, J. R., and Read, J. F., 1981, Carbonate ramp-to-deeper shale transitions of an Upper Cambrian intrashelf basin, Nolichucky Formation, Southwest Virginia Appalachians, *Sedimentology* **28**:573–597.

Mattes, B. W., and Conway Morris, S., 1990, Carbonate/evaporate deposition in the Late Precambrian–Early Cambrian Ara Formation of Southern Oman, in: *The Geology and Tectonics of the Oman Region* (A. H. F. Robertson, M. P. Searl and A. C. Ries, eds.), *Geol. Soc. Spec. Publ.* **49**:617–636.

Mel'nikov, N. V., Kilina, L. I., Krinin, V. A., and Khomenko, A. V., 1991, Neftegazonosnost' kembriyskikh rifov Suringdakonskogo svoda [Oil–gas capacity of the Cambrian reefs of the Suringdakon Arch], in: *Teoreticheskie i Regional'nye Problemy Geologii Nefti i Gaza* [Theoretical and Regional Problems in the Oil and Gas Geology] (V. S. Sukhov and A. E. Kontorovich, eds.), Nauka, Novosibirsk, pp. 180–189 [in Russian].

Monty, C. L. V., 1973, Precambrian background and Phanerozoic history of stromatolitic communities, an overview, *Ann. Soc. géol. Belg.* **96**:585–624.

Moreno-Eiris, E., 1987, Los monticulos arrecifales de Algas y Arqueociatos del Cámbrico Inferior de Sierra Morena. III: Microfacies y diagénesis, *Bol. Geol. Minero* **98**:591–621.

Morgan, N., 1976, The Montenegro bioherms: Their paleoecology, relation to other archaeocyathid bioherms and to Early Cambrian sedimentation in the White and Inyo Mountains, California, in: *Depositional Environments of Lower Paleozoic Rocks in the White-Inyo Mountains, Inyo County, California, Pacific Coast Paleogeography Field Guide 1* (J. N. Moore and A. E. Fritsche, eds.), Pacific Sec., SEPM, Los Angeles, CA, pp. 13–17.

Myrow, P. M., 1995, *Thalassinoides* and the enigma of Early Paleozoic open-framework burrow systems, *Palaios* **10**:58–74.

Newell, N. D., 1972, The evolution of reefs, *Sci. Am.* **226**:54–65.

Nikitin, I. F., Gnilovskaya, M. B., Zhuravleva, I. T., Luchinina, V. A., and Miagkova, E. I., 1974, Anderkenskaya biogermnaya gryada i istoriya ee obrazovaniya [The Anderken Bioherm Belt and the history of its development], in: *Sreda i Zhizn' v Geologicheskom Proshlom: Problemy Paleoekologii* [Environment and Life in the Geological Past: Problems of Paleoecology] (O. A. Betekhtina and I. T. Zhuravleva, eds.), *Tr. Inst. Geol. Geofiz. Sibirsk. Otd. Akad. Nauk SSSR* **169**:122–158 [in Russian].

Nitecki, M. H., and Mutvei, H., 1996, Nature and mineralogy of receptaculitid skeleton, *Bull. Inst. Océanogr. Monaco No. Spéc.* **14**:287–294.

Pegel, T. V., 1982, Kharakter raspredeleniya trilobitovykh soobshchestv v Diringdinskom rifovom massive (kembriy Yugo-Zapadnogo Prianabar'ya) [The character of distribution of trilobite communities in the Diringde Reef Massif (Cambrian of the Southwestern Prianabar'e)], in: *Stratigrafiya i Fatsii Osadochykh Basseynov Sibiri* [Stratigraphy and Facies in Sedimentary Basins of Siberia] (V. A. Astashkin, ed.), Sib. Sci.-Res. Inst. Geol., Geophys. Miner. Res., Novosibirsk, pp. 82–89 [in Russian].

Pillola, G. L., 1996, The trilobite *Giordanella* Bornemann, 1891 from Lower Cambrian of Sardinia, (Italy): A discussion on its morphology and possible mode of life, *Bull. Soc. Paleont. Ital.* (Spec. Vol.) **3**:145–158.

Playford, P. E., and Cockbain, A. E., 1989, Devonian reef complexes, Canning Basin, Western Australia: A review, in: *Fossil Cnidaria 5* (P. A. Jell and J. W. Pickett, eds.), *Mem. Assoc. Austral. Palaeontol.* **8**:401–412.

Pratt, B. R., 1989, Deep-water *Girvanella–Epiphyton* reef on a mid-Cambrian continental slope, Rockslide Formation, Mackenzie Mountains, Northwest Territories, in: *Reefs, Canada and Adjacent Area* (H. H. J. Geldsetzer, N. P. James, and E. Tebbutt, eds.), *Can. Soc. Petrol. Geol. Mem.* **13**:161–164.

Pratt, B. R., 1991, Lower Cambrian reefs of the Mural Formation, southern Canadian Rocky Mountains, in: *Geol. Assoc. Canada, Annu. Meeting* **16**:102.

Pratt, B. R., 1995, The origin, biota and evolution of deep-water mud-mounds, *Spec. Publ. Int. Assoc. Sediment.* **23**:49–123.

Ramenskiy, L. G., 1935, O printsipial'nykh ustanovkakh, osnovnykh ponyatiyakh i terminakh proizvodstvennoy tipologii zemel', geobotaniki i ekologii [On the principal orientations, main concepts, and terms of the production typology of soils, geobotany, and ecology], *Sovet. Botanika* **4**:25–42 [in Russian].

Read, B. C., 1980, Lower Cambrian archaeocyathid buildups, Pelly Mountains Yukon, *Geol. Surv. Can. Pap.* **78–18**:1–54.

Read, J. F., and Pfeil, R. W., 1983, Fabrics of allochthonous reefal blocks, Shady Dolomite (Lower to Middle Cambrian), Virginia Appalachians, *J. Sediment. Petrol.* **53**:761–778.

Rees, M. N., and Robison, R. A., 1989, Cambrian stratigraphy and paleontology of the central House Range and Drum Mountains, Utah, in: *Cambrian and Early Ordovician Stratigraphy and Paleontology of the Basin and Range Province, Western United States, Field Trip Guidebook T125* (M. E. Taylor, ed.), American Geophysical Union, Washington, DC, pp. 59–72.

Rees, M. N., Pratt, B. R., and Rowell, A. J., 1989, Early Cambrian reefs, reef complexes, and associated lithofacies of the Shackleton Limestone, Transantarctic Mountains, *Sedimentology* **36**:341–361.

Reitner, J., and Neuweiler, F., 1995, Mud mounds: A polygenetic spectrum of fine-grained carbonate buildups, *Facies* **32**:1–70.

Repina, L. N., 1977, Biofatsii trilobitov tarynskogo urovnya nizhnego kembriya Sibiriskoy platformy [Trilobite biofacies of the Lower Cambrian Taryn level on the Siberian Platform], in:

Sreda i Zhizn' v Geologicheskom Proshlom: Fatsii i Organizmy [Environment and Life in the Geological Past: Facies and Organisms] (O. A. Betekhtina and I. T. Zhuravleva, eds.), *Tr. Inst. Geol. Geofiz. Sibirsk. Otd. Akad. Nauk SSSR* **302:**51–74 [in Russian].

Repina, L. N., and Zharkova, T. M. 1974, Ob usloviyakh obitaniya trilobitov v rannekembriyskom basseyne Sibiri [On the environmental conditions of trilobites in the Early Cambrian basin of Siberia], in: *Sreda i Zhizn' v Geologicheskom Proshlom: Problemy Paleoekologii* [Environment and Life in the Geological Past: Problems of Paleoecology] (O. A. Betekhtina and I. T. Zhuravleva, eds.), *Tr. Inst. Geol. Geofiz. Sibirsk. Otd. Akad. Nauk SSSR* **84:**100–108 [in Russian].

Riding, R., 1991, Classification of microbial carbonates, in: *Calcareous Algae and Stromatolites* (R. Riding, ed.), Springer-Verlag, New York, pp. 21–51.

Riding, R., 1992, Temporal variation in calcification in marine cyanobacteria, *J. Geol. Soc. Lond.* **149:**979–989.

Riding, R., and Zhuravlev, A. Yu., 1995, Structure and diversity of oldest sponge–microbe reefs: Lower Cambrian, Aldan River, Siberia, *Geology* **23:**649–652.

Riding, R., Awramik, S. M., Winsborough, B. M., Griffin, K. M., and Dill, R. F., 1991, Bahamian giant stromatolites: Microbial composition of surface mats, *Geol. Mag.* **128:**227–234.

Rosen, B. R., 1981, The tropical high diversity enigma—the corals'-eye view, in: *Chance, Change and Challenge. The Evolving Biosphere* (P. L. Forey, ed.), British Museum (Nat. Hist.) and Cambridge University Press, London, pp. 103–129.

Rowland, S. M., 1984, Were there framework reefs in the Cambrian? *Geology* **12:**181–183.

Rowland, S. M., and Gangloff, R. A., 1988, Structure and paleoecology of Lower Cambrian reefs, *Palaios* **3:**111–135.

Rowland, S. M., and Savarese, M., 1988, The hypothesis of algal symbiosis in Archaeocyatha, *Geol. Soc. Am.* **20:**226.

Rozanov, A. Yu., 1979, Nekotorye problemy izucheniya drevneyshikh skeletnykh organizmov [Some problems in study of ancient skeletal organisms], *Byull. Mosk. Obshch. Ispyt. Prir. Otd. Geol.* **54**(3):62–69 [in Russian].

Rozanov, A. Yu., and Sayutina, T. A., 1982, Microstructures of certain algae, archaeocyathids and cribricyathids, *Proc. III N. Am. Paleont. Convent.* **2:**453–455.

Runnegar, B., Pojeta, J., Jr., Taylor, M. E., and Collins, D., 1979, New species of the Cambrian and Ordovician chitons *Matthevia* and *Chelodes* from Wisconsin and Queensland: Evidence for the early history of polyplacophoran mollusks, *J. Paleont.* **53:**1374–1394.

Sandberg, P. A., 1983, An oscillating trend in Phanerozoic non-skeletal carbonate mineralogy, *Nature* **305:**19–22.

Savarese, M., 1995, Functional significance of regular archaeocyathan central cavity diameter: A biomechanical and paleoecological test, *Paleobiology* **21:**356–378.

Savarese, M., Mount, J. F., Sorauf, J. E., and Bucklin, L., 1993, Paleobiologic and paleoenvironmental context of coral-bearing Early Cambrian reefs: Implications for Phanerozoic reef development, *Geology* **21:**917–920.

Schmitt, M., and Monninger, W., 1977, Stromatolites and thrombolites in Precambrian/Cambrian boundary beds of the Anti-Atlas, Morocco: Preliminary results, in: *Fossil Algae* (E. Flügel, ed.), Springer-Verlag, Berlin, pp. 80–85.

Schubert, J. K., and Bottjer, D. J., 1992, Early Triassic stromatolites as post-mass extinction disaster forms, *Geology* **20:**883–886.

Scrutton, C. T., 1997, The Palaeozoic corals, I: origins and relationships, *Proc. Yorkshire Geol. Soc.* **51:**177–208.

Semikhatov, M. A., Komar, Vl. A., and Serebryakov, S. N., 1970, Yudomskiy kompleks stratotipicheskoy mestnosti [The Yudoma Complex of the Stratotype Area], *Tr. Geol. Inst. Akad. Nauk SSSR* **210:**1–207 [in Russian].

Shabanov, Yu. Ya., Astashkin, V. A., Pegel, T. V., Egorova, L. I., Zhuravleva, I. T., Pelman, Yu. L., Sundukov, V. M., Stepanova, M. V., Sukhov, S. S., Fedorov, A. B., Shishkin, B. B., Vaganova, N. V., Ermak, V. I., Ryabukha, K. V., Yadrenkina, A. G., Abaimova, G. P., Lopushinskaya, T. V., Sychev, O. V., and Moskalenko, T. A., 1987, *Nizhniy Paleozoy Yugo-Zapadnogo Sklona Anabarskoy Anteklizy (po materialam bureniya)* [The Lower Paleozoic of the Southwestern

Slope of the Anabar Anteclise (According to Borehole Data], Nauka, Novosibirsk [in Russian].

Shishkin, B. B., Pegel, T. V., and Fedorov, A. B., 1978, Stroenie sredne-verkhnekembriyskogo rifogenno-akkumulyativnogo kompleksa yugo-zapadnogo Prianabar'ya [The structure of the Middle-Upper Cambrian reef-accumulative complex in the southwestern Prianabar'e], in: *Aktual'Nye Voprosy Regional'Noy Geologii Sibiri* [Actual Problems of the Regional Geology of Siberia] (V. I. Krasnov, ed.), *Tr. Sibirsk. Nauch.-Issled. Inst. Geol. Geofiz. Miner. Syr'ya* **258:** 34–44 [in Russian].

Shunula, J. P., and Ndibalema, V., 1986, Grazing preferences of *Diadema setosum* and *Heliocidaris erythrogramma* (echinoderms) on an assortment of marine algae, *Aquatic Botany* **25:**91–95.

Spencer, L. M., 1981, Paleoecology of a Lower Cambrian archaeocyathid interreef fauna from southern Labrador, in: *Short Papers for the Second International Symposium on the Cambrian System* (M. E. Taylor, ed.), *U.S. Geol. Surv. Open-File Rep.* **81-743:**215–218.

Spincer, B. R., 1996, The palaeoecology of some Upper Cambrian microbial-sponge-eocrinoid reefs, central Texas, *Paleont. Soc. Spec. Publ.* **18:**367.

Stanley, G. D., Jr., and Swart, P. K., 1995, Evolution of coral-zooxanthellae symbiosis during the Triassic: A geochemical approach, *Paleobiology* **21:**179–199.

Stenek, R. S., 1986, The ecology of coralline algal crust: Convergent pattern and adaptive strategies, *Annu. Rev. Ecol. Syst.* **17:**273–303.

Stepanova, M. V., 1986, Zavisimost' sistematicheskogo sostava vodoroslevykh soobshchestv na primere nizhnego kembriya stratotipicheskogo razreza Sibirskoy platformy [Dependence of the taxonomic composition of algal communities on facies environments on the example of the Lower Cambrian stratotype section of the Siberian Platform], in: *Paleoekologicheskiy i Litologo-Fatsial'nyy Analizy Dlya Obosnovaniya Detal'Nosti Stratigraficheskikh Shkal* [Paleoecological and Lithological-Facies Analyses for the Grounds of Scrutiny of the Stratigraphic Charts] (V. D. Krasnov, ed.), Sibirsk. Sci.-Res. Inst. Geol., Geophys. Miner. Res., Novosibirsk, pp. 22–30 [in Russian].

Stinchcomb, B. L., 1975, Paleoecology of two new species of Late Cambrian *Hypseloconus* (Monoplacophora) from Missouri, *J. Paleont.* **49:**416–421.

Stitt, J. H., 1976, Functional morphology and life habits of the Late Cambrian trilobite *Stenopilus pronus* Raymond, *J. Paleont.* **50:**561–576.

Sukhov, S. S., and Pegel, T. V., 1986, Lito-i biofatsial'nyy analiz srednekembriyskikh otlozheniy vostoka Sibirskoy platformy dlya rekonstruktsii usloviy karbonatonakopleniya [Lithological and biofacies analysis of the Middle Cambrian strata on the East of the Siberian Platform for the carbonate deposition environment reconstruction], in: *Paleoekologicheskiy i Litologo-Fatsial'nyy Analizy Dlya Obosnovaniya Detal'Nosti Stratigraficheskikh Shkal* [Paleoecological and Lithological-Facies Analyses for the Grounds of Scrutiny of the Stratigraphic Charts] (V. D. Krasnov, ed.), Sibirsk. Sci.-Res. Inst. Geol., Geophys. Miner. Res., Novosibirsk, pp. 35–50 [in Russian].

Surge, D. M., Savarese, M., Dodd, J. R., and Lohmann, K. C., 1997, Carbon isotopic evidence for photosynthesis in Early Cambrian oceans, *Geology* **25:**503–506.

Sychev, O. V., 1986, Paleogeografiya rannego ordovika Sibirskoy platformy [Early Ordovician paleogeography of the Siberian Platform], in: *Paleoekologicheskiy i Litologo-Fatsial'nyy Analizy Dlya Obosnovaniya Detal'nosti Stratigraficheskikh Shkal* [Paleoecological and Lithological-Facies Analyses for the Grounds of Scrutiny of the Stratigraphic Charts] (V. D. Krasnov, ed.), Sibirsk. Sci.-Res. Inst. Geol., Geophys. Miner. *Res.*, Novosibirsk, pp. 50–58 [in Russian].

Talent, J. A., 1988, Organic reef-building: episodes of extinction and symbiosis? Senkenberg, *Lethaia* **69:**315–368.

Thayer, C. W., 1983, Sediment-mediated biological disturbance and the evolution of marine benthos, in: *Biotic Interactions in Recent and Fossil Benthic Communities* (M. J. S. Tevesz and P. L. McCall, eds.), Plenum Press, New York, pp. 479–625.

Toomey, D. F., and Nitecki, M. H., 1979, Organic buildups in the Lower Ordovician (Canadian) of Texas and Oklahoma, *Fieldiana New Ser.* **2:**1–181.

Tucker, R. D., and McKerrow, W. S., 1995, Early Paleozoic chronology: A review in light of new U-Pb zircon ages from Newfoundland and Britain, *Can. J. Earth Sci.* **32:**368–379.

Turner, E. C., James, N. P., and Narbonne, G. M., 1997, Growth dynamics of Neoproterozoic calcimicrobial reefs, Mackenzie Mountains, northwest Canada, *J. Sediment. Petrol.* **67**:437–450.

Usychenko, O. N., 1988, Biofatsial'naya zonal'nost' v nizhnem kembrii Nepsko-Botuobinskoy anteklizy [Biofacies zonation in the Lower Cambrian of the Nepa-Botuoba Anteclise], in: *Izvestkovye Vodorosli i Stromatolity* (sistematika, biostratigrafiya, fatsial'nyy analiz) [Calcareous Algae and Stromatolites (Systematics, Biostratigraphy, Facies Analysis)] (V. N. Dubatolov and T. A. Moskalenko, eds.), Nauka, Novosibirsk, pp. 85–93 [in Russian].

Van Treeck, P., Schuhmacher, H., and Paster, M., 1996, Grazing and bioerosion by herbivorous fishes—key processes structuring coral reef communities, in: *Globale und Regionale Steuerungsfaktoren Biogener Sedimentation. I. Riff-Evolution. DFG-Schwerpunktprogr.* (J. Reitner, F. Neuweiler, and F. Gunkel, eds.), *Göttinger Arbeiten Geol. Paläont. Sonderband* **2**:133–137.

Vermeij, G. J., 1987, *Evolution and Escalation: An Ecological History of Life*, Princeton University Press, Princeton, NJ.

Walker, K. R., and Alberstadt, L. P., 1975, Ecological succession as an aspect of structure in fossil communities, *Paleobiology* **1**:238–257.

Waters, B. B., 1989, Upper Cambrian *Renalcis–Girvanella* framestone mounds, Alberta, in: *Reefs, Canada and Adjacent Area* (H. H. J. Geldsetzer, N. P. James, and E. Tebbutt, eds.), *Can. Soc. Petrol. Geologists Mem.* **13**:165–169.

Webby, B. D., 1999, Early to earliest Late Ordovician reef development, in: *Quo vadis Ordovician? Short Papers of the 8th International Symposium on the Ordovician System* (Prague, June 20–25, 1999) (P. Kraft, and O. Fatka, eds.), *Acta Univ. Carolinae, Geol.* **43**:425–428.

Westrop, S. R., 1989, Facies anatomy of an Upper Cambrian grand cycle: Bison Creek and Mistaya formations, southern Alberta, *Can. J. Earth Sci.* **26**:2292–2304.

Westrop, S. R., 1996, Temporal persistence and stability of Cambrian biofacies: Sunwaptan (Upper Cambrian) trilobite faunas of North America, *Palaeogeogr. Palaeoclimatol. Palaeoecol.* **127**:33–46.

Wickstrom, C. E., and Castenholz, R. W., 1985, Dynamics of cyanobacteria–ostracod interactions in an Oregon hot spring, *Ecology* **66**:1024–1041.

Wilkinson, C. R., and Cheshire, A. C., 1989, Patterns in the distribution of sponge populations across the central Great Barrier Reef, *Coral Reefs* **8**:127–134.

Wilson, M. A., Palmer, T. J., Guensburg, T. E., Finton, C. D., and Kaufman, L. E., 1992, The development of and Early Ordovician hardground community in response to rapid sea-floor calcite precipitation, *Lethaia* **25**:19–34.

Wood, R., 1993, Nutrients, predation and the history of reef-building, *Palaios* **8**:526–543.

Wood, R., 1999, *Reef Evolution*, Oxford University Press, Oxford, England.

Wood, R., Zhuravlev, A. Yu., and Debrenne, F., 1992, Functional biology and ecology of Archaeocyatha, *Palaios* **7**:131–156.

Wood, R., Zhuravlev, A. Yu., and Chimed Tseren, A., 1993, The ecology of Lower Cambrian buildups from Zuune Arts, Mongolia: Implications for early metazoan reef evolution, *Sedimentology* **40**:829–858.

Wray, J. L., and Playford, P. E., 1970, Some occurrences of Devonian reef-building algae in Alberta, *Can. Soc. Petrol. Geol. Bull.* **18**:544–555.

Yochelson, E. L., and Stinchcomb, B. L., 1987, Recognition of *Macluritella* (Gastropoda) from the Upper Cambrian of Missouri and Nevada, *J. Paleont.* **61**:56–61.

Zadorozhnaya, N. M., 1974, Rannekembriyskie organogenny postroyki vostochnoy chasti Altae-Sayanskoy skladchatoy oblasti [Early Cambrian organogenous buildups of the eastern part of the Altay Sayan Foldbelt], *Tr. Inst. Geol. Geofiz. Sibirsk. Otd. Akad. Nauk SSSR* **84**:158–186 [in Russian].

Zadorozhnaya, N. M., Osadchaja, D. V., Zhuravleva, I. T., and Luchinina, V. A., 1973, Rannekembriyskie organogenny postroyki territorii Tuvy (Sayano-Altayskaya skladchataya oblast') [Early Cambrian organogenous buildups on the territory of Tuva (Sayan-Altay Foldbelt)], in: *Sreda i Zhizn' v Geologicheskom Proshlom: Problemy Paleoekologii* [Environment and Life in the Geological Past: Problems of Paleoecology] (O. A. Betekhtina and I. T. Zhuravleva, eds.), *Tr. Inst. Geol. Geofiz. Sibirsk. Otd. Akad. Nauk SSSR* **169:**53–65 [in Russian].

Zamarreño, I., 1977, Early Cambrian algal carbonates in southern Spain, in: *Fossil Algae* (E. Flügel, ed.), Springer-Verlag, Berlin, pp. 360–365.

Zherikhin, V. V., 1997, Phylogenesis and phylocoenogenesis, in: *Evolution of the Biosphere* (A. Yu. Rozanov, P. Vickers-Rich, and C. Tassell, eds.), *Rec. Queen Victoria Mus. Art Gallery, Launceston* **104:**57–63.

Zhuravlev, A. Yu., 1993, A functional-morphological approach to the biology of the Archaeocyatha, *N. Jb. Geol. Paläont. Abh.* **190:**315–327.

Zhuravlev, A. Yu., 1995, Preliminary suggestions on the global Early Cambrian zonation, *Beringeria Spec. Issue* **2:**147–160.

Zhuravlev, A. Yu., 1996, Reef ecosystem recovery after the Early Cambrian extinction, in: *Biotic Recovery from Mass Extinction Events* (M. B. Hart, ed.), *Geol. Soc. Spec. Publ.* **102:**79–96.

Zhuravlev, A. Yu., 1999, Modul'nost' i stanovlenie kembriyskoy rifovoy ekosistemy [The modularity and development of Cambrian reef ecosystem], *Zh. Obshch. Biol.* **60:**29–40 [in Russian].

Zhuravlev, A. Yu., Biotic diversity and structure during the Neoproterozoic/Ordovician transition, in: *Ecology of the Cambrian Radiation* (A. Yu. Zhuravlev and R. Riding, eds.), Columbia University Press, New York, pp. 173–199.

Zhuravlev, A. Yu., and Gravestock, D. I., 1994, Archaeocyaths from Yorke Peninsula, South Australia and archaeocyathan Early Cambrian zonation, *Alcheringa* **18:**1–54.

Zhuravlev, A. Yu., and Maidanskaya, I. D., 1998, Skhodstvo faun i dinamika plit v rannem kembrii [Faunal similarities and plate tectonics in the Early Cambrian], in: *Paleogeografiya Venda–Rannego Paleozoya Severnoy Evrazii* [The Vendian-Early Paleozoic Paleogeography of Northern Eurasia] (V. A. Koroteev and A. V. Maslov, eds.), Uralian Branch, Russian Academy of Sciences, Ekaterinburg, pp. 166–171 (in Russian).

Zhuravlev, A. Yu., and Wood, R., 1995, Lower Cambrian reefal cryptic communities, *Palaeontology* **38:**443–470.

Zhuravlev, A. Yu., and Wood, R. A., 1996, Anoxia as the cause of the mid-Early Cambrian (Botomian) extinction event, *Geology* **24:**311–314.

Zhuravlev, A. Yu., Hamdi, B., and Kruse, P. D., 1996, IGCP 366: Ecological aspects of Cambrian radiation—Field meeting, *Episodes* **19:**136–137.

Zhuravleva, I. T., 1966, Rannekembriyskie organogenny postroyki territorii Sibirskoy platformy [Early Cambrian organogenous buildups on the territory of the Siberian Platform], in: *Sreda i Zhizn' v Geologicheskom Proshlom* [Organism and Environment in the Geological Past] (R. T. Hecker, ed.), Nauka, Moscow, pp. 61–84 [in Russian].

Zhuravleva, I. T., and Zelenov, K. K., 1955, Biogermy pestrotsvetnoy svity reki Leny [Bioherms from the Pestrotsvet Formation of the Lena River], *Tr. Paleont. Inst. Akad. Nauk SSSR* **56:**57–78 [in Russian].

Chapter 5

Biologically Induced Carbonate Precipitation in Reefs through Time

GREGORY E. WEBB

1. Introduction

Modern reefs are constructed largely by scleractinian corals and coralline red algae. However, through geological time, reef-building communities have varied in terms of biotic composition, community structure, and the mechanisms of reef construction. Entire groups of organisms that do not build reefs today were prominent reef builders in the past. Most studies of reef history

GREGORY E. WEBB • School of Natural Resource Sciences, Queensland University of Technology, Brisbane, QLD 4000 Australia.

The History and Sedimentology of Ancient Reef Systems, edited by George D. Stanley Jr., Kluwer Academic/Plenum Publishers, New York, 2001.

have emphasized the role of skeletal organisms in reef building (Newell, 1972; James, 1983; Fagerstrom, 1987; James and Bourque, 1992; Kauffman and Fagerstrom, 1993), but for most of geological time (i.e., Precambrian time) reefs lacked skeletal organisms altogether (Grotzinger, 1989c), and even Phanerozoic reefs that contained skeletal reef builders typically also contained nonskeletal constructional fabrics (Heckel, 1974; Pratt, 1982a,b; Webb, 1996). Nonskeletal (nonenzymatic *sensu* Webb, 1996) constructional reef fabrics result predominantly from biologically induced (*sensu* Lowenstam, 1981) carbonate precipitation and include microbialites and biologically localized marine cements.

Reefs generally are defined as autochthonous, biologically constructed, rigid structures that are capable of maintaining relief above the surrounding seafloor, even in the face of high energy conditions. The synsedimentary rigidity and relief that characterize reefs are here considered functions of reef framework. Reef framework commonly is defined in terms of large and/or densely packed, intergrown skeletal organisms, and some authors restrict the reef framework concept entirely to skeletal metazoan structures (e.g., Pickard, 1996). However, synsedimentary relief and rigidity that result from biological activity are here considered the defining attributes of reef framework; skeletons are not requisite. Therefore, rigid, relief-bearing nonenzymatic carbonates must be included in the reef framework concept (Grotzinger, 1989b,c; Webb, 1996), and buildups that meet all of the requirements for consideration as reefs, except that they are dominated by nonenzymatic frameworks rather than by skeletal metazoans (e.g., "reef mounds": James, 1983; "biogenic mounds": James and Bourque, 1992; "framework-free reefs": Kuznetsov, 1996), are here considered reefs. Recognition of the role played by biologically induced precipitation in reefs is important, not just for semantic reasons, but because nonenzymatic carbonates have major implications for the way we recognize and view ancient reefs and reef history.

Modern coral reefs are restricted to shallow, tropical, oligotrophic waters owing to the ecological requirements of scleractinian corals, their photoautotrophic symbionts, and coralline algae. Ancient reefal organisms were controlled by similar types of parameters, but the physicochemical parameters that controlled the distribution of nonenzymatic carbonates may or may not have had similar recognizable effects on accompanying skeletal biota. The physicochemical parameters that affect nonenzymatic carbonate distribution are related to a complex series of cyclic (e.g., Wilkinson *et al.*, 1985; Veizer, 1994) and unidirectional (e.g., Veizer, 1994; Webb, 1996), evolving tectonic, physiographical, geochemical, and biological systems. The temporal succession of reefs does not represent the long-term evolution of a single metazoan community or set of communities punctuated by occasional extinction-related setbacks. Instead, the geological history of reefs represents an amalgam of related and unrelated communities whose capacity to construct reefs during particular time intervals was controlled by a sequence of overlapping physicochemical and biological parameters that may have created a unique set of reef-building conditions for each interval of earth history (Webb, 1996).

This chapter is a brief survey of the role of nonenzymatic reef framework components in the geological history of reefs. Our very concept of what constitutes a reef community depends partly on factors that govern biologically induced calcification through time. Hence, the long-term temporal succession of reefs cannot make sense without reference to nonenzymatic reef framework components.

2. Biological Induction of Marine Carbonate Precipitation

Carbonate geochemistry and biomineralization have been discussed widely (e.g., Morse and Mackenzie, 1990; Doumenge, 1994). Carbonate precipitation in marine environments represents a continuum of processes that can be divided into two or three convenient categories with more or less indistinct boundaries. Abiotic mineralization, wherein spontaneous precipitation results from ambient physicochemical conditions, differs from biomineralization, where there is a biological influence on precipitation. Biomineralization can be subdivided into enzymatic (biologically controlled) biomineralization, where precipitation is completely controlled by an organism (i.e., skeletogenesis), and nonenzymatic (biologically induced) mineralization, where precipitation is a by-product of the metabolic activity or the presence or decay of an organism or organic matter (Lowenstam, 1981; Mann, 1989) (Fig. 1).

The distinction between enzymatic and nonenzymatic mineralization is relatively simple in the case of matrix-mediated skeletons, wherein crystal morphology and orientation are controlled completely by the organism. However, the exact processes that govern calcification in microbes are poorly understood (Pentecost and Riding, 1986; Riding, 1991, 2000), and enzymatic calcification in some green algae (Borowitzka, 1989) is very similar to biologically induced calcification in the sheaths of some microbes. Hence, the

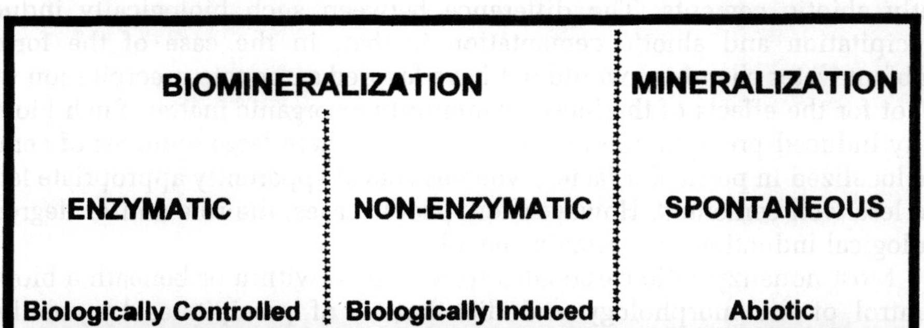

FIGURE 1. Gradational relationship between biomineralization and abiotic mineralization. Nonenzymatic carbonates fall between enzymatic (skeletal) and spontaneous abiotic end members.

boundary between enzymatic and nonenzymatic biomineralization is not distinct (Addadi and Weiner, 1989). Regardless, enzymatic biomineralizers tend to be obligate calcifiers, expending energy to effect carbonate precipitation in some cases even against unfavorable extrinsic physicochemical saturation gradients. Nonenzymatic biomineralizers (inducers) are incapable of calcifying in unfavorable physicochemical settings and are not obligate calcifiers (Riding, 1992). Whether or not an organism is an obligate or nonobligate calcifier may provide an indication of whether the calcification process was enzymatic or nonenzymatic. However, the fossil record is biased against noncalcified organisms; calcified organisms are preserved preferentially.

The distinction between biologically induced carbonate precipitation and abiotic cementation is equally problematic. Where microcrystalline nonenzymatic carbonates occur within or around the community that caused their precipitation, they may preserve biotic morphology (i.e., calcimicrobes) or form localized structures with benthic accretionary morphologies that suggest biological mediation (e.g., stromatolites). A strong case for biological induction can be made in such cases, although controversy may still exist (e.g., Fairchild, 1991; Grotzinger and Rothman, 1996). However, organisms and dead organic matter (e.g., Neuweiler, 1993) mediate water chemistry in a variety of ways and to varying degrees. Organomineralization, wherein precipitation is controlled by nonliving organic macromolecules, has been demonstrated in modern environments (Trichet and Défarge, 1995) and in ancient buildups (Neuweiler *et al.*, 1999). In open, shallow-water settings, where water volumes are large and well mixed, biologically induced precipitation may be confined to organic matter or protected pore space below the inducing community. However, within sediment or cavity systems, fluids may not be well mixed and fluid volumes are smaller. In such conditions (i.e., a relatively closed system), a community might affect the water chemistry so as to favor precipitation in the entire localized fluid volume, and the resulting carbonate precipitation might occur relatively far from the inducing community. Where such precipitates were not directly controlled by a particular organic substrate (living or dead), they might be morphologically indistinguishable from apparently abiotic cements. The difference between such biologically induced precipitation and abiotic cementation is that, in the case of the former, ambient water chemistry would not have favored carbonate precipitation were it not for the effects of the biotic community or organic matter. Such biologically induced precipitation may be indicated where large volumes of cement are localized in particular facies, whereas coeval apparently appropriate facies are less well cemented. However, for specific cases, the role and/or degree of biological induction may remain speculative.

Most nonenzymatic carbonates were formed within or beneath a biofilm. Control of the morphology and distribution of precipitates by calcifying biofilms may allow recognition of the nonenzymatic precipitates. For example, whereas an abiotic precipitate (e.g., a cement crust) may grow concentrically from all surfaces of a cavity, a biologically mediated precipitate may have a heterogeneous distribution because it is restricted to the portion of a cavity where the biofilm occurred. Biofilms consist of thin or thick layers of organic matter that may contain a variety of living and/or dead microbial communities, with or without eukaryotes such as algae and sponges, contained within a matrix of degrading organic matter and mucilage. Community structure may be simple with low diversity or highly complex with diverse communities occupying vertical zones where differing metabolic pathways are favored. Biologically induced calcification of organic matter within a biofilm may preserve recognizable organic structures (e.g., mucilagenous sheaths of cyanobacteria) or may occur as structureless microcrystalline carbonate (i.e., micrite). Although biofilms themselves may be very thin (e.g., a few micrometers), calcification of successive biofilms or calcification beneath a continually accreting biofilm may result in the production of lithified structures of considerable thickness (e.g., many meters).

2.1. Parameters Governing Nonenzymatic Carbonate Precipitation

Primary extrinsic factors controlling nonenzymatic and abiotic precipitation of carbonate minerals appear to be related to: (1) the saturation state of ambient water with respect to carbonate minerals; (2) presence of appropriate nucleation sites; and (3) absence of specific inhibitors of nucleation. The chemistry of the natural carbonic acid system in marine waters is highly complex (e.g., Morse and Mackenzie, 1990; Wollast, 1994), and for the purposes of this chapter, the concepts of saturation state, carbonate alkalinity, and nucleation are briefly introduced to enable discussion of the major parameters that govern the biological induction of carbonate precipitation.

2.1.1. Saturation State and Nucleation

The saturation state of a fluid with respect to calcite or aragonite is defined as the ratio of the free ion activities (which are related to concentration) of Ca^{2+} and CO_3^{2-} to the solubility product of the mineral in question. The solubility product is a constant that reflects the amount of dissociated mineral that can be accommodated in the fluid under a particular set of boundary conditions, the most important of which are temperature and pressure (Robertson, 1982; Morse and Mackenzie, 1990). Alkalinity can be defined as the acid neutralization capacity of the solution. Carbonate alkalinity, which reflects the contributions from CO_3^{2-} and HCO_3^-, accounts for the largest part of total seawater alkalinity (Morse and Mackenzie, 1990). Because seawater contains a very high Ca^{2+} concentration compared to alkalinity, relatively small changes in CO_3^{2-} concentration have large effects on the saturation state, but changes to Ca^{2+} concentration are unlikely to have been large enough to cause short-term fluctuations in saturation (Kempe and Kazmierczak, 1994). Nonenzymatic carbonate precipitation will not occur in a solution that is undersaturated with respect to carbonate minerals (i.e., ion activity product $<$ solubility product), but spontaneous abiotic precipitation will not necessarily occur if the solution is supersaturated (i.e., ion activity product > solubility product) owing to the kinetic activation energy barrier and various poorly understood inhibitors that impede nucleation (Morse, 1983; Wollast, 1994). The activation energy barrier represents the difference in free energy resulting from energy released by bonding of ions into the crystal lattice as a whole and the energy required to form the new solid–liquid interface. When the newly formed nucleus is very small, the energy balance favors the dissolution of the crystal lattice rather than continued growth, because the surface area is large relative to the volume (i.e., total number of bonds in the lattice) (Mann, 1989). Because of the activation energy barrier, spontaneous abiotic precipitation will occur only when saturation is much higher than the equilibrium state (e.g., Wollast, 1971; Kempe and Kazmierczak, 1990, 1994).

Modern shallow marine waters are generally supersaturated with respect to both calcite and aragonite, but deeper cold waters are generally undersaturated owing to lower temperatures, higher pressures, and increased CO_2 concentration. Modern shallow marine waters will typically support biologically induced carbonate precipitation and ancient marine supersaturation states were apparently higher for most or all of geological history (e.g., Kempe and Kazmierczak, 1990; Grotzinger and Knoll, 1995).

2.1.2. Abiotic Effects on Saturation States

A number of extrinsic parameters locally increase carbonate saturation, thereby presumably influencing the spatial distribution of nonenzymatic carbonates. (1) Solar heating of shallow waters, particularly where ponding isolates water from normal marine circulation (a common phenomenon in many reefs), causes decreased carbonate solubility while CO_2 degassing raises the pH. Increased pH (i.e., low levels of H^+ ions) increases carbonate alkalinity by promoting continued dissociation of H_2CO_3 to yield additional CO_3^{2-}. (2) Agitation of waters in littoral environments causes CO_2 degassing. (3) Evaporation in shallow settings increases ionic concentrations. (4) Weathering of silicate minerals by carbonic acid-mediated hydrolysis (i.e., "Urey reactions": Urey, 1951) increases carbonate alkalinity by producing excess CO_3^{2-} and HCO_3^- ions.

2.1.3. Biological Effects on Saturation States

Biological induction of carbonate precipitation may occur where a biotic process causes localized supersaturation to reach a critical level in the presence of appropriate nucleation sites and in the relative absence of inhibitors (for reviews, see Ehrlich, 1990; Castanier *et al.*, 1999; Riding, 2000). The saturation state is increased by processes that increase carbonate alkalinity by producing excess CO_3^{2-} and HCO_3^- or by increasing pH (Table 1).

Photosynthesis is the most extensively studied autotrophic calcification-inducing process (e.g., Borowitzka, 1989; Merz, 1992; Merz-Preiss and Riding,

Table 1. Microbial Processes Implicated in the Induction of Calcium Carbonate

Process	Sample equation and effect	Reference
Photosynthesis	$106CO_2 + 16HNO_3 + 2H_2SO_4 + H_3PO_4 + 120H_2O \rightarrow C_{106}H_{263}O_{110}N_{16}S_2P + 141O_2$ Increases pH	Morse and Mackenzie, 1990
Methanogenesis	$4H_2 + CO_2 \rightarrow CH_4 + 2H_2O$ Increases pH	Ehrlich, 1990
Deamination of amino acids	$R\text{-}CHCOOHNH_2 + 2H_2O + NAD^+ \rightarrow NADH_2 + RCOCOOH + NH_4^+ + OH^-$ Increases pH	Ehrlich, 1990
Hydrolysis of urea	$NH_2CONH_2 + 2H_2O \rightarrow 2NH_4^+ + CO_3^{2-}$ Increases carbonate alkalinity and pH	Ehrlich, 1990
Nitrate reduction	$NO_3^- + 9H^+ + 8e^- \rightarrow 2H_2O + NH_4^+ + OH^+$ Increases pH	Ehrlich, 1990
Sulfate reduction	$SO_4^{2-} + 2(CH_2O) \leftrightarrow H_2S + 2HCO_3^- \leftrightarrow H_2S + CO_2 + CO_3^{2-} + H_2O$ Increases carbonate alkalinity	Ehrlich, 1990

1999). The photosynthetic removal of CO_2 by prokaryotes or eukaryotes increases the pH and leads to a localized increase in CO_3^{2-} concentration. Photosynthetic uptake of HCO_3^- also increases the pH where accompanied by OH^- efflux (Borowitzka, 1989). Autotrophic methanogenesis by anaerobic bacteria entails the reduction of CO_2, or less commonly other organic compounds, to produce methane, with an accompanying increase in pH (Atlas and Bartha, 1993).

A variety of heterotrophic microbial processes favor the precipitation of nonenzymatic carbonates. However, microbial communities may be very complex with fine spatial chemical gradients and a variety of different metabolic processes occurring simultaneously or in succession in different parts of the community or biofilm. For instance, metabolic processes within the same community may raise carbonate alkalinity but also produce significant quantities of organic acids that inhibit carbonate precipitation (Ehrlich, 1990). Hence, microbial induction of carbonate precipitation depends not just on the types of metabolic processes that occur but on environmental gradients and the fates of particular metabolites. The most important heterotrophic carbonate inducing processes are: (1) ammonification (deamination and hydrolysis) of amino acids or urea; (2) dissimilatory reduction of nitrates; and (3) dissimilatory sulfate reduction (Castanier *et al.*, 1999). Deamination of amino acids by aerobic bacteria releases ammonia, which raises pH, thereby favoring precipitation. Bacterial hydrolysis of urea produces ammonia and CO_3^{2-} ions. However, in the presence of largely autotrophic nitrifying bacteria, the ammonia may be oxidized, forming nitric acid and thereby lowering pH and inhibiting precipitation (Ehrlich, 1990). The dissimilatory reduction of nitrates by anaerobic heterotrophic bacteria causes oxidation of organic matter, thereby increasing carbonate alkalinity (Atlas and Bartha, 1993). Anaerobic reduction of sulfates liberates H_2S, CO_3^{2-}, HCO_3^-, and OH^- ions, thus increasing both carbonate alkalinity and pH, provided that the H_2S is discharged from the system (Ehrlich, 1990). However, if H_2S is not removed from the system or used to reduce ferric iron to form FeS_2 (Schumann-Kindel *et al.*, 1997), it lowers pH, thereby causing carbonate dissolution (Morse and Mackenzie, 1990; Castanier *et al.*, 1999). Ammonification (e.g., Berner, 1968; Krumbein, 1979; Reitner, 1993) and sulfate reduction (e.g., Krumbein, 1979; Lyons *et al.*, 1984; Schumann-Kindel *et al.*, 1997) have been implicated most commonly in the heterotrophic induction of marine carbonate precipitation.

2.1.4. Biological Effects on Nucleation

Biological processes that increase carbonate supersaturation are clearly important in the induction of carbonate precipitation (and also in enzymatic biomineralization). However, the rarity of spontaneous abiotic carbonate precipitation in shallow marine environments, despite high levels of supersaturation, implies that other factors are also important. The rarity of abiotic precipitation has been attributed to a variety of organic and inorganic inhibitors that reduce precipitation rates by: (1) chemically poisoning the crystal lattice; (2) physically covering available nucleation sites; and/or (3) increasing the activation energy barrier to nucleation (Morse and Mackenzie, 1990).

In enzymatic biomineralization, organic macromolecules (organic matrix) control the location and orientation of nucleation sites and provide constraints on crystal development by acting as inhibitors. Nucleation is promoted where the substrate effectively lowers the activation energy and concentrates the appropriate ions (Mann, 1989). Various models for the effects of Ca^{2+}-binding macromolecules on carbonate lattice structure have been proposed (e.g., Simkiss, 1986; Addadi and Weiner, 1989; Mann, 1989), and Mitterer and Cunningham (1985) discussed the role of Ca^{2+}-binding aspartic acid-rich macromolecules in nonenzymatic carbonate precipitation. The carboxyl groups of such molecules are similar in structure to the carbonate anions in calcium carbonate minerals and where carboxyl groups share the lattice spacing of carbonate groups in the particular mineral, nucleation may be promoted. Reitner (1993) summarized previous work and proposed that macromolecules rich in aspartic and glutamic acids provide substrates for enhanced carbonate nucleation in modern calcified sponge–microbial biofilms. Extracellular polymeric substances (Decho, 1990; Pinckney and Reid, 1997) and degrading organic matter that contain Ca^{2+}-binding macromolecules are abundant in metazoan–microbial biofilms. Where such biofilms contain heterotrophic bacteria that increase carbonate alkalinity and/or pH, they may induce carbonate precipitation. However, biofilms are ubiquitous in shallow marine settings (Fairchild, 1991), and because nonenzymatic carbonates are not forming everywhere, many biofilms must act as inhibitors to precipitation. Westbroek *et al.* (1994) suggested that specific organic inhibitors were evolved by some organisms so as to keep their exterior membranes free of excess induced precipitation. Important recent reviews of the roles of microbes and organic matter in inducing the precipitation of carbonate minerals can be found in Camoin (1999), Riding (2000), and Riding and Awramik (2000).

3. Reef Framework Construction

3.1. Framework Construction

Reef framework can be evaluated in different ways. However, if the defining characteristics of reefs are biologically constructed, synsedimentary rigidity and relief, then reef framework should be defined and reef builders should be evaluated primarily on the basis of these two criteria. Synoptic relief can be biologically constructed in two main ways. Direct upward growth of reef builders produces relief, provided that they remain in growth position. Such relief is amplified if reef builders grow upon each other and/or generate large carbonate volume by producing large skeletons. Relief also results from localized, large-scale carbonate production, even if sediment does not remain in growth position (i.e., a sediment pile). In both cases, carbonate volume typically but not invariably is a significant component of relief and by definition both generation and maintenance of relief must be biological in nature. Hence, a sediment pile must be biologically produced and biologically bound into place to qualify as biologically produced relief; a hydrodynamically produced sandpile does not qualify.

Synsedimentary rigidity can be acquired in several ways. Large, well-skeletonized metazoans have limited inherent rigidity, but production of rigid framework requires rigid attachment to the substrate or to each other by means of their own skeletal growth or by encrustation by binders. A close relationship exists between the generation of relief and its long-term maintenance in a high-energy environment by synsedimentary rigidity.

A variety of nonenzymatic carbonates provided relief and rigidity in ancient reefs. The most widely known nonenzymatic carbonates are microbialites, which Burne and Moore (1987) defined as "organosedimentary deposits that have accreted as a result of a benthic microbial community trapping and binding detrital sediment and/or forming the locus of mineral precipitation" (pp. 241–242). Microbialites are broadly classified on the basis of macroscopic morphology (Riding, 1991, 2000). Common reef-building microbialites include: (1) stromatolites, which are laminated; (2) thrombolites, which have clotty textures; and (3) dendrolites, which have bushy textures (Riding, 1991). Stromatolites are further subdivided into: (1) agglutinated forms, where trapping and binding of detrital sediment dominates; (2) precipitate stromatolites, where biologically induced precipitates are more important than detrital sediment; and (3) skeletal stromatolites, where laminae are formed by recognizable calcified microbial fossils (calcimicrobes) (Riding, 1991). Differentiation of fine precipitates from fine agglutinated carbonate is controversial in some Precambrian stromatolites (e.g., Fairchild, 1991), but both types are considered nonenzymatic framework in this survey.

Various coccoid bacteria and cyanobacteria generally are considered the constructors of microbialites, but a variety of metazoans and nonliving organic matter also play important roles in the formation of microcrystalline carbonate deposits that resemble microbialites (e.g., Neuweiler, 1993; Reitner, 1993; Riding and Awramik, 2000). Automicrite is microcrystalline carbonate produced by biologically induced precipitation irrespective of the inducing community (Reitner *et al.*, 1995a,b). Automicrites result from precipitation induced by microbial communities (i.e., microbialites), metazoans (Reitner *et al.*, 1995a,b), or by dead organic matter (Neuweiler, 1993; Trichet and Défarge, 1995). Calcimicrobes include a variety of consistent morphological forms that more or less closely reflect specific microbial taxa and/or associations (i.e., *Renalcis*, *Girvanella*). However, some calcimicrobes grade into structureless automicrites depending on the initial rate of calcification (Merz-Preiss, 1997) and/or later diagenesis (Pratt, 1995, Fig. 62; Turner *et al.*, 2000). Hence, the morphology of nonenzymatic reef frameworks is intimately related to taphonomy. A variety of other carbonates that represent biologically induced precipitation also were important in ancient reef frameworks, but in many cases the community that is responsible for inducing the precipitation may not be preserved, or may not be identified with certainty.

Webb (1996) defined four major reef framework components (Figs. 2–4): (1) metazoan skeletons, (2) calcimicrobes, (3) microbialites, excluding calcimicrobes, and (4) biologically localized cements. The term "microbialite"

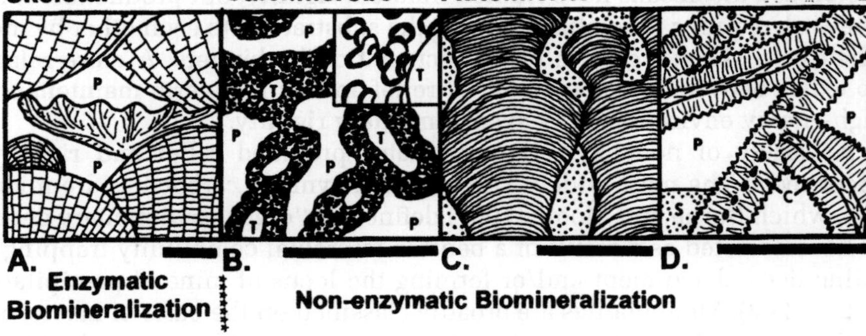

FIGURE 2. Four major reef framework components. (A) Skeletal framework, here represented by corals, results from enzymatic biomineralization. (B) Calcimicrobe framework, here represented by *Renalcis*, a common Late Devonian reef builder, results from nonenzymatic biomineralization, but the border between it and enzymatic (skeletal) biomineralization is indistinct. (C) Automicrite framework, here represented by stromatolites. (D) Biocementstone framework, here represented by cement crusts on fenestrate and ramose bryozoans, results from nonenzymatic biomineralization, but grades into abiotic mineralization. Scale bars = 10 cm, except for D, where it is 5 cm. P, pore space; T, stromatoporoid; S, sediment; C, cement.

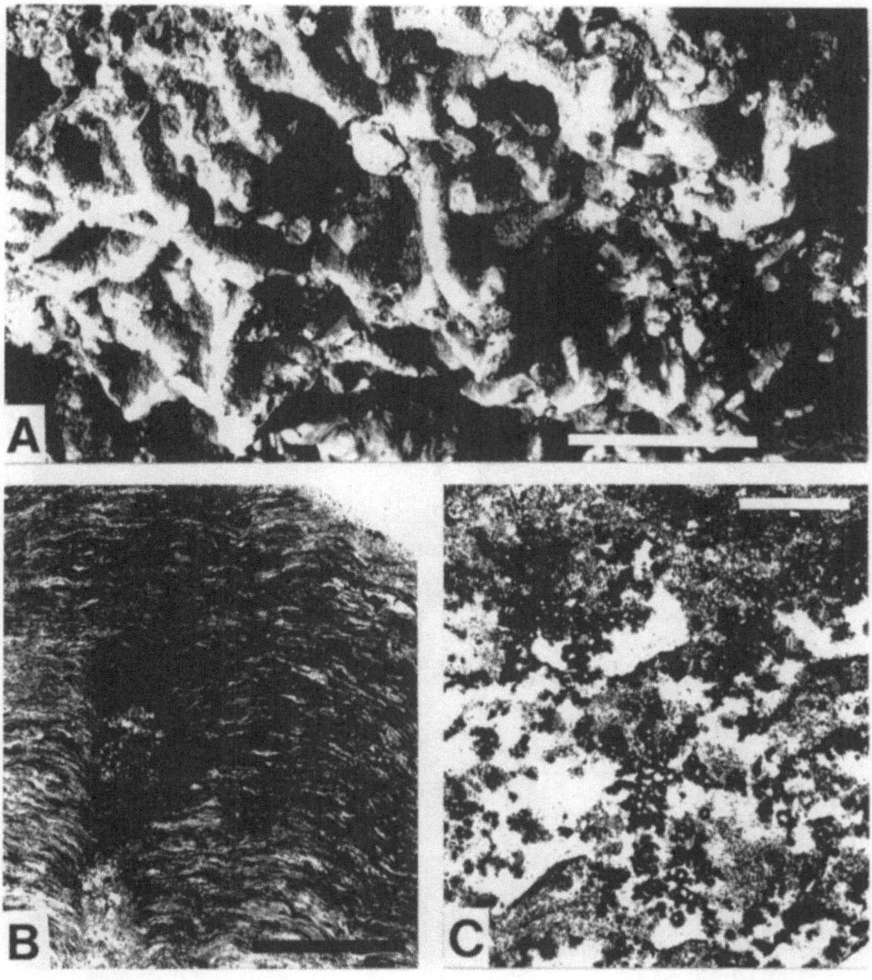

FIGURE 3. Reef framework components. (A) Skeletal reef framework mostly represented by branching *Acropora* corals in Holocene reefrock dredged from Heron Reef, Great Barrier Reef. Scale bar = 20 cm. (B) Calcimicrobe reef framework represented by *Rothpletzella* from the Frasnian (Late Devonian) of the Canning Basin, Western Australia. Scale bar = 1 cm. (C) Calcimicrobe reef framework represented by *Renalcis* from the Frasnian (Late Devonian) of the Canning Basin, Western Australia. Scale bar = 0.25 cm.

henceforth is replaced by the more inclusive term "automicrite" for structures composed of microcrystalline carbonate, so as to include not only microbial fabrics but fabrics that were precipitated as a result of induction by metazoans and nonliving organic matter. Although the components commonly occur intermixed, skeletal, calcimicrobe, and automicrite frameworks may occur as relatively pure end members. Biocementstone frameworks are defined as frameworks wherein synsedimentary rigidity is provided by cement crusts, but

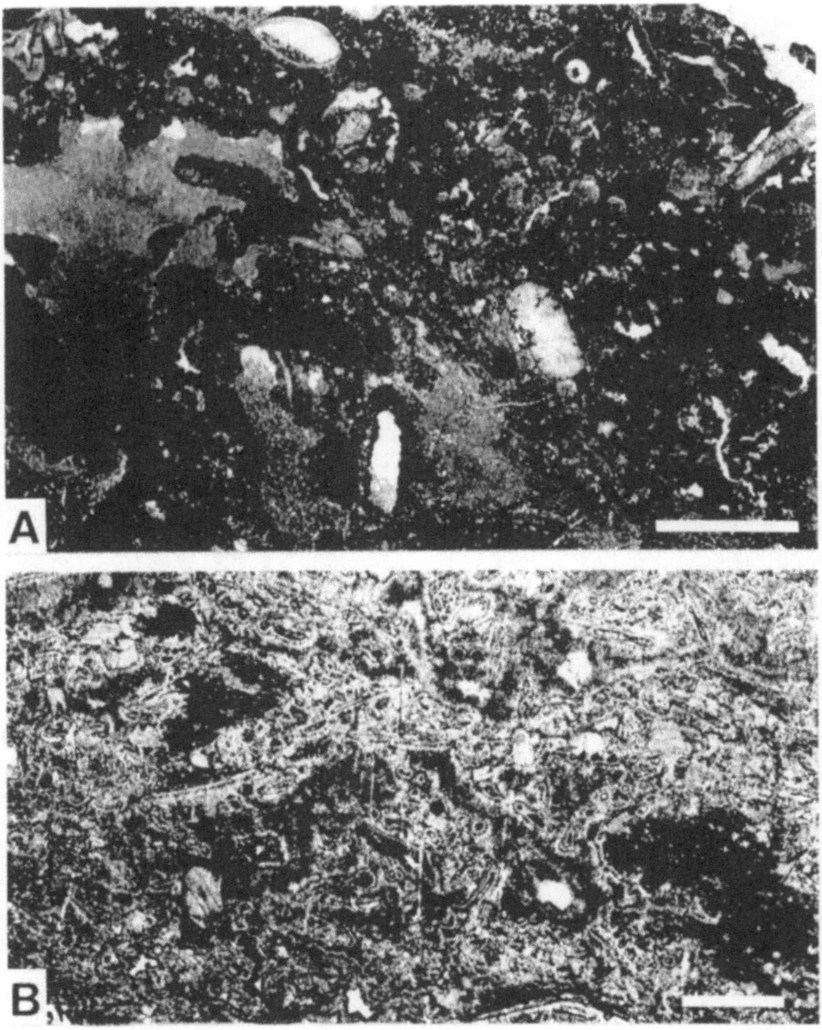

FIGURE 4. Reef framework components. (A) Automicrite reef framework represented by thrombolite (dark) from Lower Carboniferous patch reefs in the Lion Creek Limestone, Queensland, Australia. Scale bar = 0.5 cm. (B) Biocementstone reef framework represented by synsedimentary cement crusts on fenestrate bryozoan fronds from the upper part of Muleshoe Mound, Lake Valley Formation (Lower Carboniferous), Sacramento Mountains, New Mexico. Scale bar = 0.5 cm.

relief generally is provided by skeletal metazoans that may be incapable of forming rigid framework on their own. Nonmicritic, precipitate-dominated stromatolites with coarse cementlike internal textures occurred during the Precambrian and may be analogous to Phanerozoic biocementstones in the absence of larger, erect metazoans.

3.2. Reef Framework and Guild Structure

Skeletal framework has received the most attention in Phanerozoic reef studies. Fagerstrom (1987, 1991, 1994) applied the guild concept to reefs in order to investigate the role of skeletal metazoans in reef communities. The guilds of most interest for reef studies are the "reef-constructing guilds" (i.e., constructors, bafflers, and binders), which are defined largely on the basis of skeletal morphology and growth habit (Fagerstrom,1991, Table 1). Constructors are recognized primarily on the basis of (1) erect growth, (2) massive skeletons (3) large size, and (4) occurrence in growth position. Binders are recognized on the basis of (1) lateral, sheetlike growth, (2) skeletonization, and (3) occurrence in growth position as an encrustation on a substrate. Bafflers are erect taxa with smaller, less massive, less robust skeletons (Fagerstrom, 1991, Table 1).

Although the reef community guild concept applies primarily to skeletal metazoans, the concept easily accommodates nonenzymatic reef builders. The massive nature and rigidity of hemispherical stromatolites places them within the constructor guild. However, the microbial biofilms responsible for stromatolite accretion lack the erect growth habit of constructors, having instead the lateral, reptant growth habit that typifies the binder guild. Although this can be accommodated as "guild overlap" (Fagerstrom, 1987, 1994), the thin living film on an accreting stromatolite conceptually is not very different from the thin layer of living tissue on the surface of a large coral head or stromatoporoid. Other biofilms may induce precipitation of thin carbonate crusts on metazoan skeletons, therefore fitting comfortably within the binder guild. Hence, the thin, encrusting nature of the calcifying biofilm is less important to guild structure than the morphology of the rigid carbonate body that it produces; the overall automicrite structure, or cement crust, is analogous to the metazoan skeleton (e.g., Grotzinger, 1989b). More difficult to accommodate directly in the guild concept are the products of microbial calcification that occur below the sediment–water interface. In some cases, automicrite precipitation occurred interstitially within coarse or fine sediment, thereby rapidly lithifying the sediment on the seafloor. Such synsedimentary lithification of reef debris produces masses of rigid framework that lack surficial expression and therefore cannot be directly related to skeletons on the seafloor.

Nonenzymatic reefal carbonates functioned largely in the same way as skeletal organisms in framework construction. Automicrite and calcimicrobes generate relief by direct upward accretion and volume by forming massive, rigid structures or piles of fine sediment. Additionally, because calcimicrobes and many automicrites are rigid precipitates at the sediment–water interface, they are important binders, effectively lithifying the framework. As binders, nonenzymatic framework components have a major influence on the constructional roles of co-occurring skeletal metazoans by providing additional rigidity to delicate organisms. Skeletal organisms with morphologies typical of the baffler guild (i.e., erect, small, delicate) provide considerable relief to reefs where they are heavily encrusted by automicrite, calcimicrobes, or cement crusts. Delicate organisms, such as fenestrate bryozoans in "Waulsortian mounds" (Webb, 1996), palaeoberesellid algae in Late Carboniferous reefs (Horbury, 1992), and bryozoans and sponges in the Permian Capitan reef of New Mexico (Grotzinger and Knoll, 1995; Kirkland *et al.*, 1998), could not form rigid framework on their own but provided the scaffolds for rapid generation of relief when encrusted by nonenzymatic carbonates. Hence, scaffold constructors in biocementstones function more like conventional constructors than bafflers, despite their more delicate morphologies.

4. Nonenzymatic Reef Frameworks through Time

4.1. The Oldest Reefs: Precambrian Reefs

Precambrian reef history is poorly understood owing to (1) the scarcity of synthetic studies, (2) the lack of appropriate proxies for reef abundance, (3) difficulties in dating particular reef occurrences, and (4) increasing loss of the record with increasing age. However, stromatolite reefs were major contributors to the Precambrian carbonate rock record (Grotzinger, 1989b,c). The oldest described reefs are Late Archean in age (Grotzinger, 1989b), but stromatolites occurred during Early Archean time (Lowe, 1980; Walter *et al.*, 1980). Early and Middle Archean reefs have not been identified. Stable carbonate ramps were forming by Middle Archean time, but stable carbonate platforms did not form until the Late Archean, when larger masses of cratonic continental lithosphere had been assembled (Grotzinger, 1989b). The large stromatolite reef-rimmed carbonate platform of the Campbellrand Subgroup, South Africa (Beukes, 1987), which previously was considered to be Paleoproterozoic in age but subsequently was dated as Late Archean (i.e., 2521 $\pm$ 3 Ma) (Sumner and Bowring, 1996), represents one of the oldest and largest known Precambrian carbonate platforms. Hence, stromatolite reefs were intimately associated with the first large carbonate platforms. Late Archean reefs also fringed oceanic volcanoes (Grotzinger, 1989b).

Reefs were apparently more abundant during Proterozoic time or at least they are more commonly preserved. Extensive barrier reef complexes attained thicknesses of 1 km and lengths of 600 km in the Canadian Paleoproterozoic (2.1–1.88 Ga) (Grotzinger, 1989a; Hoffman, 1989; Hoffman and Grotzinger, 1989). Shelf margin reefs consisted of stacked, elongate stromatolite mounds, and deep-water pinnacle reefs occurred on ramps (e.g., Grotzinger, 1986; Jackson, 1989). Like the older Late Archean Campbellrand platform (Beukes, 1987), Paleoproterozoic reef-rimmed shelves share many sedimentological characteristics with Phanerozoic reef-rimmed shelves, including bypass margins and backstepping relationships, and the dimensions of Precambrian reef-rimmed shelves equal some of the largest Phanerozoic carbonate shelves (Grotzinger, 1989a,b). Stromatolites dominated Paleoproterozoic reef building, but thrombolites occur in cyclic, shallow-water shelf carbonates of the Rocknest Formation of Canada and also possibly in shelf-edge barrier reefs (Kah and Grotzinger, 1992). The stromatolites were precipitate dominated rather than agglutinated and cements also were important (e.g., >10 % of Pethei Platform reefs: Sami and James, 1996), but the relative importance of biologically induced precipitation versus abiotic cementation is problematical, and abiotic sea floor precipitates are known (e.g., Fairchild, 1991; Kah and Grotzinger, 1992; Grotzinger and Rothman, 1996; Sami and James, 1996).

The most extensive Mesoproterozoic carbonates occur in Siberia and China, but relatively little has been published on the reefs of those sequences (Grotzinger, 1994). In North America, barrier reef complexes were less abundant during Mesoproterozoic time (Geldsetzer *et al.*, 1989) despite increasing stromatolite diversity and abundance (Walter and Heys, 1985; Awramik, 1991). Isolated, deep-water pinnacle reefs and platform reefs were relatively common (Geldsetzer *et al.*, 1989; Narbonne and James, 1996). Despite the lack of known reef-rimmed shelves, Mesoproterozoic reefs attained large dimensions. A stromatolite platform reef in the Belcher Group of east-central Hudson Bay attained a thickness of 244 m and an aerial extent over 2500 km^2 (Ricketts and Donaldson, 1989). Deep-water pinnacle reefs of Arctic Canada (Narbonne and James, 1996) attained depositional relief greater than 75 m, and like other deep-water Mesoproterozoic reefs (e.g., Kerans and Donaldson, 1989) show complex facies architecture and zonation relating to local eustasy.

Neoproterozoic carbonate environments were dominated by ramps; rimmed shelves were very rare (Grotzinger, 1989b). However, a variety of reefs occurred during Neoproterozoic time (e.g., Geldsetzer *et al.*, 1989). Aside from typical Proterozoic, isolated, deep-water stromatolite reefs, large, latest Neoproterozoic (Ediacaran) stromatolite-constructed platform reefs occurred in Alberta (Teitz and Mountjoy, 1989). The reefs reached thicknesses of 400 m and had lateral dimensions of tens of kilometers. More unusual for the Proterozoic were the early Neoproterozoic Little Dal reefs of northwestern Canada (Aitken, 1989; Turner *et al.*, 1993, 1997), which were constructed primarily by calcimicrobes similar to Phanerozoic reef-building *Renalcis* and *Girvanella* (Turner *et al.*, 2000). Little Dal reefs also contained abundant cements, stromatolites, and thrombolites (Aitken and Narbonne, 1989). The reefs were isolated pinnacle structures with final depositional relief of 100 m. Aggradational and progradational relationships with surrounding strata correspond to local eustasy in the same way as in many Phanerozoic skeletal reefs (Turner *et al.*, 1997). However, despite representing a new and apparently successful style of reef framework development, calcimicrobe reefs are not known to have occurred again until the Cambrian Period, over 200 My later.

In summary, Precambrian reefs consisted exclusively of nonenzymatic frameworks, including some cements that may or may not have been the result of biologically induced precipitation. Late Archean stromatolite fringing reefs predated well-developed carbonate platforms and reef-building stromatolites were the primary carbonate producers and architects of Proterozoic carbonate platforms (Grotzinger, 1989b). Stromatolites (mostly automicrite framework) dominated virtually all Precambrian reefs, but calcimicrobe frameworks developed in the early Neoproterozoic Little Dal reefs. Precipitate-dominated Precambrian reefal stromatolites may be analogous to younger biocementstone fabrics in the absence of skeletal metazoans.

4.2. The Rise of Skeletal Reef Builders: Cambrian to Jurassic Reefs

The oldest unequivocal enzymatic (skeletal) reef builders are archaeocyaths and spiculate sponges in early Tommotian (Early Cambrian) reefs in the Pestrotsvet Formation of Siberia (Riding and Zhuravlev, 1995). However, most Cambrian reefs, including the oldest Cambrian reefs in Nemakit-Daldynian strata of Siberia (Kruse *et al.*, 1995), were constructed primarily by calcimicrobes. The calcimicrobe *Renalcis* was an important binder in the oldest Pestrotsvet reefs, but archaeocyaths were the primary framebuilders (Riding and Zhuravlev, 1995). The problematic, early, skeletal, tube-building organism *Anabarites* occurred in older (Nemakit–Daldynian) *Renalcis*-dominated reefs in Siberia (Luchinina, 1985), but *Anabarites* may have played no role in framework construction (Riding and Zhuravlev, 1995). Other Lower Cambrian reefs had archaeocyath-dominated frameworks (Zhuravlev, 1996), but Early Cambrian skeleton-dominated reef frameworks were the exception not the rule (e.g., Rowland and Gangloff, 1988).

Calcimicrobes continued to construct reefs during the Middle Cambrian, following the extinction of reefal archaeocyaths near the end of the Early Cambrian, and Cambrian reef construction reached its peak abundance during the early Middle Cambrian with calcimicrobe frameworks (Zhuravlev, 1996). Although rare skeletal organisms occurred in Middle Cambrian reefs, Zhuravlev (1996) concluded that benthic reefal calcimicrobes largely excluded skeletal organisms. Erect, colonial growth habits allowed some archaeocyaths to successfully compete with benthic calcimicrobes in Early Cambrian reefs, but solitary skeletal organisms were susceptible to overgrowth by calcimicrobes, thereby limiting their roles in reef construction. Calcimicrobes declined in reefs by the end of the Middle Cambrian, and Late Cambrian reefs were dominated by stromatolites and thrombolites (Zhuravlev, 1996). Similar framework compositions continued into the Early Ordovician, but new calcimicrobes and skeletal organisms (e.g., corals, bryozoans, lithistid sponges) began to inhabit reefs. By Middle Ordovician time, skeletal organisms were becoming more important, and by latest Ordovician time, bryozoans, corals, and stromatoporoid sponges were the dominant skeletal reef builders (Webby, 1984; Copper, 1997). The role of automicrites in Late Ordovician reefs is less well known (Webb, 1996).

Skeletal organisms continued to construct reefs during the Silurian, although the relative contributions of skeletal versus calcimicrobe and automicrite frameworks are controversial. Copper and Brunton (1991) and Brunton *et al.* (1997) considered stromatoporoids, corals, and bryozoans to be the dominant framework constructors of shallow-water reefs, with stromatoporoids being most important; but Silurian automicrite-dominated reefs have been described (e.g., Soja, 1994), and Bourque (1989) considered nonenzymatic reef builders to dominate many Silurian reefs. Algae and stromatolites may have been most important in inner shelf reefs during the Early Silurian (Brunton *et al.*, 1997). Kano (1994) determined that stromatoporoids were the dominant reef-building organisms in Silurian reefal and "reeflike" limestones of Gotland, Sweden, although they made up less than 50% of the facies in most cases. However, Kano did not describe the nature of the fine "matrix," which volumetrically dominates most of the facies. Kershaw (1997) considered the Gotland stromatoporoid "reefs" to be biostromes with little relief or rigidity, because the stromatoporoids did not firmly encrust the substrate and were not cemented together.

Stromatoporoids and corals also have been considered the primary Devonian reef builders (e.g., Fagerstrom, 1987). However, the relative contribution of skeletal and nonenzymatic frameworks in Devonian reefs are controversial (Webb, 1996). Skeletal organisms were clearly important constructors in many Devonian reefs and stromatoporoids in particular are among the primary carbonate producers across entire carbonate platforms, but in the Frasnian (Late Devonian) of Western Australia at least calcimicrobes and synsedimentary cements may have been the dominant providers of framework rigidity (e.g., Playford, 1980; Kerans *et al.*, 1986; Webb, 1996). North American and European Frasnian reefs may have had greater proportions of skeletal framework than Western Australian reefs, where calcimicrobes may have dominated, but quantitative studies are lacking in both regions. Regardless, whereas reefs were virtually decimated at the Frasnian/Famennian (F/F) extinction event in North America and Europe, loss of most reef-building corals and stromatoporoids did not terminate reef construction in the Famennian of Western Australia, South China, or Russia. Large Famennian reefs were constructed in these regions by stromatolites and/or calcimicrobes (Webb, 1996; Shen *et al.*, 1997). In the Canning Basin of Western Australia a short gap in reef development may have occurred immediately following the F/F boundary (Becker *et al.*, 1993; George *et al.*, 1997). However, a lowstand at the F/F boundary (George and Powell, 1997) was followed by rapid transgression (George *et al.*, 1997). Hence, continuous reef growth across the boundary would not be expected at one location and the "reef gap" is difficult to constrain.

Stromatoporoids contributed to small Famennian patch reefs in North America (e.g., Stearn *et al.*, 1987), but increasing Famennian stromatoporoid diversity does not correlate with an increase in reef building. By latest Famennian (Strunian) time, stromatoporoid diversity had partly rebounded, but reefs were apparently absent worldwide. Diverse, Strunian shallow-water stromatoporoid–coral communities in western Europe constructed biostromes, not reefs (Herbig and Weber, 1996). Interestingly, those biostromes lack automicrites and calcimicrobes (Webb, 1998).

Lower Carboniferous shallow-water reefs (Webb, 1994, 1999) were composed largely of automicrite framework with a diverse suite of skeletal organisms, such as corals, bryozoans, and calcareous algae. True skeletal frameworks composed of corals occurred by late Visean time (Webb, 1989) and scaffold constructors were important in a variety of algal and bryozoan biocementstone frameworks (e.g., Shinn *et al.*, 1983; Horbury, 1992). Biocementstone frameworks continued to be important in Late Carboniferous reefs (e.g., Davies and Nassichuk, 1989), but small, skeletal patch reefs were constructed by the sponge *Chaetetes* early in the Late Carboniferous (West, 1988).

Permian reefs include a variety of framework types (Flügel, 1994), but most reefs and particularly larger reefs had cement-rich nonenzymatic frameworks (e.g., Grotzinger and Knoll, 1995). Important Permian binders, such as the problematic *Archaeolithoporella*, probably represent nonenzymatic precipitates, but their exact mode of formation is not clear. Regardless, automicrites, biologically induced cements, and possibly even abiotic cements played important roles in Late Permian reef construction (Grotzinger and Knoll, 1995; Kirkland *et al.*, 1998).

Reefs were absent during earliest Triassic time following the Permo-Triassic extinction event(s) (Flügel, 1994; Stanley, 1997). However, reestablished reefs were in many ways similar to Late Permian reefs, being characterized by abundant biocementstone frameworks, problematic encrusters, and sponges. By the Late Triassic, skeletal framework proportions increased as scleractinian corals played a greater role in reef construction. Stanley (1988) suggested that reefs first began to take on a "modern aspect" in the Late Triassic. Reefs were absent during earliest Jurassic time, following the end-Triassic extinction event, but they rebounded with increasing abundance and diversity in direct relationship to global eustatic patterns (Leinfelder *et al.*, 1994). Jurassic reefs were dominated by thrombolite frameworks containing sponges and/or scleractinian corals. Although stromatolites and thrombolites dominated deep-water reefs (Pratt, 1995), they also were responsible for rigidity and relief in shallow-water reefs where skeletal metazoans were volumetrically more abundant (Leinfelder *et al.*, 1994). By latest Jurassic time, corals and stromatoporoids dominated the skeletal component of reefs (Stanley, 1997).

4.3. The Decline of Nonenzymatic Reef Framework: Cretaceous to Holocene Reefs

The increasing importance of Jurassic framework-constructing scleractinian corals, combined with the decline of calcimicrobe and biocementstone frameworks since the Triassic, greatly changed the complexion of global reef frameworks. Early Cretaceous reef framework basically represented a continuation of Late Jurassic forms despite a minor extinction-related retrenchment in reef abundance. However, various rudistid bivalves began to occupy positions in reefs along with corals, and by the end of the Early Cretaceous, they dominated many shallow-water shelf "reefal" settings (Kauffman and Johnson, 1988; Stanley, 1997). Shallow-water automicrite reefs were rare by the end of the Early Cretaceous (e.g., Scott and Brenckle, 1977), but automicrite reefs containing volumetrically insignificant sponges occurred in deeper slope environments (Neuweiler, 1993). Rudistid bivalve biostromes dominated shallow "reefal" environments for the remainder of the Cretaceous. The role of rudistid bivalves as reef constructors has been debated widely (e.g., Gili *et al.*, 1995; Schumann, 1996; Kauffman and Johnson, 1997). Many rudistids were clearly copious producers of carbonate sediment (Schumann, 1996), but as a group they were bound into rigid frameworks only very rarely (Kauffman and Johnson, 1988; Gili *et al.*, 1995). Regardless, a variety of Late Cretaceous buildups containing rudistid bivalves, scleractinian corals, stromatoporoids, and calcareous algae were densely bound by microbialites and/or red algae (e.g., Höfling *et al.*, 1996), and construction of vertical relief was achieved in some lower-energy areas owing to early cementation (Kauffman and Johnson, 1997). Hence, although most rudistid buildups do not qualify as reefs under the semantic constraints adopted in this chapter, scattered rigid Late Cretaceous reef frameworks were constructed and automicrites played a binding role in some of them.

Tertiary reefs are mostly poorly known owing to a lack of adequate exposure. Palaeocene reefs were relatively rare and relatively low in diversity with scleractinian corals and coralline algae acting as the primary framework constructors (Bryan, 1991; Schuster, 1996). The nature of nonenzymatic frameworks during the Palaeocene is unknown, but deep-water reefs had abundant "micritic cements" (Bryan, 1991). Eocene reefs were better developed and extensive shelf-edge scleractinian coral–coralline algal reefs occurred by the Oligocene (James, 1983; Copper, 1989; Bosellini and Russo, 1992). The role of nonenzymatic framework components has not been investigated in Oligocene reefs. Following a retrenchment in reef communities at the Oligocene–Miocene boundary, similar reef communities bounced back in the Miocene (Copper, 1989).

Miocene reefs are becoming better known owing to extensive work on well-exposed sections in the Mediterranean and western Tethys (e.g., Franseen *et al.*, 1996). Most Miocene reef communities were similar in composition to those of the Oligocene, but a large variety of other, probably anomalous reef types occurred in the Mediterranean–Paratethys region. Typical coral–algal reefs occurred globally from Early to early Late Miocene time (Esteban, 1996; Pomar *et al.*, 1996). Coral diversity was high in Early Miocene Mediterranean reefs (Cahuzac and Chaix, 1996), but by the early Messinian, *Porites* and *Tarbellastraea* were the dominant Mediterranean reef corals (Pomar *et al.*, 1996). A variety of encrusters, including coralline red algae, also were important in Miocene reefs and red algal rhodolith buildups were important structures (Esteban, 1996). During the Middle Miocene red algal–serpulid reefs occurred in the Carpathian foredeep from Poland to Moldavia (Pisera, 1985, 1996). In the same region, slightly younger Middle to earliest Late Miocene (Sarmatian) thrombolite–serpulid reefs occurred (Jasionowski, 1996; Pisera, 1996). These nonenzymatic framework reefs apparently resulted from abnormal, restricted water chemistry (i.e., high or low salinity) (Pisera, 1996).

Nonenzymatic reef frameworks became increasingly abundant in latest Miocene (Messinian) time in the western Mediterranean. A variety of coral–stromatolite, stromatolite, and thrombolite reefs have been described (e.g., Riding *et al.*, 1991; Martín and Braga, 1994; Esteban, 1996; Feldmann and McKenzie, 1997). The best known reefs had frameworks constructed by *Porites* and stromatolites with minor amounts of red and green algae and bryozoans (e.g., Riding *et al.*, 1991; Esteban, 1996). Feldmann and McKenzie (1997) suggested that most thrombolites and backreef stromatolites postdated the primary *Porites* reefs; however, *Porites* was clearly encrusted and supported by stromatolite crusts during growth (Riding *et al.*, 1991). The degree of "abnormality" or "normality" of the waters associated with the Messinian microbialite reefs is controversial. Messinian strata in the western Mediterranean were clearly associated with intervals of abnormal water chemistry (i.e., "salinity crisis," "Messinian events": Esteban, 1996), but normal marine organisms occur throughout at least some of the reefs suggesting normal levels of salinity during reef accretion (Riding *et al.*, 1991). Pedley (1996) suggested that eutrophication, not salinity, may have been the cause of atypical reef development.

While microbialite–coral reefs were forming in shallow settings in the western Mediterranean, algal mounds, termed "*Halimeda* segment reefs" by G. R. Orme and R. Riding (in Braga *et al.*, 1996), were forming in midslope environments. These rigid structures consisted of locally produced, disarticulated *Halimeda* segments, which were lithified by synsedimentary peloidal automicrite and cement crusts. The volume of automicrite approaches that of the *Halimeda* segments (Braga *et al.*, 1996; Martin *et al.*, 1997). Hence, whatever factors favored microbialite formation in shallow waters also favored the biological induction of carbonate precipitation down the slope. Regardless, the dominance of nonenzymatic frameworks in Messinian reefs appears to be a phenomenon restricted largely to the western Mediterranean.

Pliocene and Pleistocene coral reefs continued with typical, "modern" coral–coralline algae-dominated frameworks, although the dominant types of corals changed near the Plio-Pleistocene boundary (James and Bourque, 1992). The role of nonenzymatic frameworks has not been thoroughly investigated in Pliocene or Pleistocene reefs, but it is assumed to be similar to that in modern reefs. Although corals and coralline red algae dominate modern reefs, magnesium–calcite automicrites are minor to major components of several high-energy reef frameworks (Fig. 5) (Montaggioni and Camoin, 1993; Reitner, 1993; Zankl, 1993; Camoin and Montaggioni, 1994; Webb and Jell, 1997; Camoin *et al.*, 1999) and many of the abundant reefal magnesium–calcite "micrite cements" probably resulted from biologically induced precipitation, thereby making automicrites important contributors to modern reef rigidity (Webb, 1996; Webb *et al.*, 1998). Microbial automicrites, including agglutinated stromatolites, also occur in scattered modern fringing reefs (Reid *et al.*, 1995; Macintyre *et al.*, 1996) and in areas with abnormal salinity (Rasmussen *et al.*, 1993). Modern deepwater *Halimeda* mounds (e.g., Orme *et al.*, 1978; Roberts *et al.*, 1987) are not known to be lithified or to contain automicrite crusts, so they are not entirely analogous to Messinian *Halimeda* segment reefs.

FIGURE 5. Holocene cryptic reefal automicrite (thrombolite and dendrolite) forming part of a coral–algal framework, Heron Reef, Great Barrier Reef. Scale bar = 1 cm.

5. Reef History as a Tool for Reconstructing Earth History

The temporal distribution of reefs generally has been considered a function of biological history and in particular the extinction–radiation history of skeletal reef-building organisms (e.g., Fagerstrom, 1987). However, nonenzymatic reef frameworks are controlled partly by extrinsic parameters, mostly reflecting marine chemistry. Therefore, secular trends and short-term perturbations in the composition and distribution of reef framework components may serve as a proxy for extrinsic parameters, such as marine chemistry.

5.1. Secular Trends in the Volume of Nonenzymatic Carbonates

Nonenzymatic reef frameworks underwent a first-order decline in abundance, accompanied by a shift from entirely benthic to cryptic settings, from the Precambrian to the Holocene. The decline in nonenzymatic framework occurred in two steps (Fig. 6). The first step represents the rise of skeletal reef builders during the early Paleozoic. The rise of skeletal reef builders and to a lesser extent calcimicrobes also initiated the shift from benthic to cryptic settings by providing cryptic habitats, which were quickly invaded by calcifying biofilms. The second step in nonenzymatic framework decline was the rapid loss of benthic reefal automicrites (i.e., stromatolites and thrombolites) following the Jurassic. Post-Jurassic reefal automicrites, excepting anomalous

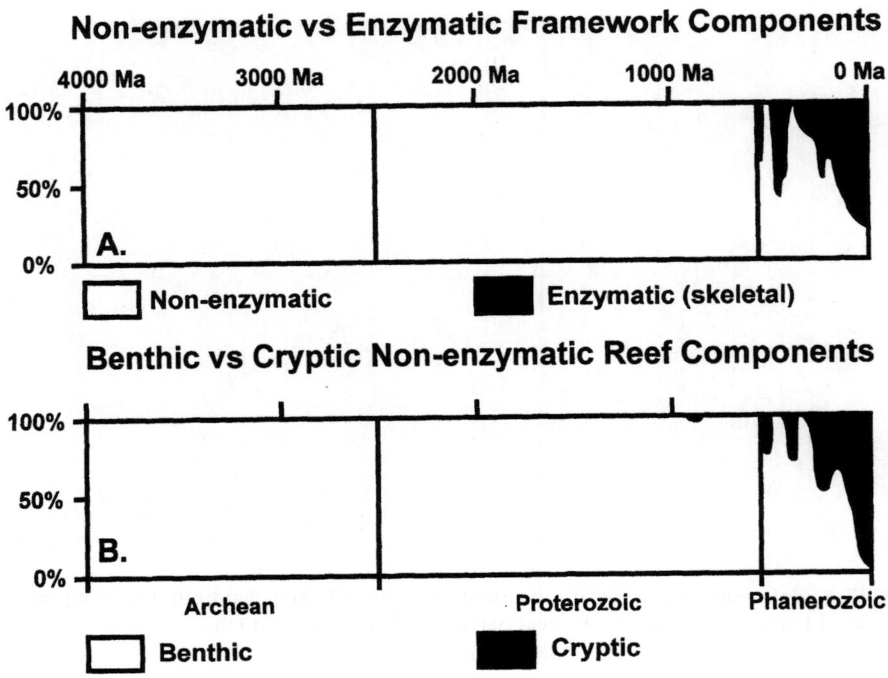

FIGURE 6. (A) Reef framework was dominated by nonenzymatic carbonates throughout the Archean and Proterozoic until skeletal framework (black) began to dominate reefs during parts of the Phanerozoic. (B) Nonenzymatic reef framework components were primarily benthic, occurring on the open seafloor, throughout most of geological time, but became dominantly cryptic by the late Phanerozoic, coincident with the rise of skeletal reef builders.

stromatolites and thrombolites in the Mediterranean Miocene, were confined largely to cryptic settings (Fig. 6).

Precambrian reefs contained temporally distinct framework types, but temporal changes in Precambrian nonenzymatic reef frameworks are difficult to constrain owing to inadequacies of the record. The dominant type of reefal stromatolites shifted from Archean and Paleoproterozoic precipitate-dominated forms to later Proterozoic agglutinated forms (Grotzinger, 1989b) and the Neoproterozoic Little Dal reefs contained the first extensive calcimicrobe frameworks.

Phanerozoic changes in reef framework composition are better constrained. Webb (1996) constructed a reef abundance curve for comparison of Phanerozoic reef framework types using the number of reference citations in the database of Flügel and Flügel-Kahler (1992) (Fig. 7). Citation data were used despite inherent bias problems, because more appropriate comprehensive databases reflecting actual reef volume or aerial distribution are lacking. The curve appears to represent a reasonable approximation of reef abudance, and bears a striking similarity to an empirical curve representing maximum

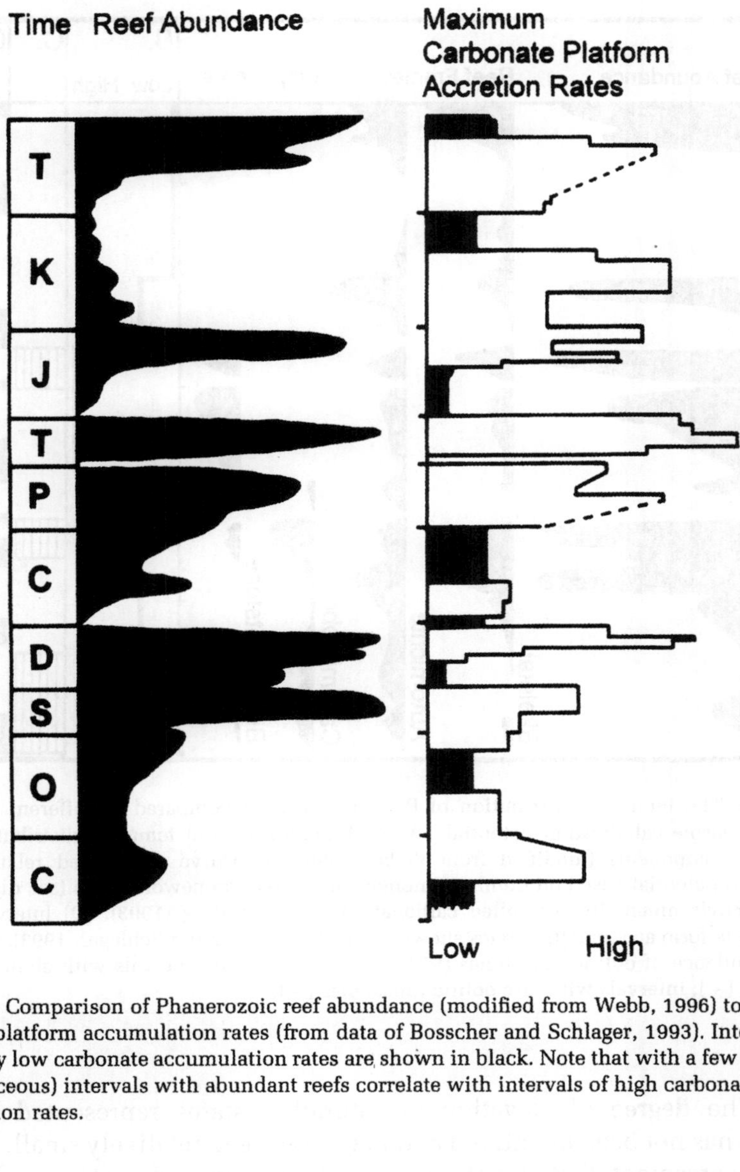

FIGURE 7. Comparison of Phanerozoic reef abundance (modified from Webb, 1996) to maximum carbonate platform accumulation rates (from data of Bosscher and Schlager, 1993). Intervals with particularly low carbonate accumulation rates are shown in black. Note that with a few exceptions (e.g., Cretaceous) intervals with abundant reefs correlate with intervals of high carbonate platform accumulation rates.

rates of carbonate platform accumulation through the Phanerozoic (Bosscher and Schlager, 1993). A more recent curve, also based on a large literature database, suggests a greater role for skeletal biota in Phanerozoic reefs (Kiessling *et al.*, 1999, Fig. 13). However, Kiessling *et al.* (1999) did not reevaluate older literature to take into account the common lack of recognition of nonskeletal framework components.

Phanerozoic intervals with extensive nonenzymatic reef frameworks presumably correlate to periods when extrinsic parameters particularly favored carbonate precipitation (e.g., elevated saturation states) (e.g., Riding, 1992,

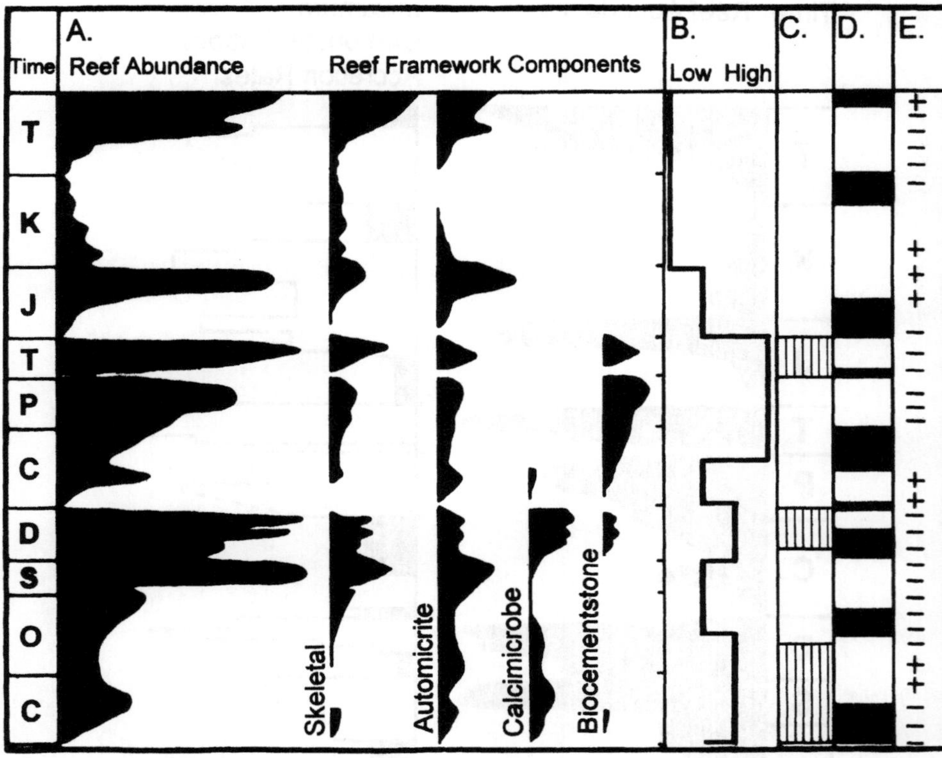

FIGURE 8. The temporal distribution of Phanerozoic reefs compared to different proxies for enhanced marine calcification potential. (A) Reef abundance and temporal distribution of reef framework components (modified from Webb, 1996). (B) Curve of inferred relative marine calcification potential based on dominant nonenzymatic reef framework types (see discussion in text). (C) Environmentally controlled carbonate events of Riding (1993). (D) Intervals of low carbonate platform accumulation rates shown in black (Bosscher and Schlager, 1993). (E) Relative global abundance of oolitic limestones (Wilkinson *et al.*, 1985). Intervals with abundant oolitic limestones (+); intervals with rare oolitic limestones (−).

1993). The degree of elevation of saturation states represented by these intervals has not been quantified and may have been relatively small, but there is wide agreement that variations in sedimentary carbonates partly reflect differing carbonate saturation through time (e.g., Sandberg, 1983; Wilkinson *et al.*, 1985; Knoll *et al.*, 1993; Riding, 1993). Although nonenzymatic reef frameworks occurred throughout the Phanerozoic, different types of framework may reflect relative differences in extrinsic parameters (Fig. 8). Skeletal frameworks are least sensitive although not insensitive (Opdyke and Wilkinson, 1990; Buddemeier and Fautin, 1996), to conditions favoring nonenzymatic carbonates. Hence, skeletal reef framework components occurred throughout most of the Phanerozoic and their first-order temporal distribution presumably reflects mostly extinction–radiation history (Newell, 1972; Fagerstrom, 1987). However, extant rapidly growing reefal skeletal biota are still largely restricted to the tropics owing to warm temperatures, high light incidence, and/or increased carbonate supersaturation (Buddemeier and Fautin, 1996), and Buddemeier and Fautin (1996) suggested that the evolution of some nonskeletal anemones from skeletal scleractinian precursors occurred during the Cretaceous and Eocene as a result of relatively low levels of carbonate saturation during those times. Regardless, biocementstone frameworks represent the other extreme, presumably requiring the most favorable extrinsic conditions, because they are under the least direct control of the community that induces precipitation.

Calcimicrobes and poorly structured automicrites may result from precipitation induced by the same organism under different conditions, or may result from subsequent diagenetic alteration (e.g., Pratt, 1995, Fig. 62; Turner *et al.*, 2000). Preservation of calcified sheaths in extant cyanobacteria requires rapid calcification in highly supersaturated waters. The same cyanobacteria if calcified more slowly produce fine carbonate sediment upon degradation (Merz-Preiss, 1997). Hence, intervals with abundant calcimicrobes may have had elevated carbonate saturation or elevated precipitation rates mediated by other parameters compared to intervals containing only automicrites (*contra* Knoll *et al.*, 1993). Whereas precipitation rates should positively correlate with the degree of saturation, other factors (i.e., presence of inhibitors) also may affect precipitation rates. Finally, benthic automicrites presumably require more favorable extrinsic conditions for calcification than cryptic automicrites, because although calcification occurs in a microenvironment in both cases, cryptic communities are bathed in relatively confined pore water (closed or semiclosed system), whereas the initial water chemistry in benthic biofilm calcification represents fully ambient seawater (open system). Hence, dominant reefal nonenzymatic carbonates suggest increasing marine calcification potential in the following order: (1) cryptic automicrite, (2) benthic automicrite, (3) calcimicrobe, and (4) biocementstone.

5.2. Proxies for Marine Calcification Potential

Although ancient marine water cannot be observed directly, a variety of proxies for marine calcification potential exist (e.g., Riding, 1992, 1993). Carbonate platform accumulation rates are related to the volume of biotic and abiotic carbonate production, both of which partly reflect parameters such as carbonate saturation state and temperature (e.g., Opdyke and Wilkinson, 1990; Buddemeier and Fautin, 1996). However, whereas upper limits to accumulation rates are controlled largely by local tectonism and subsidence (i.e., accommodation), intervals of low carbonate accumulation (Fig. 8) probably reflect low biological sediment production rates (Bosscher and Schlager, 1993). In the Mesozoic, low carbonate accumulation rates following the Permo-Triassic extinction event(s) and during the Late Cretaceous correlate well with poor reef development and low nonenzymatic reef framework content (Fig. 8). The Early Jurassic interval of low carbonate accumulation correlates with an interval of subdued reef development, but ended as automicrite reef frameworks became more abundant in the Late Jurassic. Paleozoic correlations are less consistent. The mid-Ordovician accumulation low corresponds to an interval of calcimicrobe decline, but the Early Devonian low occurred during an interval when calcimicrobes may have been relatively abundant, although less abundant than in the Late Devonian, when accumulation rates were again high. The Late Carboniferous low in carbonate accumulation occurred during an interval when biocementstone frameworks were abundant, although reef building was rather modest. Hence, platform accumulation rates correlate better with enzymatic reef framework composition than with nonenzymatic reef framework components.

Intervals of enhanced marine calcification would be expected to correlate with the temporal abundance of ooids (Wilkinson *et al.*, 1985), irrespective of whether they are biologically induced or abiotic precipitates, because ooids are clearly nonenzymatic carbonates. Peaks in ooid abundance during the Late Cambrian–Early Ordovician, Early Carboniferous, and Late Jurassic represent intervals when nonenzymatic components dominated reef frameworks (Fig. 8). However, intervals of low ooid abundance do not correspond well to intervals when nonenzymatic reef frameworks were less abundant (e.g., Permo-Triassic). Wilkinson *et al.* (1985) suggested that peaks in ooid development were correlated to intervals of changing continental emergence (i.e., Wilson Cycle-scale transgression and regression). Hence, the temporal abundance of ooids may be controlled as much by the availability of suitable shallow-shelf area as by marine chemistry.

5.3. Temporal Controls on Marine Calcification Potential

The distribution of Precambrian marine carbonates suggests that Archean waters were highly supersaturated with respect to carbonates (much higher supersaturation than modern levels) and that the levels declined through the Proterozoic reaching "near-Phanerozoic levels" near the Precambrian–Cambrian boundary (Grotzinger, 1989b; Grotzinger and Knoll, 1995). However, Precambrian seawater chemistry is the subject of much controversy. Kempe and Degens (1985) suggested that the early ocean was highly alkaline owing to an elevated sodium content (i.e., the soda ocean hypothesis). In their model, initial high alkalinity decreased gradually as sodium was transferred to newly forming continental crust, finally reaching near modern levels close to the Precambrian–Cambrian boundary (Kempe and Kazmierczak, 1994). However, most workers consider seawater chemistry to have been relatively constant, at least since the Archean (e.g., Holland, 1984; Bau and Dulski, 1996). Kasting (1987) considered elevated atmospheric levels of CO_2 to have elevated alkalinity and decreasing alkalinity resulted as inorganic carbon was tied up in dolostone and limestone on newly forming continents (Grotzinger and Kasting, 1993). Sumner and Grotzinger (1996) suggested that early high levels of marine supersaturation were maintained by high concentrations of the nucleation inhibitors Fe^{2+} and Mn^{2+}. Oxidation of these elements with rising levels of O_2 then allowed increased nucleation and formation of abundant fine carbonate sediment, with a concomitant shift from precipitate to agglutinated stromatolites and decreasing alkalinity (Sumner and Grotzinger, 1996). Apart from the large-scale trend of decreasing alkalinity, Grotzinger and Knoll (1995) suggested that intervals particularly rich in marine precipitates might have resulted when highly alkaline deep anoxic waters were intermittently mixed with Ca^{2+}-rich surface waters.

The Phanerozoic record of reefal carbonates is better constrained. The first-order post-Jurassic decline in benthic automicrites may represent a global decrease in marine carbonate supersaturation resulting from the rise of calcareous plankton in the middle Mesozoic (Fig. 9) (Gebelein, 1976; Riding 1993; Webb, 1996). The rise of calcareous plankton led to an exponential shift

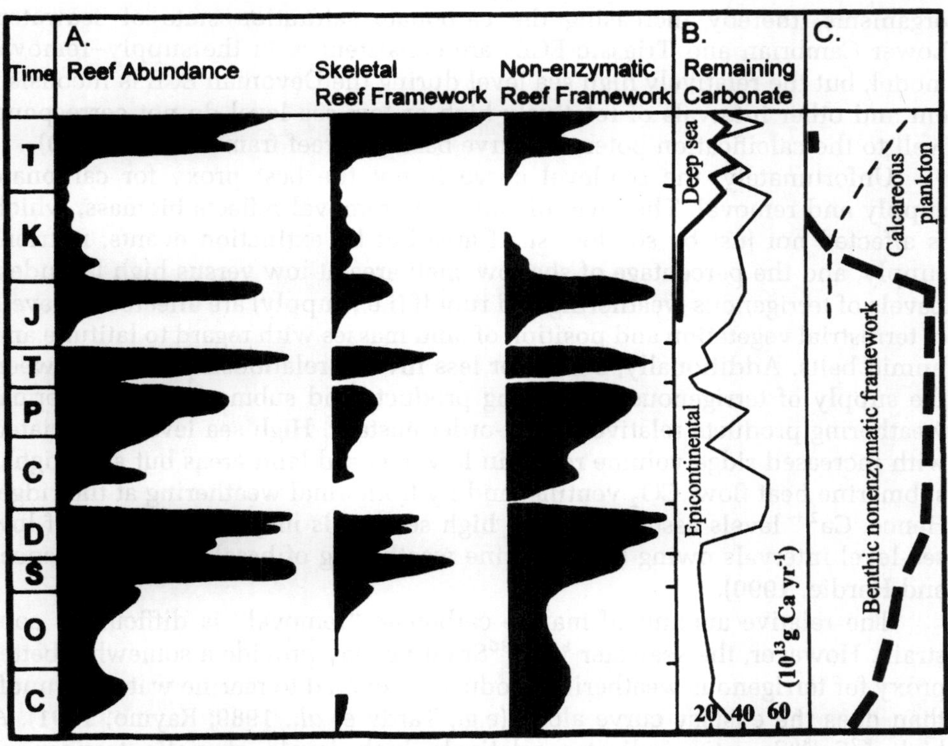

FIGURE 9. The temporal distribution of Phanerozoic reefs compared to the mass of carbonate rocks remaining from each interval. (A) Reef abundance and relative contributions to reef frameworks from enzymatic and nonenzymatic framework components. (B) Mass of Phanerozoic carbonate rocks remaining in epicontinental and deep-sea settings (Mackenzie and Morse, 1992, after Wilkinson and Walker, 1989). (C) First-order approximation of nonenzymatic carbonate abundance (heavy dashed line) through the Phanerozoic and the Mesozoic-Cenozoic exponential rise of calcareous plankton (thin dashed line).

in carbonate accumulation from shallow continental shelves to deeper pelagic settings (Mackenzie and Morse, 1992). Widely distributed calcareous plankton may have effectively "outcompeted" benthic organisms for finite carbonate resources (Wilkinson and Walker, 1989), thereby (1) sequestering carbonate in deeper settings, (2) lowering global marine carbonate saturation, and (3) disfavoring benthic nonenzymatic carbonate precipitation. Cryptic automicrites were relatively less affected, owing to their occurrence in relatively closed systems (Webb, 1996).

Short-term oscillations in Phanerozoic calcification potential have a variety of possible origins. Riding (1992, 1993) attributed "environmentally controlled carbonate events" (ECEs) (Fig. 8) primarily to a supply–removal model related to global sea level and to intervals of globally elevated temperatures. The supply–removal model suggests that increased terrigenous silicate weathering during intervals of low sea level (high continental emergence) delivers increased Ca^{2+} and HCO_3^- to shallow marine waters in runoff, while decreased shallow shelf space supports fewer benthic carbonate-removing organisms, thereby increasing the carbonate saturation state of seawater. Lower Cambrian and Triassic ECEs are consistent with the supply–removal model, but the relatively high sea level during the Devonian ECE is inconsistent and other intervals of relatively high or low sea level do not correspond well to the calcification potential curve based on reef framework (Fig. 10).

Unfortunately, the sea-level curve is not the best proxy for carbonate supply and removal. The level of carbonate removal reflects biomass, which is affected not just by shallow shelf area but by extinction events, nutrient supply, and the percentage of shallow shelf area at low versus high latitudes. Levels of terrigenous weathering and runoff (i.e., supply) are affected by levels of terrestrial vegetation and position of land masses with regard to latitude and humid belts. Additionally, a more or less inverse relationship exists between the supply of terrigenous weathering products and submarine hydrothermal weathering products relative to first-order eustasy. High sea levels associated with increased ridge volume result in low exposed land areas but also higher submarine heat flow, CO_2 venting, and hydrothermal weathering at the ridge. Hence, Ca^{2+} levels associated with high sea levels may exceed those of low sea-level intervals owing to submarine weathering of basaltic rocks (Spencer and Hardie, 1990).

The relative amount of marine carbonate "removal" is difficult to constrain. However, the seawater $^{87}Sr/^{86}Sr$ curve may provide a somewhat better proxy for terrigenous weathering products delivered to marine water as runoff than does the eustasy curve alone (e.g., Tardy *et al.*, 1989; Raymo, 1991). A high $^{87}Sr/^{86}Sr$ ratio indicates relatively high levels of sialic terrigenous weathering products, because ^{87}Sr is sourced largely from the weathering of rubidium-rich continental rocks. Although the marine $^{87}Sr/^{86}Sr$ ratio is buffered by the weathering of older carbonate rocks, increased continental weathering during eustatic lowstands combined with decreased submarine hydrothermal weathering of ^{86}Sr-rich basaltic rocks should result in a relatively high $^{87}Sr/^{86}Sr$ ratio (Faure, 1986). The $^{87}Sr/^{86}Sr$ curve indicates a general-

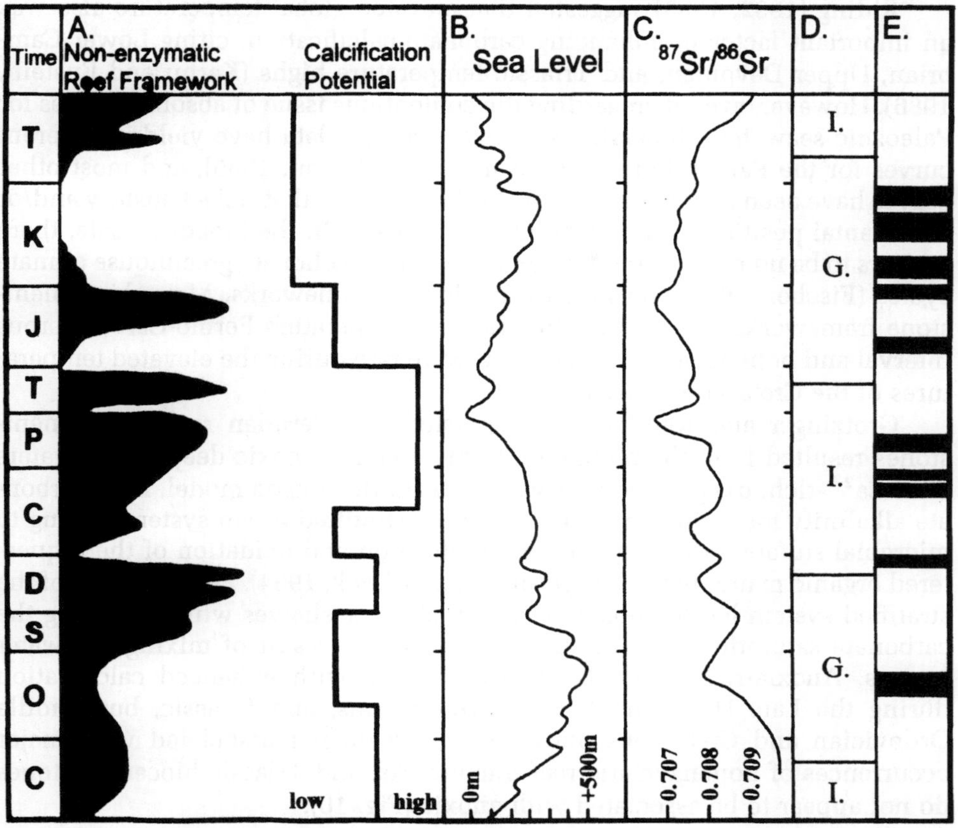

FIGURE 10. Temporal distribution of Phanerozoic reefs and inferred calcification potential compared to parameters that may affect saturation states of marine waters with respect to carbonates. (A) Reef abundance and inferred calcification potential. (B) Sea level (Hallam, 1992) where 0 m represents modern sea level. (C) $^{87}Sr/^{86}Sr$ ratios for Phanerozoic marine waters (Koepnick *et al.*, 1985) wherein higher ratios may correlate with increased terrestrial runoff. (D) First-order climatic cycles: Icehouse/Greenhouse (I/G) (Fischer, 1982). (E) Oceanic anoxic events (after Kauffman and Fagerstrom, 1993).

ized decreasing trend from the Cambrian to the Cretaceous and an increase through the Cenozoic (Koepnick *et al.*, 1985) (Fig. 10). Short-term oscillations suggest increased terrigenous supply during the Silurian, Late Devonian, Late Carboniferous, and Early Triassic. Hence, the terrigenous supply proxy is consistent with the distribution of Paleozoic nonenzymatic reef framework highs, except for the major inconsistency in the Upper Permian (Fig. 10). The $^{87}Sr/^{86}Sr$ curve is less consistent with nonenzymatic reef frameworks for the remaining part of the Phanerozoic with a major low during the Jurassic automicrite framework peak and increasing terrigenous input through the Cenozoic. However, the latter may reflect the new boundary conditions imposed by globally lower carbonate saturation states resulting from carbonate removal by calcareous plankton.

Riding (1992, 1993) suggested that elevated global temperature also was an important factor in enhancing carbonate calcification, citing Lower Cambrian, Upper Devonian, and Triassic temperature highs (Karhu and Epstein, 1986). However, even disregarding the contentious issue of absolute values for Paleozoic seawater temperatures, oxygen isotope data have yielded different curves for the Paleozoic (e.g., Hudson and Anderson, 1989), and most other curves have been constructed on the basis of models that reflect eustasy and/or continental positions (e.g., Worsley *et al.*, 1994). On the broadest scale, there appears to be no particular relationship between icehouse–greenhouse climate cycles (Fischer, 1982) and nonenzymatic reef frameworks. Major biocementstone frameworks occurred during the low temperature Permo-Carboniferous interval and nonenzymatic frameworks were rare during the elevated temperatures of the Cretaceous (Fig. 10).

Grotzinger and Knoll (1995) suggested that Permian reefal biocementstones resulted from the mixing of highly alkaline anoxic deep ocean waters with Ca^{2+}-rich, oxygenated surface waters. In the anoxia model, high carbonate alkalinity forms in deep anoxic parts of stratified ocean systems owing to microbial sulfate reduction and ammonification and oxidation of the sequestered organic matter (e.g., Kempe and Kazmierczak, 1994). Perturbations of the stratified system led to anoxia events on shallow shelves while elevating the carbonate saturation state of surface waters as a result of mixing the water masses. Anoxia-rich intervals were associated with enhanced calcification during the Late Devonian, Permo-Carboniferous, and Jurassic, but Middle Ordovician and Cretaceous anoxia events were not associated with major occurrences of nonenzymatic reef frameworks, and Triassic biocementstones do not appear to be associated with anoxia (Fig. 10).

6. Paleoecological Controls on Nonenzymatic Framework Distribution

The rise of skeletal reef builders clearly caused a decline in the relative occurrence of nonenzymatic reef frameworks, but what evidence exists for direct competition? The majority of modular skeletal organisms presumably had growth rates greater than most nonenzymatic carbonate fabrics. Therefore, they could presumably overgrow and "outcompete" nonenzymatic carbonate inducers, thereby displacing them from reefs. However, the accretion rates of most nonenzymatic reefal carbonates are poorly constrained. Lower Carboniferous reefal thrombolites accreted vertically as much as one meter during the life span of single stalked crinoids (Webb, 1987). Additionally, automicrites may be produced by nonskeletal metazoan communities (e.g., sponges) (Reitner, 1993), not just by low-relief microbial biofilms. Hence, growth rates for some automicrites may have been higher than would be expected for microbial mats and biofilm-producing communities that contained metazoans (e.g., sponges) may have efficiently competed with skeletal organisms for substrate.

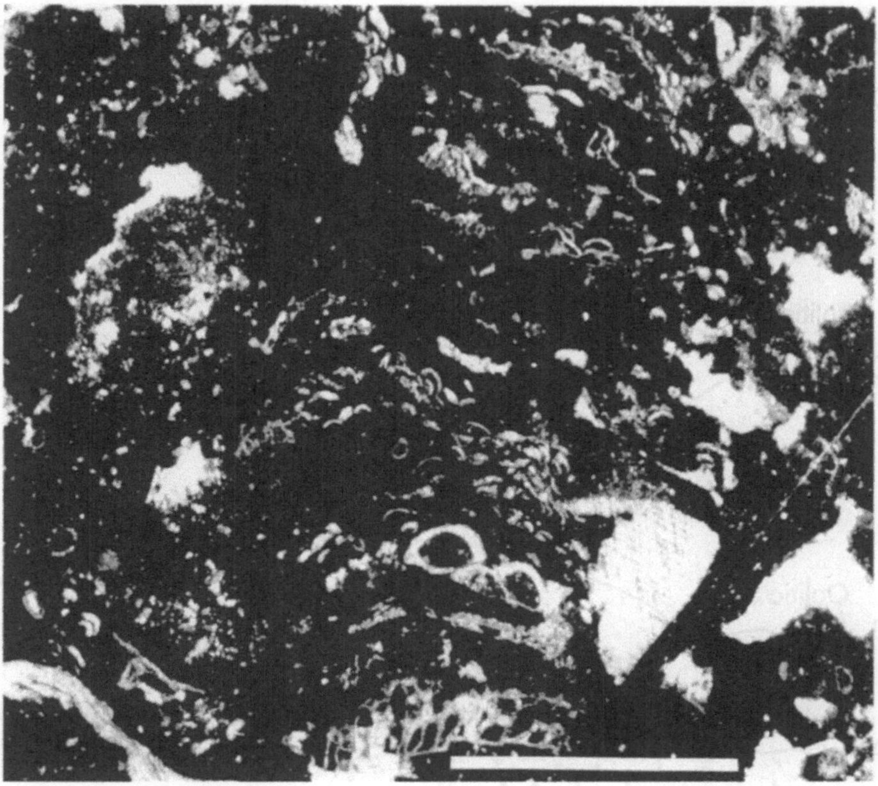

FIGURE 11. Thrombolite (automicrite) reef framework from Tournaisian (Lower Carboniferous) patch reefs in the Gudman Formation, east-central Queensland, Australia. Note that encrusting skeletal organisms outline sequential accretionary surfaces of the thrombolite. Scale bar = 0.2 cm.

Repeated, irregular encrustation of thrombolite surfaces by encrusting skeletal organisms (Fig. 11) suggests periodic disturbance (e.g., possibly by grazing) wherein the biofilm was removed, exposing the underlying lithified substrate for a sufficient time for settlement by skeletal metazoans. However, the calcifying biofilm invariably reestablished itself over the skeletal encrusters, suggesting that it was an able competitor within the community.

Regardless of accretion rates, benthic automicrite was absent from many skeletal framework-dominated Phanerozoic reefs, and conversely some automicrite-rich reefs appear to have resulted from the exclusion of skeletal organisms by temporal (e.g., extinction events) or spatial (i.e., environmental) parameters. Late Cambrian stromatolite-dominated reefs presumably reflect the absence of competitive skeletal reef builders following the extinction of archaeocyaths, and stromatolite- or thrombolite-dominated reefs in the Famennian and Early Carboniferous have been attributed widely to the loss of skeletal reef builders at the Frasnian–Famennian extinction event (e.g., West,

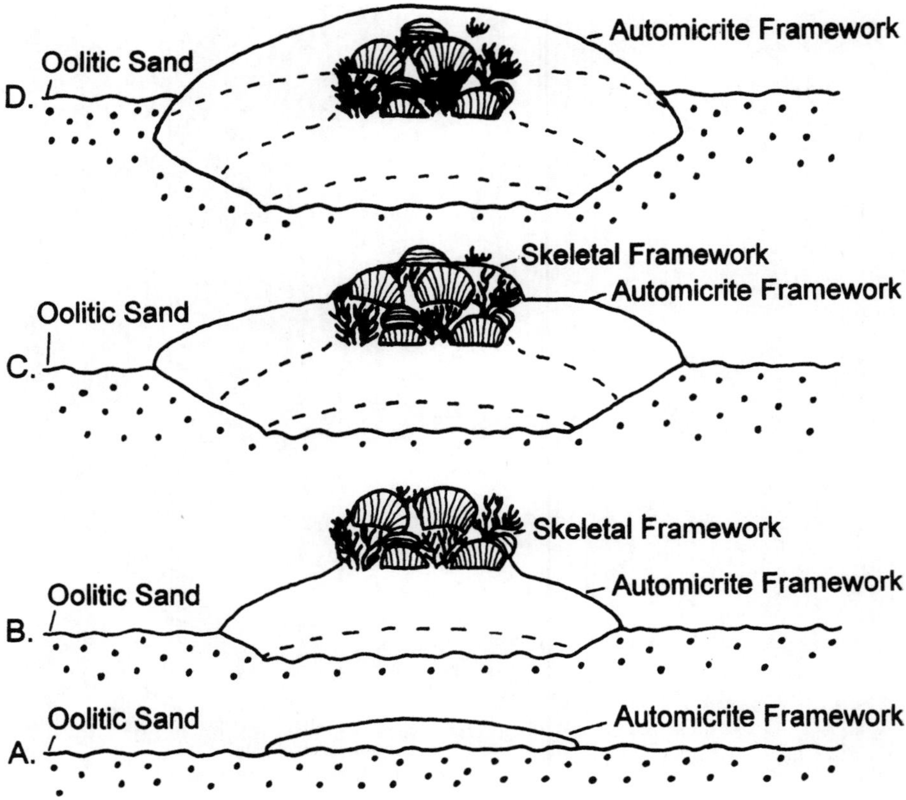

FIGURE 12. Depositional sequence of late Visean thrombolite-coral reefs in the Lion Creek Limestone, east-central Queensland, Australia (after Webb, 1989). (A) Time one: Mobile oolitic sands were initially stabilized by thrombolite (automicrite framework). (B) Time two: As sufficient relief was established on the seafloor by thrombolites, abundant corals (skeletal framework) colonized the reef crests. (C) Time three: A thrombolite buffer zone was maintained between mobile, flanking sands and the coral framework. (D) Time four: Skeletal framework was overgrown by thrombolite as the reef retracted and ooids began to encroach on the reef flanks.

1988). On the basis of such evidence, Schubert and Bottjer (1992) suggested that most Phanerozoic marine stromatolites were postextinction "disaster" forms. However, automicrites were abundant in reefs through much of the Phanerozoic, not just during postextinction intervals (e.g., Soja, 1994; Pratt, 1995; Webb, 1996). Hence, automicrite-inducing communities continuously occupied reefal environments, but only dominated reefs volumetrically when faster-growing skeletal organisms were temporarily excluded; they were not merely opportunistic communities that invaded newly vacated habitats following extinction events.

Lower Carboniferous (Visean) reefs in Queensland, Australia provide evidence of the direct interaction between automicrite producers and skeletal organisms (Fig. 12) (Webb, 1989, 1999). Shifting ooid sand substrates were

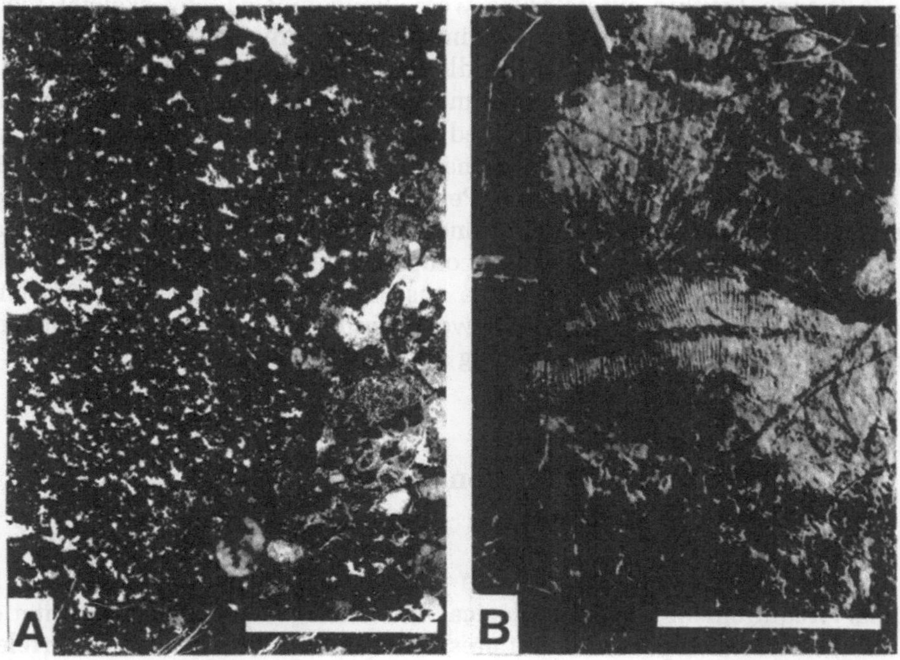

FIGURE 13. Primary reef framework types in Visean reefs in Lion Creek Limestone. (A) Thrombolite dominates the automicrite facies of Lion Creek reefs. Thin section; scale bar = 1.0 cm. (B) Lithostrotionoid and syringoporoid corals dominate the skeletal facies of Lion Creek reefs. Scale bar = 10 cm.

initially stabilized by thrombolites containing lithistid sponges and volumetrically insignificant skeletal organisms (Fig. 13A). However, when a certain area of substrate was stabilized and a certain relief above the surrounding ooid sands had been acquired, the crests of the buildups were colonized by a dense growth of colonial corals, resulting in skeletal framework (Fig. 13B). The skeletal framework persisted until such time as the mobile sands began to encroach on the marginal thrombolites. The corals were then displaced by the thrombolite-forming community, prior to being buried by sand. Hence, both stratigraphically and ecologically, the coral framework community was buffered from mobile sands by the thrombolite-forming community.

Skeletal and automicrite frameworks also coexisted in Jurassic reefs. Shallow reefs were volumetrically dominated by skeletal organisms, but owed their rigidity and relief to co-occurring automicrites (Leinfelder *et al.*, 1994). Contemporary automicrite-dominated reefs were attributed to the exclusion of most skeletal metazoans by dysoxic conditions (Leinfelder *et al.*, 1996). Hence, although Jurassic skeletal reef builders volumetrically dominated some reefs, they did not exclude automicrite-inducing communities and could not have built rigid structures without them.

Although benthic automicrites were "outcompeted" by skeletal metazoans in many environments, calcimicrobes competed somewhat better. Zhuravlev (1996) suggested that Middle Cambrian calcimicrobe communities effectively excluded solitary skeletal metazoans from reefs. Abundant benthic calcimicrobes (e.g., *Renalcis*) occurred in Frasnian reefs that also contained skeletal organisms, and nonenzymatic carbonate encrusters (e.g., *Archaeolithoporella*) were abundant in Permian and Triassic reefs that contained skeletal metazoans. Hence, some nonenzymatic carbonates must have had relatively high accretion rates and/or competed well for substrate with skeletal organisms. Additionally, as was the case for automicrites in Queensland Visean reefs, the nonenzymatic framework components of Permo-Triassic reefs provided substrates for accompanying skeletal organisms.

7. Nonenzymatic Reefal Carbonates and Global Change: Summary

In general, supersaturation with respect to carbonate minerals is the primary requirement for the biological induction of carbonate precipitation and high levels of supersaturation presumably would be expected to correlate positively with the volume of nonenzymatic carbonates. However, the types of nonenzymatic carbonates and possibly volume to some extent also may reflect the type and amount of controlling organic matter (Castanier *et al.*, 1999) and the rate of precipitation (e.g., Merz-Priess, 1997). Although the rate of precipitation is positively correlated with the level of supersaturation, other factors including temperature and the presence and concentration of organic and inorganic inhibitors also may affect precipitation rates. Although inhibitors such as Fe^{2+} and Mn^{2+} would not have been important after global O_2 levels became sufficiently high to oxidize them (Sumner and Grotzinger, 1996), Mg^{2+}, organic phosphates, and various organic acids may have mediated global or regional nonenzymatic carbonate fluxes to some extent.

Competition between calcifying and noncalcifying or carbonate-inhibiting biofilm communities through time is particularly difficult to address. The majority of modern marine biofilms appear to inhibit precipitation, despite the elevated levels of carbonate supersaturation in shallow tropical waters. Hence, the ratio of calcifying to noncalcifying biofilms through time could have partly controlled the abundance of nonenzymatic carbonates. However, it seems unlikely that the types of organic matter that favor carbonate nucleation and metabolic processes that increase alkalinity and/or pH would be preferentially excluded from shallow reefal environments over long periods of time. Hence, changing global calcification potential as recorded by nonenzymatic reefal carbonates most likely reflects changing extrinsic parameters, chief among them being the level of supersaturation.

Future work is needed to better map the distribution of and characterize the different types of organic matter and communities in calcifying and noncalcifying biofilms. Toward that aim, fundamental issues remain regarding even the recognition of biogenicity of some ancient nonenzymatic carbonates (e.g., Webb *et al.*, 1998; Riding, 2000). New biogeochemical methods for the identification of the various products of biologically induced precipitation are required. Secular trends in marine carbonate saturation levels should become better known as pCO_2 data become better constrained and new global carbon budget models are developed, allowing the relationship of carbonate saturation and nonenzymatic carbonates to be better tested. Carbonate saturation also may be independently investigated in terms of its effects on biotic evolution (e.g., Buddemeier and Fautin, 1996). Finally, much additional research is needed to understand the global flux and inhibitory effects of the various natural calcification inhibitors.

8. Conclusions

The organisms responsible for nonenzymatic reef carbonates (i.e., largely microbes) played important roles in reef construction throughout geological history, only being relegated to predominantly cryptic habitats during the Cenozoic. The temporal distribution and abundance of different nonenzymatic reef framework components were controlled by competitive interactions with skeletal metazoans and by a variety of extrinsic physicochemical parameters that directly affected the calcification potential of marine waters. Temporal variations in the level of marine supersaturation with respect to carbonate minerals may have controlled major aspects of the distribution of nonenzymatic reef frameworks. The controls on marine carbonate saturation states are many and varied, but the temporal distribution of nonenzymatic reef carbonates may serve as a proxy for fluctuations in the relative levels of global carbonate supersaturation.

The recognition and understanding of nonenzymatic reef carbonates are imperative to an understanding of the evolution of reef communities. Although skeletal metazoans dominated many reefs since the Cambrian, nonenzymatic carbonate-producing communities were important reef builders in their own right, and through direct interaction with skeletal metazoans they greatly affected the character, composition, and distribution of skeletal reef-building communities. Additionally, microbial metabolism plays a large role in reefal carbonate production even in skeletal organisms (e.g., zooxanthellae in scleractinian corals). Hence, microbes, some of the smallest organisms on earth, have been crucial in the construction of reefs, perhaps the largest biological structures produced through time. Therefore, the temporal distribution of reef communities cannot be understood only in terms of the biological history of one or more distinct skeletal metazoan communities. Extrinsic fluctuations in marine chemistry have changed important "ground rules" for reef building by regulating the occurrence, types, and accretion rates of nonenzymatic reef framework components and thereby allowing different skeletal metazoan communities to participate in the active construction of rigid, relief-bearing structures. Not only did the organisms that built reefs change through time, but the requisite characteristics of reef-building organisms changed through time as a direct result of interaction with the products of biologically induced carbonate precipitation.

ACKNOWLEDGMENTS: The author wishes to thank G. D. Stanley, Jr., for the invitation to contribute this chapter and D. Y. Sumner and M. T. Whalen for thoughtful reviews of the manuscript. The chapter owes much to conversations with numerous colleagues and the author's continued association with IGCP Project 380: Biosedimentology of Microbial Buildups.

References

Addadi, L., and Weiner, S., 1989, Stereochemical and structural relations between macromolecules and crystals in biomineralization, in: *Biomineralization: Chemical and Biochemical Perspectives* (S. Mann, J. Webb, and R. P. J. Williams, eds.), VCH Verlagsgesellschaft, Weinheim, pp. 133–156.

Aitken, J. D., 1989, Giant "algal" reefs, Middle/Upper Proterozoic Little Dal Group (>770, <1200 Ma), Mackenzie Mountains, N.W.T., Canada, in: *Reefs, Canada and Adjacent Areas*, Vol. 13 (H. H. J. Geldsetzer, N. P. James, and G. E. Tebbutt, eds.), Canadian Society of Petroleum Geologists, Calgary, pp. 13–23.

Aitken, J. D., and Narbonne, G. M., 1989, Two occurrences of Precambrian thrombolites from the Mackenzie Mountains, Northwestern Canada, *Palaios* **4**:384–388.

Atlas, R. M., and Bartha, R., 1993, *Microbial Ecology: Fundamentals and Applications*, 3rd ed., Benjamin/Cummings Pub. Co., Redwood City, CA.

Awramik, S. M., 1991, Archaean and Proterozoic stromatolites, in: *Calcareous Algae and Stromatolites* (R. Riding, ed.), Springer-Verlag, Berlin, pp. 289–304.

Bau, M., and Dulski, P., 1996, Distribution of yttrium and rare-earth elements in the Penge and Kuruman iron-formations, Transvaal Supergroup, South Africa, *Precamb. Res.* **79**:37–55.

Becker, R. T., House, M. R., and Kirchgasser, W. T., 1993, Devonian goniatite biostratigraphy and timing of facies movements in the Frasnian of the Canning Basin, Western Australia, in: *High Resolution Stratigraphy*, Geol. Soc. Lond. Spec. Publ. Vol. 70 (E. A. Hailwood and R. B. Kidd, eds.), The Geological Society of London, Bath, pp. 293–321.

Berner, R. A., 1968, Calcium carbonate concretions formed by the decomposition of organic matter, *Science* **159**:195–197.

Beukes, N. J., 1987, Facies relations, depositional environments and diagenesis in a major early Proterozoic stromatolitic carbonate platform to basinal sequence, Campbellrand Subgroup, Transvaal Supergroup, southern Africa, *Sediment. Geol.* **54**:1–46.

Borowitzka, M. A., 1989, Carbonate calcification in algae-initiation and control, in: *Biomineralization: Chemical and Biochemical Perspectives* (S. Mann, J. Webb, and R. J. P. Williams, eds.), VCH Verlagsgesellschaft, Weinheim, pp. 63–94.

Bosellini, F., and Russo, A., 1992, Stratigraphy and facies of an Oligocene fringing reef (Castro Limestone, Salento Peninsula, southern Italy), *Facies* **26**:145–165.

Bosscher, H., and Schlager, W., 1993, Accumulation rates of carbonate platforms, *J. Geol.* **101**:345–355.

Bourque, P.-A., 1989, Silurian reefs, in: *Reefs, Canada and Adjacent Areas*, Can. Soc. Petrol. Geol. Mem. Vol. 13 (H. H. J. Geldsetzer, N. P. James, and G. E. Tebbutt, eds.), Canadian Society of Petroleum Geologists, Calgary, pp. 245–250.

Braga, J. C., Martin, J. M., and Riding, R., 1996, Internal structure of segment reefs: *Halimeda* algal mounds in the Mediterranean Miocene, *Geology* **24**:35–38.

Brunton, F. R., Copper, P., and Dixon, O. A., 1997, Silurian reef-building episodes, *Proc. 8th Int. Coral Reef Symp., Panama* **2**:1643–1650.

Bryan, J. R., 1991, A Paleocene coral-algal-sponge reef from southwestern Alabama and the ecology of early Tertiary reefs, *Lethaia* **24**:423–438.

Buddemeier, R. W., and Fautin, D. G., 1996, Saturation state and the evolution and biogeography of symbiotic calcification, in: *Biomineralization 93, 7th International Symposium on Biomineralization,* Bull. Inst. Océanogr. Monaco, Num. Spec. 14(4), (D. Allemand and J. P. Cuif, eds.), Musée Océanographique, Monaco, pp. 23–32.

Burne, R. V., and Moore, L. S., 1987, Microbialites: Organosedimentary deposits of benthic microbial communities, *Palaios* **2**:241–254.

Cahuzac, B., and Chaix, C., 1996, Structural and faunal evolution of Chattian–Miocene reefs and corals in western France and the northeastern Atlantic Ocean, in: *Models for Carbonate Stratigraphy from Miocene Reef Complexes of Mediterranean Regions,* Concepts in Sedimentology and Paleontology Vol. 5 (E. K. Franseen, M. Esteban, W. C. Ward, and J. M. Rouchy, eds.), SEPM, The Society of Sedimentary Geology, Tulsa, pp. 105–127.

Camoin, G. F. (ed.), 1999, Microbial mediation in carbonate diagenesis, *Sed. Geol.* **126**:340.

Camoin, G. F., and Montaggioni, L. F., 1994, High energy coralgal–stromatolite frameworks from Holocene reefs (Tahiti, French Polynesia), *Sedimentology* **41**:655–676.

Camoin, G. F., Gautret, P., Montaggioni, L. F., and Cabioch, G., 1999, Nature and environmental significance of microbialites in Quaternary reefs: The Tahiti paradox, *Sed. Geol.* **126**:271–304.

Castanier, S., Le Métayer-Levrel, G., and Perthuisot, J.-P., 1999, Ca-carbonates precipitation and limestone genesis—The microbiogeologists point of view, *Sed. Geol.* **126**:9–23.

Copper, P., 1989, Enigmas in Phanerozoic reef development, *Mem. Assoc. Australas. Palaeont.* **8**:371–385.

Copper, P., 1997, Reefs and carbonate productivity: Cambrian through Devonian, *Proc. 8th Int. Coral Reef Symp. Panama* **2**:1623–1630.

Copper, P., and Brunton, F. R., 1991, A global review of Silurian reefs, in: *The Murchison Symposium: Proceedings of an International Conference on the Silurian System,* Spec. Pap. Palaeont. Vol. 44 (M. G. Basett, P. D. Lane, and D. Edwards, eds.), The Palaeontological Association, London, pp. 225–259.

Davies, G. R., and Nassichuk, W. W., 1989, Upper Carboniferous tubular algal boundstone reefs in the Otto Fiord Formation, Canadian Arctic Archipelago, in: *Reefs, Canada and Adjacent Areas,* Can. Soc. Petrol.Geol. Mem. Vol. 13 (H. H. J. Geldsetzer, N. P. James, and G. E. Tebbutt, eds.), Canadian Society for Petroleum Geologists, Calgary, pp. 649–657.

Decho, A., 1990, Microbial exopolymer secretions in ocean environments: Their role(s) in food webs and marine processes, *Oceanogr. Mar. Biol. Annu. Rev.* **28**:73–154.

Doumenge, F. (ed.), 1994, *Past and Present Biomineralization Processes: Consideration about the Carbonate Cycle,* Bull. Inst. Oceanogr., Monaco, Num. Spec. 13, Monaco.

Ehrlich, H. L., 1990, *Geomicrobiology,* 2nd ed., Marcel Dekker, New York.

Esteban, M., 1996, An overview of Miocene reefs from Mediterranean areas: General trends and facies models, in: *Models for Carbonate Stratigraphy from Miocene Reef Complexes of Mediterranean Regions,* Concepts in Sedimentology and Paleontology Vol. 5 (E. K. Franseen, M. Esteban, W. C. Ward, and J. M. Rouchy, eds.), SEPM, The Society of Sedimentary Geology, Tulsa, pp. 3–53.

Fagerstrom, J. A., 1987, *The Evolution of Reef Communities,* Wiley, New York.

Fagerstrom, J. A., 1991, Reef-building guilds and a checklist for determining guild membership, *Coral Reefs* **10**:47–52.

Fagerstrom, J. A., 1994, The history of Devonian-Carboniferous reef communities: Extinctions, effects, recovery, *Facies* **30**:177–192.

Fairchild, I. J., 1991, Origins of carbonate in Neoproterozoic stromatolites and the identification of modern analogues, *Precamb. Res.* **53**:281–299.

Faure, G., 1986, *Principles of Isotope Geology,* 2nd ed., Wiley, New York.

Feldmann, M., and McKenzie, J. A., 1997, Messinian stromatolite-thrombolite associations, Santa Pola, SE Spain: An analogue for the Palaeozoic? *Sedimentology* **44**:893–914.

Fischer, A. G., 1982, Long-term climatic oscillations recorded in stratigraphy, in: *Climate in Earth History* (W. Berger, ed.), Natl. Res. Council Studies in Geophys., Washington, DC, pp. 97–104.

Flügel, E., 1994, Pangean shelf carbonates: Controls and paleoclimatic significance of Permian and Triassic reefs, in: *Pangea: Paleoclimate, Tectonics, and Sedimentation during Accretion, Zenith, and Breakup of a Supercontinent*, Geol. Soc. Am. Spec. Pap. Vol. 288 (G. D. Klein, ed.) Geological Society of America, Boulder, CO, pp. 247–266.

Flügel, E., and Flügel-Kahler, E., 1992, Phanerozoic reef evolution: Basic questions and data base, *Facies* **26**:167–278.

Franseen, E. K., Esteban, M., Ward, W. C., and Rouchy, J.-M. (eds.), 1996, *Models for Carbonate Stratigraphy from Miocene Reef Complexes of Mediterranean Regions*, SEPM, Soc. Sediment. Geol., Concepts in Sedimentology and Paleontology, Vol. 5, Tulsa.

Gebelein, C. D., 1976, The effects of the physical, chemical and biological evolution of the Earth, in: *Stromatolites*, Developments in Sedimentology, Vol. 20 (M. R. Walter, ed.), Elsevier, Amsterdam, pp. 499–515.

Geldsetzer, H. H. J., James, N. P., and Tebbutt, G. E. (eds.), 1989, *Reefs, Canada and Adjacent Areas*, Can. Soc. Petrol. Geol. Mem. Vol. 13, Canadian Society for Petroleum Geologists, Calgary.

George, A. D., and Powell, C. McA., 1997, Paleokarst in an Upper Devonian reef complex of the Canning Basin, Western Australia, *J. Sediment. Res.* **67**:935–944.

George, A. D., Playford, P. E., Powell, C. McA., and Tornatora, P. M., 1997, Lithofacies and sequence development on an Upper Devonian mixed carbonate–siliciclastic fore-reef slope, Canning Basin, Western Australia, *Sedimentology.* **44**:843–867.

Gili, E., Masse, J.-P., and Skelton, P. W., 1995, Rudists as gregarious sediment-dwellers, not reef-builders, on Cretaceous carbonate platforms, *Palaeogeogr. Palaeoclimatol. Palaeoecol.* **118**:245–267.

Grotzinger, J. P., 1986, Evolution of Early Proterozoic passive-margin carbonate platform, Rocknest Formation, Wopmay Orogen, Northwest Territories, Canada, *J. Sediment. Pet.* **56**:831–847.

Grotzinger, J. P., 1989a, Construction of Early Proterozoic (1.9 Ga) barrier reef complex, Rocknest Platform, Northwest Territories, in: *Reefs, Canada and Adjacent Areas*, Can. Soc. Petrol. Geol. Mem.Vol. 13 (H. H. J. Geldsetzer, N. P. James, and G. E. Tebbutt, eds.), Canadian Society of Petroleum Geologists, Calgary, pp. 30–37.

Grotzinger, J. P., 1989b, Facies and evolution of Precambrian carbonate depositional systems: emergence of the modern platform archetype, in: *Controls on Carbonate Platform and Basin Development*, SEPM Spec. Publ. Vol. 44 (P. D. Crevello, J. L. Wilson, J. F. Sarg, and J. F. Read, eds.), Society of Economic Paleontologists and Mineralogists, Tulsa, pp. 79–106.

Grotzinger, J. P., 1989c, Introduction to Precambrian reefs, in: *Reefs, Canada and Adjacent Areas* (H. H. J. Geldsetzer, N. P. James, and G. E. Tebbutt, eds.) *Can. Soc. Petrol. Geol. Mem.* **13**:9–12.

Grotzinger, J. P., 1994, Trends in Precambrian carbonate sediments and their implications for understanding evolution, in: *Early Life on Earth* (S. Bengtson, ed.), Columbia University Press, New York, pp. 245–258.

Grotzinger, J. P., and Kasting, J. F., 1993, New constraints on Precambrian ocean composition, *J. Geol.* **101**:235–243.

Grotzinger, J. P., and Knoll, A. H., 1995, Anomalous carbonate precipitates: Is the Precambrian the key to the Permian? *Palaios* **10**:578–596.

Grotzinger, J. P., and Rothman, D. H., 1996, An abiotic model for stromatolite morphogenesis, *Nature* **383**:423–425.

Hallam, A., 1992, *Phanerozoic Sea-Level Changes*, Columbia University Press, New York.

Heckel, P. H., 1974, Carbonate buildups in the geologic record: A review, in: *Reefs in Time and Space*, SEPM Spec. Publ. Vol. 18 (L. F. Laporte, ed.), Society of Economic Paleontologists and Mineralogists, Tulsa, pp. 90–154.

Herbig, H. G., and Weber, H. M., 1996, Facies and stromatoporoid biostromes in the Strunian (latest Devonian) of the Aachen region, Germany, in: *Global and Regional Controls on Biogenic Sedimentation. I. Reef Evolution. Research Reports*, Göttinger Arb. Geol. Paläont. Vol. Sb2 (J. Reitner, F. Neuweiler, and F. Gunkel, eds.), Geologische Institute, Georg-August-Universität Göttingen, Göttingen, pp. 359–364.

Hoffman, P. F., 1989, Pethei reef complex (1.9 Ga), Great Slave Lake, N. W. T., in: *Reefs, Canada and Adjacent Areas*, Can. Soc. Petrol. Geol. Mem. Vol. 13 (H. H. J. Geldsetzer, N. P. James, and G. E. Tebbutt, eds.), Canadian Society of Petroleum Geologists, Calgary, pp. 38–48.

Hoffman, P. E., and Grotzinger, J. P., 1989, Abner/Denault reef complex (2.1 Ga), Labrador Trough, N. E. Quebec, in: *Reefs, Canada and Adjacent Areas*, Can. Soc. Petrol. Geol. Mem. Vol. 13 (H. H. J. Geldsetzer, N. P. James, and G. E. Tebbutt, eds.), Canadian Society of Petroleum Geologists, Calgary, pp. 49–54.

Höfling, R., Moussavian, E., and Götz, S., 1996, Development of Cretaceous and Paleogene reef communities in the Alpine-Mediterranean Realm—Selected case studies, in: *Global and Regional Controls on Biogenic Sedimentation. I. Reef Evolution. Research Reports*, Göttinger Arb. Geol. Paläont. Vol. Sb2 (J. Reitner, F. Neuweiler, and F. Gunkel, eds.), Geologische Institute, Georg-August-Universität Göttingen, Göttingen, pp. 179–183.

Holland, H. D., 1984, *The Chemical Evolution of the Atmosphere and Oceans*, Princeton University Press, Princeton, NJ.

Horbury, A. D., 1992, A late Dinantian peloid cementstone-palaeoberesellid buildup from North Lancashire, England, *Sediment. Geol.* **79**:117–137.

Hudson, J. D., and Anderson, T. F., 1989, Ocean temperatures and isotopic compositions through time, *Trans. R. Soc. Edinb. Earth Sci.* **80**:185–192.

Jackson, M. J., 1989, Lower Proterozoic Cowles Lake foredeep reef, N. W. T., Canada, in: *Reefs, Canada and Adjacent Areas*, Can. Soc. Pet. Geol. Mem. Vol. 13 (H. H. J. Geldsetzer, N. P. James, and G. E. Tebbutt, eds.), Canadian Society of Petroleum Geologists, Calgary, pp. 64–71.

James, N. P., 1983, Reef environment, in: *Carbonate Depositional Environments*, Am. Assoc. Petrol. Geol. Mem. Vol. 33 (P. A. Scholle, D. G. Bebout, and C. H. Moore, eds.), American Association of Petroleum Geologists, Tulsa, pp. 346–440.

James, N. P., and Bourque, P.-A., 1992, Reefs and mounds, in: *Facies Models: Response to Sea Level Change* (R. G. Walker and N. P. James, eds.), Geol. Assoc. Canada, St. John's, Newfoundland, pp. 323–347.

Jasionowski, M., 1996, Budowle serpulowo-mikrobialitowe sarmatu na Roztoczu: Niezwykle joint-venture, *Przegl. Geol.* **44**:1044–1048.

Kah, L. C., and Grotzinger, J. P., 1992, Early Proterozoic (1.9 Ga) thrombolites of the Rocknest Formation, Northwest Territories, Canada, *Palaios* **7**:305–315.

Kano, A., 1994, Quantitative compositions and reef development of the Silurian limestones of Gotland, Sweden, *Cour. Forschungsinst. Senckenb.* **172**:141–146.

Karhu, J., and Epstein, S., 1986, The implications of the oxygen isotope records in coexisting cherts and phosphorites, *Geochim. Cosmochim. Acta* **50**:1745–1756.

Kasting, J. F., 1987, Theoretical constraints on oxygen and carbon dioxide concentrations in the Precambrian atmosphere, *Precamb. Res.* **34**:205–229.

Kauffman, E. G., and Fagerstrom, J. A., 1993, The Phanerozoic evolution of reef diversity, in: *Species Diversity in Ecological Communities* (R. E. Rickfels, and D. Schluter, eds.), University of Chicago Press, Chicago, pp. 315–329.

Kauffman, E. G., and Johnson, C. C., 1988, The morphological and ecological evolution of Middle and Upper Cretaceous reef-building rudistids, *Palaios* **3**:194–216.

Kauffman, E. G., and Johnson, C. C., 1997, Ecological evolution of Jurassic–Cretaceous Caribbean reefs, *Proc. 8th Int. Coral Reef Symp. Panama* **2**:1669–1676.

Kempe, S., and Degens, E. T., 1985, An early soda ocean? *Chem. Geol.* **53**:95–108.

Kempe, S., and Kazmierczak, J., 1990, Calcium carbonate supersaturation and the formation of *in situ* calcified stromatolites, in: *Facets of Modern Biogeochemistry* (V. Ittekkot, S. Kempe, W. Michaelis, and A. Spitzy, eds.), Springer-Verlag, Berlin, pp. 255–278.

Kempe, S., and Kazmierczak, J., 1994, The role of alkalinity in the evolution of ocean chemistry, organization of living systems, and biocalcification processes, in: *Past and Present Biomineralization Processes: Considerations of the Carbonate Cycle*, Bull. Inst. Océanogr. Monaco Num. Spec. 13 (F. Doumenge, ed.), Musée Océanographique, Monaco, pp. 61–117.

Kerans, C., and Donaldson, J. A., 1989, Deep-water conical stromatolite reef, Sulky Formation (Dismal Lakes Group), Middle Proterozoic, N.W.T., in: *Reefs, Canada and Adjacent Areas*,

Can. Soc. Petrol. Geol. Mem. Vol. 13 (H. H. J. Geldsetzer, N. P. James, and G. E. Tebbutt, eds.), Canadian Society of Petroleum Geologists, Calgary, pp. 81–88.

Kerans, C., Hurley, N. F., and Playford, P. E., 1986, Marine diagenesis in Devonian reef complexes of the Canning Basin, Western Australia, in: *Reef Diagenesis* (J. H. Schroeder and B. H. Perser, eds.), Springer-Verlag, Berlin, pp. 357–380.

Kershaw, S., 1997, Palaeoenvironmental change in Silurian stromatoporoid reefs, Gotland, Sweden, *Bol. R. Soc. Esp. Hist. Nat. Secc. Geol.* **91**:329–342.

Kiessling, W., Flügel, E., and Golonka, J., 1999, Paleoreef maps: Evaluation of a comprehensive database on Phanerzoic reefs, *Amer. Assoc. Petrol. Geol. Bull.* **83**:1552–1587.

Kirkland, B. L., Dickson, J. A. D., Wood, R. A., and Land, L. S., 1998, Microbialites and microstratigraphy: The origin of encrustations in the middle and upper Capitan Formation, Guadalupe Mountains, Texas and New Mexico, USA, *J. Sediment. Res.* **68**:956–969.

Knoll, A. H., Fairchild, I. J., and Swett, K., 1993, Calcified microbes in Neoproterozoic carbonates: Implications for our understanding of the Proterozoic/Cambrian transition, *Palaios* **8**:512–525.

Koepnick, R. B., Burke, W. H., Denison, R. E., Hetherington, E. A., Nelson, H. F., Otto, J. B., and Waite, L. E., 1985, Construction of the seawater $^{87}Sr/^{86}Sr$ curve for the Cenozoic and Cretaceous: supporting data, *Chem. Geol.* **58**:55–81.

Krumbein, W. E., 1979, Photolithotropic and chemoorganotrophic activity of bacteria and algae as related to beachrock formation and degradation (Gulf of Aqaba, Sinai), *Geomicrobiol. J.* **1**:139–203.

Kruse, P. D., Zhuravlev, A. Yu., and James, N. P., 1995, Primordial metazoan–calcimicrobial reefs: Tommotian (Early Cambrian) of the Siberian Platform, *Palaios* **10**:291–321.

Kuznetsov, V. G., 1996, Framework-free reefs: Types, structure, and position in the geological sequence: Communication 1. Problem statement. Cyathiform organism buildups, *Lithol. Miner. Resour.* **31**:493–501.

Leinfelder, R. R., Krautter, M., Laternser, K., Nose, M., Schmid, D. U., Schweigert, G., Werner, W., Keupp, H., Brugger, H., Herrmann, R., Rehfeld-Keifer, U., Schroeder, J. H., Reinhold, C., Koch, R., Zeiss, A., Schweizer, V., Christmann, H., Menges, G., and Luterbacher, H., 1994, The origin of Jurassic reefs: Current research developments and results, *Facies* **31**:1–56.

Leinfelder, R. R., Werner, W., Nose, M., Schmid, D. U., Krautter, M., Laternser, K., Takacs, M., and Hartmann, D., 1996, Paleoecology, growth parameters and dynamics of coral, sponge and microbialite reefs from the Late Jurassic, in: *Global and Regional Controls on Biogenic Sedimentation. I. Reef Evolution. Research Reports*, Göttinger Arb. Geol. Paläont. Vol. Sb2 (J. Reitner, F. Neuweiler, and F. Gunkel, eds.), Geologische Institute, Georg-August-Universität Göttingen, Göttingen, pp. 227–248.

Lowe, D. R., 1980, Stromatolites 3,400-Myr old from the Archean of Western Australia, *Nature* **284**:441–443.

Lowenstam, H. A., 1981, Minerals formed by organisms, *Science* **211**:1126–1131.

Luchinina, V. A., 1985, Algal buildups of the early Paleozoic in the north of the Siberian Platform, *Akad. Nauk SSSR Sibirskoye Otdel. Tr. Inst. Geol. Geofiz.* **628**:45–50.

Lyons, W. B., Long, D. T., Hines, M. E., Gaudete, H. E., and Armstrong, P. B., 1984, Calcification of cyanobacterial mats in Solar Lake, Sinai, *Geology* **12**:623–626.

Macintyre, I. G., Reid, R. P., and Steneck, R. S., 1996, Growth history of stromatolites in a Holocene fringing reef, Stocking Island, Bahamas, *J. Sediment. Res.* **66**:231–242.

Mackenzie, F. T., and Morse, J. W., 1992, Sedimentary carbonates through Phanerozoic time, *Geochim. Cosmochim. Acta* **56**:3281–3295.

Mann, S., 1989, Crystallochemical strategies in biomineralization, in: *Biomineralization: Chemical and Biochemical Perspectives* (S. Mann, J. Webb, and R. P. J. Williams, eds.), VCH Verlagsgesellschaft, Weinheim, pp. 35–62.

Martín, J. M., and Braga, J. C., 1994, Messinian events in the Sorbas Basin in southeastern Spain and their implications in the recent history of the Mediterranean, *Sediment. Geol.* **90**:257–268.

Martín, J. M., Braga, J. G., and Riding, R., 1997, Miocene *Halimeda* algal–microbial segment reefs in the marginal Mediterranean Sorbas Basin, Spain, *Sedimentology* **44**:441–456.

Merz, M. U., 1992, The biology of carbonate precipitation by cyanobacteria, *Facies* **26:** 81–101.

Merz-Preiss, M. U., 1997, Calcifying cyanobacteria and cyanobacterial micrite production: Environmental interpretation, in: *Microbial Mediation in Carbonate Diagenesis. International Workshop (I.A.S.–A.S.F.–I.G.C.P. 380), Chichilianne, 22–24 September, 1997—Abstract Book, Publ. Assoc. Sedimentol. Francais* **26:**45,46.

Merz-Preiss, M. U., and Riding, R., 1999, Cyanobacterial tufa calcification in freshwater streams: Ambient environment, chemical thresholds and biological processes, *Sed. Geol.* **126:**103–124.

Mitterer, R. M., and Cunningham, R., Jr., 1985, The interaction of natural organic matter with grain surfaces: Implications for calcium carbonate precipitation, in: *Carbonate Cements*, S.E.P.M Spec. Publ. Vol. 36 (N. Schneidermann and P. M. Harris, eds.), Society of Economic Paleontologists and Mineralogists, Tulsa, pp. 17–31.

Montaggioni, L. F., and Camoin, G. F., 1993, Stromatolites associated with coralgal communities in Holocene high-energy reefs, *Geology* **21:**149–152.

Morse, J. W., 1983, The kinetics of calcium carbonate dissolution and precipitation, in: *Reviews in Mineralogy*, Vol. 11 (R. J. Reeder, ed.), Mineralogical Society of America, Washington, DC, pp. 227–264.

Morse, J. W., and Mackenzie, F. T., 1990, *Geochemistry of Sedimentary Carbonates*, Elsevier, Amsterdam.

Narbonne, G. M., and James, N. P., 1996, Mesoproterozoic deep-water reefs from Borden Peninsula, Arctic Canada, *Sedimentology* **43:**827–848.

Neuweiler, F., 1993, Development of Albian microbialites and microbialite reefs at marginal platform areas of the Vasco-Cantabrian Basin (Soba Reef area, Cantabria, N. Spain), *Facies* **29:**231–250.

Neuweiler, F., Gautret, P., Thiel, V., Lange, R., Michaelis, W., and Reitner, J., 1999, Petrology of lower cretaceous carbonate mud mounds (Albian, N. Spain): Insights into organomineralic deposits of the geological record, *Sedimentology* **46:**837–859.

Newell, N. D., 1972, The evolution of reefs, *Sci. Am.* **226:**54–65.

Opdyke, B. N., and Wilkinson, B. H., 1990, Paleolatitude distribution of Phanerozoic marine ooids and cements, *Palaeogeogr. Palaeoclim. Palaeoecol.* **78:**135–148.

Orme, G. R., Flood, P. G., and Sargent, C. E. G., 1978, Sedimentation trends in the lee of outer (ribbon) reefs, northern region of the Great Barrier Reef province, *Phil. Trans. R. Soc. Lond. A* **291:**85–99.

Pedley, M., 1996, Miocene reef distributions and their associations in the central Mediterranean region: An overview, in: *Models for Carbonate Stratigraphy from Miocene Reef Complexes of Mediterranean Regions*, S.E.P.M Soc. Sediment. Geol. Concepts in Sedimentology and Paleontology Vol. 5 (E. K. Franseen, M. Esteban, W. C. Ward, and J.-M. Rouchy, eds.), S.E.P.M., The Society of Sedimentary Geology, Tulsa, pp. 73–87.

Pentecost, A., and Riding, R., 1986, Calcification in cyanobacteria, in: *Biomineralization in Lower Plants and Animals*, Syst. Assoc. Spec. Vol. 30 (B. S. C. Leadbeater and R. Riding, eds.), Clarendon Press, Oxford (for the Systematics Association), pp. 73–90.

Pickard, N. A. H., 1996, evidence for microbial influence on the development of Lower Carboniferous buildups, in: *Recent Advances in Lower Carboniferous Geology*, Geol. Soc. Lond. Spec. Publ. Vol. 107 (P. Strogen, I. D. Somerville, and G. Ll. Jones, eds.), The Geological Society of London, Bath, pp. 65–82.

Pinckney, J. L., and Reid, R. P., 1997, Productivity and community composition of stromatolitic microbial mats in the Exuma Cays, Bahamas, in: *Biosedimentology of Microbial Buildups, IGCP Project No. 380, Proceedings of 2nd Meeting Göttingen/Germany 1996* (F. Neuweiler, J. Reitner, and C. Monty, eds.), *Facies* **36:**204–207.

Pisera, A., 1985, Diagenesis of the Middle Miocene algal-vermetid reefs from Poland: An example of the local late diagenetic continental control, *Proc. 5th Int. Coral Reef Symp. Tahiti* **3:**277–282.

Pisera, A., 1996, Miocene reefs of the Paratethys: A review, in: *Models for Carbonate Stratigraphy from Miocene Reef Complexes of Mediterranean Regions*, S.E.P.M. Soc. Sediment. Geol. Concepts in Sedimentology and Paleontology Vol. 5 (E. K. Franseen, M. Esteban, W. C. Ward, and J. M. Rouchy, eds.), S.E.P.M., The Society of Sedimentary Geology, Tulsa, pp. 97–194.

Playford, P. E., 1980, Devonian "Great Barrier Reef" of Canning Basin, Western Australia, *Bull. Am. Assoc. Pet. Geol.* **64**:814–840.

Pomar, L., Ward, W. C., and Green, D. G., 1996, Upper Miocene reef complex of the Llucmajor area, Mallorca, Spain, in: *Models for Carbonate Stratigraphy from Miocene Reef Complexes of Mediterranean Regions*, S.E.P.M. Soc. Sediment. Geol. Concepts in Sedimentology and Paleontology Vol. 5 (E. K. Franseen, M. Esteban, W. C. Ward, and J.-M. Rouchy, eds.), S.E.P.M., The Society of Sedimentary Geology, Tulsa, pp. 191–225.

Pratt, B. R., 1982a, Stromatolite decline—A reconsideration, *Geology* **10**:512–515.

Pratt, B. R., 1982b, Stromatolitic framework of carbonate mud-mounds, *J. Sediment. Pet.* **52**:1203–1227.

Pratt, B. R., 1995, The origin, biota and evolution of deep-water mud-mounds, in: *Carbonate Mud-Mounds: Their Origin and Evolution*, Int. Assoc. Sediment. Spec. Publ. Vol. 23 (C. L. V. Monty, D. W. J. Bosence, P. H. Bridges, and B. R. Pratt, eds.), International Association of Sedimentologists, Oxford, pp. 49–123.

Rasmussen, K. A., Macintyre, I. G., and Prufert, L., 1993, Modern stromatolite reefs fringing a brackish coastline, Chetumal Bay, Belize, *Geology* **21**:199–202.

Raymo, M. E., 1991, Geochemical evidence supporting T. C. Chamberlin's theory of glaciation, *Geology* **19**:344–347.

Reid, R. P., Macintyre, I. G., Browne, K. M., Steneck, R. S., and Miller, J., 1995, Modern marine stromatolites in the Exuma Cays, Bahamas: Uncommonly common, *Facies* **33**:1–18.

Reitner, J., 1993, Modern cryptic microbialite/metazoan facies from Lizard Island (Great Barrier Reef, Australia): Formation and concepts, *Facies* **29**:3–40.

Reitner, J., Gautret, P., Marin, F., and Neuweiler, F., 1995a, Automicrites in a modern marine microbialite. Formation model via organic matrices (Lizard Island, Great Barrier Reef, Australia), in: *Biomineralization 93, 7th International Symposium on Biomineralization*, Bull. Inst. Océanogr. Monaco, Num. Spec. 14(2) (D. Allemand and J. P. Cuif, eds.), Musée Océonographique, Monaco, pp. 237–263.

Reitner, J., Neuweiler, F., and Gautret, P., 1995b, Part II, Modern and fossil automicrites: Implications for mud mound genesis, in: *Mud Mounds: A Polygenetic Spectrum of Fine-grained Carbonate Buildups* (J. Reitner, and F. Neuweiler, eds.), *Facies* **32**:4–17.

Ricketts, B. D., and Donaldson, J. A., 1989, Stromatolite reef development on a mud-dominated platform in the Middle Precambrian Belcher Group of Hudson Bay, in: *Reefs, Canada and Adjacent Areas*, Can. Soc. Petrol. Geol. Mem. Vol. 13 (H. H. J. Geldsetzer, N. P. James, and G. E. Tebbutt, eds.), Canadian Society of Petroleum Geologists, Calgary, pp. 113–119.

Riding, R., 1991, Classification of microbial carbonates, in: *Calcareous Algae and Stromatolites* (R. Riding, ed.), Springer-Verlag, Berlin, pp. 21–51.

Riding, R., 1992, Temporal variation in calcification in marine cyanobacteria, *J. Geol. Soc. Lond.* **149**:979–989.

Riding, R., 1993, Phanerozoic patterns of marine $CaCO_3$ precipitation, *Naturwissen* **80**: 513–516.

Riding, R., 2000, Microbial carbonates: The geological record of calcified bacterial–algal mats and biofilms, *Sedimentology* **471**(Suppl. 1):179–214.

Riding, R., and Awramik, S. M., 2000, *Microbial Sediments*, Springer-Verlag, Heidelberg.

Riding, R., and Zhuravlev, A. Yu., 1995, Structure and diversity of oldest sponge–microbe reefs: Lower Cambrian, Aldan River, Siberia, *Geology* **23**:649–652.

Riding, R., Martin, J. M., and Braga, J. C., 1991, Coral–stromatolite reef framework, Upper Miocene, Almería, Spain, *Sedimentology* **38**:799–818.

Roberts, H. H., Phipps, C. V., and Effeudi, L., 1987, *Halimeda* bioherms of the eastern Java Sea, Indonesia, *Geology* **15**:371–374.

Robertson, W. G., 1982, The solubility concept, in: *Biological Mineralization and Demineralization* (G. H. Nancollas, ed.), Springer-Verlag, Berlin, pp. 5–21.

Rowland, S. M., and Gangloff, R. A., 1988, Structure and paleoecology of Lower Cambrian reefs, *Palaios* **3**:111–135.

Sami, T. T., and James, N. P., 1996, Synsedimentary cements as Paleoproterozoic platform building blocks, Pethei Group, northwestern Canada, *J. Sediment. Res.* **66**:209–222.

Sandberg, P. A., 1983, An oscillating trend in Phanerozoic non-skeletal carbonate mineralogy, *Nature* **305**:19–22.

Schubert, J. K., and Bottjer, D. J., 1992, Early Triassic stromatolites as post-mass extinction disaster forms, *Geology* **20**:883–886.

Schumann, D., 1996, Upper Cretaceous rudist reefs of central Oman, in: *Global and Regional Controls on Biogenic Sedimentation. I. Reef Evolution. Research Reports*, Göttinger Arb. Geol. Paläont. Vol. Sb2 (J. Reitner, F. Neuweiler, and F. Gunkel, eds.), Geologische Institute, Georg-August-Universität Göttingen, Göttingen, pp. 193–197.

Schumann-Kindel, G., Bergbauer, M., Manz, W., Szewzyk, U., and Reitner, J., 1997, Aerobic and anaerobic microorganisms in modern sponges: A possible relationship to fossilization processes, in: *Biosedimentology of Microbial Buildups*, IGCP Project No. 380, Proceedings of 2nd meeting Göttingen/Germany 1996 (F. Neuweiler, J. Reitner, and C. Monty, eds.), *Facies* **36**:268–272.

Schuster, F., 1996, Paleocene coral reefs and related facies associations, Kharga Oasis, Western Desert, Egypt, in: *Global and Regional Controls on Biogenic Sedimentation. 1. Reef Evolution.* Research Reports, Göttinger Arb. Geol. Paläont. Vol. Sb2 (J. Reitner, F. Neuweiler, and F. Gunkel, eds.), Geologische Institute, Georg-August-Universität Göttingen, Göttingen, pp. 169–174.

Scott, R. W., and Brenckle, P. L., 1977, Biotic zonation of a Lower Cretaceous coral–algal–rudist reef, Arizona, *Proc. 3rd Int. Coral Reef Symp. Miami* **2**:183–189.

Shen, J., Yu, C., and Bao, H., 1997, A Late-Devonian (Famennian) *Renalcis–Epiphyton* reef at Zhaijiang, Guilin, South China, *Facies* **37**:195–210.

Shinn, E. A., Robin, D., Lidz, B., and Hudson, H., 1983, Influence of deposition and early diagenesis on porosity and chemical compaction in two Paleozoic buildups: Mississippian and Permian age rocks in the Sacramento Mountains, New Mexico, in: *Carbonate Buildups: A Core Workshop*, S.E.P.M. Core Workshop Vol. 4 (P. M. Harris, ed.), Society of Economic Paleontologists and Mineralogists, Tulsa, pp. 182–223.

Simkiss, K., 1986, The processes of biomineralization in lower plants and animals—An overview, in: *Biomineralization in Lower Plants and Animals,* Syst. Assoc. Spec. Vol. 30 (B. S. C., Leadbeater and R. Riding, eds.), Clarendon Press, Oxford (for the Systematics Association), pp. 19–38.

Soja, C. M., 1994, Significance of Silurian stromatolite–sphinctozoan reefs, *Geology* **22**:355–358.

Spencer, A. J., and Hardie, L. A., 1990, Control of seawater composition by mixing of river waters and mid-ocean ridge hydrothermal brines, in: *Fluid–Mineral Interactions: A Tribute to H. P. Eugster*, Geochem. Soc. Publ. Vol. 19 (R. J. Spencer and I. M. Chou, eds.), Geochemical Society, University Park, PA, pp. 409–419.

Stanley, G. D., Jr., 1988, The history of early Mesozoic reef communities: A three-step process, *Palaios* **3**:170–183.

Stanley, G. D., Jr., 1997, Evolution of reefs of the Mesozoic, *Proc. 8th Int. Coral Reef Symp. Panama* **2**:1657–1662.

Stearn, C. W., Halim-Dihardja, M. K., and Nishida, D. K., 1987, An oil-producing stromatoporoid patch reef in the Famennian (Devonian) Wabamun Group, Normandville Field, Alberta, *Palaios* **2**:560–570.

Sumner, D. Y., and Bowring, S. A., 1996, U-Pb geochronologic constraints on deposition of the Campbellrand Subgroup, Transvaal Supergroup, South Africa, *Precamb. Res.* **79**:25–35.

Sumner, D. Y., and Grotzinger, J. P., 1996, Were kinetics of Archean calcium carbonate precipitation related to oxygen concentration? *Geology* **24**:119–122.

Tardy, Y., N'Kounkou, R., and Probst, J. J., 1989, The global water cycle and continental erosion during Phanerozoic time (570 my), *Am. J. Sci.* **289**:455–483.

Teitz, M. W., and Mountjoy, E. W., 1989, The Late Proterozoic Yellowhead carbonate platform west of Jasper, Alberta, in: *Reefs, Canada and Adjacent Areas*, Can. Soc. Petrol. Geol. Mem. Vol. 13 (H. H. J. Geldsetzer, N. P. James, and G. E. Tebbutt, eds.), Canadian Society of Petroleum Geologists, Calgary, pp. 129–134.

Trichet, J., and Défarge, C., 1995, Nonbiologically supported organomineralization, in: *Biomineralization 93, 7th International Symposium on Biomineralization, Bull. Inst. Océanogr. Monaco Num. Spec.* **14**(2):203–236.

Turner, E. C., Narbonne, G. M., and James, N. P., 1993, Neoproterozoic reef microstructure from the Little Dal Group, northwestern Canada, *Geology* **21**:259–262.

Turner, E. C., James, N. P., and Narbonne, G. M., 1997, Growth dynamics of Neoproterozoic calcimicrobial reefs, Mackenzie Mountains, northwest Canada, *J. Sediment. Res.* **67**:437–450.

Turner, E. C., James, N. P., and Narbonne, G. M., 2000, Taphonomic control on microstructure in Early Neoproterozoic reefal stromatolites and thrombolites, *Palaios* **15**:87–111.

Urey, H. C., 1951, The origin and development of the Earth and other terrestrial planets, *Geochim. Cosmochim. Acta* **1**:209–277.

Veizer, J., 1994, The Archean–Proterozoic transition and its environmental implications, in: *Early Life of Earth* (S. Bengtson, ed.), Columbia Univ. Press, New York, pp. 208–219.

Walter, M. R., and Heys, G. R., 1985, Links between the rise of the Metazoa and the decline of stromatolites, *Precamb. Res.* **29**:149–174.

Walter, M. R., Buick, R., and Dunlop, J. S. R., 1980, Stromatolites 3400–3500 Myr old from the North Pole area, Western Australia, *Nature* **284**:443–445.

Webb, G. E. 1987, Late Mississippian thrombolite bioherms from the Pitkin Formation of northern Arkansas, *Bull. Geol. Soc. Am.* **99**:686–698.

Webb, G. E., 1989 [1991], Late Visean coral–algal bioherms from the Lion Creek Formation of Queensland, Australia, *C. R. 11th Int. Congr. Strat. Geol. Carbonif.* **3**:282–295.

Webb, G. E., 1994, Non-Waulsortian Mississippian bioherms: A comparative analysis, in: *Pangea: Global Environments and Resources*, Can. Soc. Petrol. Geol. Mem. Vol. 17 (A. F. Embry, B. Beauchamp, and D. J. Glass, eds.), Canadian Society of Petroleum Geologists, Calgary, pp. 701–712.

Webb, G. E., 1996, Was Phanerozoic reef history controlled by the distribution of nonenzymatically secreted reef carbonates (microbial carbonate and biologically induced cement)? *Sedimentology* **43**:947–971.

Webb, G. E., 1998, Earliest known Carboniferous shallow-water reefs, Gudman Formation (Tn1b), Queensland, Australia: Implications for Late Devonian reef collapse and recovery, *Geology* **26**:951–954.

Webb, G. E., 1999, Youngest Early Carboniferous (late Visean) shallow-water patch reefs in eastern Australia (Rockhampton Group, Queensland): Combining quantitative micro- and macroscale data, *Facies* **41**:111–140.

Webb, G. E., and Jell, J. S., 1997, Cryptic microbialite in subtidal reef framework and intertidal solution cavities in beachrock, Heron Reef, Great Barrier Reef, Australia: Preliminary observations, in: *Biosedimentology of Microbial Buildups*, IGCP Project No. 380, Proceedings of 2nd Meeting Göttingen/Germany 1996 (F. Neuweiler, J. Reitner, and C. Monty, eds.), *Facies* **36**:219–223.

Webb, G. E., Baker, J. C., and Jell, J. S., 1998, Inferred syngenetic textural evolution in Holocene cryptic reefal microbialites, Heron Reef, Great Barrier Reef, Australia, *Geology* **26**:355–358.

Webby, B. B., 1984, Ordovician reefs and climate: A review, in: *Aspects of the Ordovician System*, Vol. 295 (D. L. Bruton, ed.), Palaeont. Contrib. Univ., Oslo, pp. 87–98.

West, R. R., 1988, Temporal changes in Carboniferous reef mound communities, *Palaios* **3**:152–169.

Westbroek, P., Buddemeier, B., Coleman, M., Kok, D. J., Fautin, D., and Stal, 1994, Strategies for the study of climate forcing by calcification, in: *Past and Present Biomineralization Processes: Considerations of the Carbonate Cycle*, Bull. Inst. Océanogr. Monaco Num. Spec. 13 (F. Doumenge, ed.), Musée Océanographique, Monaco, pp. 37–60.

Wilkinson, B. H., and Walker, J. C. G., 1989, Phanerozoic cycling of sedimentary carbonates, *Am. J. Sci.* **289**:525–248.

Wilkinson, B. H., Owen, R. M., and Carroll, A. R., 1985, Submarine hydrothermal weathering, global eustasy, and carbonate polymorphism in Phanerozoic marine oolites, *J. Sediment. Pet.* **55**:171–183.

Wollast, R., 1971, Kinetic aspects of the nucleation and growth of calcite from aqueous solutions, in: *Carbonate Cements* (O. P. Bricker, ed.), Johns Hopkins University Press, Baltimore, pp. 264–273.

Wollast, R., 1994, The relative importance of biomineralization and dissolution of $CaCO_3$ in the global carbon cycle, in: *Past and Present Biomineralization Processes: Considerations about the Carbon Cycle*, Bull. Inst. Océanogr., Monaco, Num. Spec. 13 (F. Doumenge, ed.), Musée Océanographique, Monaco, pp. 13–35.

Worsley, T. R., Moore, T. L., Fraticelli, C. M., and Scotese, C. R., 1994, Phanerozoic CO_2 levels and global temperatures inferred from changing paleogeography, in: *Pangea: Paleoclimate, Tectonics, and Sedimentation During Accretion, Zenith, and Breakup of a Supercontinent*, Geol. Soc. Amer. Spec. Pap. Vol. 288 (G. D. Klein, ed.), Geological Society of America, Boulder, CO, pp. 57–73.

Zankl, H., 1993, The origin of high-Mg-calcite microbialites in cryptic habitats of Caribbean coral reefs—their dependence on light and turbulence, *Facies* **29**:55–59.

Zhuravlev, A. Yu., 1996, Reef ecosystem recovery after the Early Cambrian extinction, in: *Biotic Recovery from Mass Extinction Events*, Geol. Soc. Lond. Spec. Publ. Vol. 102 (M. B. Hart, ed.), The Geological Society of London, Bath, pp. 79–96.

Chapter 6

A Half Century Later

The Permian Guadalupian Reef Complex of West Texas and Eastern New Mexico

NORMAN D. NEWELL

1. Introduction

It was a compliment for me to be invited by George Stanley to write a chapter on the Guadalupe reefs. I was one of several pioneers on this subject long ago and I pondered seriously about whether or not I could contribute anything of substance at this time. The Guadalupian reef complex is one of the most studied reefs in the world and serves as a model for understanding fossil reefs elsewhere. The application of the microscope to the organic fabric removed much uncertainty about this famous reef barrier. Although I visited this reef many times since and attended a field conference in Alpine in 1991, my interests have been directed to other fields (e.g., the Great Bahama Bank, Pacific atolls, and the systematics of fossil bivalves). I concluded that it might be useful to start by writing of my own experience in the Guadalupes and to review a few of the diverse viewpoints that have since emerged on this complex subject.

NORMAN D. NEWELL • Paleontology Division (Invertebrates), The American Museum of Natural History, New York, New York 10024.

The History and Sedimentology of Ancient Reef Systems, edited by George D. Stanley Jr., Kluwer Academic/Plenum Publishers, New York, 2001.

In 1977, a Field Conference Guidebook on the Guadalupe Complex (Hileman and Mazzullo, 1977) appeared. It was an excellent hardback, profusely illustrated, quarto volume of more than 500 pages, touching on a wide range of studies, with a selected bibliography of more than 100 published works. My present purpose is not to critique this work, which would be a staggering undertaking, but to approach the subject from a more personal point of view.

2. Early Work in the Guadalupes

Early in the 20th century it had become evident that buried fossil reefs worldwide frequently were traps for petroleum. Consequently, much of the globe had been and still is continuously explored for ancient reefs with favorable geologic conditions for petroleum reservoirs using outcrop data, seismic soundings, and test drilling.

It was in the late 1930s that I first became interested in the Guadalupe region, especially Pine Springs (Fig. 1), as the source of well-preserved, beautiful silicified invertebrate fossils, easily extractable with acid. Following the pioneer geologic work of G. H. Girty (1902), E. R. Lloyd (1929), P. B. King and Robert E. King (1929), G. Arthur Cooper (Cooper and Grant, 1972) had been making important collections of fossils from the marine Permian rocks in the mountains around the Delaware Basin in West Texas.

In the following decade I visited the Guadalupes several times, during which I made the acquaintance of Mr. Wallace E. Pratt, a retired Humble Oil and Refining Company geologist, executive, and landowner in the Guadalupes. He told us about favorable oil prospects in the country to the east of the Guadalupe Mountains and suggested that I might undertake a survey of the outcropping reef complex as basic research in oil geology. With Wallace Pratt's influence, the Humble Oil Company gave me enthusiastic and generous financial support for comparisons between modern carbonate enviroments in the Bahama Islands and the rocks of the Guadalupe Mountains.

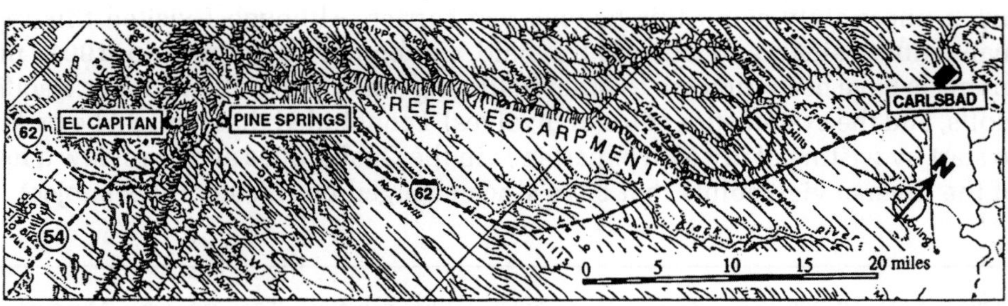

FIGURE 1. Map of the Guadalupe Mountains, West Texas, and southern New Mexico, a magnificent sponge–algal reef complex of Middle Permian age. The fore-reef facies of the Capitan forms the escarpment rim of the deep-water Delaware Basin on the right. The back-reef and ancient shore are in the background. (After King, 1942.)

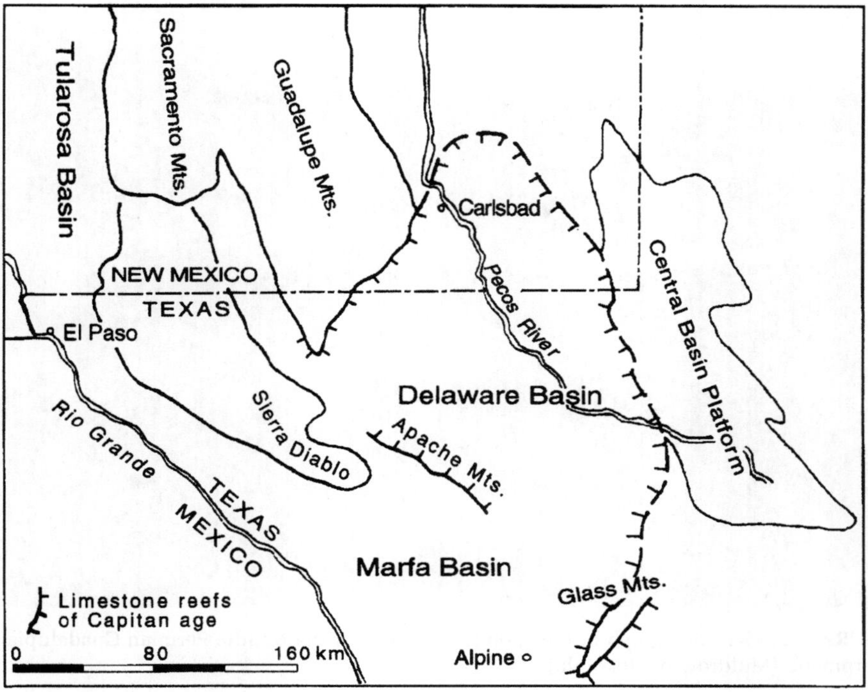

FIGURE 2. The Guadalupe reef complex: Limestone reefs of Permian age, around the Delaware Basin, showing the reef escarpment and the other topographic features. Compare with map in Fig. 1. (After King, 1934.)

Starting in the 1940s, I organized teams of university students from Lawrence, Kansas, Madison, Wisconsin, and Columbia, New York City, to undertake comparative studies of living coral reefs in the Bahama Islands and Permian sponge–algal reefs in the Guadalupe Mountains. All this work, financed by the Humble Oil Company, was published as a book (Newell *et al.*, 1953), republished in 1972, and a paper (Newell and Rigby, 1957). The book attracted wide attention and eventually led Congress to establish the Guadalupe Mountains National Park, the Capitan reef complex being the most studied of any fossil reefs. The work on its geology and fossils continues to the present (Fig. 2).

3. The Guadalupe Reef Barrier

There are parallels between the Capitan complex and modern reefs that indicate that the massive limestone around the adjacent Delaware Basin is indeed a shelf edge barrier (Fig. 3). In the Bahamas we learned that the optimum situation for attached living calcareous algae and corals is at shelf

FIGURE 3. Aerial photograph of the sponge–algal reef of the Middle Permian Guadalupian reef escarpment. (Muldrow photograph.)

edge and at a depth of average wave base in clear waters. Sediments derived from erosion of this reef front are carried away in occasional gaps, becoming progressively finer in the relatively quieter waters of the back reef lagoon. The result is a textural and concurrent biological gradient away from the reef front. Talus from the reef front advancing into deeper waters accumulates as a major part of the reef mass. Most of the bulk of the reef complex thus is composed of detritus.

In the Guadalupe complex, the rock facies agree in their distribution with those of the Bahamas. During growth, the sponge–algal solid reef sheds sediments in both directions: the deep waters of the Delaware Basin and the shallow platform to the north. Algal nodules of both regions are characteristic of the rear margin of the reefs.

The Capitan, unlike modern coral reefs, is almost wholly lacking in corals, their place being taken by calcareous sponges, bryozoans, fusulinid foraminifers, and calcareous algae.

In spite of repeated descriptions based on surface observations and borings, a persistent misconception remains that organic reefs, including the Guadalupian, are frames of cemented *in situ* skeletons of reef builders. Instead, they are better viewed as structural complexes, in which usually more than nine tenths of the volume is composed of detritus and fragments which were quickly converted to limestone during early exposure to fresh water or air in the vadose zone.

Rugose and tabulate corals are extinct forms, unrelated to living reef-building taxa, which at times were quite important in middle Paleozoic reefs. These corals, however, became sparse in the late Paleozoic time, finally becoming extinct toward the end of the Permian. The scleractinian corals, which played such an important role in later reef construction from the Late Triassic to the present (Stanley, 1988), were long considered to have descended from the Paleozoic *Rugosa*, though this is no longer considered likely (Oliver, 1996). Strangely, corals of any sort are rare in the Capitan complex and unknown in the Lower Triassic, where no transitional forms have yet been found.

The Rugosa and Scleractinia morphologically are quite dissimilar. The Rugosa skeletons are composed of calcium carbonate, calcite, while those of the Scleractinia are composed of the polymorph calcium carbonate mineral, aragonite. The purported scleractinian coral, named *Numidiaphyllum* (Flügel, 1976; Ezaki, 1997), was discovered by Gillian Newell in the 1970s, in the Middle Permian of Tunisia, on one of our expeditions there. The failure to find preserved scleractinians in the Lower Triassic is a mystery, but does not signify that ancestors were not living during that time.

Scrutton and Clarkson (1991) have suggested that early Paleozoic scleractinianlike fossils may have had separate histories as offshoots of soft-bodied anemones. Stanley (1988) has suggested that ancestors of the Scleractinia were most likely around during the Early Triassic as soft-bodied anemones and he even cited some Lower Triassic trace fossils attributed to anemones. Romano (1996) summarized molecular dates gathered on living scleractinian corals to suggest that these corals extend back to the Carboniferous, over 300 million years ago.

The fossil reefs of the Guadalupe Mountains share diverse faunas with neighboring low mountains: the Sierra Diablo, Glass Mountains, and other nearby areas. The reef core and back reef carbonates are dolomitized and many of the fossils have been diagenetically destroyed. Decades of paleontologists have etched exquisite silicified invertebrate shells from the lower marine Permian limestone talus, in the deeper part of the Delaware Basin (Newell *et al.*, 1953). Calcareous shells, sponges, algae, fusulinids, and bryozoans are plentiful in the uppermost reef talus, but unfortunately there is no silicification.

4. Changing Ideas about the Capitan Complex

The Capitan complex indeed is difficult to interpret and many of the early conclusions about its nature and origin are controversial. The reef complex provides an instructive lesson on the pitfalls between data gathering and the interpretation of those data. Much remains to be done and learned. After decades of research, one may ask: How much information is enough?

The great Guadalupian escarpment (Fig. 4) has long been regarded by many workers as an exhumed fossil barrier reef developed around the Delaware Basin (Lloyd, 1929). But others (Baars, 1964; Achauer, 1969) thought that

FIGURE 4. Aerial photograph of cliff exposure, west side of Guadalupe Mountain, showing basinward dip of the fore reef talus. Area between peaks is talus. (Newell photograph.)

it is an ancient bank deposit of organic detritus. They were unable to envision reef structures and frame-building organisms in the Capitan.

In my early thinking about the history of the Capitan complex (Newell *et al.*, 1953), I did not consider the effect of significant changes in sea level. Modern studies of the Earth show that in a few years orbital excentricities in the solar system, coupled with changes in ice volume at the north and south poles, cause frequent oscillations in the level of the ocean worldwide. The seafloor also is unstable, as Darwin showed in 1842.

Studies of sinking coral atolls in the Pacific show that there are thousands of drowned volcanoes at depths in this region, which have been mapped and are well known. Accordingly, many features of reef texture and structure in the Guadalupes and elsewhere may be evaluated as effects of dissolution and cementation by fresh waters at the vadose zone during emergence.

5. More Recent Work in the Guadalupes

This generalization prompted a few investigators to look for evidence of one or several exposures of Capitan rocks during their formation. Pioneers in this line of studies were Price *et al.* (1946), who compared Capitan pisolites with algal pisolites of the late Cenozoic fluviatile cover on the Great Plains.

Two geologists, R. J. Dunham and C. M. Thomas (Dunham, 1969), mapped extensive pisolite deposits of the Capitan shelf edge (Fig. 5) and interpreted

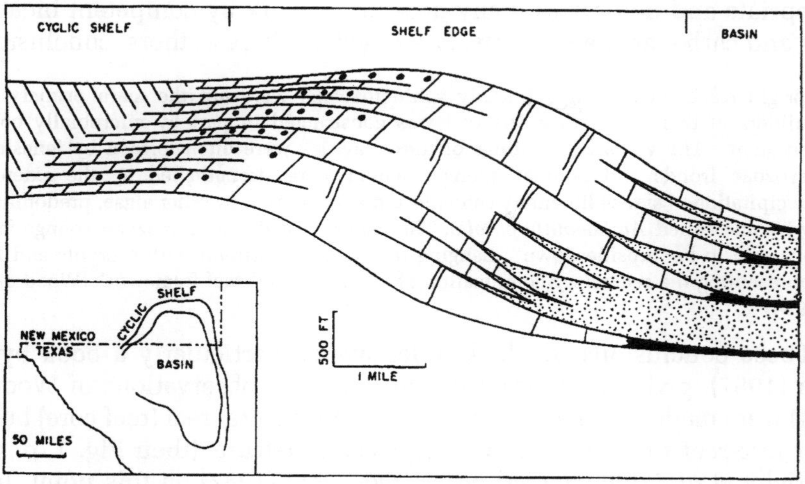

FIGURE 5. Diagrammatic relationships across the Guadalupian reef complex, showing vadose pisolite facies, indicated by dots, at the summit. (After Dunham, 1969.)

them as *vadose caliche* deposits formed during the emergence of the Capitan. Unlike the Great Plains pisolite, which are irregular encrustations of sand grains, the Guadalupian features are small, spherical concretions, deposited in pre-existing cavities in a dolomitic carbonate.

The restricted numbers of larger frame builders, as compared with the less conspicuous small ones, probably explains the prolonged controversy over the reef–nonreef nature of the Capitan. The use of microscopic sections shows that the microbiota constitutes 57–96% of the reef mass (Weidlich and Fagerstrom, 1998). Ubiquitous cavities in the carbonate were more or less filled with early diagenetic cements or quartz-rich sediment of diverse origins. These partially filled cavities are abundant in the Capitan reef.

Wood *et al.* (1996) have shown that many of the openings are primary within a loose framework of frondose bryozoa and platy sponges, such as *Gigantospongia disciformis* (Rigby and Senowbari-Daryan, 1996). These reached as much as 2 m across and formed ceilings of large cavities that supported a diverse community of smaller animals and algae, called a crypt community. Although the Capitan reef was basically fragile, it remained intact by a postmortem cementation and by ingrowth of botryoidal algae and fine calcareous sediment.

Sparse vertical sheets of laminated calcite occur as fissure fillings in the basinward part of the reef (Dunham, 1972). They trend parallel to the shelf margin, are inches to several feet in width, and can be traced tens to hundreds of feet laterally. Their formation clearly involved fracturing of the host rock and their encrusting biota, demonstrating a submarine origin.

Wood *et al.* (1994) published a study of the Capitan reef complex, under the title "Turning the Capitan Reef Upside Down." This title is not really appropriate and denigrates many decades of work by competent biostratigraphers and carbonate petrologists. It is based on these authors' conclusion that:

> upright calcareous sponges (mainly sphinctozoans) were not the main primary reef builders of this reef: rather, the reef was constructed in part by microbially bound sediments. The walls and ceilings of these cavities were often formed by fenestellid bryozoan fronds and laminar sponges which were strengthened by the extensive precipitation of syn-sedimentary cements and encrustation by other algae, predominately by one called *Archaeolithoporella*. The majority of the sphinctozoan sponge fauna was said to be "upside down"; hanging in a pendant fashion within crypts and thus representing the secondary colonization of an established reef framework (Wood *et al.*, 1994).

These conclusions overlook older works, particularly a book by Fagerstrom (1987), giving a different conclusion. The observations of Wood *et al.* (1994) were made on what I consider to be not central reef (reef core) but rather upper fore-reef slope, most of it composed of detritus (their Fig. 1, p. 423).

Rigby *et al.* (1998) agreed with Fagerstrom (1987) on this point. In 1996, Wood *et al.* again tried to redefine the Guadalupian reef, but in my opinion their conclusions are superficial and not based on enough substantial field work to support the claims. More recently, a new paper by Fagerstrom and Weidlich (1999) addresses with great care the previously controversial issues about the reef versus nonreef nature of the Capitan reef, without changing their earlier opinion.

6. Late Permian Mass Extinctions and Their Effect on the Reef

The Guadalupes in the Ochoan stages contain evaporites of gypsum, salt, and dolomite, filling the deeper part of the Delaware Basin and overlying the fossiliferous Capitan reef carbonate. Deep drill corings show that the Ochoan evaporites reach almost 4500 feet in thickness (Adams *et al.*, 1939).

While poor exposures and lack of fossils have turned away biostratigraphers from studying these strata, the Ochoan rocks themselves tell of an episode of marked aridity followed presumably by a paraconformity. Similar types of lithofacies underlie the Upper Permian in England, the North Sea, and Germany.

In the past one and one-half billion years of reef evolution, the Late Permian was the time of the most sweeping annihilation of life. The Guadalupian reef extended only into the Middle Permian, but it was followed by climatic and other changes leading to a mass extinction. I view this mass extinction event as starting in the Middle Permian and culminating in the Upper Permian (Ochoan). I suggest that the Ochoan evaporites point to an episode of aridity indicating a contributing cause of that mass extinction.

Bowring *et al.* (1998) have shown from uranium–lead ratios in China that the boundary between the Triassic and the Permian, occurred 221.4 ($\pm$0.3) million years ago. Erwin (1993) and Bowring *et al.* (1998), in addition to others, have shown convincing evidence that the mass extinction at the end of the Permian was indeed rapid rather than slow and gradual. Many workers, including this author, have changed their minds about how slow the extinction was. Based on new data the extinction event probably lasted less than one million years.

Several generations of paleontologists have pondered possible causes of the dramatic worldwide change in life over this period of time. Peter Ward (1998) thinks the tip of South Africa, where fossil vertebrates of several kinds are abundant and occur in simple, flat-lying stratigraphic sequence, may provide a solution to the mystery. The Permian–Triassic boundary is marked by an abrupt break in the vertrebrate faunas and sudden changes from drab colors below to red beds above. Ward thinks the way to make sense of this situation is to adopt an ingenious theory of Andrew Knoll and several colleagues at Harvard. Knoll *et al.* (1996) gave an explanation that I also favor for the terminal extinction at the end of the Paleozoic Era. It also could apply to the Guadalupian–Ochoan extinction. During extinction periods, there were no ice caps on Earth and the oceans were stagnant, thus becoming charged with the poisonous gas carbon dioxide. During the Guadalupian–Ochoan extinction period, a great flood of lava simultaneously was rising to the surface in Siberia and discharging massive amounts of carbon dioxide into the atmosphere. This simultaneous charging of the waters and the air with carbon dioxide made it lethal for both marine and terrestrial animals alike. Of course, this scenario would have been especially disastrous for shallow-water organisms living on reefs.

Ward (1998) describes evidence from South Africa that corroborates this theory. Although his study deals with a different time span, it indicates that at the end of the extinction there was not only a complete turnover in the fauna but also an abrupt change in color of the sediments from drab to red, indicating the oxidation by enriched oxygen in the postulated carbon dioxide atmosphere.

7. Significance of the Guadalupian Reef Complex and Future Directions of Research

Was the Capitan reef a bank deposit of unlithified sediment? Our photographs (Newell *et al.*, 1953), especially plates 9 (Figs. 1 and 2), 14 (Figs. 1 and 2, 15), and plate 16 (Fig. 3) should have answered the question. We demonstrated in that publication that the upper part of the reef was already lithified while it was forming and that the submerged fore-reef talus slides consisted of cobbles of hard limestone and dolomite, embedded in a mud matrix. We felt that this talus should be studied further to ascertain the kinds of reef limestone involved. As far as I know, this has not yet been done.

Altogether, the evidence indicates that the Guadalupian Reef Complex, one of the best studied reefs in the world, is of extraordinary importance in the overall story of ancient reefs. It also is crucial to the study of global geology.

After reviewing a large number of recent publications on the Capitan complex, it became clear that most of the controversies about its nature and significance have been resolved. The greatest advances came with the application of quantitative petrographic micro-observations of carbonate rock thin sections by R. J. Dunham and C. M. Thomas (Dunham, 1969). Their work showed that most of the reef is composed of microscopic biota not apparent on surface inspection. The needed taxonomy of many reef organisms has yet to be worked out, and it may be that this deficiency has promoted confusion and controversy in the past. Going back to the work of Lloyd and the King brothers in 1929, over 70 years ago, the famous Guadalupe Mountains Capitan reef complex still remains one of the most important and relevant reef models. It will likely remain for decades to come as a centerpiece for any future discussion on fossil reefs.

ACKNOWLEDGMENTS: In the preparation of this brief chapter, I have been aided by several reef experts, especially George Stanley, Keith Rigby, Sr., and Al Fagerstrom. In addition, I would like to thank the following people for their assistance at the American Museum of Natural History: Walter Elvers, who has volunteered many hours in library and other research and assistance with the illustrations, and Gillian Newell, who has ably assisted at every stage in the preparation of the manuscript.

References

Achauer, C. W., 1969, Origin of Capitan Formation, Guadalupe Mountains, New Mexico and Texas, *Am. Assoc. Petroleum Geol. Bull.* **53**(11):2314–2323.

Adams, J. E., Cheney, M. G., DeFord, R. K., Dickey, R. I., Dunbar, C. O., Hills, J. M., King, R. E., Lloyd, E. R., Miller, A. K., and Needham, C. E., 1939, Standard Permian section of North America, *Am. Assoc. Petroleum Geol. Bull.* **23**:1673–1681.

Baars, D. L., 1964, Modern carbonate sediments as a guide to old limestones, *World Oil* **158**(5):95–100.

Bowring, S. A., Erwin, D. H., Jin, Y. G., Martin, M. W., Davidek, K., Wang, W., 1998, U/Pb zircon geochronology and tempo of the End–Permian mass extinction, *Science* **280**:1039–1045.

Cooper G. A., and Grant R. E., 1972, Permian brachiopods of West Texas, I., in: *Smithsonian Contributions to Paleobiology*, Smithsonian Inst. Press, Washington, DC, pp. 231.

Darwin, C., 1842, *The Structure and Distribution of Coral Reefs.* (First part of the geology of the Voyage of the Beagle.) London, Reprinted University of California Press, 1962.

Dunham, R. J., 1969, Vadose pisolite in the Capitan Reef (Permian), New Mexico and Texas, in: *Depositional Environments in Carbonate Rocks* (G. M. Friedman, ed.), SEPM Spec. Pub. 14, pp. 182–191.

Dunham, R. J., 1972, Capitan reef New Mexico and Texas: Facts and questions to aid interpretation and group discussion, *Permian Basin Sect., SEPM, Pub.* 72–15 (no consecutive page numbers).

Erwin, D. H., 1993, *The Great Paleozoic Crisis*, Columbia University Press, New York.

Ezaki, Y., 1997, The Permian coral *Numidiaphyllum:* New insights into Anthozoan phylogeny and Triassic Scleractinian origins, *Palaeontology* **40**(1):1–14.

Fagerstrom, J. A., 1987 (1988), *The Evolution of Reef Communities*, Wiley, New York.

Fagerstrom, J. A., and Weidlich, O., 1999, The origin of the Upper Capitan–Massive Limestone

(Permian), Guadalupe Mountains, New Mexico–Texas: Is it a reef? *Geol. Soc. Am. Bull.* **111**(2):159–176.

Flügel, H. W., 1976, Numidiaphyllidae-eine neue Familie der Rugosa aus dem Ober-Perm von Süd-Tunis, *Neues Jahrb. Geol. Paläontol, Monatsh.* **9**:54–64.

Girty, G. H., 1902, The upper Permian in western Texas, *Am. J. Sci.* (4th ser.) **14**:363–368.

Hileman, L. E., and Mazzullo, S. J. (eds.), 1977, Upper Guadalupian Facies, Permian Reef Complex, Guadalupe Mountains, New Mexico and West Texas, in: *1977 Field Conference Guidebook*, Permian Basin Section, SEPM, pp. 77–16.

King, P. B., 1934, Permian stratigraphy of trans-Pecos Texas, *Geol. Soc. Am. Bull.* **45**:697–798.

King, P. B., 1942, Permian of west Texas and southeastern New Mexico, *Am. Assoc. Petroleum Geol. Bull.* **26**:535–763.

King, P. B, and King, R. E, 1929, Stratigraphy of outcropping Carboniferous and Permian rocks of trans-Pecos Texas, *Am. Assoc. Petroleum Geol. Bull.* **13**:907–926.

Knoll, A. H., Bambach, R. K, Canfield, D. E., and Grotzinger, J. P., 1996, Comparative earth history and Late Permian mass extinction, *Science* **273**: 452–457.

Lloyd, E. R., 1929, Capitan limestone and associated formations of New Mexico and Texas, *Am. Assoc. Petroleum Geol. Bull.* **13**:645–657.

Newell, N. D., and Rigby, J. K., 1957, Geologic studies on the Great Bahama Bank, in: *SEPM Spec. Publ.*, Regional Aspects of Carbonate Deposition, **5**:15–79, Tulsa.

Newell, N. D., Rigby, J. K., Fischer, A. G., Whiteman, A. J., Hickox, J. E., and Bradley, J. S., 1953, *The Permian Reef Complex of the Guadalupe Mountains Region, Texas and New Mexico*, W. H. Freeman, San Francisco. Republished 1973, Hafner Pub. Co. NY.

Oliver, W. A., Jr., 1996, Origins and relationships of Paleozoic coral groups and the origin of the Scleractinia, in: *Paleobiology and Biology of Corals*, Vol. 1 (G. D. Stanley, Jr., ed.), Paleontological Society, Pittsburgh, pp. 107–135.

Price, W. A., Elias, M. K., and Frye, J. C., 1946, Algae reefs in cap rock of Ogallala formation of Llano Estacado plateau, New Mexico and Texas, *Am. Assoc. Petrol. Geol. Bull.* **30**:1742–1746.

Rigby, J. K., and Senowbari-Daryan, B., 1996, *Gigantospongia*, new genus, the largest known Permian sponge, Capitan Limestone, Guadelupe Mountains, New Mexico, *J. Paleontol.* **70**(3):347–355.

Rigby, J. K., Senowbari-Daryan, B., and Liu, H., 1998, Sponges of the Permian Upper Capitan Limestone, Guadalupe Mountains, New Mexico and Texas, *Brigham Young Univ. Geol. Stud.* **43**:19–117.

Romano, S. L., 1996, A molecular perspective on the evolution of scleractinian corals, in: *Paleobiology and Biology of Corals*, Vol. 1 (G. D. Stanley, Jr. ed.), Paleontological Society, Pittsburgh, pp. 39–59.

Scrutton, C. T., and Clarkson, E. N. K., 1991, A new scleractinian-like coral from the Ordovician of the Southern Uplands, Scotland, *Palaeontology* **34**:179–194.

Stanley, G. D., Jr., 1988, The history of Early Mesozoic reef communities: A three-step process, *Palaios* **3**:170–183.

Ward, P. D., 1998, The greenhouse extinction, *Discover* **8**:54–58.

Weidlich O., and Fagerstrom, J. A., 1998, Evolution of the Upper Capitan-Massive (Permian), Guadalupe Mountains, New Mexico, *Brigham Young Univ. Geol. Stud.* **43**:167–187.

Wood, R., Dickson, J. A. D., and Kirkland-George, B., 1994, Turning the Capitan Reef upside down: a new appraisal of the ecology of the Permian Capitan Reef, Guadalupe Mountains, Texas and New Mexico, *Palaios* **9**:422–427.

Wood, R., Dickson, J. A. D., and Kirkland-George, B., 1996, New observations on the ecology of the Permian Capitan Reef, Texas and New Mexico, *Palaeontology* **30**(3):733–762.

Chapter 7

Triassic Reefs of the Tethys

ERIK FLÜGEL and BABA SENOWBARI-DARYAN

1. Introduction: What Do We Know about Triassic Reefs?

The evolution of Triassic reefs started with a long-lasting global crisis of the metazoan reef ecosystem after the Permian–Triassic mass extinction (about 12 Ma), followed by a relatively rapid recovery during the Middle Triassic. Reef systems were differentiated during the Upper Triassic but were severely affected by a global crisis at the Triassic–Jurasic boundary. The present contribution is focused on the biological controls of Triassic reefs, particularly in the Tethyan realm, and on the major changes in reef ecosystems recorded

ERIK FLÜGEL and BABA SENOWBARI-DARYAN • Institute of Paleontology, University Erlangen-Nürnberg, D-91054 Erlangen, Germany.

The History and Sedimentology of Ancient Reef Systems, edited by George D. Stanley Jr., Kluwer Academic/Plenum Publishers, New York, 2001.

by differences in reef types and reef biota. The term "reef" as used in this chapter refers to bioconstructions characterized by (1) biological control during the formation of the structure (predominantly by sessile benthic organisms), (2) a laterally restricted topographic relief, and (3) (inferred) rigidity of the structure.

1.1. Current State of Research

Triassic reefs are known from Europe, Asia, northwestern Australia as well as western North America and western South America (Fig. 1). Mud and reef mounds as well as biostromes and frame-built metazoan reefs have been described from the Middle Triassic (middle and upper Anisian, Ladinian) and Upper Triassic (Carnian, Norian, and Rhaetian). The Lower Triassic "reef gap" interval (Flügel and Stanley, 1984) characteried by the global absence of true ecological reefs and potential reef builders (no coralline sponges, no scleractinian corals, no calcareous algae) comprises a time interval of more than 12 my, including the Scythian and the lower part of the Anisian. During this interval only microbial carbonate buildups were formed. Reef development stopped at the Triassic–Jurassic boundary in the Tethys and in Panthalassa except for very few Panthalassan terranes (Stanley, 1994a).

Scientific study of Triassic reefs started in the last century in Europe. The spectacular mountain ranges of the Dolomites, a part of the Southern Alps in South Tyrol (northern Italy), and the Northern Calcareous Alps in Austria and Bavaria were the regions where the origin of huge Middle and Upper Triassic

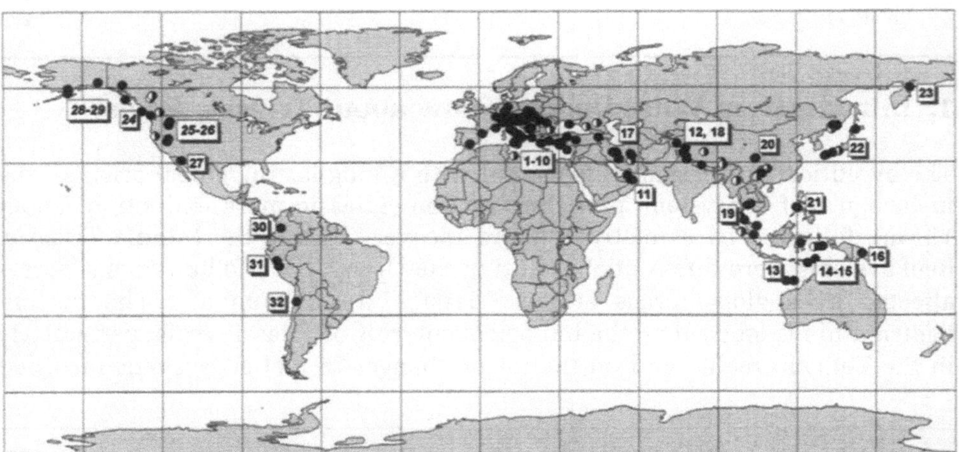

FIGURE 1. Recent distribution of Upper Triassic (Norian and Rhaetian) reefs. Note the concentration of reefs in the Alpine–Mediterranean region that was the western part of the former Tethys, extending to the east until to modern Papua New Guinea. Locations see Fig. 3.

carbonate masses as reefs has been recognized at first (Richthofen, 1860; Mojsisovics, 1879). Subsequent to the still classical investigations in the time prior to World War II, research on Triassic reefs was intensified in the 1960s starting with studies in the Alps followed by investigations in the Mediterranean area and North America. Many studies were concerned with the systematic description of paleontological data aiming to an understanding of reef paleoecology and the significance of various reef "types." Research on the importance of Triassic reefs as "carbonate factories" and reef diagenesis started relatively late perhaps because of the minor importance of Triassic reefs as potential reservoir rocks. Most scientists currently working on Triassic reefs are located in Austria, China, Germany, Italy, Poland, Slovakia, Slovenia, Russia, and the United States. Current research goals are threefold: (1) Analysis of reef biota with respect to biogeographical, climatic, and oceanographical aspects of Triassic oceans; (2) investigations of the impact of microbes, organic crusts and cryptic biota on reef formation; and (3) examination of reef geometries within the frame of sequence stratigraphy and the evaluation of global sea-level curves.

Several review articles summarize the knowledge and the problems involved with the analysis of Triassic reefs (Flügel, 1981, 1982; Stanley, 1988; Flügel and Senowbari-Daryan, 1996). Triassic reefs are relatively well-studied as compared with reefs from other Phanerozoic systems. The "Erlangen reef bibliography" (Flügel and Flügel-Kahler, 1992; at the moment comprehending more than 9000 references on Phanerozoic reefs and reef organisms) yields about 1500 references dealing with Triassic reefs and reef biota. Although the paleontological knowledge of Triassic reef builders (particulary sponges and corals) has significantly increased during the last 20 years, there is still a strong demand for thorough studies of reef biota and facies. Similar to modern reef organisms, only a small percentage of the total reef diversity has been recognized as exemplified by the high number of new sponge, coral, or foraminiferal taxa discovered per year. Preservation biases, inconsistencies of systematic classifications of reef-building fossils, as well as strong variations in the intensity of investigations make the interpretation of distributional patterns of Triassic reefs more difficult. These objections, however, can be overcome by the evaluation of large data sets (Flügel, 2001).

1.2. Distribution of Triassic Reefs

The "classical" reefs of the Alps originated at the western end of the Tethys, a huge western bay of Panthalassia (the precursor of the present Pacific Ocean). The Triassic Tethys expanded from Spain in the western Mediterranean to Timor and Papua New Guinea in the east. Current paleogeographical models indicate the differentiation of the Triassic Tethys into a northern and a southern branch and one or several microcontinents located between these branches (Dercourt *et al.*, 1993; Philip *et al.*, 1996; Golonka and Ford, 2000). Triassic reefs were built both off the shelves and separated platforms of the northern, northwestern, southwestern, and southern coasts of the Tethys (Flügel, 2001). Upper Triassic reefs formed in the northern Tethys are known from the northwestern Caucasus and the Crimea, as well as Bulgaria and Romania. Reefs of the northwestern part of the Tethys were recorded from the Carpathian Mountains and the Northern Alps. These reefs were formed in a shelf position. In contrast, Upper Triassic reefs of the Southern Alps in Northern Italy, Slovenia, and in the southwestern part of the Tethys (Apennines, Sicily, Dinarides, Greece, southern Turkey) originated on or at the margins of separated carbonate platforms bordered by deeper-water zones. Upper Triassic reefs formed off the coasts of the southern Tethys are known from Oman, Central Asia (Indian and Nepalesian Himalayas, southern Tibet), offshore northwestern Australia, Timor, parts of Indonesia, and Papua New Guinea. The differently named midcontinent (Cimmerian continent, Transional Plate, Kreios) comprises Upper Triassic reefs in Central Iran and in parts of Central Asia (Karakorum, northern Tibet, Pamir Range).

Upper Triassic reefs known from southern China, Burma–Malaya, Singapore, Thailand, Vietnam, Borneo, and the northern Philippines originated on shelves of various microcontinents (terranes) within the western Panthalassan Ocean. Some of these terranes (e.g., Sichote-Alin and Karkaren–Primorje region in Far Eastern Russia; Japan) were later transported and accreted to South Asia. Shallow-marine carbonates with and without reefs also were formed within volcanic island settings of the eastern ancient Pacific. These terranes moved northward before or after collision with one another and became amalgamated with the North American plate. Today, they form parts of the Western North American Cordillera (Stanley, 1979, 1994a; Reid and Ginsburg, 1986). The interpretation of Upper Triassic coral- and sponge-bearing shallow-water sediments occurring in Chile, Peru, and Bolivia is controversial. Some authors consider the strata as sediments deposited on the South American craton, others as parts of amalgamated eastern Pacific terranes (Stanley, 1994b; Prinz-Grimm, 1995).

2. Permian, Triassic, and Lower Jurassic Reef Types

Modern reefs are easily classified according to their dominating reef builders. Most recent tropical and subtropical reefs correspond to coral reefs and coral–algal reefs constructed predominantly by scleractinian corals or corals and coralline red algae. Ancient reefs can be categorized in a similar way using the relative amount of principal organic contributors to reef formation (cf. Fagerstrom, 1987).

The discussion of the evolution of Triassic reefs also should consider Permian and Lower Jurassic reefs. Biota used in categorizing Permian to Lower Jurassic reefs include microbes, calcareous algae, corals, sponges, bryozoans, brachiopods, serpulid worms, bivalves, crinoids, and many microproblematica (small fossils that cannot readily be placed in established systematic groups;

	Late Permian	Triassic						Early Jurassic
		Early	Middle		Late			
		Scythian	Anisian	Ladinian	Carnian	Norian	Rhaetian	
my		6.5	7.4	6.9	6.7	11.1	3.9	
Microbial Reefs								
Thrombolites	●	●	●	●	●	●	●	
Microbial Crusts	●			●		●		
Stromatolites	●					●		
'Tubiphytes'	●		●	●				
Sponge Reefs								
Coralline Sponges ('Calcisponges')	●		●	●	●	●	●	●
Siliceous Sponges	●				●			●
Coral Reefs								
Rugose Corals	●							
Scleractinian Corals			●	●	●	●	●	●
Algal Reefs								
Dasycladaleans				●		●		
Bryozoan Reefs	●							
Brachiopod Reefs	●							●
Pelecypod Reefs			●	●				
Serpulid Reefs						●	●	

FIGURE 2. Major Triassic reef types, characterized by the principal reef-building groups. Note the high number of Ladinian and Norian reefs in contrast to Anisian and Rhaetian reef types. There exist no metazoan reefs during the time of the Lower Triassic reef gap.

e.g., *Tubiphyte*). A subdivision of these taxonomic groups into systematic categories (e.g., phylloid algal vs. dasycladalean algal reefs) allows further differentiation of more specific reef types. Modifying earlier proposals (Flügel and Stanley, 1984; Davies *et al.*, 1988; Flügel, 1994; Kawamura and Machiyama, 1995; Weidlich and Fagerstrom, 1998; Shen *et al.*, 1998), eight major reef categories are distinguished defined by the quantitative predominance of microbes, sponges, corals, calcareous algae, bryozoans, brachiopods, bivalves, or serpulids. Microbial reefs include four subtypes, sponge, and coral reefs each two subtypes. The consideration of nonpaleontological criteria, e.g., the contribution of synsedimentary carbonate cement to reef formation, allows furthergoing categorizations. If necessary, cement-rich microbial crust reefs or *Tubiphytes* reefs can be specifically designed (microbial crust–cement reef). Many reefs are characterized by a combination of two or more quantitatively important groups of sessile reef builders. It may be reasonable to express these combinations using terms as "serpulid–microbe reefs" for reefs composed of serpulid aggregates, encrusted by microbes, or "serpulid–microbe–cement reefs" for reefs formed by a framework of serpulids and microbial crusts, supported by large amounts of synsedimentary marine cements. Understanding the temporal and spatial development of reef communities may require the indication of principal reef-building taxa (e.g., Middle Triassic *Placunopsis* bivalve reefs).

Microbial, sponge, coral, and algal reefs are known both from the Permian and from the Triassic. This long-range occurrence, however, should not be overrated: A more detailed look reveals differences not only at high taxonomic levels (rugose coral reefs vs. scleractinian coral reefs), but also at lower taxonomic levels: Late Permian *Tubiphytes* reefs and Triassic *Tubiphytes* reefs differ in taxonomic, systematic, and ecological criteria of this encrusting enigmatic fossil. A similar situation exists for Permian and Triassic coralline sponge reefs.

The spatial distribution of the major reef types seems to be related to paleoclimate, at least in Late Permian times (Kawamura and Machiyama, 1995). Triassic distributional patterns are more complicated because of a latitudinal expansion of Triassic reefs through time (Flügel, 2001). Most Middle Triassic Tethyan reefs were formed north of the paleoequator. Starting with the upper Carnian, Triassic reefs were built within a broad belt between about 40°N (Caucasus) and approximately 40°S (Papua New Guinea). Most Upper Triassic Tethyan reefs, however, originated in the northwestern and southwestern Tethys between 10° and 30°N.

3. Reef Biota

3.1. Reef Builders

Most Triassic reefs were built by coralline sponges and corals. In addition, microbes and algae as well as serpulids are of importance. Microbial crusts and biogenic encrustations upon and between the primary reefbuilders contributed to the enhancement of rigidity and stability of the reef structures.

3.1.1. Sponges

Coralline sponges are the dominant constituents of many Middle and Upper Triassic reefs. Siliceous sponges are rare (Senowbari-Daryan and Wurm, 1994) but they may be of reef-building importance as shown by hexactinellid sponge reef mounds in upper Carnian of southern China (Wu, 1989; Wendt *et al.*, 1989) and the Norian of Central Iran (Senowbari-Daryan, 1996; Senowbari-Daryan and Hamadani, 1999). In contrast, Triassic coralline sponges are widely distributed and highly diversified. The poorly defined term "calcisponge" formerly has been used for systematically different, polyphyletic groups of coralline sponges, characterized by a nonspiculate calcareous skeleton. Coralline sponges include segmented sphinctozoid and nonsegmented inozoid sponges as well as chaetetid sponges (which partly have been designated as tabulozoans in some earlier papers). Many spongiomorphs and disjectoporids, formerly included within the hydrozoans, are now interpreted as sponges. The taxonomic knowledge of Triassic sphinctozoid coralline sponges is significantly greater than that of inozoid sponges, and Triassic chaetetid sponges are in need of more detailed studies. Spongiomorphs and disjectoporids have not yet been studied in the context of their relationships with sponges.

Sphinctozoid sponges comprise currently more than 60 genera and about 200 species (Senowbari-Daryan, 1990). They increase in diversity from Middle Triassic to Upper Triassic (Riedel and Senowbari-Daryan, 1991). The total number of genera increases from upper Anisian to Ladinian and Carnian and reaches a climax during Norian–Rhaetian times.

The contribution of coralline sponges to the formation of Triassic reefs varied in time: No records are known from the Lower Triassic. Sponges reported from Scythian deposits of the Alps (Flügel and Stanley, 1984) are too poorly preserved for an undisputable taxonomic determination. Anisian sphinctozoid sponges are predominantly small encrusting species. Ladinian and Carnian organic buildups yield highly diverse sponge faunas consisting predominantly of sphinctozoids. Only a few sphinctozoids known from Anisian reefs occur also in Ladinian and early Carnian reefs, but the most common Ladinian species originated in the Ladinian. Sphinctozoids with a magnesium–calcite skeleton appeared for the first time during the Ladinian and became important constituents of Carnian reef biota. New aragonitic and magnesium–calcitic sphinctozoids as well as inozoids and chaetetid sponges originated during the middle and late Carnian. The taxonomic composition of middle and late Carnian sphinctozoid faunas is similar to that of Ladinian and Cordevolian faunas but there are differences with regard to a differentiation into geographically restricted associations. A major taxonomic change in the composition of sphinctozoid faunas occurred during an interval between late Carnian to early Norian time. The change is characterized by the extinction of about 95% of the Carnian species and the predominance of aragonitic sphinctozoids from middle Norian to Rhaetian time. Late Carnian and early Norian reefs in southern Turkey exhibit marked differences in the ranking of the main groups of coralline sponges during time (Riedel, 1990; Senowbari-Daryan, 1994). New sphinctozoid and inozoid taxa originated during the Norian and Rhaetian. The diversity (species richness) of Norian sponge faunas is distinctly higher than that of Ladinian and Carnian reefs. Norian and Rhaetian sphinctozoids do not differ taxonomically. Nearly all sphinctozoid species disappeared at the Triassic–Liassic boundary.

The species diversity gradient increases from the northwestern to the southwestern part of the Tethys (Northern Calcareous Alps vs. Sicily and southern Turkey). As compared with Norian reefs in the Alps, reefs in Sicily are characterized by a high number of endemic sphinctozoid taxa (about 50%) and the abundance of nonsphinctozoid sponges (Senowbari-Daryan and Schäfer, 1986). Marked compositional differences also are evident in comparing Norian–Rhaetian reef sponges from the western Tethys with those from the

Cimmerian plate (Senowbari-Daryan, 1996; Senowbari-Daryan and Hamadani, 1999; Senowbari-Daryan *et al.*, 1997).

Ladinian coralline sponges occurred in close association with microbial and other biogenic crusts. Many Norian and Rhaetian coralline sponges acted as primary framebuilders forming the substrate for many epibionts. As compared with Ladinian and Carnian reefs, the sphinctozoid sponges of Norian reefs were distinctly larger in size and were often more crowded and covered areas of several square meters. Smaller coralline sponges were important parts of cryptic communities.

3.1.2. Corals

Middle to Late Triassic reef corals have been studied since the last century, but our knowledge of them is still insufficient due to preservational biases and discrepancies of systematics treatment. Current studies aim to a standardization of scleractinian taxonomy (e.g., Roniewicz, 1989, 1995).

No corals are known from the Lower Triassic. Scleractinian corals appeared at fairly diverse levels during middle Anisian time. They both appeared in nonreef and in reef environments of the Middle Triassic in the western Tethys and also in southern China. Ezaki (1998), however, claimed an origination of the Scleractinia much earlier during the Paleozoic.

Several intervals and turning points have been distinguished in the evolution of Triassic corals (Roniewicz, 1989; Roniewicz and Morycowa, 1989, 1993; Riedel, 1991; Melnikova, 1994; Stanley and Swart, 1995). Interestingly, four time intervals, characterized by changes in relative coral importance within higher taxonomic groups and their diversity, coincide roughly to those reflected by sphinctozoid sponge faunas: (1) Middle Anisian to early Ladinian, (2) late Ladinian to early Carnian, (3) upper Carnian to early Norian and (4) mid to late Norian–Rhaetian.

3.1.3. Microbes and Organic Crusts

Middle and Upper Triassic reef limestones yield a wealth of differently organized crusts growing on reef builders, between them, or on the sediment. Different names are used to characterize these crusts including spongiostromate crusts, stromatolites, micrite crusts, or microbialites. The term "microbes" as used in the study of ancient reefs includes noncalcified or calcified eubacteria and cyanobacteria, algae, and fungi as well as small protozoan and metazoan organisms associated with the organic crusts. Tiny tubular fossils interpreted as remains of microbes have been called "calcimicrobes" or "porostromate algae."

Microbes were ubiquitous on the extended shelves in various parts of the early Triassic world. They were able to form mound-shaped and biostromal carbonate buildups during the long-lasting early Triassic reef gap, which was characterized by the lack of framework reefs built by metazoans (see Section 5.2).

Biogenic crusts were of different importance for the formation of Middle and Upper Triassic reefs. Some reefs are formed almost exclusively by microbial crusts (e.g., Anisian buildups in South China: Enos *et al.*, 1997; Ladinian mud mounds in Spain: Calvet and Tucker, 1995; Norian reefs in southern Spain and the Apennines: Braga and Lopez-Lopez, 1989; Climaco *et al.*, 1997). In other reefs built by sponges and/or corals organic crusts acted as secondary frame builders. Structure, composition, and quantitative importance of these crusts varied through time: Micrite crusts associated with low-growing organisms occur in Anisian reefs of the western Tethys (Senowbari-Daryan *et al.*, 1993). Common Anisian encrusters were sponges, bryozoans, porostromate algae, foraminifers, serpulids, and microproblematica. Microbial encrustions are rare as compared with later reefs. Ladinian and Lower Carnian (Cordevolian) reefs are characterized by abundant and morphologically diverse crust types. Thin microbial crusts, enigmatic encrusting organisms (e.g., *Bacinella*, *Tubiphytes*), and large amounts of synsedimentary carbonate cement were responsible for the formation of many Ladinian reefs in the Dolomites (Brandner *et al.*, 1991; Harris, 1993, 1994a,b). Microbially controlled carbonate crusts building ledges on Ladinian fore-reef blocks show striking textural similarities to laminar micrite crusts formed in modern deeper fore-reef environments (Brachert and Dullo, 1994). Common Ladinian and Carnian encrusters are microbes, *Tubiphytes*, and in places also sphinctozoid sponges. Middle and Upper Carnian crusts are characterized by a cavernous structure and were formed by low-growing encrusting organisms with segmented walls in association with micrite layers and porostromate algae. Norian and Rhaetian micritic spongiostromate crusts developed predominantly on coralline sponges (Wurm, 1982). The number of Norian–Rhaetian organisms contributing to the formation of biogenic crusts is high and includes foraminifers, coralline inozoid and chaetetid sponges, solenoporacean algae, and many microproblematica (Wurm, 1982). Current studies point to taxonomic differences in the composition of the crusts occurring in Upper Triassic reefs in the western Tethys and crusts developed in Upper Triassic reefs formed on the shelves of the Iranian Plate (Senowbari-Daryan, 1996).

3.1.4. Other Organisms

3.1.4a. Algae and Microproblematica. Calcareous algae: Green algae (represented by udoteacean and dasycladacean algae) and red algae (solenoporaceans) are important constituents of Triassic reefs. Udoteaceans and solenoporaceans built frameworks and acted as bafflers. Solenoporacean algae are of particular interest because of their volumetrically important role as reef builders in some Norian reefs of the southern Tethys (Oman, Timor). Dasycladceans are characteristic elements of Middle and Upper Triassic back-reef and lagoonal environments but also occur within the central reef facies of Upper Rhaetian reefs. In a few places, dense growth of dasycladacean algae and rapid synsedimentary cementation resulted in the formation of meter-sized biostromes (e.g., Ladinian and Norian of Southern Spain). Distributional and diversity patterns of dasycladacean algae allow the differentiation of Middle Triassic far-reef, near-reef, and platform environments (Zorn, 1976; Ott, 1967).

The temporal distribution of Triassic dasycladacean algae is similar to that of Triassic sponges and corals: Almost no records from the Lower Triassic, first assemblages in the Middle Anisian, and significant changes of the taxonomic composition near the Anisian–Ladinian, lower–upper Ladinian, Ladinian–early Carnian to middle–upper Carnian boundary, and during late Norian time.

Microproblematica: Millimeter-sized, calcified sessile organisms of unknown systematic position are important constituents of Middle and Upper Triassic reefs. The taxonomic composition of these microproblematica varies in time (Senowbari-Daryan, 1984; Flügel and Senowbari-Daryan, 1996). Some taxa exhibit strong environment controls and assist in the recognition of reef zonations and subenvironments (Schäfer and Senowbari-Daryan, 1981). Fossils, commonly assigned to *Tubiphytes* (now *Shamovella*, cf. Riding, 1993) were active in the formation of Anisian frameworks (Lehrmann *et al.*, 1998) and Ladinian and Carnian biocementstones and boundstones (Henrich, 1982). Common encrusting microproblematica in Ladinian and Carnian reefs of the western Tethys are *Ladinella* and *Plexoramea*.

3.1.4b. Bivalves and Serpulids. Bivalve reefs are represented by Anisian and Ladinian Muschelkalk bioherms known from France, southern Germany, and Poland (e.g., *Placunopsis* reefs: Klotz and Lukas, 1988; Hüssner, 1993). Upper Triassic serpulid reefs and serpulid–cement reefs seem to be restricted to a relatively small region in the central western Tethyan zone (Spain, Italy: Braga and Lopez-Lopez, 1989; Berra and Jadoul, 1996).

3.2. Reef Destroyers: Micro- and Macroborers

The importance of microborers (bacteria, algae, fungi, and others) seems to increase from Anisian to Upper Triassic time. Quantitative differences in the association of boring thallophytes may be caused by differences in the substrate rather than by changes of boring organisms during time. In the Upper Triassic the infestion by endolithic microborers is high in mollusk and brachiopod shells as well as in sponges but relatively low in corals. Interestingly, there is no change in the overall composition of microborer associations from Upper Permian to Middle Triassic despite the significant changes in the composition of reef builders (Schmidt, 1992; Balog, 1996). The ecological niches used by the Triassic principal boring thallophytes were established in the Permian and can be traced from the Mesozoic until today. Microborer associations have proved very useful as indicators of paleobathymetry in Phanerozoic reefs (Vogel *et al.*, 1999). Using microendoliths as bathymetric indicators, sponges and corals of the central reef area of the Norian Hoher Göll reef indicate that the deepest part of the shallower euphotic zone and the fore-reef debris in the deeper euphotic zone roughly correspond to maximum water depths of a few tens of meters (Balog, 1996). Carnian sponge–coral reef mounds of the Upper Cassian Formation in the southern Alps and Ladinian bivalve reefs in southwestern Germany yield similar microboring assemblages indicating water depths between 20 and 30 m (Glaub and Schmidt, 1994; Vogel *et al.*, 1995).

Macroborings in bryozoans, sponges, and brachiopod shells excavated by bivalves and cirrepedians are common in Anisian reef mounds (Senowbari-Daryan *et al.*, 1993). Many Upper Triassic coralline sponges and corals and also brachiopods were bored by bivalves, gastropods, and polychaetes. In Upper Rhaetian reefs of the Alps, the percentage of corals affected by macroborers is generally high (about 30%) (Stanton and Flügel, 1989).

4. Reef Paleoecology

Middle Triassic and some Upper Triassic reef-building biota are characterized by low-growing communities that acted as bafflers and binders rather than as true constructors.

4.1. Guild and Community Structures

The application of the "reef guild concept" allows comparisons of the role of sponges, corals, algae, microbes, and other organisms in framework structures of different Triassic reefs. Criteria necessary for the assignment of reef biota to one of the reef-building guilds are growth habit, skeletonization, growth form, preservation in growth position or *in situ,* as well as skeletal packing and density (Fagerstrom and Weidlich, 1999). A quantitative survey of 183 Triassic reefs exhibits a number of characteristics: (1) a distinct predominance of the baffler guild as compared with the binder–encruster and constructor guilds, and (2) a change from binder-dominated reef communities in the Anisian to baffler-dominated reef communities in the Norian-Rhaetian (Flügel, 2001). Most Anisian reef builders (sphinctozoid sponges, solenoporacean algae) were small, low-growing organisms belonging predominantly to the baffler guild or binder–encruster guilds. Macrobial encrusting associations (sponges, bryozoans, foraminifers, serpulids, cyanobacterial crusts) were probably of greater importance in reef building during the Anisian than during Ladinian and Late Triassic time. Open-surface communities predominated. Cryptic communities are limited to very small, centimeter-sized constructional cavities. Guild structures of many Ladinian reefs are dominated by bafflers (sponges and corals) and binders–encrusters. A diversification of the guild structures during the Carnian is related to an increase of the relative abundance of the constructor guild (represented mostly by scleractinian corals). This change is recorded in early Carnian (Cordevolian) platform reefs (Turnšek *et al.*, 1984).

Using a "community" definition as a group of quantitatively dominating, skeletonized organisms living together and characterized by a distinctive composition, the analysis of Triassic reef communities reveals the following patterns:

Anisian communities consists of both low-diverse and high-diverse associations. Low-diverse communities are represented by bivalve reefs of the Muschelkalk built by *Placunopsis* or *Entostrotion*, the *Olangocoelia* and *Celyphia* reefs of the Southern Alps built by only s few sphinctozoid sponges (Ott *et al.*, 1980), and the microbial–calcimicrobial mounds described from southern China. High-diverse communities composed of sphinctozoid and inozoid sponges as well as corals occur in Anisian reefs in the Southern Alps and southern Spain. The number of co-occurring communities is high from the very first beginning of reef growth. The late Anisian reef mounds in the Dolomites yield about ten communities composed of coralline sponges, bryozoans, corals, algae, and microbes (Fois and Gaetani, 1984; Senowbari-Daryan *et al.*, 1993).

Ladinian communities known both from the Northern and Southern Alps exhibit distinct distribution patterns along water energy and depth gradients. Many communities are low in diversity and are defined by the predominance of one or only few taxa (e.g., Wetterstein reefs of the Northern Calcareous Alps) (Henrich, 1982). The total number of Ladinian reef communities is significantly higher than the number of Anisian communities and it seems to increase during Early Carnian (Cordevolian) time as shown by Cordevolian reefs formed on the Julian platform in Slovenia characterized by a wealth of well-defined communities. A similar situation existed in Slovenia during the formation of late Carnian carbonate platforms and reefs (Turnšek *et al.*, 1987; Turnšek and Buser, 1989).

Norian and Rhaetian Tethyan reefs include a great number of specific communities many of which have a pan-Tethyan distribution. These communities are characterized by the abundance of a few dominant taxa, distinct associations of primary reef builders and associated epibionts, and the occurrence of diagnostic foraminiferal assemblages. Upper Triassic Dachstein and Upper Rhaetian reefs in the Northern Calcareous Alps in Austria and Bavaria are characterized by six standard communities some of which are more geographically widespread, while others are rarer and not as widely distributed (Stanton and Flügel, 1987). In Upper Rhaetian reefs, the communities typically occur in patches only a few meters in diameter. Major controls on the community structures in the Upper Triassic reefs studied so far were water energy and depth, competition for space, as well as sediment input (Bernecker *et al.*, 1999). Absence of steep environmental gradients (e.g., constant water depths) resulted in homogenous community compositions within some patch reefs (Schäfer, 1979).

Many Norian–Rhaetian reef communities differ not only in taxonomic composition but also in a distinct spatial separation of low or high growth forms. Examples are Upper Rhaetian reefs of the Northern Calcareous Alps, the Norian Pin-Spiti reefs of the Himalaya region (Bhargava and Bassi, 1985), and Rhaetian reefs located off the northwestern Australian shelf (Sarti *et al.*, 1992).

The taxonomic composition of Norian–Rhaetian communities seems to have been controlled by the physiographic setting of the patch reefs. Reef communities of platform reefs, platform edge reefs, and reefs formed in downslope positions differ in amount and diversity of coralline sponges, corals, and solenoporacean algae (Zankl, 1969; Senowbari-Daryan, 1990; Satterley, 1994). Another possible control on community composition is the paleolatitude. Norian reefs in Oman differ in community composition and abundance of specific communities from coeval reefs in the northwestern part of the Tethys (Bernecker and Weidlich, 1994; Bernecker, 1996).

4.2. Reef Successions

Successions in Triassic reefs are recorded by sequences of encrusters or by vertical changes among reef communities. Regular successions of encrusters have been described from some Carnian and Norian reefs (Zankl, 1969; Fürsich and Wendt, 1977) but many Upper Triassic reefs exhibit no distinct patterns of encrustation (Wurm, 1982).

The sponge → coral succession: This type of succession corresponds to a vertical community sequence composed of a basal coralline sponge-dominated reef fauna followed by a coral-dominated fauna. This sequence already occurs in the Anisian (Fois and Gaetani, 1984; Martin and Braga, 1987; Bodzioch, 1994) but is more common in many Norian and Rhaetian reefs of the western and southern Tethys as well as in northwestern Australia and the western Cordillera of North America (Stanley, 1979; Flügel, 1981; Röhl *et al.*, 1991; Sarti *et al.*, 1992). It is known at different scales within a vertical distance of less than one to several tens of meters. Generally, there is no gradual transition but rather there is an abrupt change. One possible explanation for this pattern is a decrease in water depth connected with a change from low to high-energy conditions.

5. Testimonies of Tethyan Triassic Reefs

The study of Triassic reefs of the Tethys offers the possibility to follow long-term changes in the composition of reef biota over a time span of about 42 million years. Using the timescale of Gradstein and Ogg (1996), we discuss the consequences of the dramatic Permian–Triassic mass extinction and speculate on the Permian heritage of Triassic reef organisms. We also examine the question of whether or not late Triassic reefs' development can be regarded as the dawn of modern reefs.

5.1. Changes through Time

Triassic reefs exhibit distinct changes in (1) biotic composition, (2) abundance, size, and paleogeographic setting, and (3) the importance of synsedimentary carbonate cements for the formation of reef framework. These changes characterize four stages in the development of the Triassic reef ecosystems: (1) Anisian (and early Ladinian?), (2) Ladinian and lower Carnian (Cordevolian), (3) Middle and upper Carnian (to lower Norian), and (4) Norian–Rhaetian. The biotic changes between 1 to 2 and between 2 and 3 reefs correspond with a gradual replacement of taxa. The changes within the Carnian to Norian interval represent a combination of a short-lasting turnover in the composition of faunas and a rapid extinction event.

5.1.1. Anisian Reefs

Knowledge of Anisian reefs is crucial in understanding the recovery of the reef ecosystem after the end-Permian mass extinction. Middle and Upper Anisian reefs are known from Poland, Germany, southern Spain, the Dolomites and Northern Calcareous Alps, western Carpathians, Hungary, Slovenia, Bosnia-Hercegovina, and southern China. However, most of our information is based on the investigation of only a few reefs in the Dolomites (Gaetani *et al.*, 1981; Fois and Gaetani, 1984; Senowbari-Daryan *et al.*, 1993; Schafhauser, 1997) and in the Guizhou province, South China (Lehrmann *et al.*, 1998). All Anisian reefs are located north of the paleoequator within paleolatitudes from 10° to about 25°N (Flügel, 2001). There are three centers of reef development: (1) southwestern Germany and central Poland (bivalve reefs and sponge and coral reefs), (2) Dolomites, and (3) southern China (microbial reefs). Reef building started during the Middle Anisian (Pelsonian substage) and became more common in the Upper Anisian (Illyrian). The oldest reefs are known from the northern Dolomites in South Tyrol, northern Italy.

Anisian reefs comprise microbial reefs (thrombolitic reefs and *Tubiphytes* reefs, sponge reefs, and coral reefs, as well as bivalve reefs. Reef types include mud mounds, reef mounds, framework reefs, and biostromes. Most Anisian reefs are rich in carbonate mud but poor in carbonate cements. Exceptions are shelf edge framestones formed by skeletal stromatolites and encrusting microproblematica (Enos *et al.*, 1997).

5.1.1a. Size and Setting. The Southern Alpine middle and upper Anisian reefs are small structures comprising sizes between <1 and only a few meters in thickness and a lateral extension of some tens of meters. The reefs were built on carbonate or mixed carbonate–siliciclastic ramps in low-energy environments.

5.1.1b. Biota. Organisms contributing to the formation of Anisian reefs as primary or secondary reef builders are astonishingly diverse and include microbes, calcareous algae, many enigmatic organisms, coralline and siliceous sponges, several major groups of scleractinian corals, bryozoans, various bivalves, crinoids, serpulid worms, and encrusting foraminifers. Sponges, bryozoans, porostromate and solenoporacean algae, as well as various epibionts are essential constituents of Upper Anisian bioconstructions. Most Anisian reef builders were small, low-growing, centimeter-sized organisms. Lateral zonation patterns of reef builders and foraminifers have been described from the Southern Alps (Scheuber, 1990; Senowbari-Daryan *et al.*, 1993).

5.1.2. Ladinian and Early Carnian (Cordevolian) Reefs

Knowledge of Ladinian and Cordevolian reefs is better than that of Anisian reefs. There are many records from the Alpine and Carpathian region, Slovenia, Italy, and western Turkey. In Asia, Ladinian reefs are known from southern China and Thailand. The only Ladinian reefal facies occurring in North America has been described from Nevada (Stanley, 1979; Roniewicz and Stanley, 1998). Most of our information is based on five regions: the Northern Calcareous Alps, the Western Carpathians, the Dolomites of South Tyrol, the southern Alps in of Slovenia, and Ladinian buildups in the Southern Apennines, Italy. The paleolatidudinal distribution of Ladinian and Cordevolian reefs in the western Tethys corresponds roughly to that of Anisian reefs (10° to about 25°N). There are, however, more centers of reef building, including (1) southwestern Germany and central Poland (bivalve reefs and sponge and coral reefs) (Bodzioch, 1994), (2) Northern Calcareous Alps and western Carpathians, (3) Southern Calcareous Alps and the southern and central Apennines (Fois, 1981; Fois and Gaetani, 1984; Iannace *et al.*, 1998), and (4) southern Spain. South China (Guizhou) remained an important reef region. The early Ladinian (Fassanian) are represented by bivalve buildups in Germany, dasycladacean biostromes and microbial mud mounds in Spain and microbial and coralline sponge–microbe reef mounds in the Southern Calcareous Alps. In the western Tethys, areal distribution and abundance of reef growth increased significantly during the late Ladinian (Longobardian) and Cordevolian as shown by the extended Wetterstein reef complexes of the Northern Calcareous Alps and western Carpathians, and the kilometer-thick late Ladinian carbonate buildups in the Dolomites. Smaller but highly differentiated Longobardian and/or Cordevolian reefs occur in Slovenia.

Ladinian and Cordvolian reefs comprise microbial reefs (thrombolitic, microbial crust, and *Tubiphytes* reefs), dasycladacean algal reefs, coralline sponge reefs, and coral reefs as well as bivalve reefs. Reef types include mud mounds, reef mounds, framework reefs, and biostromes. Reef mounds and framework reefs, however, are more abundant than in the Anisian. Reef mounds of the Southern Calcareous Alps in northern Italy and in Slovenia correspond to low-relief microbe–cement reefs and to mounds formed by coralline sponges, corals, solenoporacean algae, and encrusting organisms.

5.1.2a. Importance of Sediment and Carbonate Cement. Larger Ladinian and Cordevolian reefs yield significantly more synsedimentary carbonate cement (developed in growth cavities as well as in form of cement crusts) than Anisian reefs. Carbonate mud is abundant in smaller patch reefs of lower slope or basinal settings.

5.1.2b. Size and Setting. Ladinian reefs of the Southern and Northern Calcareous Alps are distinctly larger structures than in Anisian examples. The size of Ladinian and Cordevolian reefs in the Northern Calcareous Alps varies between tens and hundred of square kilometers. The thickness of individual reefs bodies is within a range of decameters but the total reef facies may reach thicknesses of several hundred meters. Similar dimensions are known from the Southern Calcareous Alps. The Alpine reefs were mostly formed at the edges of extended carbonate platforms and on steep upper slopes or on lower slopes both in high- and low-energy environments. Reefs formed in the Slovenian trough were mud mounds and reef mounds and depending in their size were affected by fluctuations in siliciclastic input. Cordevolian reefs formed on the Julian Platform have lateral extensions of tens of kilometers.

5.1.2c. Biota. Reef builders in Tethyan Ladinian and Cordevolian reefs are microbes, calcareous algae (solenoporaceans, udoteaceans, dasycladaceans), many enigmatic organisms (e.g., *Tubiphytes*), coralline sponges, scleractinian corals, bivalves, crinoids, serpulid worms, and encrusting foraminifers. As compared with Anisian reefs, the taxonomic composition of Upper Ladinian and Cordevolian reef builders differs from that of Anisian faunas. Most Anisian sphinctozoid sponges became extinct; the origination rate is high. The first sphinctozoid sponges with a magnesium–calcite skeleton appeared during the Ladinian. Only about one third of the coral taxa known from the Anisian continued into Ladinian and Cordevolian time. Anisian and Ladinian encrusting biota exhibit distinct differences. Microbial crusts and small epibionts as well as coralline sponges are essential constituents of Ladinian reefs. Many Ladinian reef builders were low growing, but high-growing communities (sponges, corals) also existed. Reefs formed within the Slovenian trough and on the Julian platform are characterized by the association of coralline sponges and corals and by the abundance of the latter group. Ladinian reefs of the southern Apennines exhibit many affinities with Ladinian reefs in the Northern Alps with regard to biotic composition and the quantitative abundance of microbe–cement crusts. Distinct lateral zonation patterns of reefbuilders and foraminifers seem to have increased in Ladinian and Cordevolian reef complexes.

5.1.3. Middle-Upper Carnian Reefs and the Carnian–Norian Turnover

Late Carnian Tethyan reefs are known from the Alps and the Carpathians, Sicily, Slovenia, Romania, Greece, southern Turkey, and Oman. Additional records come from Sichuan, southern China, and the Russian Far East. As compared with Ladinian and Cordevolian reefs, the paleolatitudinal distributional pattern of Middle and Upper Carnian reefs is wider, comprising reefs formed both north and south of the paleoequator. Concerning the temporal range, Middle and/or Upper Carnian reefs exhibit three common patterns: (1) reefs continuing from the Lower Carnian to the Middle and even to the Upper Carnian (e.g., western Carpathians); (2) reefs restricted to a specific part of the Carnian (e.g., the biostromal Leckkogel reefs of the Northern Calcareous Alps); and (3) Carnian reefs continuing their growth during early Norian time (Hydra Island, Greece; western Taurus, Turkey; Sikhote-Alin, Far East Russia). These reefs exhibit a mixture of "Carnian" and "Norian" reef fossils both in the Carnian and Norian parts of the respective sections.

Middle and Upper Carnian examples include microbial reefs, coralline sponge reefs, siliceous sponge reefs, and scleractinian coral reefs. Reef mounds seem to have been more common than framework reefs and mud mounds.

5.1.3a. Size and Setting. As compared with Ladinian and Cordevolian reefs, Middle and Upper Carnian reefs were smaller. Reefs were formed in various settings including carbonate ramps, mixed carbonate–siliciclastic environments (e.g., Southern Calcareous Alps), on carbonate platforms and on shoals within intraplatform basins.

5.1.3b. Biota. Coralline sponges, microproblematica (*Tubiphytes*), biogenic crusts, and scleractinian corals are still the prevailing constituents of the fauna in reefs of this age. By and large the composition of most Middle and Upper Carnian (Julian and Tuvalian) reef-building associations known from the western Tethys is similar to that of Ladinian and Cordevolian reefs. There are, however, some differences: The numerical diversity seems to increase during the Carnian and the composition of the organic crusts as well as the association of microproblematica and reef foraminifers differ from those of Ladinian reefs. Another aspect as compared with Ladinian and Cordevolian reefs is the development of distinct differences in taxonomic composition between the northwestern and the southwestern parts of the Tethys (Slovenia, southern Turkey) and southern China. Differentiation within the western Tethys is expressed in the composition of sphinctozoid associations, the importance of nonsphinctozoid sponges, and the frequency of *Tubiphytes*. The Late Carnian reef mounds in southern China are unique in the predominance of hexactinellid sponges. These mounds growing in deeper water formed a reef belt that extended laterally about 20 km (Wu, 1989; Wendt *et al.*, 1989).

5.1.3c. Late Carnian–Early Norian Biotic Turnovers and Extinction Events. Several localities in the western Tethys allow the discussion of the mode of biotic changes, which were partly connected with extinction events (Riedel, 1990, 1991). The late Carnian to Norian Pantokrator Limestone on the Greek island of Hydra consists of a vertical succession with a *Tubiphytes* association, a coral–coralline sponge association, a coralline sponge association, and a coral-dominated association. The sphinctozoid sponges and corals include taxa known from the Ladinian reefs of the Northern Calcareous Alps and from Carnian reefs of the Southern and Northern Calcareous Alps. Late

Carnian (Tuvalian) and early Norian (Lacian) reef biotas of the Dereköy Unit and the Kasimler Formation near Antalya and Isparta, southwestern Turkey, differ significantly in the relative abundance of principal reef builders: In the Tuvalian, sphinctozoids are dominating, inozoids common, chaetetids rare; in the Lacian chaetetids and inzoids are dominating, corals are abundant, and sphinctozoids common. There is a mixed taxonomic composition of the coral fauna (about 40% of Lacian coral taxa already are known in Carnian reefs). Lacian sphinctozoid sponge genera are characterized by high endemism (Senowbari-Daryan, 1994).

Interestingly, sponges and corals were both affected by significant changes in taxonomic diversity, faunal composition, and degree of skeletal construction during Late Carnian–Early Norian time: Sponges and corals as well as microproblematica all exhibit high rates of extinction during the Upper Carnian. About 95% of sphinctozoid sponge species disappear within the Upper Carnian interval (Riedel, 1990). Estimates of specific and generic extinction rates among Upper Carnian and Lower Norian biotas do not yield equal results (Riedel, 1990; Melnikova, 1994) but all authors agree that considerable differences exist in the taxonomic composition between Ladinian–Carnian and Upper Carnian–Norian coral faunas not only in the western Tethys but also in the shallow shelves of the Cimmerian Plate (Pamir region) (Melnikova, 1975, 1994). Coral diversity decreased near the Carnian–Norian boundary and increased in the Middle Norian (Roniewicz and Morycowa, 1993). Changes also are reflected in the coral growth forms which dominate the fauna (Riedel, 1991) as well as in the skeletal mineralogy of sphinctozoid sponges (Senowbari-Daryan, 1990). Most Ladinian and Carnian sphinctozoids had skeletons composed of calcite or aragonite and some taxa were composed of high magnesium–calcite. Early Norian sphinctozoans exhibit mostly aragonite and high magnesium–calcite skeletons. Aragonite skeletons dominated from Middle Norian to Rhaetian times. These changes in skeletal mineralogy were variously interpreted as responses to fluctuations in sea-water chemistry (Railsback and Anderson, 1987; Riedel, 1990; Reitner, 1992; Stanley and Hardie, 1998).

There are different opinions concerning the rates and importance of the proposed Carnian extinction events and biotic turnovers (Benton, 1986, 1991; Hallam, 1990). A mid-Carnian event was discussed for several marine invertebrate group including bivalves, crinoids, echinoids, and bryozoans and a nonsynchronous end-Carnian event was proposed for terrestrial vertebrates. Sea-level fluctuations and changes in climate as well as changes in nutrient input have been discussed being major controls on Carnian and end-Carnian biotic changes (Ciarapica *et al.*, 1990; Seffinga, 1988; Simms and Ruffell, 1989, 1990; Riedel, 1990).

5.1.4. Norian–Rhaetian Reefs

Our information about Norian and Rhaetian reefs is considerably greater than our knowledge of Middle Triassic or Carnian reefs. Possibly this could

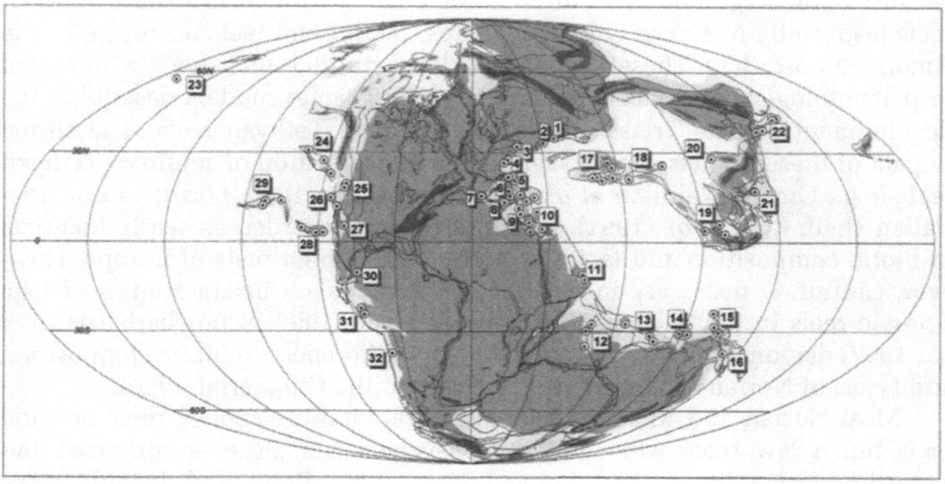

FIGURE 3. Paleo-distribution of Upper Triassic reefs. Reefs are concentrated in the western and southern part of the Tethys, on the shelves of the Cimmerian Continent, and on various microplates in Asia as well as in the eastern parts of Panthalassa. Locations: (1) Caucasus, (2) Crimea, (3) Carpathians, (4) northern Calcareous Alps, (5) southern Calcareous Alps, (6) Apennines (Italy), (7) southern Spain, (8) Sicily, (9) Greece, (10) southern Turkey, (11) Oman, (12) Spiti/Himalaya, (13) off NW Australia, (14) Timor, (15) Ceram and Sulawesi, (16) Papua New Guinea, (17) Iran, (18) Pamir Mountains, (19) Thailand and Malaysia, (20) southern China, (21) northern Phillipines, (22) Japan and far east Russia, (23) Kamchatka, (24) British Columbia, (25) Oregon and Idaho, (26) California and Nevada, (27) Mexico, (28) Vancouver, (29) Alaska, (30) Columbia, (31) central Peru, (32) northern Chile. Note that the latitudinal and even more the longitudinal position of many Panthalassan terranes with reefs is a matter of ongoing discussion (e.g., Kamchartka, or the western north American terranes).

reflect the much wider distribution of late Upper Triassic reefs in the western and southern Tethys and in Panthalassia. In contrast to Middle Triassic reefs, Norian and Rhaetian reefs were formed both in the western and central Tethys and in Panthalassa. They grew in a much wider paleolatitudinal zone stretching from 35°S to about 40°N (Fig. 3).The original position of some high-latitude reefs is a matter of debate (e.g., Russian Far East, Japan). Interestingly, only few reefs occur close to the paleoequator and more than two thirds of the reefs were located in the northern hemisphere. Many late Upper Triassic reefs are not dated precisely and therefore the designation Norian–Rhaetian reef is rather vague. However, there are reefs in the Northern and Southern Calcareous Alps, in southern Anatolia, Turkey, as well as in the Pamir Range whose age has been proven by macro- and microfossils as being Lower, Middle, or Upper Norian. Many so-called Norian–Rhaetian reefs in the western Tethys are Upper Norian (Sevatian) in age. Well-dated Rhaetian reefs are known from the Alps and the Carpathians.

Norian–Rhaetian reefs differ from older Triassic reefs in distributional patterns, reef types, and taxonomic composition of reef-building organisms from older reefs. Our knowledge of Norian and Rhaetian reefs is biased because of the large amount of information dealing with northwestern Tethyan reefs (especially Northern Calcareous Alps, Sicily) and reefs occurring in the American Cordillera. These regions have been studied intensively with regard to paleontological and palecological data. Most facies models describing the development of Late Triassic reefs are based on Tethyan reefs. The strong impact of these models can be seen in the interpretation of southern Tethyan reefs (e.g., Oman: Weidlich *et al.*, 1993, Bernecker, 1996; northwestern Australian shelf: Röhl *et al.*, 1991), which have been regarded as nearly identical in biotic composition and facies patterns with Alpine reefs of Europe. However, caution is necessary in the transfer of models. Current studies of Late Triassic reefs in Central Iran (Senowbari-Daryan, 1996; Senowbari-Daryan *et al.*, 1997) demonstrate the existence of many differences in biotic composition and types of Norian–Rhaetian reefs formed on the Cimmerian Plate.

Most Norian and Rhaetian examples are coralline sponge reefs or coral reefs but a few reefs were built by dasycladacean algae or microbes and serpulid worms (Braga and Lopez-Lopez, 1989; Berra and Jadoul, 1996; Climaco *et al.*, 1997) (Fig. 3).

5.1.4a. Size and Setting. Many Norian and Rhaetian reefs were platform margin reefs bordering deep or shallow open-marine basins (e.g., Northern Alps, Sicily, Pamir). Other reefs were built on upper slopes, at the margins of intraplatform sags, on carbonate and siliciclastic ramps, as well as on shallow attached or oceanic carbonate platforms. The dimensions vary widely ranging from meter- and decameter-sized biostromes and reef mounds to huge reef complexes reaching several hundred meters in thickness and lateral extensions of tens to hundreds of kilometers. Smaller reef structures are common on ramps and in settings influenced by siliciclastic sedimentation.

5.1.4b. Biota. Major reef builders comprise various groups of coralline sponges and scleractinian corals as well as solenoporacean algae and a wealth of encrusting organisms (calcimicrobes, foraminifers, microproblematica). The taxonomic composition of Norian and Rhaetian sponges, corals, calcareous algae, foraminifers, and microproblematica is highly different from those of Ladinian and most Carnian reefs. Such differences reflect the strong impact of the late Carnian to early Norian biotic change. Sponges are the most important organisms in many Norian reefs. They are represented by highly differentiated sphinctozoids, inozoids, chaetetids, and also hexactinellid silicisponges. More than 50% on generic level and more than 95% on specific level of Norian–Rhaetian sphinctozoid taxa originated during the Norian, mostly during the late Norian (Sevatian). Interestingly, sphinctozoid sponges exhibit a high extinction rate and a very low origination rate during the Rhaetian. Chaetetid sponges (tabulozoans of older descriptions) increased in importance during the Norian. Most Norian and Rhaetian coralline sponges acted as primary frame builders forming the substrate for many epizoans and epiphytes and stabilizing the reef framework (Senowbari-Daryan, 1990). As compared with the Ladinian and Carnian, Norian sphinctozoid sponges are distinctly larger in size and cover a greater surface area. Smaller sponges are important components of cryptic communities within reefs.

Scleractinian corals are common both in Norian and Rhaetian reefs but increase in abundance in late Rhaetian reefs of the Northern Calcareous Alps and the Carpathians. As compared with Carnian reefs, the generic and specific diversity of Norian corals is very high, reflecting high origination rates, the differentiation into faunal provinces (Melnikova, 1994), and an increase in the number of reef communities. Diversity in Rhaetian reefs is still high but decreases dramatically at the Triassic–Liassic boundary. There is a marked increase in the number of cerioid coral growth forms during the Norian and Rhaetian, possibly reflecting changes in nutrient levels and competion for space with sponges (Riedel, 1991).

Many Norian and Rhaetian reefs exhibit lateral zonation or vertical succession characterized by the dominance of sponges or corals within distinctly separated units or by the occurrence of specific taxa within particular reef settings (Satterley, 1994). Sponges are common in low-growing communities of deeper-water or protected settings (e.g., lower slope), whereas corals often flourished on shallow upper slope, reef crest, and reef flat environments. There are many exceptions from this generalized pattern (downslope and deeper-water coral patch reefs, mixed sponge–coral reef mounds; cf. Satterley, 1994), however, demonstrating broad adaptions to varying environmental conditions even for taxonomically rather uniform coral communities. The success of corals in latest Triassic reefs might have been controlled by the invention of coral/zooxanthellate symbioses (Stanley, 1981; Stanley and Swart, 1995) but it is still difficult to use morphological and facies criteria without contradiction in the recognition of zooxanthelllate and nonzooxanthellate features of Late Triassic scleractinians (Stanton and Flügel, 1987, 1989).

5.2. The Lower Triassic Reef Gap

The present state of knowledge points to the existence of a global reef gap in the early Triassic (Flügel and Stanley, 1984; Flügel, 1994) characterized by the lack of true metazoan reefs. No indisputable records of potential skeletonized reef builders (sponges, corals) are known so far from the Scythian time interval. Sessile calcified benthic metazoans are only represented by bryozoans (Schäfer, 1994) occurring in nonreef environments of predominantly high-latitudinal temperate shelves (Canada, Spitsbergen, Greenland, Siberia, New Zealand) (Schäfer and Grant-Mackie, 1998).

However, microbes and calcimicrobes, already important in the formation of latest Permian reefs (Guo and Riding, 1992), were ubiquituous on the carbonate platforms of the wide shelves in various parts of the early Triassic world (Schubert and Bottjer, 1995; Sano and Nakashima, 1997; Baud *et al.*, 1997). Small- to medium-scaled biogenic structures formed by thrombolitic and stromatolitic microbialites are known from Lower Triassic shallow shelf carbonates (Great Basin, northwestern United States; Antalya region, southwestern Turkey) as well as from open-marine carbonate ramps (Transcaucasia Mountains, South Armenia; Abadeh region, Central Iran, and northeast Iran).

Carbonate structures recognized in the deep subsurface of the eastern Precaucasian area west of the Caspian Sea (Nazarevich *et al.*, 1986) are of particular interest. These oil- and gas-bearing massive algal limestones are composed of stromatolites and oncoids yielding cyanobacteria, *Tubiphytes obscurus*, and in other levels also dasycladacean algae and brachiopods. The 500- to 800-m thick buildups have been interpreted as banks and reefs within a shallowing-upward sequence. The age of the carbonate buildups occurring in the upper part of the Kumanskaya series of the Precaucasian region is middle to late Lower Triassic by conodonts, foraminifers, and mollusks. Lower Triassic microbialites described from the United States are of late Lower Triassic (Spathian) age. Well-dated records of lowermost early Triassic (Griesbachian) microbial buildups are known from the western Tethys (southern Turkey: Baud *et al.*, 1997), from a seamount of the Panthalassan ocean (southwest Japan: Sano and Nakashima, 1997), and from southern China (Lehrmann, 1999). These records demonstrate the overwhelming significance of bacteria, cyanobacteria, and fungi in microbially controlled carbonate production following the end-Permian mass extinction. The rapid and wide conquest of early Triassic seas by microbes was facilitated by absence or low diversity of metazoans, absence of predation and intense bioturbation, decreased competition for space and substrate colonization, low sedimentation rates, and the extraordinary capacity of microbes to proliferate in strongly fluctuating physicochemical conditions. The time span of the worldwide metazoan–reef gap after the Permian/Triassic crisis comprises the duration of the Scythian plus the duration of the early part of the Anisian; that is, more than 10 million years.

5.3. The Permian–Triassic Connection

There is a strong consent to the statement that Middle Triassic and Carnian reef-building organisms are virtually very similar or even identical with Permian reef biota and that there are many coincidences between Permian and Triassic reefs (Flügel and Stanley, 1984; Stanley, 1988). These assumptions are based (1) on the comparison of tectonosedimentary setting, gross composition, and diagenetic sequences (shelf-margin reefs; high amount of synsedimentary cement) of late Permian reefs, particularly the Permian Reef Complex in United States, with the Ladinian and early Carnian Wetterstein reefs in the Northern Alps and the Carpathians (Ott, 1967; Wilson, 1975; Mazzullo and Lobitzer, 1988; Lobitzer *et al.*, 1990), and (2) on factual (or apparent) taxonomic coincidences. The latter argument includes the necessity to differentiate between "Lazarus" from "Elvis" taxa (Erwin and Droser, 1993).

Both the comparison of reef types and the comparison of biotic characters, depend strongly on the scale used: Comparisons of Late Permian and Middle Triassic reefs on the high level of groups exhibit apparent similarities because of the overall importance of coralline sponges, *Tubiphytes*, and various biogenic crusts. Differentiation of these groups into taxonomic subgroups, however, reveals distinct differences between Late Permian, Anisian, and Ladinian–Cordevolian reefs. These differences are substantiated by the complete absence of some important Permian reef organisms in Middle Triassic rocks (e.g., phylloid algae, rugose corals). Comparisons on generic levels exhibit correspondence for bryozoans (Schäfer, 1994), several dasycladacean algae (Flügel, 1985), and a few Anisian and Ladinian sphinctozoid sponges (Senowbari-Daryan *et al.*, 1993). Morphological similarities of the sponges, however, also can be explained by homeomorphy. There is no indisputable correspondence at the species level. *Tubiphytes obscurus* from the Middle Triassic must be separated taxonomically from late Paleozoic *T. obscurus* and is a good example of an Elvis taxon (Senowbari-Daryan and Flügel, 1993).

5.3.1. Arrested Reappearance of Permian Taxa during the Late Triassic?

Interestingly, most Permian Lazarus genera found in reefs reappeared during the Late Triassic (Norian) rather than in Middle Triassic time: coralline and hexactinellid sponges, gastropods and strophomenid brachiopods (Hallam, 1991; Senowbari-Daryan and Flügel, 1996), and udoteacean algae (Reid, 1986). The sphinctozoid genus *Neoguadalupia*, described from Late Permian reefs of Southern China reappeared in the Pamir region and in central Iran (Boiko *et al.*, 1991; Senowbari-Daryan and Hamadani, 1999). The inozoid sponge *Radiofibra*, known from Late Permian reefs in Tunisia reappeared in Late Triassic reefs of Central Iran (Senowbari-Daryan *et al.*, 1997), and *Discosiphonella* (= *Cystauletes*), a common sphinctozoid sponge in Permian reefs reappeared in Norian reefs in southern Turkey (Senowbari-Daryan and Link, 1998) as well as in the Northern Calcareous Alps (Zankl, 1969). The Permian species of these Lazarus genera are endemic and not cosmopolitan taxa that are generally believed to be better candidates for taxonomic survival after mass extinctions. Stanley (1994a) stressed the possibility of preferential survival during mass extinctions on volcanic island refuges somewhere in the ancient Pacific Ocean. These late Paleozoic to early Mesozoic terranes subsequently have been accreted to the Americas. An interesting explanation for the delayed migration of Triassic reef biota from Panthalassan refuges to the western Tethys was offerred by Kozur (1998) assuming long-lasting oceanic superanoxia in the shallow waters of the Tethyan shelves.

5.3.2. Microbiota

Similarities and coincidences of microbial or algal crusts are difficult to assess because of the lack of distinctive morphological criteria. *Archaeolithoporella*, a common organism of Permian reefs, also has been described from Triassic reefs, but the correspondence may be superficial. The only microbiota occurring both in Late Permian and Triassic reefs are specific associations of microborers (Vogel *et al.*, 1999). Interestingly enough, these cyanobacterial, algal, and fungal borings exhibit no morphological differences subsequent to the Permian/Triassic crisis, and thus must be regarded as possibly representing identical organisms.

5.4. Norian–Rhaetian Reefs: The Dawn of Modern Reefs?

Fagerstrom (1987) stressed the major differences between Cenozoic and pre-Cenozoic reefs. In comparison with Cenozoic reefs, pre-Cenozoic reefs are characterized by (1) different taxonomic composition of reef communities, (2) lower diversity of reef communities, (3) overlap of binder and baffler guilds with the constructor guild, (4) abundance of less well-skeletonized reefbuilders, (5) less-developed zonation, (6) smaller dimensions of reef structures, and (7) lower relief of the reefs. In addition, associations of patch reef and bioclastic sediment dominate over true framework reefs.

Modern shallow-marine reefs are rigid framework structures built predominantly by zooxanthellate scleractinian corals and calcareous algae. Essential biological criteria of recent reefs are (1) the abundance of sessile, gregarious organisms, (2) the dominance of organisms with calcareous skeletons, (3) rapid growth and high coloniality, (4) the domination of constructor guilds, (5) biogenic sedimentation between and around corals, (6) a high proportion of cryptic habitats as compared with open surface habitats, and (7) strong bioerosion. Nonbiological criteria are (1) the dominance of tropical–subtropical settings, (2) occurrence in well-lit, stenohaline warm-water environments, and (3) location of reefs in both open ocean settings and on continental shelves.

Considering the criteria of late Upper Triassic reefs, some but not all of the Upper Rhaetian coral-dominated reefs in the Northern Calcareous Alps exhibit several "modern" features including an abundance of sessile, gregarious, highly calcified, and high-growing colonial corals. A quantitative study of the famous Adnet "coral reef" near Salzburg (Austria) revealed further ecological features that can be compared with those of modern coral thickets (Bernecker *et al.*, 1999): The key role of branching corals at Adnet is the formation of a dense coral thicket, the segregation patterns of coral colonies, indicating avoidance of colony contacts, and decreasing areal coverage from what is interpreted as sheltered to shallow, stressed areas. There are, however, important differences between the Rhaetian coral thickets and modern coral reefs: Modern coral reefs are characterized by the predominance of constructor guilds, predominance of branching colonial corals, association of corals and coralline algae, a high proportion of cryptic habitats as compared with open surface habitats, and strong bioerosion. These criteria cannot be derived from the Adnet study. At Adnet the dominating *Retiophyllia* communities belong to baffler and binder guilds rather than to the constructor guild, exhibit predominantly phaceloid–dendroid growth forms, and the amount of cryptic habitats is very small and bioerosion of the coral thickets is very low.

5.5. The End-Triassic Reef Crisis

In addition to the disappearance of many invertebrate taxa, the end-Triassic mass extinction is also marked in the marine realm by the collapse of the reef ecosystem at the end of the Rhaetian (Hallam and Goodfellow, 1990). Tabulation of survivorship and ranges of Triassic corals and coralline sponges emphasizes the abrupt nature of the end-Triassic extinction that terminated reef building in the Tethys (Stanley, 1988; Stanley and Beauvais, 1994). Less than 1% of all Triassic coral species (14 of 321) and 8.6% of sphinctozoid coralline sponge genera (5 of 58) survived into Liassic time. Only two coral groups survived. The most successful Late Triassic coral group—the distichophyllids—became extinct (Roniewicz and Morycowa, 1989).

On a global scale records of earliest Jurassic (Hettangian) coral reefs are extremely rare (Far East Russia: Krasnov, 1997). An example of a Tethyan-type patch reef is known from the Lower Jurassic (Sinemurian) Hazelton Group of the Stikine terrane in British Columbia (Stanley and Beauvais, 1994). This small coral bioherm developed on interbedded carbonate–volcaniclastic sediments and consists of three superimposed biohermal mounds. Reef growth was succeeded by submarine erosion and volcanic flows. The occurrence of common Upper Triassic coral species that escaped the end-Triassic extinction supports the assumption that oceanic islands around the ancient Pacific might have served as refugia and centers of renewed reef development.

Most Liassic corals correspond to solitary corals or small colonies scattered within a nonreef sediment (Beauvais, 1984). An exception is the Middle Liassic (Pliensbachian) coral fauna of the Moroccoan reefs (DuDresnay, 1977; Beauvais, 1977, 1984). This Jurassic fauna includes 35% endemic taxa. The greatest part of the fauna, however, consists of taxa that flourished during the Late Triassic among Tethyan coral faunas. These Triassic genera disappeared in the late Liassic where a significant radiation of new genera and families took place, as documented in the Upper Sinemurian to Upper Plienbachian reefs of the High Atlas Mountains, Morocco. These reefs are coral–algal patch reefs, sponge biostromes, and bivalve mounds (DuDresnay, 1979; Warme, 1988; Kenter and Campbell, 1991). Full recovery of the reef ecosystem was not complete until Mid-Jurassic time when coral, sponge, and hydrozoan reefs emerged in different parts of the Tethys.

Most authors explain the end-Triassic disappearance of reefs as a result of sea-level fluctuations or changes in climate: Based on geochemical data, a late Triassic to early Jurassic global regression followed by an early Jurassic transgression and associated anoxic events have been proposed by Hallam and Goodfellow (1990) to explain the disappearance of the Rhaetian reefs of the Northern Calcareous Alps. Repeated phases of karstification at the very end of the Triassic are known from some Upper Rhaetian reefs of the Northern Alps

(Stanton and Flügel, 1989; Bernecker *et al.*, 1999). Climate cooling as a cause of mass extinction of reefs at the Triassic–Liassic boundary (Fabricius *et al.*, 1970) is not supported by isotope data (Hallam and Goodfellow, 1990). Other explanations for the disturbance of the reef ecosystem are local salinity reduction (Hallam and El Shaarawy, 1982; McRoberts and Newton, 1995) or increased volcanism during the breakup of Pangaea (Dickins, 1993).

6. Conclusions

1. Two-step development: The evolution of Triassic reefs was characterized by a two-step development before the Lower Triassic to early Anisian metazoan reef gap. Anisian–Ladinian–Carnian reefs and Norian–Rhaetian reefs differ in biotic composition, ecological structures, prevailing reef types as well as abundance, dimensions, and paleolatidudinal distributional patterns.

2. Mass extinction and reef evolution: Triassic reefs were affected by the end-Permian, Carnian, and end-Triassic extinction events. Some Permian-type Lazarus taxa reappeared in the Norian, indicating a long-lasting existence of unknown Panthalassan refuge areas. Recovery of reef ecosystems after the Permian–Triassic event occurred in the western Tethys and on the shelves of the South China Plate. High extinction rates of coralline sponges, corals, calcareous algae, foraminifers, and microproblematica during the time interval from Late Carnian to early Norian were compensated by high origination rates within the same time interval. Upper Triassic (Rhaetian) reefs disappeared more or less synchronously in the western Tethys and the southern Tethys. Disappearance of the latest Triassic reefs did not affect all Upper Triassic reef builders. Some corals and microencrusters continued into lower and middle Liassic time.

3. Distributional patterns: Paleolatitudinal distribution of Triassic reefs expanded during Anisian to Upper Triassic time. Anisian and Ladinian to early Carnian (Cordevolian) reefs were restricted to the Northern Hemisphere, while late Carnian to Rhaetian reefs occur in both hemispheres. Maximum reef distribution occurred in the Norian. Norian–Rhaetian Tethyan reefs were concentrated in the western Tethys, along the northern coast of Gondwana, on the shelves of the Cimmerian continent, and around some microplates. Late Triassic (Norian) reefs of the Tethys occurred in large, separated domains and they reached dimensions of hundreds to several thousand of square kilometers.

ACKNOWLEDGMENTS: Our studies were supported by the Deutsche Forschungsgemeinschaft (Priority Program "Global and Regional Controls on Biogenic Sedimentation–Reef Evolution," project Fl 42/69, "Controls on Triassic reef evolution"). We are indebted to Wolfgang Kiessling (Berlin) and Michaela Bernecker (Erlangen) for valuable discussions. The chapter benefited from reviews by George D. Stanley, Jr. and Reinhold Leinfelder.

References

Balog, S.-J., 1996, Boring thallophytes in some Permian and Triassic reefs: Bathymetry and bioerosion, *Göttinger Arbeiten Geol. Paläont. Sonderband* **2**:305–308.

Baud, A., Cirilli, S., and Marcoux, J., 1997, Biotic response to mass extinction: The lowermost Triassic microbialites, *Facies* **36**:238–242.

Beauvais, L., 1977, Main characters of the Liassic coral fauna from Morocco, *Proceedings of the 3rd International Coral Reef Symposium, Miami* **2**:375–378.

Beauvais, L., 1984, Evolution and diversification of Jurassic Scleractinia, *Palaeontogr. Am.* **54**:219–224.

Benton, M. J., 1986, One than one event in the Late Triassic mass extinction, *Nature* **321**:857–861.

Benton, M. J., 1991, What really happened in the Late Triassic, *Hist. Biol.* **5**:257–278.

Bernecker, M., 1996, Upper Triassic reefs of the Oman Mountains: Data from the South Tethyan margin, *Facies* **34**:41–76.

Bernecker, M., and Weidlich, O., 1994, Attempted reconstruction of Permian and Triassic skeletonization from reefbuilders (Oman, Turkey): Quantitative assessment with digital analysis, *Abhandlungen Geol. Bundesanst. Wien.* **50**:31–56.

Bernecker, M., Weidlich, O., and Flügel, E., 1999, Response of Triassic reef coral communities to sea-level fluctuations, storms and sedimentation: Evidence from a spectacular outcrop (Adnet, Austria), *Facies* **40**:229–280.

Berra, F., and Jadoul, F., 1996, Norian serpulid and microbial bioconstructions: Implications for the platform evolution in the Lombardy Basin (Southern Alps, Italy), *Facies* **35**:143–162.

Bhargava, O. N., and Bassi, U. K., 1985, Upper Triassic coral knoll reefs: Middle Norian, Spiti-Kinnaur, Himachal Himalaya, India, *Facies* **12**:219–242.

Bodzioch, A., 1994, Paleoecology of hexactinellid sponges from the epicontintal Triassic of Poland, in: *Sponges in Time and Space* (R. W. M. van Soest, T. M. G. van Kempen, and J. C. Braekman, eds.), Balkema, Rotterdam, pp. 35–44.

Boiko, E. V., Belyaeva, G. V., and Zhuravleva, T., 1991, Phanerozoic sphinctozoans from the territory of Russia, Acad. Sci. USSR, Siberian Department, Inst. Geol. Geophys., Acad. Sci. Tajikistan, (in Russian), Dushanbe (Akademiya Nauk), pp. 223.

Brachert, T. C., and Dullo, W. C., 1994, Micrite crusts on Ladinian foreslopes of the Dolomites seen in the light of a modern scenario from the Red Sea, *Abhandlungen Geol. Bundesanst. Wien* **50**:57–68.

Braga, J. C., and Lopez-Lopez, J. R., 1989, Serpulid bioconstructions at the Triassic-Liassic boundary in Southern Spain, *Facies* **21**:1–10.

Brandner, R., Flügel, E., and Senowbari-Daryan, B., 1991, Microfacies of carbonate slope boulders: Indicators of the source area (Middle Triassic: Rifugio Molignon Cliff, Western Dolomites), *Facies* **25**:279–296.

Calvet, F., and Tucker, M., 1995, Mud mounds with reefal caps in the upper Muschelkalk (Triassic), eastern Spain, in: *Carbonate Mud mounds. Their Origin and Evolution*, Vol. 23 (C. L. V. Monty, D. W. J. Bosence, P. H. Bridges, and B. R. Pratt, eds.), Blackwell Science, Oxford, pp. 311–333.

Ciarapica, G., Cirilli, S., Martini, R., Rettori, R., Zaninetti, L., and Salvini-Bonnard, G., 1990, Carbonate buildups and associated facies in the Monte Facito Formation (Southern Apennines), *Boll. Soc. Geol. Ital.* **109**:151–164.

Climaco, A., Boni, M., Iannace, A., and Zamparelli, V., 1997, Platform margins, microbial/serpulid bioconstructions and slope-to-basin sediments in the Upper Triassic of the "Verbicaro Unit" (Lucania and Calabria, Southern Italy), *Facies* **36**:37–56.

Davies, G. R., Richards, B. C., Beauchamp, B., and Nassichuk, W. W., 1998, Carboniferous and Permian reefs in Canada and adjacent areas, in: *Reefs, Canada and Adjacent Areas* (H. H. J. Geldsetzer, N. P. James and G. E. Tebbutt, eds.), Mem. Canadian Society Petroleum Geologists, Calgary, pp. 565–574.

Dercourt, R., Ricou, L. E., and Vrielynck, B. (eds.), 1993, Atlas Tethys palaeoenvironment maps. Late Murghabian (266–264 Ma); Late Anisian (237–234 Ma), 9–33, maps, Gauthier-Villars, Paris.

Dickins, J. M., 1993, Climate of the Late Devonian to Triassic, *Palaeogeogr. Palaeoclimat. Palaeoecol.* **100**:89–94.

DuDresnay, R., 1977, Le milieu récifal fossile du Jurassique inférieur (Lias) dans le domaine des Chaines atlasiques du Maroc, *Bull. Bur. Rech. Gol. Min. Mem.* **89**:296–312.

DuDresnay, R., 1979, Recent data on the geology of the Middle Atlas (Morocco), *Lect. Notes Earth Sci.* **15**:293–320.

Enos, P., Jiayong, W., and Yangji, Y., 1997, *Facies* distribution and retreat of Middle Triassic platform margin, Guizhou province, south China, *Sedimentology* **44**:563–584.

Erwin, D. H., and Droser, M. L., 1993, "Elvis taxa," *Palaios* **8**(6): 623–624.

Fabricius, F. H., Friedrichsen, H., and Jacobshagen, V., 1970, Paläotemperaturen und Paläoklima in Obertrias und Lias der Alpen, *Geol. Rundschau* **59**:805–826.

Fagerstrom, J. A., 1987, *The Evolution of Reef Communities*, Wiley, New York.

Fagerstrom, J. A., and Weidlich, O., 1999, Strengths and weaknesses of the reef guild concept and quantitative data: Application to the Upper Capitan-massive community (Permian), Guadalupe Mountains, New Mexico-Texas, *Facies* **40**:131–156.

Flügel, E., 1981, Paleoecology and facies of Upper Triassic reefs in Northern Calcareous Alps, *Soc. Econ. Paleont. Min. Spec. Publ.* **30**:291–359.

Flügel, E., 1982, Evolution of Triassic reefs: current concepts and problems, *Facies* **6**:297–328.

Flügel, E., 1985, Diversity and environments of Permian and Triassic Dasycladacean algae, in: *Paleoalgology* (D. F. Toomey and M. H. Nitecki, eds.), Springer, Berlin, pp. 344–351.

Flügel, E., 1994, Pangean shelf carbonates: Controls and paleoclimatic significance of Permian and Triassic reefs, *Geol. Soc. Am. Spec. Paper* **288**:247–266.

Flügel, E., 2001, Triassic reef patterns, in: *Phanerozoic Reef Patterns* (E. Flügel, W. Kiessling, and J. Golonka, eds.), *Soc. Econ. Paleont. Min. Spec. Publ.*, Tulsa, OK, pp. 000–000.

Flügel, E., and Bernecker, M., 1996, Upper Triassic reefs of the South Tethyan Margin (Oman), *Göttinger Arbeiten Geol. Paläont. Sonderband* **2**:273–277.

Flügel, E., and Flügel-Kahler, E., 1992, Phanerozoic reef evolution: Basic questions and database, *Facies* **26**:167–278.

Flügel, E., and Senowbari-Daryan, B., 1996, Evolution of Triassic reef biota: State of the art, *Göttinger Arbeiten Geol. Paläont. Sonderband* **2**:285–294.

Flügel, E., and Stanley, G. D., 1984, Reorganization, development and evolution of post-Permian reefs and reef organisms, *Palaeontogr. Am.* **54**:177–186.

Fois, E., 1981, The Sass da Putia carbonate buildup (Western Dolomite): Biofacies succsession and margin development during the Ladinian, *Riv. Ital. Paleont. Strat.* **87**:565–598.

Fois, E., and Gaetani, M., 1984, The recovery of reef-building communities and the role of cnidarians in carbonate sequences of the Middle Triassic (Anisian) in the Italian Dolomites, *Palaeontogr. Am.* **54**:191–200.

Gaetani, M., Fois, E., Jadoul, F., and Nicora, A., 1981, Nature and evolution of Middle Triassic buildups in the Dolomites (Italy), *Marine Geol.* **44**:25–57.

Glaub, I., and Schmidt, H., 1994, Traces of endolithic microboring organisms in Triassic and Jurassic bioherms, *Kaupia* **4**:103–112.

Golonka, J., and Ford, D., 2000, Pangean (Late Carboniferous–Middle Jurassic) paleoenvironment and litho facies, *Palaeogeogr., Palaeoclimatol, Palaeoecol.* **161**:1–34.

Gradstein, F. M., Agterberg, F. P. Ogg, J. G., Hardenbol, J., van Veen, P., Thierry, J., and Huang, Z., 1994, A Mesozoic time scale, *J. Geophys. Res.* **24**:51–74.

Guo, L., and Riding, R., 1992, Microbial micritic carbonates in uppermost Permian reefs, Sichuan Basin, southern China: some similarities with Recent travertines, *Sedimentology* **39**:37–53.

Hallam, A., 1990, The end-Triassic mass extinction event, *Geol. Soc. Am. Spec. Papers* **247**:577–583.

Hallam, A., 1991, Why was there a delayed radiation after the end-Paleozoic extinctions? *Hist. Biol.* **5**:257–262.

Hallam, A., and El Shaarawy, Z., 1982, Salinity reduction of the end-Triassic sea from the Alpine region into northwestern Europe, *Lethaia* **2**:169–178.

Hallam, A., and Goodfellow, W. D., 1990, Facies and geochemical evidence bearing on the end-Triassic disappearance of the Alpine reef ecosystem, *Hist. Biol.* **4**:131–138.

Harris, M. T., 1993, Reef fabric, biotic crusts and syndepositional cements of the Latemar reef margin (Middle Triassic, northern Italy), *Sedimentology* **40**:383–401.

Harris, M. T., 1994a, The foreslope and toe-of-slope facies of the Middle Triassic Latemar buildup (Dolomites, northern Italy), *J. Sed. Res.* **B64**:132–145.

Harris, M. T., 1994b, A volume-based approach to reef productivity and submarine erosion rates—A case-study of a Middle Triassic reef margin (Latemar buildup, northern Italy), *J. Geol.* **102**:603–610.

Henrich, R., 1982, Middle Triassic carbonate margin development: Hochstaufen–Zwieselmassif, Northern Calcareous Alps, Germany, *Facies* **6**:85–106.

Hüssner, H., 1993, Rifftypen im Muschelkalk Süddeutschlands, in: *Muschelkalk* (H. Hagdorn and A. Seilacher, eds.), Schöntaler Symposium der Gesellschaft für Naturkunde in Württemberg, pp. 261–269.

Iannace, A., Radoicic, R., and Zamparelli, V., 1998, A new dasycladalean alga from the Middle Triassic carbonates of southern Calabria (Southern Italy), *Facies* **39**:67–74.

Kawamura, T., and Machiyama, H., 1995, A Late Permian coral reef complex, South Kitakami Terrane, Japan, *Sed. Geol.* **99**:135–150.

Kenter, J. A. M., and Campbell, A. E., 1991, Sedimentation on a Lower Jurassic carbonate platform flank: Geometry, sediment fabric and related depositional structures (Djebel Bou Dahar, High Atlas, Morocco), *Sed. Geol.* **72**:1–34.

Klotz, W., and Lukas, V., 1988, Bioherme im Unteren Muschelkalk (Trias) Südosthessens, *Neues Jahrbuch fur Geologie und Paläontologie, Monatshefte* **11**:661–669.

Kozur, H. W., 1998, Problems for evaluation of the scenario of the Permian–Triassic boundary biotic crisis and of its causes, *Geol. Croatica* **51**:135–162.

Krasnov, E. V., 1997, Early Mesozoic reef-like coral communities development in the Russian Far-East, *Bol. R. Soc. Esp. Hist. Nat. Sec. Geol.* **92**:61–64.

Lehrmann, D. J., 1999, Early Triassic calcimicrobial mounds and biostromes of the Nanpanijang basin, south China, *Geology* **27**:359–362.

Lehrmann, D. J., Wei, J., and Enos, P., 1998, Controls on facies architecture of a large Triassic carbonate platform: the great bank of Guizhou, Nanpanjiang Basin, South China, *J. Sed. Res.* **B68**:311–326.

Lobitzer, H., Mandl, G. W., Mazzullo, S. J., and Mello, J., 1990, Comparative study of Wetterstein carbonate platforms of eastern Northern Calcareous Alps and West Carpathian Mountains: Preliminary results, in: *Thirty Years of Geological Cooperation between Austria and Czechoslovakia* (D. Minarikova and H. Lobitzer, eds.), Geologische Bundesanstalt, Wien, pp. 136–158.

Martin, J. M., and Braga, J. C., 1987, Bioconstrucciones del Anisiense–Ladiniense en el Trias Alpujarride, *Cuadernos Geol. Iberica* **11**:421–444.

Mazzullo, S. J., and Lobitzer, H., 1988, Facies and diagenesis of Triassic (Austrian Alps) and Upper Permian (New Mexico-Texas) platform-margin carbonates: A comparison, *Am. Assoc. Petrol. Geol. Bull.* **72**:219.

McRoberts, C. A., and Newton, C. R., 1995, Selective extinction among end-Triassic European bivalves, *Geology* **23**:102–104.

Melnikova, G. K., 1975, Pozdnetriasovie skleraktinii Jugo-Vostochnogo Pamira, *Akademiya Nauk Tadzhikistanski, SSR, Institut Geologii (Donit)*, Donit, Dushanbe, pp. 236.

Melnikova, G., 1994, Triassic corals, their stratigraphic significance and geographic distribution, *Courier Forschungsinstitut Senckenberg* **172**:35–41.

Mojsisovics, M., 1879, Die Dolomit-Riffe von Südtirol und Venetien: Beiträge zur Bildungsgeschichte der Alpen, Hölder, Wien.

Nazarevich, B. P., Nazarevich, I. A., and Boiko, N. I., 1986, Uloviya formirovaniya i osobennosti razmetsennya nizhnetriasovikh iskopaemikh orgạnogennykh postroek vostochnogo Predkavkazya, in: *Fanerozoiskie rify i korally SSSR* (B. S. Sokolov, ed.), Otdel. Geol. Geofiz. Geochim. i Gornykh Nauk, Moskva, pp. 161–166.

Ott, E., 1967, Segmentierte Kalkschwämme (Sphinctozoa) aus der alpinen Mitteltrias und ihre Bedeutung als Riffbildner im Wettersteinkalk, *Bayer. Akad. Wiss. Math.-Naturwiss. Kl. Abh. N.F.* **131**:1–96.

Ott, E., Pisa, G., and Farabegoli, E., 1980, *Celyphia zoldana* sp. n., a reef building sphinctozoan sponge in Anisian limestones of the southeastern Dolomites, *Riv. Ital. Paleont. Strat.* **85**:892–942.

Philip, J., Masse, J. P., and Camoin, G., 1996, Tethyan carbonate platforms, in: *The Ocean Basins and Margins*, Vol. 8 (A. E. M. Nairn, L. E. Ricou, B. Vrielnyck, and J. Dercourt, eds.) Plenum Press, New York, pp. 239–266.

Prinz-Grimm, P., 1995, Triassische Korallen der südlichen Zentral-Anden, *Geol. Palaeontol.* **29**:233–243.

Railsback, L. B., and Anderson, T. F., 1987, Control of Triassic seawater chemistry and temperature on the evolution of post-Paleozoic aragonite-secreting faunas, *Geology* **15**:1002–1005.

Reid, R. P., 1986, Discovery of Triassic phylloid algae; Possible links with the Paleozoic, *Can. J. Earth Sci.* **23**:2068–2071.

Reid, R. P., and Ginsburg, R. N., 1986, The role of framework in Upper Triassic patch reefs in the Yukon (Canada), *Palaios* **1**:590–600.

Reitner, J., 1992, "Coralline Spongien." Der Versuch einer phylogenetisch-taxonomischen Analyse, *Berliner Geowiss. Abhandlungen, Reihe E* **1**:1–352.

Richthofen, F. V., 1860, Geognostische Beschreibung der Umgebung von Predazzo, St.Cassian und der Seiser Alpe in Süd-Tyrol, Perthes, Gotha.

Riding, R., 1993, *Shamovella obscura*: The correct name for *Tubiphytes obscurus* (Fossil), *Taxon* **42**:71–73.

Riedel, P., 1990, Riffbiotope im Karn und Nor (Obertrias) der Tethys: Entwicklung, Einschnitte und Diversitätsmuster, Ph.D. Thesis, Riedel-Palaontologisches Institut Universität Erlangen.

Riedel, P., 1991, Korallen in der Trias der Tethy: Stratigraphische Reichweiten, Diversitätsmuster, Entwicklungstrends und Bedeutung als Rifforganismen, *Mitteilungen Gesellschaft Geol. Bergbaustudenten Österreich* **37**:97–118.

Riedel, P., and Senowbari-Daryan, B., 1991, Pharetronids in Triassic reefs, in: *Fossil and Recent Sponges* (J. Reitner and H. Keupp, eds.), Springer, Berlin, pp. 465–476.

Röhl, U., Dumont, T., von Rad, U., Martini, R., and Zaninetti, L., 1991, Upper Triassic Tethyan carbonates off Northwest Australia (Wombat Plateau, ODP Leg 122), *Facies* **25**:211–252.

Roniewicz, E., 1989, Triassic scleractinian corals of the Zlambach Beds, Northern Calcareous Alps, Austria, *Denkschriften Österr. Akad. Wiss. Math.-Naturw. Klasse* **126**:1–152.

Roniewicz, E., 1995, Upper Triassic solitary corals from the Gosankamm and other North Alpine regions, *Sitzungsberichte der österreichischen Akademie der Wissenschafien, mathematisch-naturwissen schaftliche Klasse, Abteilung I* **202**:3–41.

Roniewicz, E., and Morycowa, E., 1989, Triassic Scleractinia and the Triassic/Liassic boundary, *Mem. Assoc. Australas. Palaeontol.* **8**:347–354.

Roniewicz, E., and Morycowa, E., 1993, Evolution of the Scleractinia in the light of microstructural data, *Cour. Forschungsinstitut Senckenberg* **164**:233–240.

Roniewicz, E., and Stanley, G. D., 1998, Middle Triassic cnidarians from the New Pass Range, Central Nevada, *J. Paleontol.* **72**:246–256.

Sano, H., and Nakashima, K., 1997, Lowermost Triassic (Griesbachian): Microbial bindstone–cementstone facies, southwest Japan, *Facies* **36**:1–24.

Sarti, M., Russo, A., and Bosellini, F. R., 1992, Rhaetian strata, Wombat Plateau: Analysis of fossil communities as a key to paleoenvironmental change, *Proc. Ocean Drilling Program Sci. Results* **122**:181–195.

Satterley, A. K., 1994, Sedimentology of the Upper Triassic reef complex at the Hochkönig Massiv (Northern Calcareous Alps, Austria), *Facies* **30**:119–150.

Schäfer, P., 1979, Fazielle Entwicklung und palökologsche Zonierung zweier obertriadischer Riffstrukturen in den Nördlichen Kalkalpen ('Oberrhät'-Riff-Kalke, Salzburg), *Facies* **1**:3–245.

Schäfer, P., 1994, Bryozoen der Trias—eine Übersicht, *Abhandlungen Geol. Bundesanstalt Wien* **50**:387–397.

Schäfer, P., and Grant-Mackie, J. A., 1998, Revised systematics and paleobiogeography of some late Triassic colonial invertebrates in the Pacific region, *Alchingeria* **22**:87–122.

Schäfer, P., and Senowbari-Daryan, B., 1981, Facies development and paleoecologic zonation of four Upper Triassic patch-reefs, Northern Calcareous Alps near Salzburg, Austria, *Soc. Econ. Paleont. Min. Spec. Publ.* **30**:241–259.

Schafhauser, M., 1997, Statigraphie und Fazies in der Mitteltrias der Sudkara wanken (Kärten/Osterreich) des angrenzenden Sudalpins: Ph.D. Thesis Technische Universität Berlin, pp. 161.

Scheuber, M., 1990, Der Spitzkalk von Recoaro (Vicentinische Alpen, Norditalien): Sedimentologie, Paläontologie und Paläogeographie eines mitteltriassischen Sedimentationsraumes, *Facies* **23**:57–96.

Schmidt, H., 1992, Mikrobohrspuren ausgewählter Faziesbereiche der tethyalen und germanischen Trias (Beschreibung, Vergleich und bythymetrische Interpretation), *Frankfurter geowiss. Arbeiten, Serie A* **12**:1–228.

Schubert, J. K., and Bottjer, D. J., 1995, Aftermath of the Permian–Triassic mass extinction event: Paleoecology of Lower Triassic carbonates in the western USA, *Palaeogeogr. Palaeoclimat. Palaeoecol.* **116**:1–39.

Seffinga, G., 1988, Possible evidence of 'glacial' conditions during the Julian substage of the Karnian (Upper-Triassic), *Proc. Koninklijke of the Nederlandse Akademie van Wetenschappen Ser. B* **91**:91–100.

Senowbari-Daryan, B., 1984, Mikroproblematika aus den obertriadischen Riffkalken von Sizilien, *Münstersche Forschungen Geol. Paläont.* **61**:1–81.

Senowbari-Daryan, B., 1990, Die systematische Stellung der thalamiden Schwämme und ihre Bedeutung in der Erdgeschichte, *Münchner Geowiss. Abhandlungen, Reihe A, Geol. Paläont.* **21**:1–326.

Senowbari-Daryan, B., 1994, Segmentierte Schwämme ('Sphinctozoen') aus der Obertrias (Nor) des Taurus-Gebirges (S-Türkei), *Abhandlungen Geol. Bundesanstalt Wien* **50**:415–446.

Senowbari-Daryan, B., 1996, Upper Triassic reefs and reef communities of Iran, *Göttinger Arbeiten Geol. Paläont. Sonderband* **2**:299–304.

Senowbari-Daryan, B., and Flügel, E., 1993, Triassic reefs and platform carbonates in the Northern Calcareous Alps, in: *Facial Development of Algae-Bearing Carbonate Sequences in the Eastern Alps* (R. Höfling, E. Moussavian, and W. Piller, eds.), Field Trip Guidebook, International Algal Symposium, Regional Meeting, Institut für Paläontologie und Historische Geologie Universität München, München, pp. A1–A35.

Senowbari-Daryan, B., and Flügel, E., 1996, A "problematic fossil" revealed: *Pycnoporidium? eomesozoicum* Flügel, 1972 (Late Triassic, Tethys)—Not an enigmatic alga but a strophomenid brachiopod (*Gosaukammerellla* n.g.), *Facies* **34**:83–100.

Senowbari-Daryan, B., and Hamadani, A., 1999, Thalamid sponges from the Upper Triassic (Norian–Rhaetian) Nayband Formation near Wali-Abad, SE Abadeh, Central Iran, *Riv. Ital. Paleontol. Strat.* **105**:79–100.

Senowbari-Daryan, B., and Link, M., 1998, A new thalamid sponge from the Upper Triassic (Norian) reef limestones of the Antalya region (Turkey), *Acta Geol. Hungarica* **41/3**:343–354.

Senowbari-Daryan, B., and Schäfer, P., 1986, Sphinctozoen (Kalkschwämme) aus den norischen Riffen von Sizilien, *Facies* **14**:235–284.

Senowbari-Daryan, B., and Stanley, G. D., 1998, *Neoguadalupia oregonenesis* new species: Reappearance of a Permian sponge genus in the Upper Triassic Wallowa Terrane, Oregon, *J. Paleontol.* **72**:221–224.

Senowbari-Daryan, B., and Wurm, D., 1994, *Radiocella prima* n.g., n.sp., der erste segmentierte Schwamm mit tetracladinem Skelett aus den Dachstein-Riffkalken (Nor) des Gosaukammes (Nördliche Kalkalpen, Österreich), *Aghandlungen Geol. Bundesanstalt* **50**:447–452.

Senowbari-Daryan, B., and Zamparelli, V., 1999, Upper Triassic sponges from Northern Calabria (Southern Italy), *Riv. Ital. Paleontol. Strat.* **105**:145–154.

Senowbari-Daryan, B., Zühlke, R., Bechstädt, T., and Flügel, E., 1993, Anisian (Middle Triassic) buildups of the Northern Dolomites (Italy): The recovery of reef communities after the Permian/Triassic crisis, *Facies* **28**:181–256.

Senowbari-Daryan, B., Seyed-Emani, K., and Aghanabati, A., 1997, Some inozoid sponges from Upper Triassic (Norian–Rhaetian) Nayband Formation of Central Iran, *Riv. Ital. Paleontol. Strat.* **103**:293–322.

Shen, J., Kawamura, T., and Yang, W., 1998, Upper Permian coral reef and colonial rugose corals in northwest Hunan, South China, *Facies* **39**:35–66.

Simms, M. J., and Ruffell, A. H., 1989, Synchroneity of climatic change and extinctions in the late Triassic, *Geology* **17**:265–268.

Simms, M. J., and Ruffell, A. H., 1990, Climatic and biotic change in the late Triassic, *J. Geol. Soc. London* **147**:321–327.

Stanley, G. D., 1979, Paleoecology, structure, and distribution of Triassic coral buildups in Western North America, *Univ. Kansas Paleontol. Inst.* **65**:1–58.

Stanley, G. D., 1981, Early history of scleractinian corals and its geological consequences, *Geol.* **9**:507–511.

Stanley, G. D., 1988, The history of early Mesozoic reef communities: A three-step process, *Palaios* **3**:170–183.

Stanley, G. D., 1994a, Late Paleozoic and early Mesozoic reef-building organisms and paleogeography: The Tethyan–North American connection, *Courier Forschungsinstitut Senckenberg* **172**:69–75.

Stanley, G. D., 1994b, Paleontology and stratigraphy of Triassic and Jurassic rocks in the Peruvian Andes, *Palaeontographica A* **233**:1–208.

Stanley, G. D., and Beauvais, L., 1994, Corals from an Early Jurassic coral reef in British Columbia—refuge on an oceanic island reef, *Lethaia* **27**:35–47.

Stanley, G. D., and Swart, P. K., 1995, Evolution of the coral–zooxanthellate symbiosis during the Triassic: a geochemical approach, *Paleobiology* **21**:179–199.

Stanley, S. M., and Hardie, L. H., 1998, Secular oscillations in the carbonate mineralogy of reef-building and sediment-producing organisms driven by tectonically forced shifts in seawater chemistry, *Palaeogeogr. Palaeoclimatol. Palaeoecol.* **144**:85–144.

Stanton, R. J., and Flügel, E., 1987, Paleoecology of Upper Triassic reefs in the Northern Calcareous Alps: Reef communities, *Facies* **16**:157–186.

Stanton, R. J., and Flügel, E., 1989, Problems with reef models: The Late Triassic Steinplatte "reef" (Northern Alps, Salzburg/Tyrol, Austria), *Facies* **20**:1–138.

Turnšek, D., and Buser, S., 1989, The Carnian reef complex on the Pokljuka (NW Yugoslavia), *Razprave IV, Razreda SAZU* **30**:76–105.

Turnšek, D., Buser, S. , and Ogorelec, B., 1984, The role of corals in Ladinian–Carnian reef communities of Slovenia, Yugoslavia, *Palaeontogr. Am.* **54**:201–209.

Turnšek, D., Buser, S., and Ogorolec, B., 1987, Upper Carnian reef limestone in clastic beds at Perbla near Tolmin (NW Yugoslavia), *Razprave IV Razreda SAZU Ljubljana* **27**:37–64.

Vogel, K., Bundschuh, M., Glaub, I., Hofmann, K., Radtke, G., and Schmidt, H., 1995, Hard substrate ichnocoenoses and their relations to light intensity and marine bathymetry, *N. Jb. Geol. Paläont. Abhandlungen* **195**:49–61.

Vogel, K., Balog, S. J., Bundschuh, M., Gektidis, M., Glaub, J., Krutschinna, J., and Radtke, G., 1999, Bathymetrical studies in fossil reefs, with microendoliths as paleoecological indicators, *Profil* **16**:181–191.

Warme, J. E. (ed.), 1988, Evolution of the Jurassic High Atlas Rift, Morocco: Transtension, structural and eustatic controls on carbonate facies, tectonic inversion, *Amer. Ass. Petrol. Geol. Mediterranean Conference Nice Sept., 1988, Fieldtrip 9, Guidebook, Nice*, pp. 310.

Weidlich, O., and Fagerstrom, J. A., 1998, Evolution of the Upper Capitan–Massive (Permian) Guadalupe Mountains, New Mexico, *Brigham Young Univ. Stud.* **43**:169–187.

Weidlich, O., Bernecker, M., and Flügel, E., 1993, Combined quantitative analysis and microfacies studies of ancient reefs: An integrated approach to Upper Permian and Upper Triassic reef carbonates (Sultanate of Oman), *Facies* **28**:115–144.

Wendt, J., Wu, X., and Reinhardt, J. W., 1989, Deep-water hexactinellid sponge mounds from the Upper Triassic of Northern Sichuan (China), *Palaeogeogr. Palaeoclimatol. Palaeoecol.* **76**:17–29.

Wilson, J. L., 1975, Carbonate facies in geologic history, Springer, Berlin, pp. 471.

Wu, X., 1989, Carnian (Upper Triassic) sponge mounds of the Northwestern Sichuan Basin, China: Stratigraphy, facies and paleoecology, *Facies* **21**:171–188.

Wurm, D., 1982, Mikrofazies, Paläontologie und Palökologie der Dachsteinriffkalke (Nor) des Gosaukammes, Österreich), *Facies* **6**:203–296.

Zankl, H., 1969, Die Hohe Göll—Aufbau und Lebensbild eines Dachsteinkalk-Riffes in der Obertrias der nördlichen Kalkalpen, *Abh. Senckenberg. Naturforsch. Gesellschaft* **519**:1–123.
Zorn, H., 1976, Über den Lebensraum fossiler Wirtelalgen in der Trias der Alpen, *Naturwissenschaften* **63**:426–429.

Chapter 8

Jurassic Reef Ecosystems

REINHOLD R. LEINFELDER

1. Introduction

Corals and sponges from Jurassic reefs have attracted both amateur and professional paleontologists for a long time. In particular, the often beautifully preserved corals, which may look just like a coral skeleton from an extant coral, nourished the idea that Jurassic reefs were quite similar to modern representatives. Also, Jurassic reefs often are considered to have been very prolific, having outcompeted even the modern Great Barrier Reef by forming a reef belt at least 7000-km long.

This view persisted not only among amateur paleontologists but also among many geoscientists. Many Jurassic reefs contain a wealth of sponges and therefore were labeled in major revisions of Phanerozoic reef systems as "coral–sponge" reefs, at least until the last decade (e.g., James, 1983; Fager-

REINHOLD R. LEINFELDER • Institute for Palaeontology and Historical Geology, University of München, D-80333 München, Germany.

The History and Sedimentology of Ancient Reef Systems, edited by George D. Stanley Jr., Kluwer Academic/Plenum Publishers, New York, 2001.

strom, 1987; Scott, 1988). However, this was not generally considered a major difference from modern reef ecosystems, because sponges also may play prominent roles in reefs today, such as Caribbean reefs. The common implication was that Jurassic reef ecosystems were fairly similar to modern reef systems in terms of composition, structures, and ecological requirements.

Some earlier studies (e.g., Crevello and Harris, 1984) and particularly recent comparative analysis of Jurassic, chiefly Upper Jurassic reef systems (e.g., Geister and Lathuilière, 1991; Leinfelder, 1993a, 1994b; Nose, 1995; Schmid, 1996; Leinfelder *et al.*, 1996; Insalaco *et al.*, 1997; Matyszkiewicz, 1997) have shown that the reefs comprise a wide variety of different types and that comparison with modern reefs must occur with great caution.

2. Jurassic Reefs

This chapter illustrates Jurassic reef ecosystems by focusing on similarities and differences of reef fauna, reef structure, and ecological demands between modern reefs and Jurassic reefs. This contribution is based on analyses of reefs from the late epoch of the Jurassic period, because it was at that time that reefs developed most vigorously, and comparative studies can be performed best using this time interval. However, the temporal development of reef growth in the course of the Jurassic period also will be briefly discussed.

2.1. Distribution of Jurassic Reefs

In terms of the general tectonic setting, Upper Jurassic coral reefs occurred predominantly in variable positions on pure carbonate to mixed carbonate–siliciclastic, flat, or steepened ramps. Quite a few of them grew very close to the shoreline, whereas others were situated in fairly deep settings, possibly down to several hundreds of meters.

Examples of coral reefs on flat, partly near level-bottom ramps are widespread in Europe (e.g., Germany, France, Spain, Portugal, England: e.g., Ali, 1983; Aurell and Bádenas, 1997; Bertling, 1993; Errenst, 1990a,b; Fezer, 1988; Flügel *et al.*, 1993; Geister and Lathuilière, 1991; Insalaco *et al.*, 1997; Leinfelder, 1993a; Leinfelder *et al.*, 1996; Nose, 1995). Coral reefs at the margins of and behind steepened ramps occur, for example, in Switzerland (Gygi and Persoz, 1986; Pümpin, 1965; Pittet *et al.*, 1995; Takacs, in prep.) and in central and southern Portugal (Ellis *et al.*, 1990; Leinfelder, 1994b, Nose, 1995; Schmid and Jonischkeit, 1995). Coral reefs also may rim platforms with depositional or bypass margins, with examples coming from central Portugal (Leinfelder, 1992, 1994b), Austria (Steiger, 1981; Steiger and Wurm, 1980), Italy (Sartorio, 1989), the Caucasus (Scott, 1988), Morocco (Steiger and Jansa, 1984), and the US Gulf Coast (Baria *et al.*, 1982; Montgomery, 1993, 1996), among many other localities.

Sponge reefs also were widespread, with prominent occurrences in Romania (Draganescu, 1976; Herrmann, 1996), Poland (Trammer, 1988; Matyszkiewicz, 1996), southern Germany (Gwinner, 1976; Wagenplast, 1972; Meyer and Schmidt-Kaler, 1989; Brachert, 1992; Leinfelder *et al.*, 1994, 1996; Matyszkiewicz, 1997; Werner *et al.*, 1994; Keupp *et al.*, 1996; Pisera, 1997), Switzerland (Takacs, in prep.), France (Gaillard, 1983; Palmer and Fürsich, 1981), Spain (Deusch *et al.*, 1991; Krautter, 1995, 1996), but also developed in Portugal (Ramalho, 1988; Leinfelder *et al.*, 1993a), Italy (Krautter, 1996), and Morocco (Wiedenmayer, 1980; Warme *et al.*, 1988), among many other places. Microbolite reefs are particularly common in Iberia, Portugal (Leinfelder *et al.*, 1993b; Schmid, 1995), and offshore Nova Scotia (Jansa *et al.*, 1989). For a more complete list of references on occurrences of Jurassic coral, sponge, and microbolite reefs, see Leinfelder (1994a). Important occurrences are also indicated in Fig. 1.

Focusing on the plate tectonic setting of the Jurassic most reefs grew on the stable and wide shelf of the northern Tethys margin. Particularly during the Late Jurassic, sea level was about 100 to 150 m higher than today (Haq *et al.*, 1988), which in Europe resulted in many pericontinental and epicontinental shallow to moderately deep shelf seas connecting the Tethys. In a belt extending from Russia to Rumania through Poland, Germany, Switzerland, Eastern France, eastern Spain, Southern Portugal to Texas and New Mexico, reefs developed extensively, especially during the Oxfordian. However, no continuous barrier reef developed; instead, reefs occurred as isolated bodies of variable extent in a beltlike yet broad shelf area. Another seaway with reef development was the North Atlantic rift systems and adjacent epicontinental seas, which during the Late Jurassic connected the Texas realm with the Portuguese Lusitanian rift basin, western France, England, and Northern Germany. Many seaways connected the epicontinental seas adjacent to the northern Tethys with those adjacent to the North Atlantic rift system. Reefs were less widespread, yet existing on the southern Tethys shelf (see Section 3). During the Late Jurassic, warm-water coral reefs and coral associations also grew in fairly low latitudes, such as in northern England (Ali, 1983; Insalaco *et al.*, 1997) and southern Argentina (Leggareta, 1991). This is evidence of very equilibrated seawater temperatures (Leinfelder, 1994a).

2.2. General Types of Jurassic Reefs

Jurassic reefs contain various proportions of "parazoans" (sponges) and/or "true" metazoan reef building organisms (corals and others), microbial crusts, mud, and peloidal to calciclastic particles, as well as highly variable proportions of framework development or preservation. The basic types of Jurassic reefs can be grouped into the following categories, although a great variety of transitional and successional types exist (Fig. 2):

- Coral reef types
- Siliceous sponge reef types
- Pure microbolite reef types

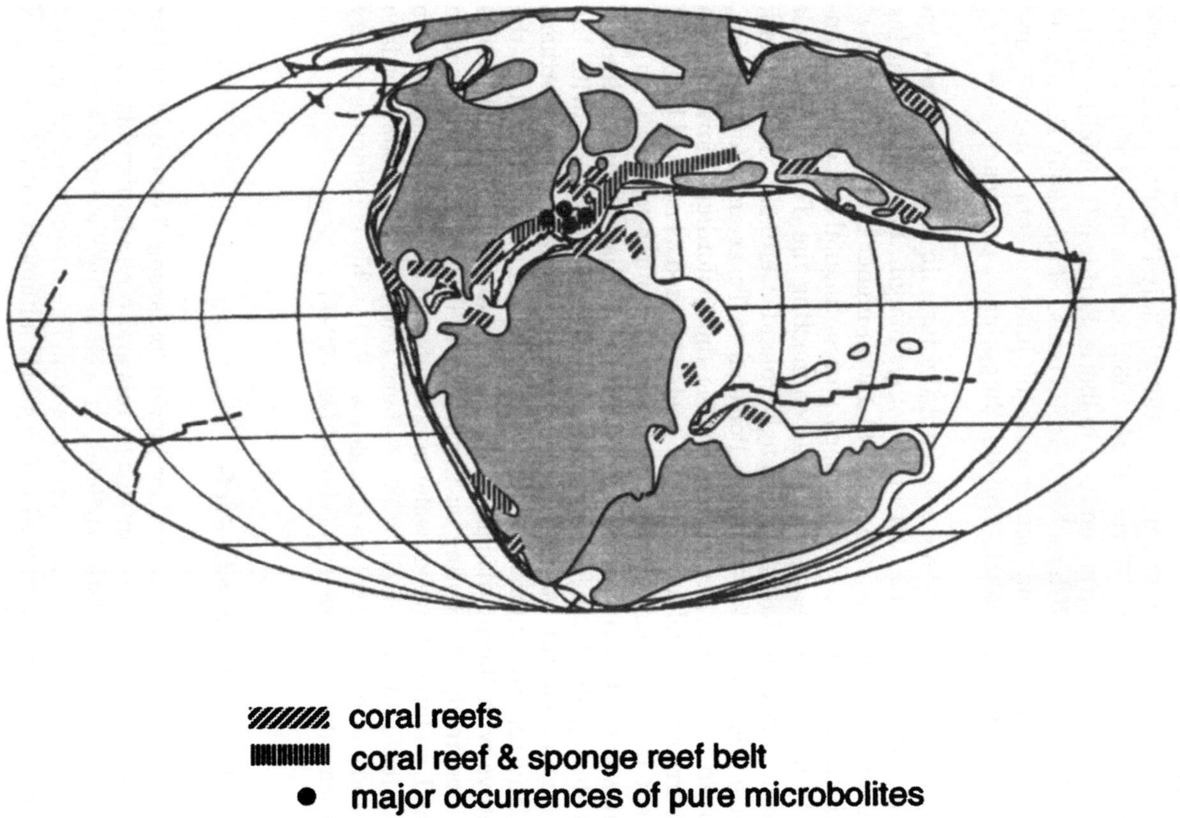

FIGURE 1. The global distribution of coral, siliceous sponge and pure microbolite reefs during the Late Jurassic. (Modified after Leinfelder, 1994a.) Pangea paleogeography after Scotese *et al.* (1993). Reefs are more frequent along the northern Tethys shelf. Note occurrence of warm-water coral reefs in high paleolatitudes, indicating equilibrated water temperatures. Reefs were less frequent during the Early and Mid-Jurassic (see sketch maps in Leinfelder, 1994a).

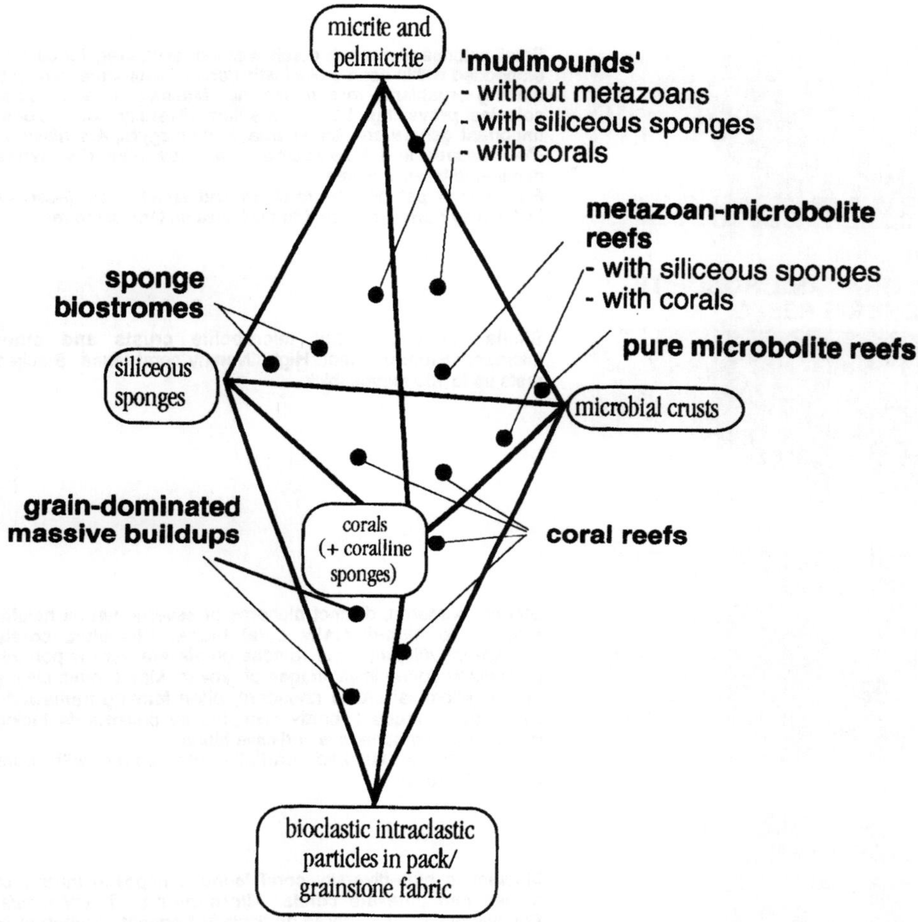

FIGURE 2. Compositional types of Jurassic reefs. The end-members coral reefs, siliceous sponge reefs, microbolite reefs, mudmounds, and bioclastic sand shoals form many transitional types. Most common types are indicated by dots. (After Leinfelder, 1993a, and Leinfelder and Keupp, 1995, modified.)

2.2.1. Coral Reefs

Recent literature on Upper Jurassic coral reef types exists (e.g., Leinfelder, 1993a, 1994a,b; Leinfelder *et al.*, 1996; Nose, 1995; Insalaco 1996a,b; Insalaco *et al.*, 1997), therefore only a short review of coral reef types is given here in tabular form (Fig. 3). The dominant fauna of these reefs are scleractinian corals that may or may not be preserved in life position. If preserved *in situ*, in rare cases they may build true framestones with massive corals growing on top of each other. Coral bushes forming bafflestones, however, are more frequent and in rare examples may grow up to 4 m. In general, dimensions are from a few

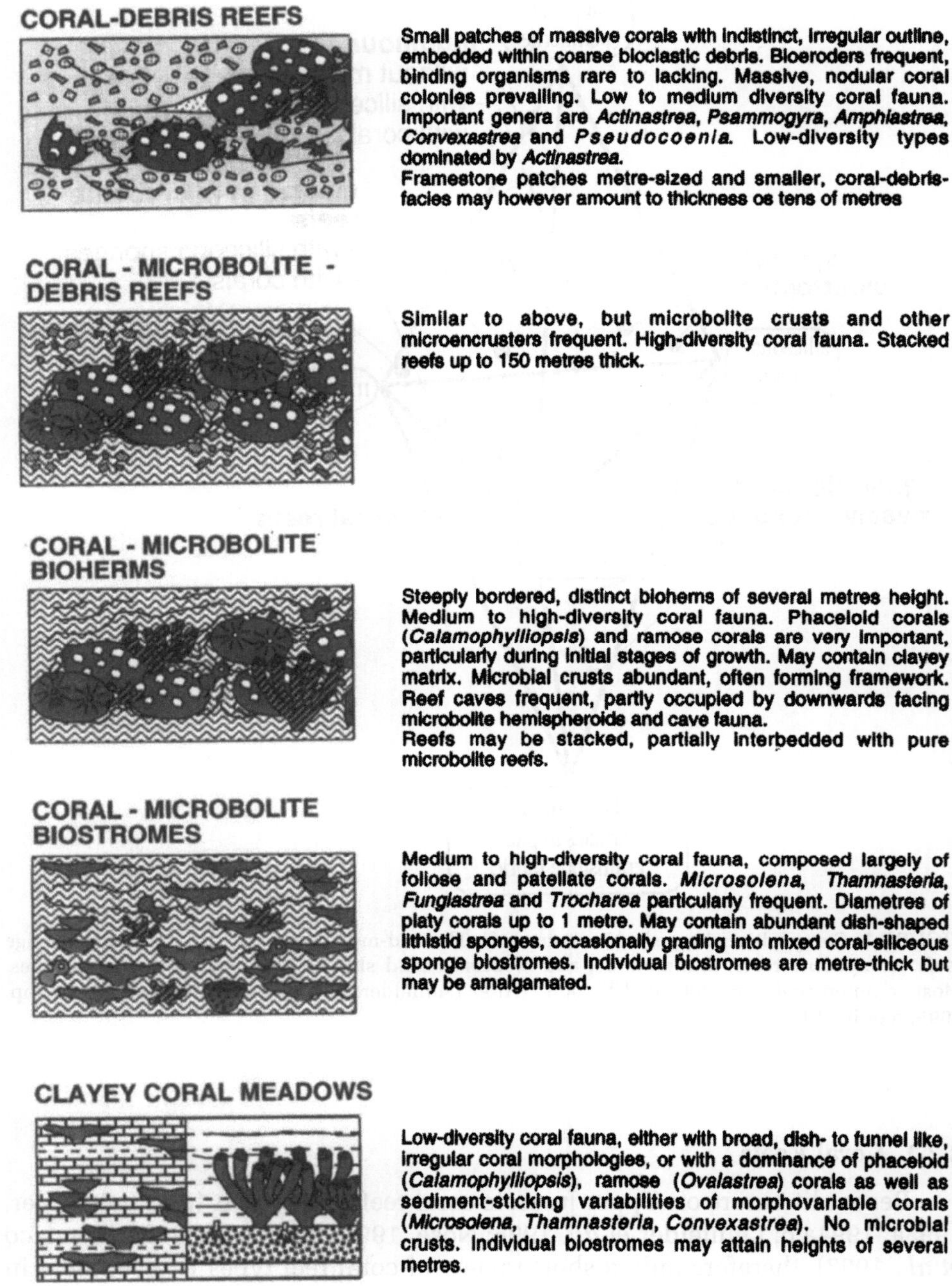

FIGURE 3. The most prominent reef types of the Late Jurassic. (Simplified after Nose and Leinfelder, submitted for publication).

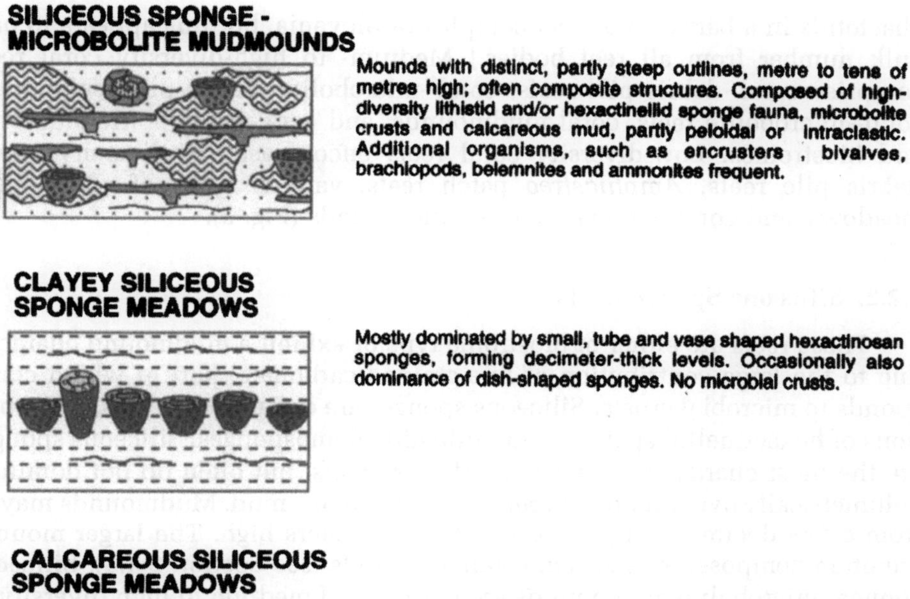

FIGURE 3. *Continued.*

decimeters height up to stacked reefs attaining the cumulative thickness of 100 m and more. Most frequent are coral reefs with thickness ranging from several meters up to 15 m. Upper Jurassic coral reefs can be typified by their faunal composition, dominance, and frequency of coral species, their sedimentological characteristics, and their general shape and dimensions. In general, medium- to high-diversity coral reefs with 40 and more coral species in one coral association should be distinguished from low-diversity coral reefs. [Turnšek *et al.* (1981), recorded 109 species of corals, hydrozoans, and chaetetids in a barrier-type reef complex of Slovenia, but this appears to be a bulk number from all reef bodies.] Medium- to high-diversity coral reefs comprise coral–debris pile reefs, coral–microbolite–debris pile reefs, marly coral–microbolite reefs, coral mudmounds, and coral–microbolite bioherms, and biostromes. Low-diversity coral reefs encompass low-diversity coral–debris pile reefs, *Amphiastrea* patch reefs, various types of marly coral meadows, and coral–stromatoporoid mudmounds (Fig. 3).

2.2.2. Siliceous Sponge Reefs

These reefs are either of biostrome type or exhibit a mudmound character due to the large contribution of fine-grained carbonate, part of which corresponds to microbial crusts. Siliceous sponges are composed of various proportions of hexactinellid sponges and "lithistid" demosponges. Siliceous sponges are the most characteristic element of these reefs, but often do not dominate volumetrically over microbial crusts and calcareous mud. Mudmounds may be from a few decimeters up to several tens of meters high. The larger mounds are often composed of stacked, smaller mounds (see Section 2.4.4). Siliceous sponge–microbolite mudmounds are normally of medium to high diversity in terms of sponge taxa, although a great deal of work is still to be done to determine sponge associations in a quantitative way at the species level. Besides the various types of sponge mudmounds, there are a lot of sponge biostromes, which in the case of the Oxfordian sponge beds from eastern Spain stretch across more than 70,000 km^2, with their original extension certainly having been much broader (Krautter, 1995). Sponge biostromes or sponge meadows are variable and mostly of low faunal diversity. Marly meadows dominated by vase- and tube-shaped sponges are distinguished from calcareous biostromes dominated by dish-shaped sponges. Abundance of sponges may be very high, with sponges being almost the only rock element, such as in the case of the Spanish sponge biostromes (see Section 2.4.4). Somewhat muddier, thick-bedded biostromes have less frequent sponges but contain a lot of microbial crusts (Fig. 3)

2.2.3. Pure Microbolite Reefs

An interesting reef type is the one that is almost completely composed of microbolite. Such reefs were widespread in some areas during the Late Jurassic and are also known from the Early Jurassic. Microbolite crusts, dominated by clotted, thrombolitic fabric, form a dense framework, building up bioherms from a few decimeters up to 30 m in height (Leinfelder *et al.*, 1993a,b). Macrofauna is either virtually absent, confined to narrow levels, or in rare cases scarcely scattered irregularly throughout the reefs in miniature forms. Whenever fauna appears, it is mostly siliceous sponges, chiefly of the hexactinellid type. Serpulids, terebellid worms, and enigmatic encrusting microorganisms may be occasionally very frequent. An interesting feature is that such pure microbolite reefs also may occur in a repetitive, stacked manner within some coral reefs (see Section 2.4.4).

2.3. Abilities and Demands of Jurassic Reef Organisms: The Key to Paleoenvironmental Reconstruction

Reef organisms are strongly dependent on each other, and thus play many different roles in reef systems, which are often referred to as "cities under water" due to the formation of preservable structures and the intense interdependence of the reef biota. In such comparisons, the reef organisms are described as belonging to different guilds, thus having different "jobs," such as chief constructors, cementers, recyclers of building material, gardeners, shopkeepers, water filterers, waste recyclers, and much more (e.g., Fagerstrom, 1987, Ginsburg, 1997; Leinfelder and Ginsburg, 1998). All jobs are necessary to keep the reef city system running and no single chain or web link can be missing. In order to accomplish their tasks, reef organisms have very special demands and abilities. In the fossil example, it is absolutely necessary to uncover those abilities and demands of extinct reef organisms in order to evaluate their role and efficiency in the fossil reef system. Only by doing so is it possible to reconstruct the factors controlling the growth of fossil reefs. This is the ultimate goal in analysis of fossil reefs and allows using fossil reefs as paleoenvironmental and paleostructural monitors (Leinfelder, 1994a; see also Section 2.4).

2.3.1. The Master Builders: Reef Corals

The major constructors of modern reefs are the scleractinian corals, which depending on taxa have variable growth rates, shapes, and fragility. The paramount feature of the great majority of modern reef corals is their perfect symbiotic relationship with zooxanthellate corals, making them largely independent of the heterotrophic feeding mode. Besides the tropical reef corals, there are nonzooxanthellate species, some of which form reeflike (though normally not preservable) meadows in deep and cold waters (Henrich *et al.*, 1996).

Jurassic scleractinians had a high and across the epochs of the Jurassic generally increasing taxonomic diversity (see Section 3). During the Late Jurassic, coral associations generally appeared in large-scale, shallowing upward successions in a broad variety of environments whose general environmental parameters in many cases can be determined by criteria independent of corals. This enables paleontologists to evaluate the abilities and demands of Jurassic corals. Once calibrated, coral associations then can be used as environmental monitors for settings that cannot be reconstructed by other criteria. In this chapter I want to particularly emphasize similarities and differences of Jurassic corals from modern corals.

2.3.1a. Bathymetry and the Photosymbiosis Question. One of the key questions is whether Jurassic reef corals already possessed a photosymbiotic relationship with algae, which in modern reef corals is paramount in determining the environmental necessities and physiological abilities of corals and in particular allows them to grow fast. Despite diagenetic alteration Jurassic corals do often show a distinct macroscopic banding of their skeletons. Microstructures of the individual bands are astonishingly similar to modern zooxanthellate corals, showing low- and high-density bands, which in modern corals are attributed to seasonal differences in illumination. Higher illumination allows for a more rapid linear growth, as it is reflected as a band of lower structural density (Allison *et al.*, 1996). The distinct development of high- and low-density banding in many Upper Jurassic corals supports a good argument over whether many of them already possessed zooxanthellae or other algal symbionts.

The following additional arguments support such interpretation: (1) Many Jurassic taxa are highly integrated forms, which means that they developed complex calical features, such as thamnasterioid and meandroid types with perforate septae, dissepiments, and pennulae (Nose, 1995). Such forms are exclusively zooxanthellate today (Coates and Jackson, 1987), whereas simpler forms may be both zooxanthellate and nonzooxanthellate. (2) Within shallowing-upward successions, Jurassic coral associations are restricted to the upper part. (3) Jurassic coral associations are commonly associated with high-energy sediments such as oolites or shallow lagoonal sediments, like dasycladacean limestones and intertidal loferite sequences. (4) Jurassic coral associations may span a larger bathymetric framework than modern reefal coral associations and even show a partial overlap with siliceous sponge associations of deeper waters. Nevertheless, distinct associations change characteristically along the bathymetric gradient, with deeper water or turbid settings characterized by a dominance of plate-shaped corals. Plate shape is a widespread adaptation toward lower illumination. There are some Jurassic morphovariable coral taxa that show nodular growth in shallow water, flattened, irregular to plate-shaped growth in low-light settings, and branching growth form under elevated sediment influx (Nose, 1995; Nose and Leinfelder, 1997). Deeper-water associations might be completely composed of plate-shaped taxa, such as the many microsolenid associations from Iberia or France (Errenst, 1990a,b; Leinfelder *et al.*, 1996; Insalaco, 1996a). In the transitional zone with lithistid demosponges, some Jurassic microsolenid corals may develop thin plate shapes measuring more than 1 m across (Leinfelder *et al.*, 1993a). However, platy shapes might also be an adaptation toward poor plankton availability or osmotic nutrition, and moreover a two-dimensional plate-shaped cross-section through a coral colony need not necessarily correspond to a similar three-dimensional plate shape (Fig. 4).

Additionally, but not unequivocally, partial support for the occurrence of photosymbionts in Mesozoic reef corals comes from carbon and oxygen isotopes. Stanley and Swart (1995) have argued that a positive correlation of these isotope ratios as well as oxygen ratios above $-6\ \Delta\ O^{18}$ are characteristic of nonzooxanthellate forms due to near-equilibrium conditions with the ambient seawater. Strong clustering of values around -3 to $+1\ \Delta\ C^{13}$ and -5 to $-3\ \Delta\ O^{18}$ is interpreted as the reflection of the vital effect of zooxanthellae.

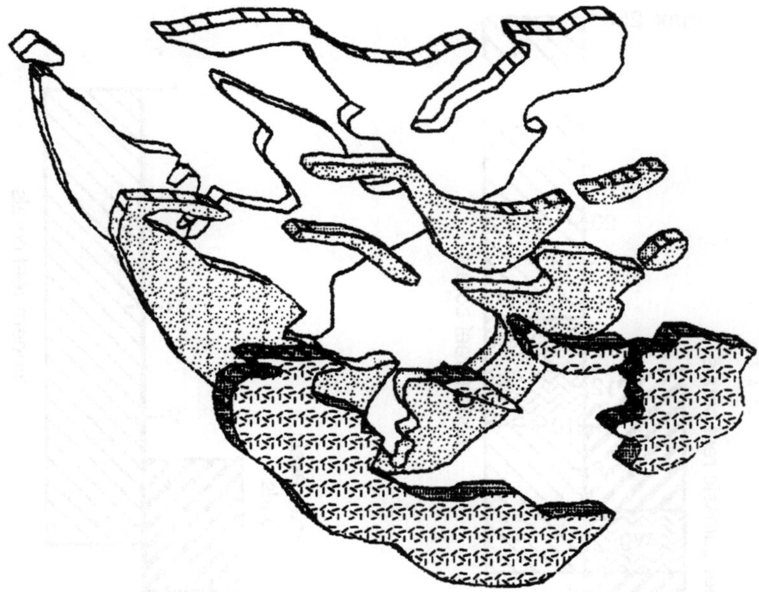

FIGURE 4. Three-dimensional reconstruction of microsolenid corals from the Swiss Liesberg Beds (Oxfordian). The irregular funnel to spiral shape may appear dish-shaped in a fragmental, two-dimensional cross-section. Growth form is interpreted as "best-fit" adaptation toward both reduced illumination and background sedimentation. (From Tacaks, unpublished data.)

The authors used Mesozoic corals that were still preserved as aragonite. However, not all these results are compatible with the above criteria. Using Polish material, the species *Thamnasteria concinna* should be a nonzooxanthellate form according to Stanley and Swart (1995), whereas this species and its close relative *Thamnasteria lobata* show clearly developed growth banding, highly integrated calices, and flattening interpreted to be from greater water depth. Isotopic values of Iberian (nonaragonitic) material show no isotope correlation and plot in the zooxanthellate field with their $\Delta\ C^{13}$ values but are depleted in oxygen, which might be a diagenetic effect (Nose and Schmid, unpublished results). This contradiction may be due to the fact that in both aragonitic and neomorphic material, diagenetic overprinting could be too intense to use this method or rather that the isotope signals of Jurassic corals were not as strong as in modern corals because the efficiency of the photosymbiotic relation was still much lower than in modern corals (Nose and Leinfelder, 1997).

There are a couple of important arguments that support the last statement. Despite showing the distinct bimodal growth bands of modern zooxanthellates, growth rates of Jurassic corals were considerably less than those of modern corals, being in the range of 3 to 5 mm/year in comparison with the modern zooxanthellate corals' average rates of 10 to 15 mm/year (Fig. 5).

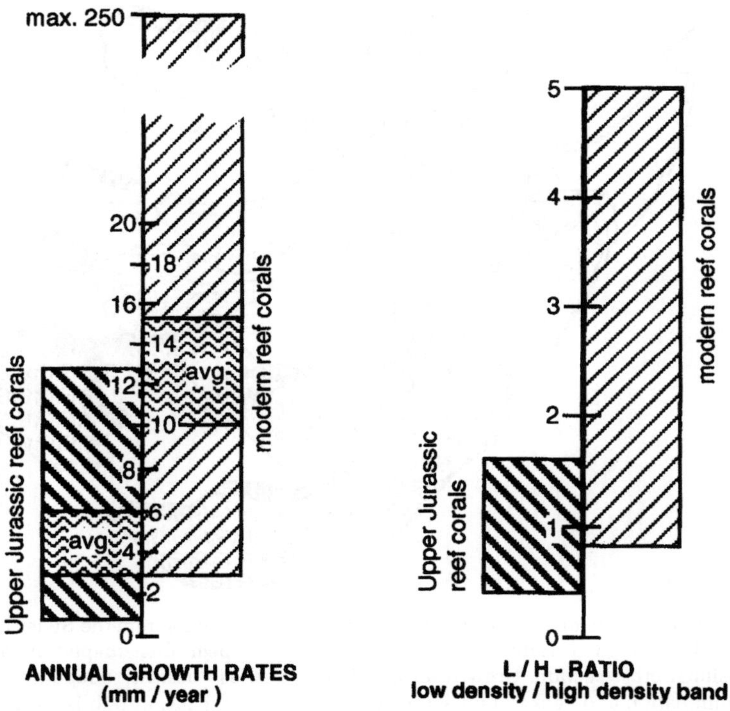

FIGURE 5. (Left) Linear growth rates of Upper Jurassic and modern reef-building scleractinians; avg, average. (Recent data after Buddemeier and Kinzie, 1976; Schuhmacher, 1976.) (Right) Ratio of low to high density band thickness in Upper Jurassic and modern reef building scleractinians (recent data after Highsmith, 1979). (Modified after Nose and Leinfelder, 1997.)

Maximum values of modern corals are 280 mm/year and more, whereas the highest growth rates of Jurassic corals measurable to date are about 13 mm/year (Bertling, personal communication, 1998). This lower growth rate is accompanied by a lower ratio of low- versus high-density band thickness, which in modern zooxanthellate reef corals ranges from 0.8 to 5, whereas in Upper Jurassic taxa the ratios average around 1 and never exceed 1.6. This shows that despite a different calcification pattern due to elevated light availability, there was no major push in growth rate, suggesting that the photosymbiotic relationship was not yet as effective as in modern forms. Corals also may have been apozooxanthellate, i.e., switching from zooxanthellate to nonzooxanthellate state (Stanley, personal communication, 1998). Another important argument is that highest species diversity of Upper Jurassic coral associations appears in settings with a highly reduced yet clearly noticeable siliciclastic influx rather than in pure carbonate settings (Section 2.4.4). Again, this indicates that the photosymbiotic relation was not yet perfected so that these corals were not yet adapted to strongly oligotrophic settings and moderately oligotrophic to mesotrophic sites were preferred for at least some Upper Jurassic coral associations.

2.3.1b. Sedimentation Rate. In contrast to the above-mentioned positive effect of a very low terrigeneous influx, raised sedimentation rates were as just a threat to Jurassic corals as they are to modern ones (Rogers, 1990; Riegl, 1995) and strongly elevated sedimentation rates could not be tolerated. Nevertheless, there was a considerable range of tolerable background sedimentation among Jurassic corals. In general, coral growth forms and calical types may be adapted to sedimentation to a variable degree. Growth form strategies are straightforward, with cylindrical solitary forms and bushy corals better adapted than massive colonies. Among massive colonies, domal forms are less vulnerable than flat, encrusting, or plate-shaped corals. Also, number of septae (as an expression of available tentacles that can be used for cleaning) and calical types of massive colonies may be good indicators of potential adaptation. Cerioid forms with deep calices and strong calical walls are poorly adapted, whereas thamnasteroid and meandroid shapes pose fewer problems in sorting out unwanted particles by tentacle activity (Hubbard and Pocock, 1972; Leinfelder, 1986, 1994b). However, there are many exceptions to such morphological concepts. Poorly adapted corals might have grown under the shelter of larger and better-adapted forms. Colonial skeletal morphologies must also express factors other than sedimentation, such as illumination. Corallum shapes are multifunctional and may represent a best-fit compromise toward different adaptations. An example are the dominantly platy shapes of many microsolenid associations interpreted as a low-light adaptation in deeper or turbid settings (Bertling, 1997b; Errenst, 1990a,b; Leinfelder *et al.*, 1993a, 1996; Insalaco, 1996a). Microsolenids have been compared with the modern zooxanthellate tabular deep-water form *Leptoseris fragilis* (Leinfelder, 1992; Insalaco, 1996a). *Leptoseris fragilis* occurs down to 150 m in the clear waters of the Red Sea by combining both a very sophisticated indirect photosymbiotic relationship and enhanced suspension feeding morphologies (Schlichter, 1992). Insalaco (1996a) presented a key study on such associations based on comparative study of occurrences from eastern France, Switzerland, and England. He also interpreted pure carbonate, low-diversity microsolenid associations with plate shape dominance as deeper water settings, whereas the "microsolenid window" was raised in areas with terrigeneous influence due to turbid waters. However, such interpretation might be biased due to two-dimensional interpretation of growth forms. Actually, detailed analysis of growth forms from the Swiss and French occurrences showed that microsolenids are not generally platy, but commonly exhibit intermittent pustular growth, enabling calical growth to keep up with episodes of sedimentation (Takacs and Stuttgart, unpublished results). Three-dimensional reconstruction based on parallel slabs shows that many platy growth forms measured from two-dimensional slabs actually refer to broad funnel-shaped to irregular colonies (Fig. 4), which represent the best adaptational fit toward (1) surface enlargement necessary because of turbidity, (2) removal possibilities of allochthonous material settling down from the water column, and (3) rapid upward growth to cope with sedimentation.

Bertling (1997b), in an Oxfordian example from northern Germany, pointed out that biostromes composed of platy corals also may develop under strong sediment stress, when sedimentation is intermittent. Growth banding shows that coral growth is limited to about 20 years of nonsedimentation, whereas strong sediment pulses are reflected by the siliciclastic matrix. In many other cases, however, microsolenids underwent more frequent, probably seasonal terrigeneous influx, triggering the irregular, broad funnel shapes described above. Moreover, coral calices also may be oriented downward, rejecting the simplified low-light theory for broadened microsolenid corals. In Switzerland and France, clayey microsolenid associations are overlain by pure carbonate microsolenid associations, both dominated by broadened, two-dimensional platy growth forms. They occur within a generally shallowing upward succession and it appears that the lower, clayey association grew in deeper waters than the compositionally distinct, superimposed, calcareous association (Laternser, 2000), which is a contrasting view to the interpretation of Insalaco (1996a). This new interpretation is corroborated by the occurrence of diagnostic shallow-water organisms, such as *Lithocodium* and *Bacinella* (see Section 2.3.4), by a dominantly bioclastic ground mass and by echinoid types indicative of elevated water energy (see Section 2.3.3).

Besides general growth and calical forms of corals, additional criteria are necessary to identify the degree of sedimentation rate in a Jurassic coral association, particularly since many modern corals have a good capability of cleaning themselves through mucus secretion. This has no direct expression in the morphology of the corallite or the general growth form. Other useful criteria are:

- Corals may show steplike rugged undersides or nonenveloping growth bands that demonstrate that corals became partially buried while growing. Rugged margins are indicative of occasional sedimentary events, whereas smooth margins with nonenveloping growth band are characteristic of continuously elevated sedimentation (Nose and Leinfelder, 1997).
- Thick microbial crusts with frequent microencrusters on coral surfaces are good indicators of very reduced background sedimentation.
- Microencrusters alone or associated with thin microbial crusts might be indicative of strongly reduced but intermittent sedimentation.
- Very low diversities of coral associations point strongly to sedimentation stress, if other stress factors, such as great depth, strong abrasion, or abnormal salinities can be ruled out.

2.3.1c. Water Energy. There are some additional differences between Jurassic and modern corals and coral associations. The dominance of distinct morphology in comparable hydraulic settings is not compatible between

Jurassic and modern corals. Jurassic high-energy settings are dominated by massive hemispherical rather than branching forms as in modern coral reefs. In the Jurassic, branching forms were largely restricted to lower energy settings. However, some species of *Calamophylliopsis*, *Stylosmilia*, and *Dermosmilia* do occur in higher-energy reefs of Portugal and Lorraine (France), which could indicate strategies similar to modern acroporoids, such as rapid regeneration potential after storms. Flat encrusting forms, which may occur in modern, highly abrasive settings, are not similarly developed in Jurassic reefs owing to the fact that such environments were normally unsuitable for colonization (see Section 2.3.5). However, loaf-shaped, broad, nonencrusting morphologies are a typical element of unstable, sand-ground, high-energy reefs and probably represent stabilization strategies by a broad, lower resting surface.

2.3.1d. Salinity. A few Jurassic coral taxa were very euryhaline. *Amphiastrea piriformis*, for instance, formed small, monospecific reef bodies up to 1 m large in an oyster–*Isognomon* association within delta embayments in Kimmeridgian and Tithonian rocks of Portugal (Fürsich, 1981; Fürsich and Werner, 1986; Leinfelder, 1986).

2.3.1e. Temperature. No Jurassic cold-water corals are known. Astonishing coral productivity occurred in fairly high paleolatitudes (e.g., Germany: Bertling, 1993, 1997b; Southern England: Insalaco *et al.*, 1997). Low coral diversities of southern Argentinean Oxfordian coral associations are interpreted as the result of siltation stress (Morsch, 1989), rather than climatic stress, since coral colonies are large and co-occur with dasycladacean algae and calcareous oolites (Morsch, 1989; Matheos and Morsch, 1990; Leggareta, 1991).

2.3.2. Multipurpose Workers: The Sponges

Most of the sponges in modern reefs are soft sponges belonging to the demosponge group. Much rarer are coralline sponges that exhibit a basal coral skeleton, are normally small, and often are restricted to reef caves. The sponges are often described as the water filter system of the reef, by their abilities to filter huge amounts of water in order to feed on bacteria, a principal food source. They also serve partly to stabilize loose substrates that subsequently become cemented by other organisms, reducing abrasive influence of sand-sized particles within the reef (Greb *et al.*, 1996). Many modern sponges live outside modern coral reefs in nearly all water depths. The home of most modern siliceous sponges (lithistid demosponges and hexactinosan sponges) are deeper waters, down to bathyal depths (Reid, 1968; compilation of additional references in Krautter, 1997). Only rarely do they form siliceous sponge meadows (Henrich *et al.*, 1992), and only a few localities are known to date where rigid hexactinosan sponges occur in deep shelf mudmounds

(Conway *et al.*, 1991; Conway and Barrie, 1997). However, the latter locality is not a direct analogue to Jurassic siliceous sponge mudmounds because the modern example is composed of terrigeneous mud, whereas the mud in the Jurassic examples is largely carbonate. Recently, Pisera (1997) pointed to some possible modern analogues for siliceous sponge settings on deeper carbonate-dominated shelves, such as the bathyal lithistid associations on the slopes of New Caledonia (Lévi and Lévi, 1988; Roux *et al.*, 1991). Detailed analysis is still lacking and further studies will show whether these occurrences could serve as models for Jurassic sponge reefs. To date, it appears that none of the known modern sponge settings are directly comparable with those of the Jurassic, particularly the spectacular sponge mounds.

Whereas not much is known about Jurassic soft sponges, those with a rigid skeleton occur frequently in Jurassic reefs. In coral reefs, stromatoporoids (now generally assigned to the coralline sponges that have an identical basal calcareous skeleton) and other sponges with calcareous skeletons (Calcarea) are common elements and occasionally even dominate to form meadows (Fürsich and Werner, 1991; Werner *et al.*, 1994). Nearly all of them are restricted to shallow water. Siliceous sponges are rare in Jurassic coral reefs but are the dominant element in the sponge reefs. In general, there may be either a dominance of lithistids with additional hexactinosan elements or a dominance of hexactinosans without lithistids. Taxa diversity ranges from high to very reduced, with distinct growth forms clearly predominating in the low-diversity associations (dish shape dominance or tube–vase shape dominance; see Section 2.4.4).

The biology of modern siliceous sponges is not fully understood. Recently, Krautter (1997) compiled available data on modern sponges and, based on both modern and Jurassic examples, developed morphological criteria to highlight demands and abilities of Jurassic sponges. Of particular interest is that the two major groups of siliceous sponges differ considerably in their biology and physiology.

2.3.2a. Energy Uptake. The normal feeding strategy of sponges is active filtering of minute planktic organic matter. Since ostia are rarely larger than 200 μm, the usable particle size is below this limit, down to 1 μm; hence, sponges largely filter on the micro-, nanno-, and picoplankton, most of which arew free living bacteria. The amount of free bacteria rapidly diminishes in greater water depth (Hobbie *et al.*, 1972; Rheinheimer, 1980), a fact for which sponges compensate by enormous water pumping activity or by additional forms of energy uptake. Many sponges of the demosponge group (including the important Jurassic lithistid sponges) cultivate enormous amounts of bacteria or cyanobacteria, which can amount to more than 80% of the soft tissue of sponges. This appears a useful strategy to deal with major strong fluctuations in nutrient availability. Whereas the symbiotic relationship with cyanobacteria is evident (Wilkinson and Trott, 1985; Wilkinson and Evans, 1989), the same is also plausible for the settlement of pure bacteria. Bacteria probably benefit from the chemical microenvironment and the waste products of the host sponge, whereas the sponge can feed on metabolic products of the bacteria or the bacteria themselves, especially in times of reduced availability of external food. This strategy enables them to inhabit environments that are depleted in external planktic food. In fossil examples, lithistid sponges with very large thick body walls appear to have used this strategy. This is particularly obvious if in morphovariable forms the same species appears thin in some environments but thickened and globoid in other settings. In lithistid-dominated associations, low nutrient availability may be indicated if:

- Thick-walled forms dominate, whereas thin-walled, morphoconstant forms are missing.
- Morphovariable forms are of the nodular to globoid, i.e., the high-volume-per-surface type.
- The diversity of the entire association is low.

Hexactinellid sponges have a completely different bauplan, and thus are considered an independent phylum by some authors (discussion in Reiswig and Mackie, 1983). The volume of their organic tissue is extremely thin, just coating the silica spicule skeleton. Their filter-feeding capacity is much poorer than in lithistid sponges, but their tissue characteristics enable them to feed largely on dissolved and colloidal organic matter by osmotrophy. Dissolved and colloidal organic matter become strongly enriched in deeper waters through the decay of sinking dead plankton and other organic particular material. Dissolved organic material in shallow water is recycled directly by the wealth of unicellular living plankton that is lacking largely in deeper water, making deeper settings the preferential site of hexactinellid growth. Even if dissolved organic material is rare due to general low productivity of the surface waters or from thermohaline stratification preventing dead material from sinking to the bottom, some hexactinellids can still adapt to such impoverished conditions by enlarging their surface-to-volume ratio. Superoligotrophic conditions are to be assumed by the following criteria:

- Strong dominance of thin-bodied, dish-shaped hexactinellid sponges.
- If nonmorphovariable, only such taxa with an original thin dish-shape are represented.
- If lithistid sponges are present, they are of the massive to nodular type (Fig. 6).
- Low- to very-low-diversity associations, with hardly any additional faunal elements.

Such a peculiar association is known from the Oxfordian of eastern Spain (see Section 2.4.4).

2.3.2.b. Sedimentation Rate. Sponges, like any other fixosessile organism, are vulnerable to elevated sedimentation rates but have a variety of adaptations that are also recognized in fossil examples and help define the ancient environment where sponge associations lived. Development of a tube shape is one of the primary adaptations toward sedimentation. Tube shape

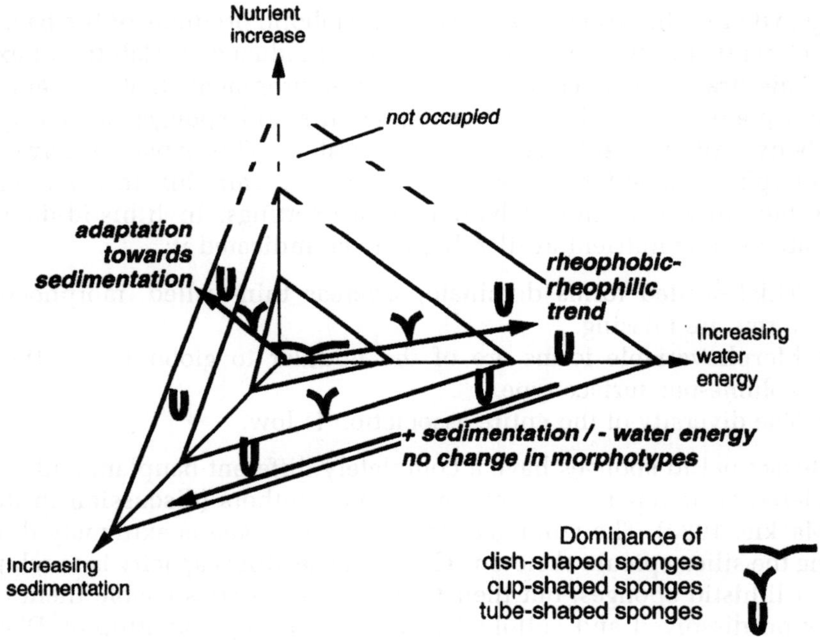

FIGURE 6. Selected environmental factors controlling morphologies of Jurassic siliceous sponges (strongly simplified). Dominance of morphologies varies from thin dishes to thick-walled knobs, to cup and tube shape along environmental gradients. Distinct trends may occur along increasing water energy and increasing sedimentation, which is normally coupled with increasing nutrient availability. Note that dominance of tube shapes may both be indicative of slightly elevated water energy and elevated sedimentation. See text for details.

helps in filtering through the "chimney" (Bernoulli) effect, provided slow horizontal currents are available. In completely quiet settings, the exhalant current becomes bundled in tube-shaped sponges, and thus prevents fine material from settling down. Associations of small, tube- and narrow vase-shaped sponges occur frequently (Leinfelder *et al.*, 1993a; Fürsich and Werner, 1991; Werner *et al.*, 1994). Dish-shaped sponges are completely unprotected. However, the sponges do have some abilities to cleanse themselves of material that actually penetrates the sponge. Demosponges have multipurpose cells, the so-called archaeocytes, which can capture unwanted particles, transport them to the outer surface of the sponge, and release them there, although this does not work well in lithistid demosponges because the denseness of their spicular skeleton strongly restricts motility of archaeocytes. Consequently, this group of demosponges is largely restricted to clear waters. Hexactinosan sponges have the ability to let particles migrate through their tissues, although the process and especially the limitations of it are not yet fully understood (Krautter, 1997). Yet, the fossil examples show that hexactinosans are often the only sponges appearing in clayey deposits that show other features of elevated sedimentation rate, such as lack of microbolite crusts (see Section 2.4.4). A couple of calcareous sponges were able to cover their lower, older inhalant pores by a secondary skeleton, hence allowing them to cope with sediment accumulation. This was detected in the Jurassic sponge *Eudea clavata* by Krautter (1994).

2.3.2c. Temperature. Another important feature of paleoenvironmental implication is the restriction of modern hexactinosan sponges to water temperatures colder than 15° Celsius (Mackie *et al.*, 1983; Dayton *et al.*, 1994), although this might be an evolutionary adaptation and not necessarily transferable to the ancient examples.

2.3.2d. Growth Rate. The growth rates of sponges, though poorly known and apparently very variable, appear to range around the average of 2 cm/year for modern siliceous sponges (M. Krautter, personal communication, 1998). Upper Jurassic siliceous sponges occasionally may be more than 2 m in diameter. Such individuals thus should have attained an age of several hundred years, making growth rates and individual life spans comparable to scleractinian corals.

2.3.3. Bioeroders: The Recyclers of Building Material

In a modern reef a wealth of organisms are continuously eroding the coral skeletons by rasping, gnawing, and biting off pieces of the surface or even by drilling into them. Most bioeroders, such as herbivorous snails, parrot fish, or many sea urchins are doing so in search of food, which is often soft algae. Prokaryotic, microscopic cyanobacteria and fungi drill into coral skeletons for protection, as do the frequent lithophagid bivalves and boring sponges. Bioeroders are important and in a healthy reef, they are in perfect equilibrium with reef growth. They remove old, sick, and dead stone corals and other calcareous skeletons before the surfaces are lost by settlement of a soft algal cover. If bioeroders attack living corals, it is particularly the old ones that already suffer from partial necrolysis, hence giving the larvae of both non-dominating and dominating taxa improved chances for settlement. By doing so, they maintain high coral diversity. Paleozoic reefs often show time-dependent, intrinsic change from a pioneer, through intermediate (high-diversity) to a dominance stage, without any apparent extrinsic environmental changes (e.g., Walker and Alberstadt, 1975). Aging of a reef, as expressed by the intrinsic change of reef associations, is largely unknown from Mesozoic and Cenozoic reefs and might be an expression of the much lower availability of bioeroders during the Paleozoic, and hence the lack of the "rejuvenescance" mechanism provided by these organisms.

Bioeroders break the reef material into particles of variable size. Cobble- and sand-sized pieces, if cemented by other organisms, become recycled by forming a hard foundation for the new settlement of larvae. Most of the smaller generated material, including silt- and mud-sized bioeroded particles, also is removed from open surfaces in the reef. Part of this material fills the open cavities of the underlying dead reef, and by doing so stabilizes the entire living reef. Another large fraction is winnowed into the open ocean or the lagoon by waves and storms, thus preventing the reef from getting choked by its own debris. This is particularly important for productive reefs growing in a nonsubsiding tectonic setting. The available ecospace can be used much longer by this strategy.

In Jurassic reefs, quite similar bioeroding organisms were active. Boring microbes were ubiquitous in coral reefs and back reef sands, as evidenced by typical microcrystalline rims with an irregular inner margin (coated grains *pro parte*). Even the oldest boring foraminifer, *Troglotella incrustans*, known from the Jurassic, developed a peculiar commensal lifestyle within the test of another foraminifer and did not contribute much to general bioerosion (Schmid and Leinfelder, 1996). Lithophagid bivalves often mined massive skeletons to their complete destruction. Bertling (1997a) demonstrated that taxonomic composition of the boring fauna also was related to the degree of sedimentation.

While clionid boring sponges were not yet as important in Mesozoic reefs as in Neogene reefs (Bertling, 1997a), the boring haplosklerid sponge *Aka* was widespread in deeper sponge reef settings (Reitner and Keupp, 1991). Sea urchins are a very frequent element in both Jurassic sponge and coral reefs. In rare cases, scratch marks from sea urchins on bivalve shells are known from the Upper Jurassic (Leinfelder, 1986), providing evidence for the existence of algal turfs or microbial coatings on these shells.

Regular echinoids are frequent in many Jurassic reefs containing microbial crusts. As in modern reefs, they appear to have had the task of keeping growth of microbial films under control. Interestingly, some coral reefs do show an upward and outward increase of microbial crusts, which is astonishingly paralleled by a relative decrease rather than an increase of regular sea urchins. Despite the potential higher food availability provided by prolific microbial growth, this fact indicates environmental deterioration at least for sea urchins, such as intervals of poor oxygenation, which many have speeded up the turnover from coral dominance to exclusive microbolite domination (see Section 2.4.1). Also, sea urchins are less frequent in monospecific coral reef meadows struggling with sedimentation. It seems that in the latter example there was too little food for the urchins because the strategy of the phaceloid to ramose corals was to have their branches partially buried by sediment (see Section 2.3.1).

To date, it is impossible to calculate mass balance budgets of reef construction versus reef destruction in Jurassic reefs, although bulk accretion rates can be determined in some cases. Nevertheless, the patterns show that bioerosion was an important factor in Jurassic reefs as well. However, some Jurassic bioeroders may be valuable indicators of specific environmental factors. For example, analysis of constructional morphology of regular Upper Jurassic echinoids particularly allows for the interpretation of water energy, oxygen availability, and water depth (Fig. 7) (Baumeister, 1997; Baumeister and Leinfelder, 1998).

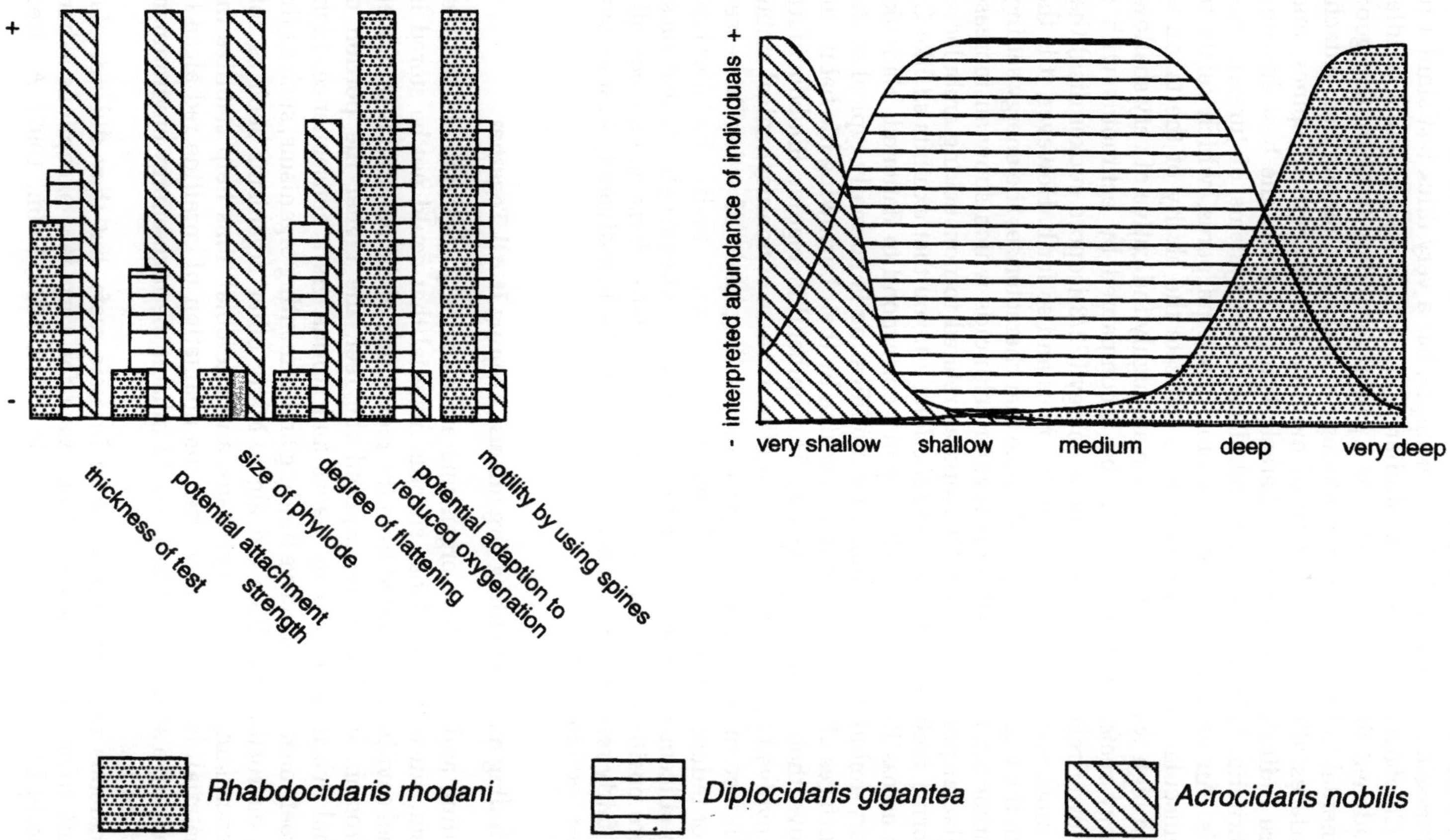

FIGURE 7. Analysis of constructional morphology of Jurassic regular echinoids for environmental interpretation. The three chosen examples represent a well-adapted and specialized deep-water form (*R. rhodani*), a largely ubiquitous, poorly specialized form of moderate water depth (*D. gigantea*), and a specialized shallow-water high-energy taxon (*A. nobilis*). (Data from Baumeister, 1997, and Baumeister and Leinfelder, 1998.)

Thickness of the test of a regular echinoid is a first indicator of water energy. Jurassic *Acropora nobilis*, for example, has a very thick test similar to modern *Colobocentrotus atratus*. Similarly, *A. nobilis* also exhibits saddle-type broadened short spines serving as a partial secondary test. Another good environmental indicator is the morphology of oral spine mammillae. Attachment surfaces allow interpretation of motility and strength of spines, and hence are indicative of substrate characteristics. Probably the best environmental criteria, however, are the ambulacral pore systems of Jurassic reef echinoids. An enormous number of oral P3/4-type pores, which relate to strong muscular ambulacral sucker feet, indicate the ability of the taxon to withstand high water energy, and hence are highly indicative of very shallow, high-energy reefs. Such forms may be accompanied by echinoids with a slightly lower resistivity toward water energy (P2/3 isopore sucker disk) that lived in more protected areas within the reefs and so forth. However, it is the form with the highest potential resistance that characterizes the energy setting.

Another interesting feature is regular echinoids which show an increase of ambulacral pore rows with P1 isopores related to nonsucking ambulacral feet or forms such as *Rhabdocidaris rhodani*, a taxon that exhibits slitlike C1 isopores across its entire test. C1 isopores correspond to flattened, bladelike respiratory ambulacral feet. Such foot modification is normally typical of the aboral surfaces of irregular echinoids, which by their burrowing life habit had to improve their respiratory system. The epibenthic regular *Rhabdocidaris rhodani* evolved this independently and the increase of nonmechanical ambulacral P1 feet in the other examples also is diagnostic of either increased activity or reduced oxygen availability of these forms. Actually, the latter type occur in siliceous sponge reefs, and independent evidence shows that these reefs were positioned in outer ramp settings. *Rhabdocidaris rhodani* actually lived in fairly restricted settings as reflected by the low-diversity host association (see Section 2.4.4).

2.3.4. Binding and Cementing Organisms: Keeping It All Together

Binding and cementing organisms are extremely important in the reef ecosystem, since they fix loose surplus material that could not be stored in lower reef cavities or exported from the reef. Much of the material is produced by bioerosion, but the more exposed the reef, the higher the portion of additional debris generated by waves and storms. As a consequence, high-energy reefs need very effective binding and cementing organisms, such as the modern encrusting coralline red algae. Many other organisms such as soft corals, encrusting sponges, bryozoans, and microbial films help stabilize the loose material, but it was only after the adaptation of coralline red algae to highly abrasive settings that reefs could inhabit the high-energy environment (see Section 4).

Calcareous red algae did exist in Jurassic reefs, but they did not play important roles in reef stabilization, save for very few exceptions where solenoporid bindstones cover coral reefs (Nose, 1995; Helm, 1997). A direct ancestor to the coralline algae, *Marinella lugeoni*, arose during the Late Jurassic and also inhabited coral reefs sporadically, but again was of no importance for reef stabilization (Leinfelder and Werner, 1993). The most important stabilizers in Jurassic reefs were microbial films and mats, which calcified as typical microbolite crust fabrics (Leinfelder *et al.*, 1993b). Microbial calcification, resulting in comparable types, also is important in modern reefs, but is largely restricted to the cavity and cave environments, probably because of competition with coralline algae (Reitner, 1993; Reitner *et al.*, 1996; Montaggioni and Camoin, 1993).

2.3.4a. Jurassic Reefal Microbolites. Jurassic, and similarly other, microbolites should be categorized under a macroscopic, mesoscopic, and microscopic scale (Schmid, 1996). A combination of these allows not only for a straightforward descriptive classification but also for genetic clues (Schmid, 1996; Leinfelder *et al.*, 1996) (Fig. 8).

The occurrence of Jurassic microbolites with similar thrombolitic fabrics at different water depths, including deep shelf settings probably about 400 m

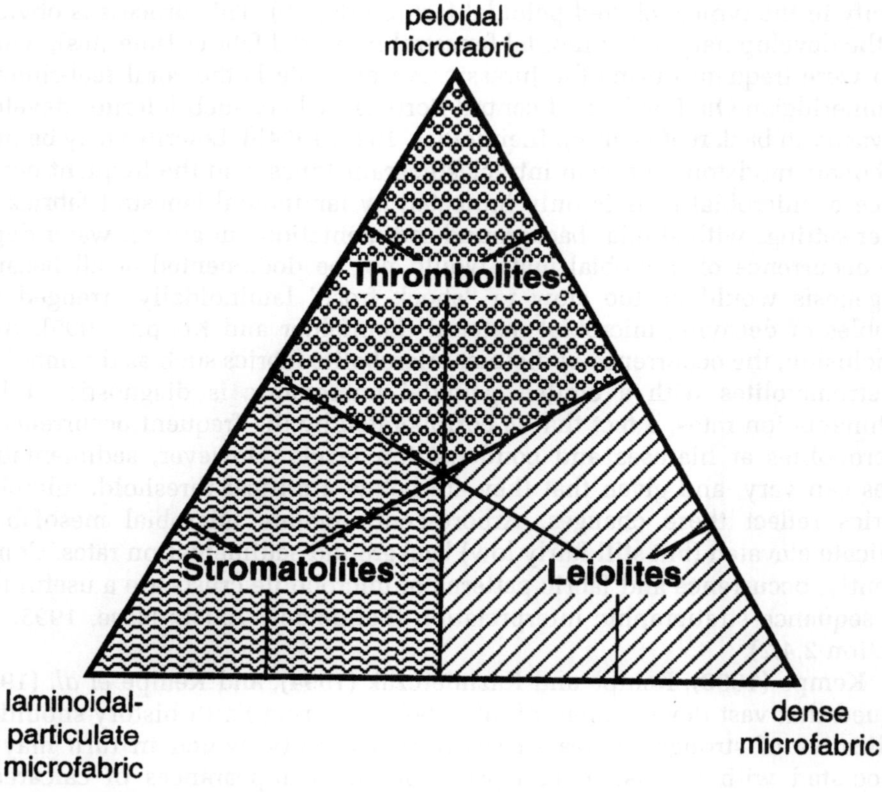

FIGURE 8. Classification of Jurassic microbolites according to their fabric. (Simplified after Schmid, 1995, and Leinfelder *et al.*, 1996.)

deep (Jansa *et al.*, 1989; Dromart *et al.*, 1994) as well as their frequent and prolific development in Jurassic reef caves (Schmid, 1995), shows that microbolites were potentially aphotic, and hence are not related solely to cyanobacterial origin. This does not necessarily mean that cyanobacteria were not involved; they may even have dominated in shallow settings. The conclusion is that microbial films and mats composed of cyanobacteria and/or eubacteria, possibly even diatoms, can result in the same typical clotted peloidal microfabric that triggers a thrombolitic mesofabric. Such microbes often produce similar polysaccaroid macromolecules that fix calcium ions and act as an organic catalyst for calcification (Reitner, 1993). Thrombolitic types are frequent from shallow to deep water, whereas stromatolitic types are largely though not exclusively related to shallow water, probably because diurnal changes of light intensities are reflected by a dominance of motile oscillatorian cyanobacteria. If occurring in deeper water, other regular changes such as background sedimentation or nutrient availability might have caused a similar though less pronounced laminated fabric (Leinfelder and Schmid, 2000).

Reduced sedimentation is the most important prerequisite for the development of microbial crusts. Microbial films also may grow under elevated background sedimentation but then trap and baffle sediment rather than calcify in the typical clotted peloidal fashion (Fig. 9). This process is obvious by the development of laminoidal fenestral intertidal fabrics (loferites), which also were frequent during the Jurassic. An example is the coral reef-rimmed Kimmeridgian Ota Platform of central Portugal, where such loferites developed vastly in back reef position (Leinfelder, 1992, 1994b). Loferites may be pure carbonate mudstones or even intraclastic grainstones, and the frequent occurrence of microbial mats is only indicated by laminoidal fenestral fabrics. In other settings with similar background sedimentation but greater water depth the occurrence of microbial mats would not be documented at all because diagenesis would be too slow to "shock frost" laminoidally arranged gas bubbles of decaying microbial material (Leinfelder and Keupp, 1995). As a conclusion, the occurrence of distinct microbolite fabrics such as thrombolites or stromatolites with a clotted peloidal microfabric is diagnostic of low sedimentation rates, a fact that is corroborated by the frequent occurrence of microbolites at hiatuses and condensation levels. However, sedimentation rates can vary, and given that they are below a critical threshold, microbial fabrics reflect these changes. Arborescent, digitate microbial mesofabrics indicate elevated (but still fairly low) background sedimentation rates. Consequently, occurrence and fabric patterns of microbolite crusts are a useful tool for sequence stratigraphic interpretation (Leinfelder, 1993b; Nose, 1995; see Section 2.4.2).

Kempe (1990), Kempe and Kazmierczak (1994), and Kempe *et al.* (1996) argued that vast development of microbolites during Earth history should be indicative of strongly increased seawater alkalinity, which in turn may be associated with ecosystem collapses and mass appearances of calcareous microbolites. During the Precambrian, this was thought to be due to the sodic nature of the oceans, whereas during the Phanerozoic alkalinity could be

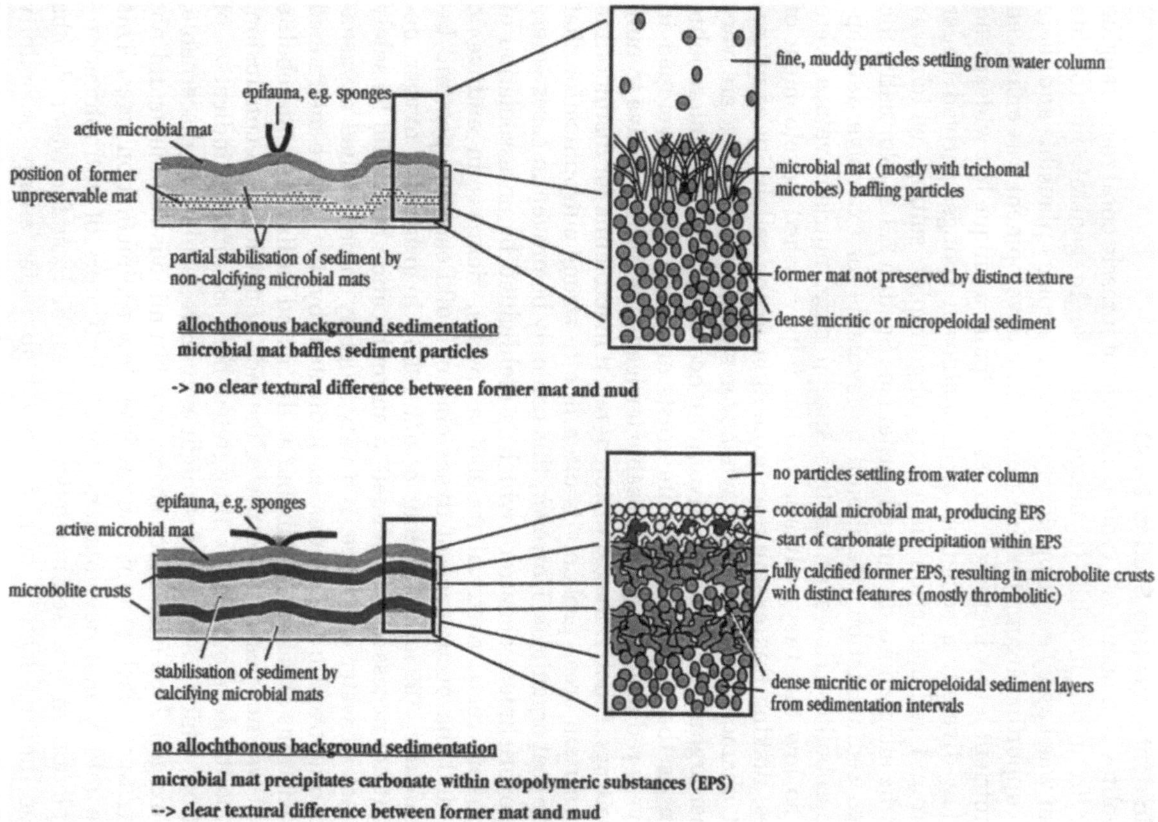

FIGURE 9. Preservation of Jurassic microbial mats. Mats may grow with or without background sedimentation. Under slow sedimentation, mat microbes were probably dominated by trichomal cyanobacteria-trapping sediment particles. Distinct microbolite structures did not develop. Under strongly reduced to zero sedimentation microbes produced large amounts of exopolymeric substances (EPS), which largely are polysaccharoids and some proteins. EPS may act as catalyst for calcification in distinct, mostly thrombolitic, microbolite patterns.

regionally increased by stagnant basins via the H_2S alkalinity pump or by increased influx of sialic weathering products. Probably Jurassic seas had a slightly higher alkalinity than today, which might have facilitated the formation of microbolite precipitation. However, the alkalinity model alone cannot explain the co-occurrence of pure thrombolite reefs and macrofauna-rich reefs at certain time intervals (see Section 2.4.4).

Microbolites also occur as a typical element of Jurassic coral reefs, ranging presumably from oligotrophic to mesotrophic, so in general, microbolite development was largely eurytrophic. However, it appears plausible and there are several supporting arguments that microbolite development was enhanced whenever nutrient levels increased. In eastern Spain, a unique but widespread Oxfordian low-diversity association of hexactinosan sponges thrived under zero background sedimentation in a very oligotrophic setting, which was possible by the adaptations discussed earlier (see Section 2.3.2). Normally, the very reduced sedimentation rate should be favorable for extensive development of microbolites. However, in this special case, nutrient values were apparently so low that they only allowed for impoverished development of microbolites, making this example an exception to the reliability of microbolites that characteristically occur whenever sedimentation rates are very low. Another argument for the positive influence of nutrients on microbolite development is that reefal microbolite crusts are often better developed in areas with very reduced though noticeable terrigeneous influx relative to pure carbonate settings. Finally, the exclusion of reefal macrofauna for distinct time intervals in certain areas probably is often due to strong eutrophication that excluded the reef organisms. However, this can only be proven in cases where oxygen impoverishment occurred, which is highlighted by an association of pure microbolite reefs, bacterial framboidal pyrite, richness in authigenic glauconite, and the occurrence of dysaerobic or poikiloaerobic pectinid bivalves (Leinfelder, 1993a; Leinfelder *et al.*, 1996). In principle, oxygen depletion might not necessarily be related to eutrophication but could be solely caused by lack of water exchange. Actually, slightly impoverished oxygenation under oligotrophic conditions is indicated by the before-mentioned sponge associations of eastern Spain, but as discussed earlier, vast microbolite development was not possible because of the lack of nutrients. In conclusion, pure microbolite development is interpreted here as directly indicative of eutrophication, which in some cases might even have resulted in discernable, occasional oxygen depletion. Black shales and bituminous sediments did not develop due to the lack of rapid burial from low sedimentation rates. This makes pure microbolite development perfect indicators of eutrophication–oxygen depletion in low sedimentation regimes, at least for water depths where other limiting factors such as very high water energy or salinity fluctuations can be ruled out. Microbolites therefore are important tools for paleoceanographic reconstructions (see Section 2.4.4).

2.3.4b. Living on Microbolite Crusts. A wealth of other encrusting organisms lived on the surfaces of Jurassic microbolite crusts. Many of these were

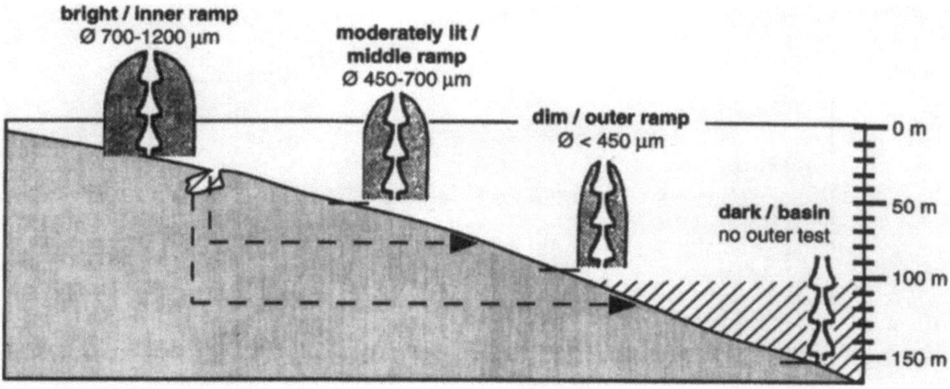

FIGURE 10. *Tubiphytes morronensis*, interpreted by Schmid (1995) as a miliolid foraminifer with a twofold test, occurs across a broad bathymetric range in Jurassic coral, sponge, and microbial reefs. The thickness of its outer test is correlated with illumination and can be used as a relative bathymetric dipstick, provided reef cave dwelling is recognized. (After Schmid, 1995, and Leinfelder *et al.*, 1996, simplified.)

considered earlier as enigmatic, alga-type organisms, but are now in part interpreted as foraminifers (Leinfelder, 1986; Schmid, 1995, 1996; Schmid and Leinfelder, 1996). Some of these, such as the loftusiid foraminifer *Lithocodium aggregatum*, the miliolid foraminifer *Tubiphytes morronensis*, or the enigmatic organism *Bacinella irregularis*, in some cases may actually contribute to binding and construction of reefs, whereas like the other microencrusters (e.g., the enigmatic *Koscinobullina socialis*, bryozoans, serpulids) they are normally just accessory organisms. However, they often are perfect indicators of environmental factors, particularly water depth: *Lithocodium* and *Bacinella* both are restricted to shallow settings; *Girvanella minuta* is frequent in coastal settings; and *Tubiphytes*, though eurybathic, is nevertheless a good indicator (Fig. 10), since the thickness of its outer wall changes with light availability, although low-light but shallow cave settings must be taken into consideration (Schmid, 1996; Leinfelder and Schmid, 2000). Using microencruster associations rather than indicator species, allows for a very refined paleobathymetric interpretation (Fig. 11).

2.3.5. Other Organisms and Infrastructure Jobs

Sponges are assisted in water filtering by an enormous crowd of other organisms. However, it would be beyond the scope of this chapter to discuss them all. Among the most common are pectinid, ostracean, and pteriacean bivalves, serpulids, crinoids, bryozoans, and brachiopods. Deeper sponge mounds also might contain burrowing bivalves, including rare nuculids and pholadomyids, Parts of which were sediment feeders.

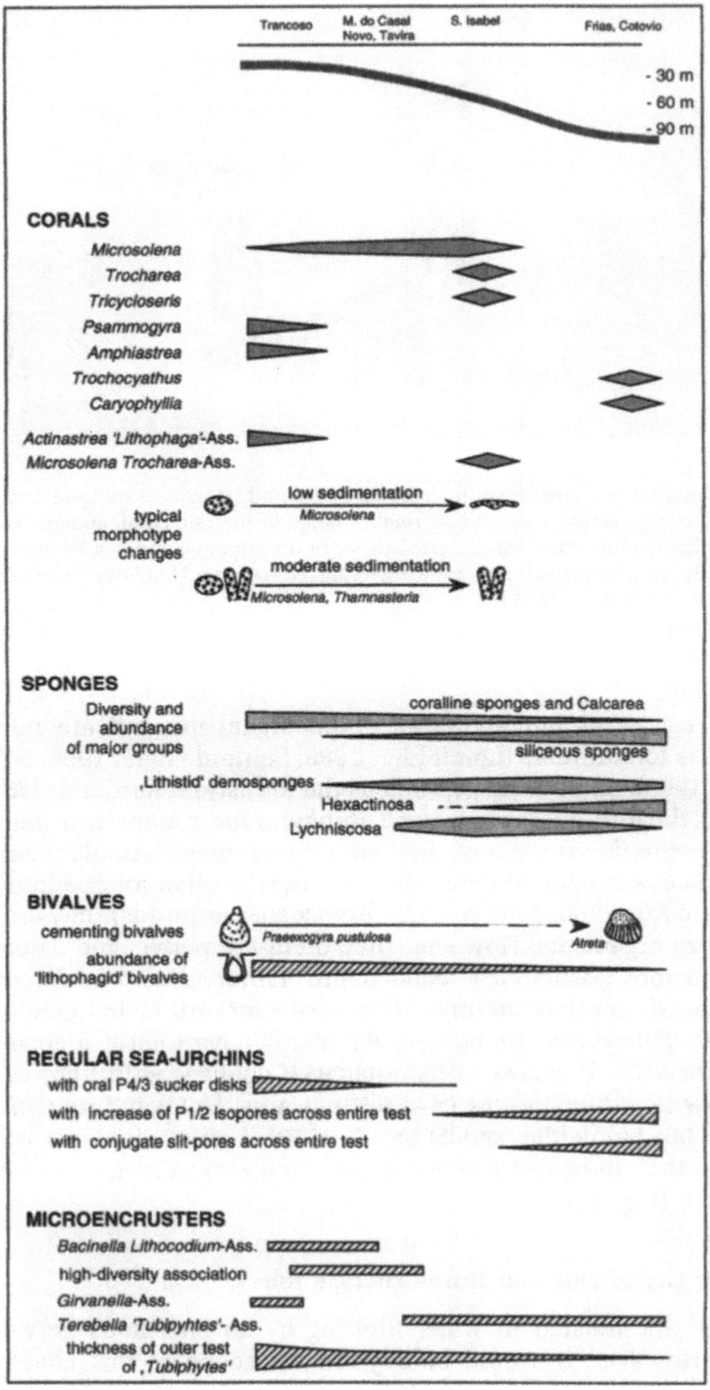

FIGURE 11. Bathymetric distribution and comparative paleoecology of Upper Jurassic reef organisms and associatons from Iberia. (Modified and expanded after Leinfelder *et al.*, 1993a,b, and Werner *et al.*, 1994).

Gastropods, including nerineids, also are widespread, but their life habit is largely unknown. Belemnite and ammonoid taxa, belonging to the predators and possibly scavengers, were frequent in sponge reefs and several taxa definitely lived within the reefs. Crustaceans, many of which belong to the "litter recycling brigade", certainly were also frequent in Upper Jurassic reef systems but are rarely preserved. A high number of taxa, however, are known from the Solnhofen, Nusplingen, and other lithographic limestones of southern Germany, which developed in close association with Upper Jurassic coral and sponge reefs. Among the vertebrates are spectacularly preserved organisms of Upper Jurassic reef ecosystems, such as "parrot-fish-like" chondrostean *Gyrodus*, which probably fed on valved benthos and corals, as well as sharks and rays, reptiles, and even *Archaeopteryx*, the first bird that lived on island in the Upper Jurassic coral seas of southern Germany (for recent findings, see, e.g., Röper *et al.*, 1996; Renesto and Viohl, 1997; Dietl *et al.*, 1997). Many of these organisms provide additional clues to the special settings of Jurassic ecosystems. Cementing bivalve taxa change along a water depth gradient (Werner *et al.*, 1994), burrowing bivalves allow for the recognition of soft muds within reef systems, crinoid types may indicate energy levels, and the general abundance of filter feeders and sediment feeders gives insight into the trophic situation. Comparative paleoecological analysis allows evaluation of differences in these patterns. It is very obvious, for example, that the general occurrence of "true" (i.e., noncyanobacterial) calcareous algae, such as dasycladaceans, occur in shallow-water coral reef settings rather than in sponge reefs. Additionally, certain taxa of dasycladaceans, foraminifers, nerineids, and ammonoids allow for paleotemperature analysis, biogeographic comparison, or biostratigraphic correlation.

2.4. Controlling Factors of Jurassic Reef Ecosystems: The Comparative Approach

Above I have discussed the role of Jurassic organisms in the reef ecosystem and highlighted some examples where functional autecological interpretation and general considerations about the biology of reef organisms give clues to many environmental parameters. This knowledge is important when interpreting qualitative and quantitative patterns of co-occurrence of organisms. Modern reef settings are best characterized by diagnostic reef associations that particularly reflect water depth and energy levels but also other factors such as nutrient availability or sedimentation rate. This approach is particularly useful in the Atlantic–Caribbean realm where general reef diversity is lower than in the Indopacific and dominance of certain taxa under given environmental parameters is obvious. Examples of useful reef associations are the *Acropora palmata* association, the *Montastrea annularis* association, the *Porites porites* association, or the *A. cervicornis* association, among many others (e.g., Geister, 1983, 1992; Greb *et al.*, 1996), all of which reflect different energy levels and water depths. Slightly elevated background sedimentation rates can be reflected in different ratios of key taxa abundance, such as *Montastrea cavernosa* versus *Montastrea annularis* (Greb *et al.*, 1996), or in the occurrence of robust associations as, for instance, in the strongly terrigeneously influenced reefs off Brazil (Leão and Ginsburg, 1997).

Although reef associations may appear already distinct in a qualitative manner, it is particularly important to use quantitative or at least semiquantitative criteria on abundance of taxa (whenever possible at the species level) in order to decipher environmental parameters in a reliable fashion. A lot of quantitative and semiquantitative faunal and floral analyses of many Upper Jurassic reefs are now available, besides new qualitative data on reef faunas, with possibly the best examples from Portuguese and Spanish coral reefs (Errenst, 1990a,b; Rosendahl, 1985; Leinfelder, 1986, 1994b; Nose, 1995; Schmid, 1996; Aurell and Bádenas, 1997; Nose and Leinfelder, 1997). A very useful method for comparative studies is the establishment of coral fauna diversities based on the "trophic nucleus" concept and the Shannon Index (Kauffman and Scott, 1976; Odum, 1983; Werner, 1986). The trophic nucleus is composed of the minimum number of species whose individuals amount to at least 80% of the entire individual number. It is plausible to use volume percent (which can be done by image analysis or by applying volume factors to percentages based on numbers of individuals). Good taxonomic data on the species level also are available from northern Germany (Bertling, 1993). Many new data also are available from the Lorraine and Swiss coral reefs (Insalaco, 1996a; Insalaco *et al.*, 1997), but taxonomic resolution mostly is only on the generic level, except for a few studies (Geister and Lathuilière, 1991; B. Lathuilière, personal communication; Laternser, 2000). As for the siliceous sponge facies, the Iberian Upper Jurassic examples are well studied (Leinfelder *et al.*, 1993a; Werner *et al.*, 1994; Krautter, 1995, 1997; M. Krautter, personal communication) down to the species level, and many new results on sponge reefs are available from the French, Swiss, German, Polish, and Rumanian Jurassic (Gaillard, 1983; Herrmann, 1996; Matyszkiewicz, 1996, 1997; Keupp *et al.*, 1996; Koch, 1996; Koch *et al.*, 1994; Pisera, 1997), although distinct sponge associations based on representative populations could only be established occasionally (Werner *et al.*, 1994; Krautter, 1995, 1997).

2.4.1. Paleoecological Gradients and Their Interplay

Leinfelder (1993a) and Leinfelder *et al.* (1993a,b) recognized the strong control of physical environmental parameters on the establishment and composition of Upper Jurassic reefs based on paleoecological, sedimentological, and sequential analysis. They stated that water depth, sedimentation rate, and nutrient–oxygen fluctuations were the dominant controlling mechanisms besides water energy, temperature, substrate, and salinity. They demonstrated (Fig. 12) that reef growth is only possible if background sedimentation drops below a critical threshold, allowing for improvement in diversity and formation of microbial crusts when substantially lowered. This is true for shallow-water and deeper shelf reef types, which change along the bathymetrical

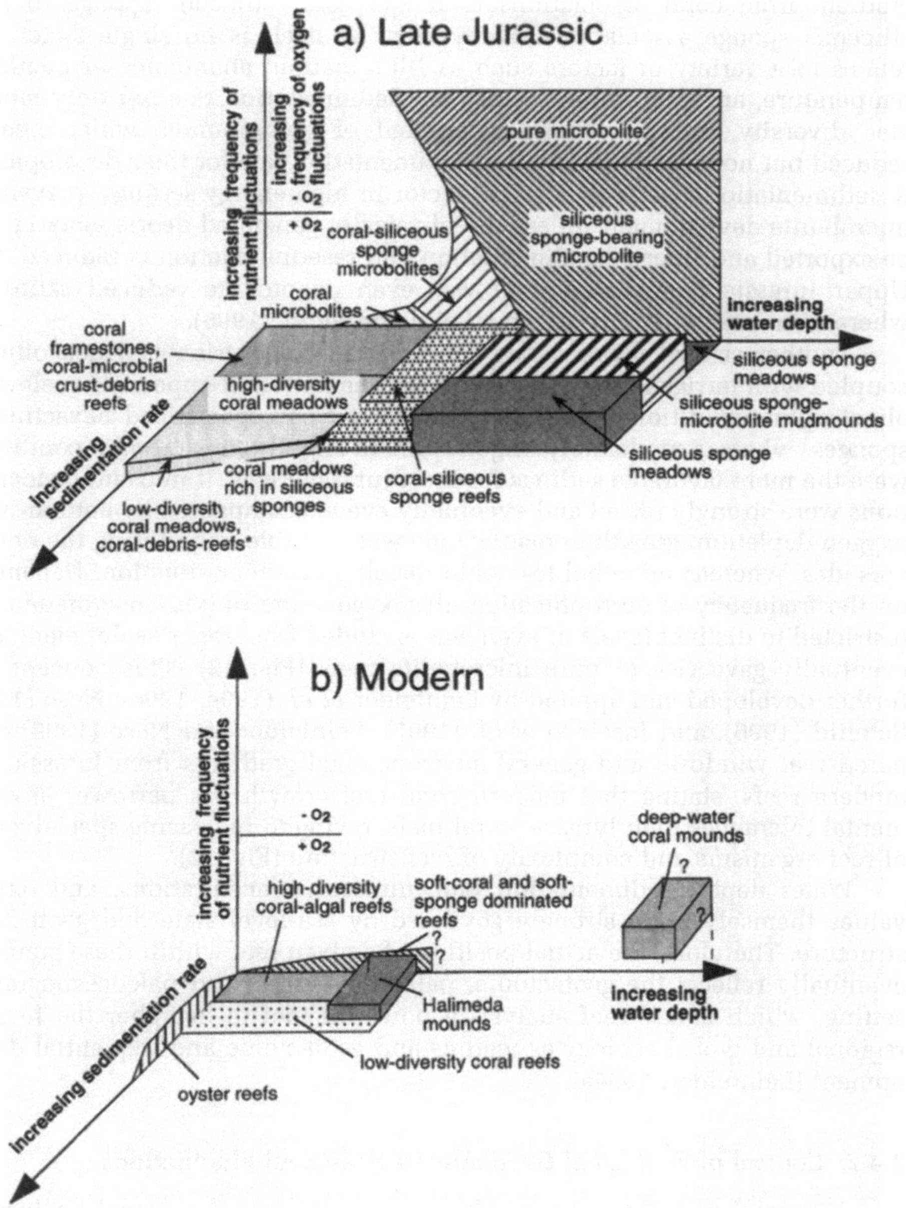

FIGURE 12. (a) The triple factor model of Upper Jurassic reefs in comparison with (b) a similar tentative model for modern reefs. Differences in background sedimentation rate, bathymetry, nutrients, and oxygen concentrations largely determines the occurrence and composition of Jurassic reefs. Environmental tolerances are much smaller in modern reefs. Note that modern deepwater coral mounds show an offset distribution, appearing in a separate "reef window" (cf. 20). See text for further explanation. (Modified after Leinfelder, 1993a, Leinfelder *et al.*, 1996, and Leinfelder and Nose, 1999.).

gradient from coral associations to mixed coral–siliceous sponge to pure siliceous sponge associations. Bathymetry by itself is no single factor, but relates to a variety of factors such as illumination, abundance of plankton, temperature, and atmospheric pressure. Sedimentation rate not only moderates diversity, but also the development of mudmounds, which need a reduced but noticeable background sedimentation rate for their development. Resedimentation also is the critical factor in high-energy settings, preventing microbolite development unless the physically generated debris material can be exported and internal sedimentation and resedimentation is reduced. One Upper Jurassic coral association was even adapted to reduced salinities, whereas the rest was fully marine (Leinfelder *et al.*, 1996).

Another critical factor is the concentration of nutrients, which often is coupled with terrigeneous background sedimentation. It appears that elevated oligotrophic conditions could be tolerated only by specialized hexactinellid sponges, whereas moderately oligotrophic to slightly mesotrophic conditions were the more favorable settings for most Jurassic reefs. If nutrient concentrations were strongly raised and eventually even accompanied by bottom-water oxygen depletion, growth of macrofauna was possible only during the normal episodes, whereas microbolites could develop in either situation. Depending on the frequency of eutrophication–dysoxygenation pulses, macrofauna was restricted to distinct levels or even was excluded from reef development, so it eventually gave rise to pure microbolite reefs (Fig. 12). This concept was further developed and applied by Leinfelder *et al.* (1994, 1996), Nose (1995), Schmid (1996), and Insalaco *et al.* (1997). Leinfelder and Nose (1999) compared reef windows and general environmental gradients from Jurassic and modern reefs, stating that modern coral reef growth has narrower environmental tolerances than Jurassic coral reefs, owing to increasing specialization of reef organisms and complexity of reef structure (Fig. 12).

Water depth, sedimentation rate, nutrient concentrations, and oxygen values themselves are strongly governed by sea-level state and geotectonic structure. Therefore, the actual position of a given reef within these gradients eventually reflects the geotectonic, paleogeographic, and paleoceanographic setting, which makes reef analysis a powerful tool to decipher the Jurassic regional and global ecology as well as and geotectonic and sequential development (Leinfelder, 1994a).

2.4.2. Control of Ecological Gradients by Sea-Level Fluctuations

A general sequence stratigraphic model has been developed for Jurassic reefs influenced by terrigeneous input (Fig. 13). Reef growth generally occurred in the company of third-order sea-level rises that widely reduced the terrigeneous influx across the shelf. Independent analysis of growth rates for reefs, based on coral and microbolite growth rates (Schmid, 1996), however, revealed that most Iberian reefs did not grow across an entire transgressive third-order cycle but occurred only during additional environmental improvement along with fourth- or fifth-order sea-level rise that eventually opened the

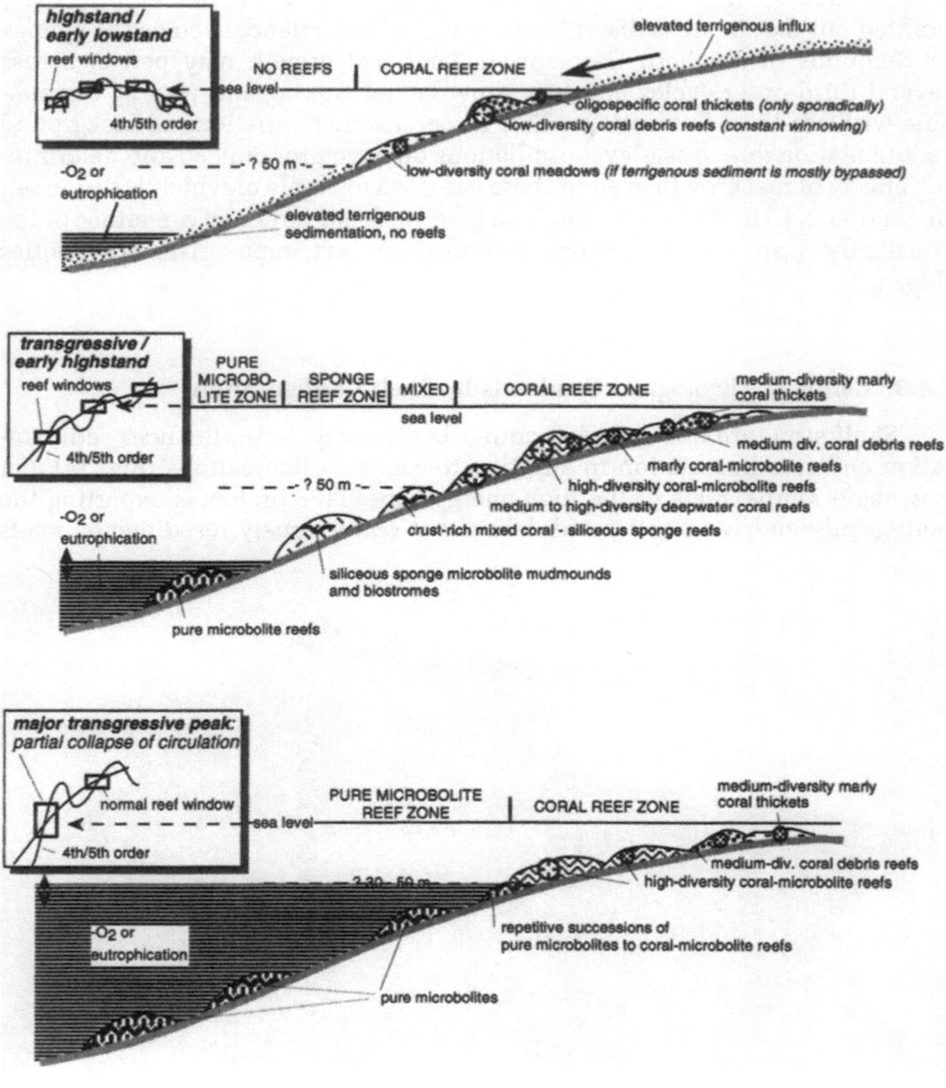

FIGURE 13. Simplified sequence stratigraphic model for Upper Jurassic reefs in terrigeneously influenced settings. During highstand and early lowstand reefs, sedimentation is elevated across the shelf so that reefs may only occur in the constantly wave-washed zone, giving rise to the typical coral–debris reefs. Low-diversity marly coral meadows also may occur sporadically. During sea-level rise, reefs of various types expand widely across the shelf. Due to very reduced sedimentation, reefs often contain a high-diversity fauna and are rich in microbolite crusts, resulting in distinct reef bodies, commonly with pronounced relief. Sea-level rise, however, might also lead to a partial collapse of shelf circulation due to additional climatic equilibration, giving rise to eutrophic or oxygen-depleted settings with pure microbolite reefs occurring up to fairly shallow waters. Most reefs grew only during fourth- and fifth-order floodings within a third-order framework.

"reef window" (Fig. 13). Leinfelder (1993a) and Leinfelder *et al.* (1994) pointed out that in areas distant to siliciclastic hinterland, such as the sponge mudmounds of southern Germany, general reef growth may persist across several third-order cycles but does show lateral waxing and waning correlatable with sea-level fluctuations. This illustrates that variations of background sedimentation due to sea-level oscillations do have pronounced and determining effects of reef growth even in these cases, an example of which is discussed in Section 2.4.4b. Highstand and lowstand reefs are rarer and are either of the constantly winnowed high-energy type or show very impoverished diversities (Fig. 13).

2.4.3. Control of Ecological Gradients by Shelf Configuration

Shelf structure not only determines bathymetry, but influences sedimentation and resedimentation in addition to sea-level fluctuations (Fig. 14). On low-angle ramps, reefs of the high-energy zone have problems exporting the biologically and wave-generated debris that continuously resediments. Reefs

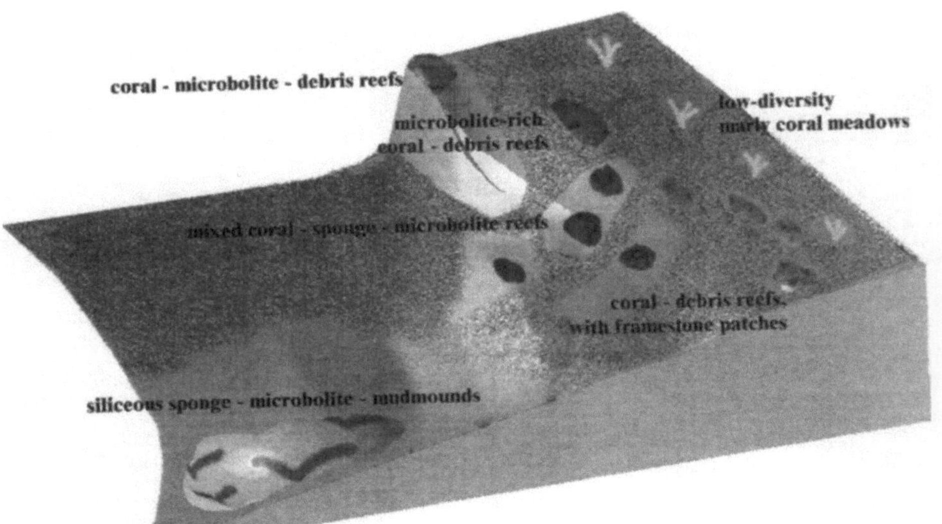

FIGURE 14. Jurassic reef types differ according to different shelf configurations and positions. Some characteristic examples are given. Homoclinal ramps (foreground) show characteristic coral–debris reefs in agitated water and siliceous sponge mounds or biostromes in the lower mid to outer ramp. Steepened ramps and rimmed shelves may suppress sponge mound development in deeper waters because of proximity to mud and sand-exporting, shallow-water carbonate factories. Steepened depositional slopes are the preferred site for crust-rich reefs of moderately deep water, whereas coral–microbolite–debris reefs are indicative of a position in close proximity to a bypass margin, enabling enhanced gravitational export of surplus calciclastics generated within the reefs.

are therefore crust-free and short-lived, since they normally are suffocated in their own debris. Such reefs must have been among the most widespread but are often overlooked since their relics are taken as allochthonous reef debris of other nonpreserved reefs. Large relics of isolated massive corals or even relics of the original framework are indicative of the existence of true reefs (coral–debris reef type). A good example of a partially preserved reef are some Oxfordian reefs at St. Ursanne, Switzerland (Takacs, unpublished results) or many reefs in Iberia, such as reef bodies of the Amaral formation (Nose, 1995).

If a slope break existed, a large proportion of the debris could be exported, giving rise to the development of coral–microbial–debris reefs. These were reefs that grew at slope edge in the constantly wave-washed zone, and hence are rich in fragments of corals and other reef organisms. Contrasting the coral–debris of high-energy ramps, these reefs contain a lot of microbial crusts, stabilizing and cementing the loose, wave-generated material. This was only possible due to facilitated winnowing and gravitational export over a steep depositional or bypass margin so that microbolite crusts could stabilize the remaining material. A case study for this type are the shelf-edge reefs of the narrow Ota platform of central Portugal (Leinfelder, 1992, 1994b), but comparable reefs occur elsewhere (e.g., Schmid and Jonischkeit, 1995).

Export of sediment across a steepened slope prevented the growth of deeper shelf sponge reefs, and it is only at the steep slopes that microbial crust-rich reefs may have developed due to the given bypass possibilities. Consequently, coral–microbial reefs and coral–sponge–microbial reefs are characteristic of very reduced sediment influx, and as such are often indicative of steepened slopes. They may, however, also occur during sea-level rise on homoclinal ramps below the fair weather wave base as a consequence of reduced sedimentation. If on a slope setting, such reefs are accompanied by allochthonous sediments, such as turbidites and debrites, or canyon development. Well-studied examples originate from central and southern Portugal and Spain (Leinfelder *et al.*, 1993a; Nose, 1995; Schmid, 1996; Baumgärtner and Reyle, 1995). The coral–microbolite–debris reef type mentioned above, however, is always diagnostic for steep slopes, even without additional sedimentological criteria, such as allochthonous sediments.

Siliceous sponge mudmounds are largely confined to lower mid to outer, often homoclinal, ramp settings that are distant from shallow-water carbonate factories, because in this position they receive suitable amounts of allochthonous muddy material for their growth. In cases where shallow-water carbonate factories are lacking and terrigeneous sediment also is not imported, no mounds can grow due to both lack of allochthonous material and superoligotrophic conditions, the latter of which prevents the pervasive growth of microbolites (Leinfelder, 1994a; Leinfelder *et al.*, 1996). Under these conditions, low-diversity sponge biostromes with enormous individual numbers may develop (Krautter, 1997; Section 2.4.4c). If influx of carbonate mud is too high for the growth of sponge mounds, sponge biostromes may still develop, which in this case are mud-rich and often contain lower numbers and different siliceous sponge morphotypes than low-sedimentation biostromes. If carbonate factories were too near because of steep shelf slopes, neither biostromes nor bioherms could develop.

Given suitable conditions, growth of sponge mounds may be pervasive, such as on the broad shelf sea of southern Germany. Mounds have the tendency to amalgamate vertically, and thus may create strong submarine relief if sponge reef growth is not interrupted by longer periods of sedimentation. During the late Kimmeridgian, mounds with submarine relief of more than 50 m are known (Gwinner, 1976). These gave rise to the punctuated occurrence of coral reefs on the highest mounds. Most of these reefs are of the coral–debris type and the coral–microbolite–debris type (unpublished results, cf. Paulsen, 1964). In this example, the homoclinal, nearly level-bottom initial outer ramp configuration was suitable for extensive growth of sponge–microbial mudmounds from the Late Oxfordian onward, which in turn changed the configuration of the shelf, and hence the controlling factors of reef growth, such as bathymetry, water energy, and sedimentation–resedimentation patterns. Similar, partially intrinsic change of reef composition and accompanying controlling factors occurred on the wide and shallow, coral-dominated Oxfordian shelf of eastern France. Here, shallowing was accompanied by reduction in terrigeneous influx which allowed for the establishment of uniform microsolenid associations substituting for each other before autodifferentiation of communities changed the entire setting (unpublished results, cf. Geister and Lathuiliére, 1991; Insalaco *et al.*, 1997).

2.4.4. The Interplay of Control Mechanisms: Selected Examples

A few examples should illustrate the interplay of processes and models outlined above.

2.4.4a. Example 1. *Bathymetry and sedimentation rates as major controllers of coral associations from the central Lusitanian Basin, Portugal.* Figure 15 shows the principal coral associations from the Kimmeridgian of the central Lusitanian Basin, which were analyzed in a semiquantitative to quantitative manner using image analysis and counts-per-area data. The reefs occur at different positions within a generally prograding succession shallowing upward from basinal and prodelta marls and clays with intercalated turbiditic sandstones to inner ramp oolitic and coraliferous carbonates (Nose, 1995). Seismic data support the interpretation of a prograding slope system and give additional bathymetric control (Leinfelder and Wilson, 1989). Similar to modern Caribbean associations these reef associations can be arranged along a bathymetric gradient using dominant coral morphologies, microencruster associations, and depth-diagnostic cementing bivalve taxa, as well as the position of associations within a shallowing upward succession and their lateral interfingering with high- or low-energy sediments (oolitic and bioclastic grainstones vs. lime wackestones and clayey muds). The deepest coral associations are dominated by platy microsolenid corals and already show a high proportion of lithistid sponges. Figure 15, however, also shows that although

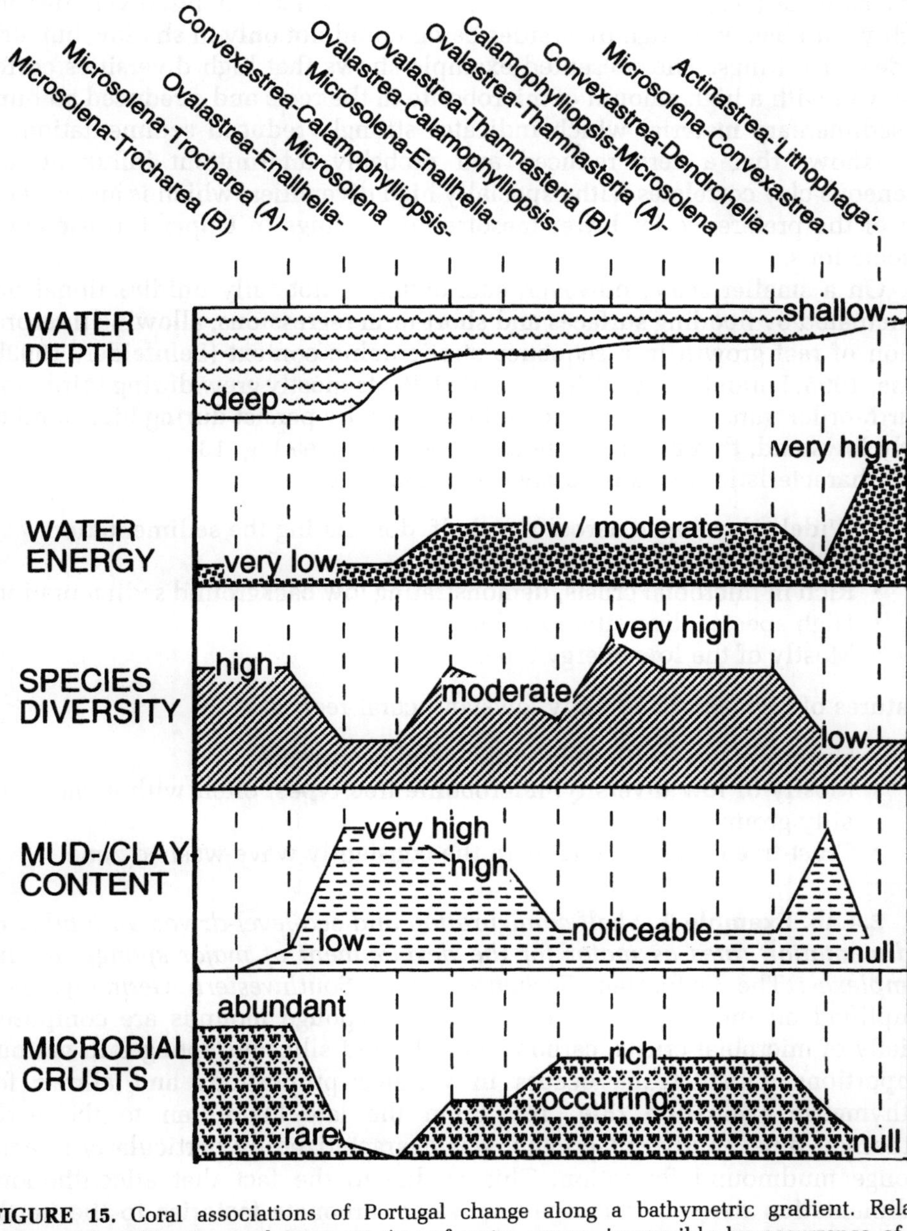

FIGURE 15. Coral associations of Portugal change along a bathymetric gradient. Relative bathymetric calibration and interpretation of water energy is possible by occurrence of reef associations within shallowing-upward succession, by sedimentological features, by morphotype interpretation of diagnostic corals, and by associated fauna (compare with 11). Note that species diversity does not change uniformly along with bathymetry, but mostly is negatively correlated with terrigeneous clay content in the ground mass and positively correlated with the amount of microbial crusts, pointing to sedimentation rate as the major modifier of diversity. Highest diversities occur, however, in associations with a low but noticeable clay content, indicating the importance of nutrient availability for Jurassic reef corals (see text for further explanation). (After Nose, 1995, and Nose and Leinfelder, 1997, modified.)

coral associations are easily lined up along a bathymetric trend, diversities are highly variable, with high diversities being found not only in shallow but also in deeper settings. The presented example shows that high diversities correlate well with a high amount of microbolite in the reefs and a reduced amount of sedimentary material which indicates strongly reduced sedimentation. It also shows that a very reduced and probably intermittent influx of terrigeneous clay correlates with especially high diversities, which is an expression of the preference for lower mesotrophic settings in Upper Jurassic coral associations.

On a smaller scale, however, shallowing is not fully unidirectional but punctuated by flooding surfaces and short-term regressions, allowing interpretation of reef growth in a sequence stratigraphic context (Leinfelder, 1993b; Nose, 1995; Leinfelder and Wilson, 1998). Reefs mostly grew during third- and fourth-order transgressions but could occasionally persist during highstand or early lowstand, though in different composition (see Fig. 13).

Characteristics of transgressive coral reefs are:

- Widely developed across the shelf, dominating the sedimentation.
- Pure carbonate.
- Rich in microbial crusts, demonstrating low background sedimentation.
- High species diversity, low dominance.
- Mostly of the low-energy type.

Features of highstand and early lowstand coral reefs are:

- Generally rare, with only punctuated occurrence.
- Mostly of low diversity, microbolite free types, often with a clayey or silty groundmass.
- Crust-free debris type reefs in the constantly wave-washed zone.

2.4.4b. Example 2. *Shelf-configuration and sea-level-driven variability of sedimentation rates as motor for the development of major sponge mound complexes:* The Oxfordian *Gosheim reef of Southwestern Germany.* In a simplified manner, Upper Jurassic siliceous sponge mounds are composed chiefly of microbial crusts, carbonate muds, and siliceous sponges in various proportions. Other fauna occurs in variable proportions and allows for bathymetric evaluation (Fig. 16). From the late Oxfordian to the early Tithonian time, southern Germany was characterized by particularly intense sponge mudmound formation. This is due to the fact that allochthonous sedimentation derived from shallow-water carbonate factories to the north, east, and northwest was low enough to allow mound formation. In the Swiss Jurassic and particularly the south Portuguese Algarve Basin, carbonate factories were much closer to the suitable depth provinces of the outer ramp due to the steepened character of the ramps, which did not or only occasionally allow for mound formation. On the other hand, a certain amount of allochthonous calcareous muds were necessary to helping a mound accumulate (Leinfelder and Keupp, 1995).

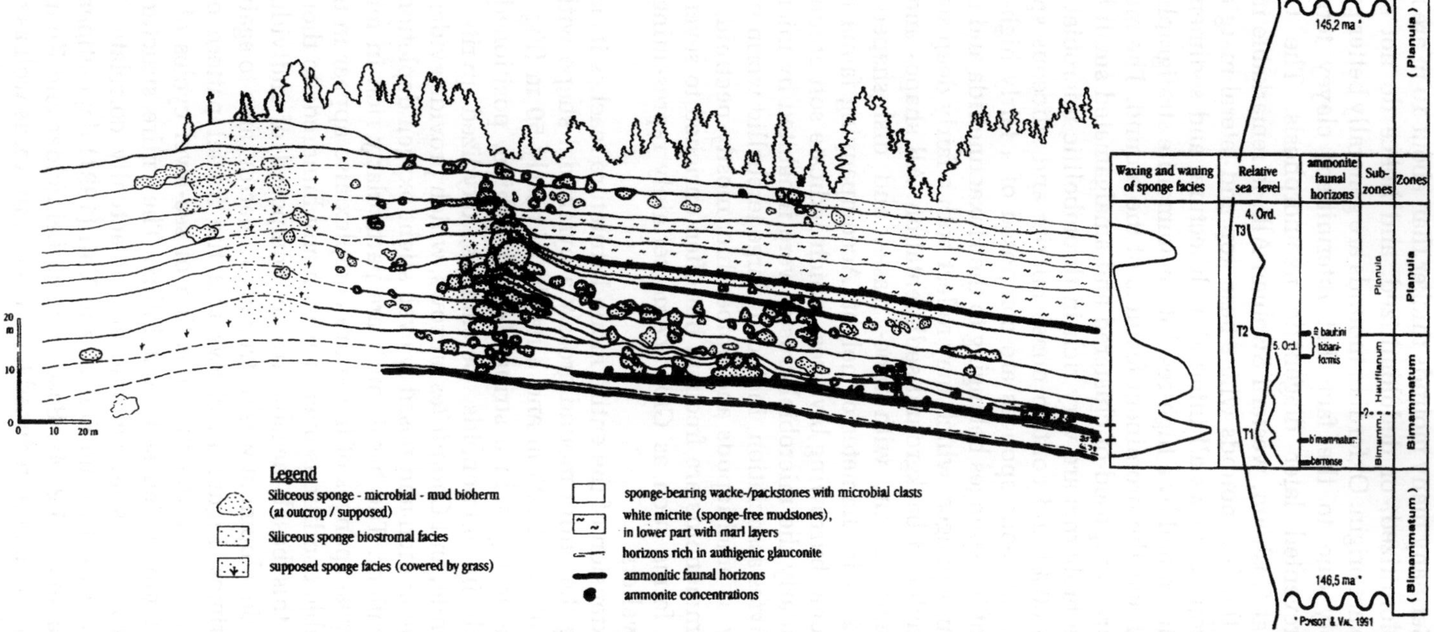

FIGURE 16. Architecture and sequential interpretation of a major composite–siliceous sponge–microbolite mudmound complex of the western Swabian Alb, southern Germany. Growth of the entire complex is related to reduction of sediment influx during a third-order sea-level rise. Episodes of lateral expansion of the composite complex are correlative with fourth- and fifth-order floodings. The (late?) highstand of T3 stopped the growth of the entire complex due to raised import of allochthonous calcarous mud from shallow-water carbonate factories. See text for details.

In southern Germany many mounds thicker than about 10 m are intensely recrystallized, dolomitized, or dedolomitized, and hence do not show any clues as to their exact origin. Oxfordian mounds are generally better preserved, which is probably due to the fact that intermittent clayey terrigeneous sedimentation prevented later diagenetic neomorphosis. The Oxfordian mound structure at Gosheim, western Swabian Alb, is remarkable in various ways. It is one of the few mounds where base, top and lateral margins are all observed in outcrop. It shows all internal architectural and sedimentological features and is constrained by a high-resolution ammonite stratigraphic framework. Figure 16 shows the prominent features of the mound. The mound is a composite structure composed of clustered and amalgamated small bioherms of meter-size. The bioherms are very rich in thrombolitic microbial crust (in some cases up to 80%) and contain dense micrite and siliceous sponges in variable amounts. Siliceous sponges are composed of a fairly high-diversity fauna, with 88% of the sponges belonging to the Hexactinellida and only 12% to the lithistid demosponges, which may reflect both a fairly deep setting and frequent though reduced background sedimentation. All shapes and sizes of sponge morphologies occur, with tube-, vase-, and dish-shaped sponges measuring up to 2 m in diameter or height. Accompanying fauna is rich as well, with occasional burrowing bivalves, which indicate soft grounds. Hard substrates, particularly the microbial crusts, were inhabited by microencrusters of the *Tubiphytes* association, including large terebellid worm tubes, and by a great variety of brachiopods and epibenthic, mostly pectinid, bivalves. Belemnites and ammonoids are frequent but often enriched to several levels, with some small forms such as *Glochiceras* certainly representing bottom-related mound dwellers.

The lateral extension of the entire mound structure reaches its maximum at its base, giving the entire mound grossly a pyramidal shape with a basal lateral extension of about 200 m and a height of nearly 50 m (Fig. 16). The basal mounds are arranged in a string-of-pearl fashion, positioned directly above a bed rich in ammonoids, belemnites, oxydized pyrite nodules, authigenic glauconite, and *Chondrites* burrows, which provide evidence of the cessation of sedimentation and possibly even slight oxygen depletion, at least in the topmost seafloor. The basal muds are particularly rich in microbolite crusts and the largest sponges of the entire complex also appear in this level, evidencing favorable conditions over at least many hundreds to thousands of years. Following this basal megabiostrome, composed of individual small bioherms, mound development was restricted to a narrow area to again expand laterally several times. Figure 16 shows that the fractal pattern of mound development is caused by the superposition of sea-level cycles of different magnitude, which modulated sediment influx. The entire structure can be referred to a third-order sea-level rise and it actually correlates with the transgressive system tract of the Ponsot and Vail (1991) depositional system between 146.5 Ma and 145.2 Ma. Based on the Leinfelder and Keupp (1995) model this transgression both reduced terrigeneous influx as well as carbonate mud export from the remote shallow-water carbonate factory, which increased again during highstand, causing burial of the mound structure by allochthonous carbonate muds (highstand shedding). This correlation suggests that the Oxfordian deep shelf sediments were largely of allochthonous origin rather than representing hemipelagic planktic ooze. Actually, relics of coccoliths are extremely rare, which possibly does not represent preservational bias. Establishment as well as pulses of lateral expansion of the mound structures correlate with ammonoid concentrations and formation of authigenic glauconite relatable to fourth- and higher-order flooding events.

The Gosheim mound thus is a key example for analyzing the origin of Jurassic sponge-rich mudmounds. It highlights the importance of strongly reduced but noticeable intermittent influx of allochthonous sedimentation. There was no obvious structural control. This also appears true for most if not all other sponge mounds in southern Germany. The lateral interfingering with bedded muddy limestones is evidence of the composite character of the mound that despite its cumulative height of about 50 m never rose above the seafloor more than a couple of meters. A composite analysis of biostratigraphic data, maturity analysis of authigenic glauconites, and independent growth rate measurements of microbolites from measurable coral reefs indicates that mound growth was around 1 mm/year, but preservation potential was only 15 to 20%.

2.4.4c. Example 3. *Low-diversity sponge biostromes as indicators of starved oligotrophic shelves.* In eastern Spain, more than 70,000 km^2 is covered by a unique and extremely uniform association of siliceous sponges of Mid-Oxfordian age. This association was recently described in detail by Krautter (1995, 1997). The laterally extensive biostromal unit is only a couple of meters thick but nevertheless spans several ammonite zones (maximum cordatum to planula zone, hence six ammonite zones). Except for a very few exceptions no relief is formed and the unit is composed almost solely of dish-shaped sponges comprising up to 90% of rock volume. Despite astronomical individual numbers, the fauna is of low diversity. Ninety percent of the fauna are represented by hexactinosan sponges comprising 24 species. Nearly all these species are known from other sponge reef localities, but the accompanying taxa are missing and three genera strongly dominate. Eighty percent of all sponges are of the dish shape. The rarity of encrusting organisms on sponges, such as serpulids, bryozoans, or oysters, is remarkable, although these forms occur sporadically. The only other elements worth mentioning are terebratulid bryozoans and the regular echinoid *Rhabdocidaris rhodani*, which is indicative of diminished oxygen availability. Similarly, levels of *Chondrites* burrows also indicate that sediment was occasionally depleted in oxygen. Judging from the long time span involved and the frequent iron oxide crusts with hardgrounds, no background sedimentation existed. Interestingly, microbolitic bioherms did not develop, although microbolitic crusts occur as thin layers. The peculiar sponge fauna of this reef was adapted to very oligotrophic conditions that relate to the lack of terrigeneous influx. Lack of nutrients kept microbolite development at a low pace, and together with the lack of allochthonous sedimentation helps explain why bioherms did not develop.

2.4.4d. Example 4. *Oceanographic changes during the hypselocyclum–divisum transgression.* One of the most puzzling features of Upper Jurassic reef development is the development of pure microbolite reefs related to eutrophication and/or oxygen depletion (see Section 2.4.1). Many of these reefs have been described in detail elsewhere (Leinfelder *et al.*, 1993a,b; Nose, 1995; Schmid, 1996). Coral and sponge reefs can be very rich in microbial crust, which here is not considered a pure microbolite reef because macroorganisms and microbolite crusts strongly interfinger. However, there are reefs composed nearly completely of pure thrombolitic microbolite or of a reef structure with upward alternation of pure microbolite and coral–microbolite. Above it was stated that pure microbolite development might be due to a variety of nonbiological factors, such as increased or decreased salinity or intertidal position. We focus here only on examples where such mechanisms can be excluded and where criteria for oxygen depletion, such as clusters of dysaerobic epibenthic bivalves, authigenic glauconite, framboidal pyrite, or secondary gypsum are available. The most interesting episode was that of the late Hypselocyclum–early Divisum zones. Here such reefs developed widely in Iberia and correlate elsewhere with other peculiar reefs. This time was characterized by a strong transgression (Haq *et al.*, 1988; Hantzpergue, 1988; Ponsot and Vail, 1991; Leinfelder, 1993b) promoting vigorous reef growth, despite the fact that in many areas the Lower Kimmeridgian and early Upper Kimmeridgian stages were dominated by terrigeneous deposits. However, many Iberian shelf sections do not show normal trends for reefs like during the major Oxfordian or late Kimmeridgian transgressions. Shallow water coral reefs were not accompanied by deeper-water sponge mounds but rather by pure microbolites. The pure microbolites occasionally exhibit thin horizons of siliceous sponges. These microbolites occurred both in considerably deep waters as well as in moderately deep slope settings. The clusters of meter-scaled club-shaped microbolites at Cotovio (Oxfordian) or the 30-m thick Lower Kimmeridgian Rocha thromobolite of southeastern Portugal (Fig. 17d) are good examples (Leinfelder *et al.*, 1993a). However, even coral reef growth in some examples became interrupted by pure microbolite growth showing characteristics of oxygen depletion. Examples are the Tormon reef of eastern Spain (Fig. 17a) or the Serra Isabel reefs of central Portugal (Fig. 17c) (Leinfelder *et al.*, 1993a,b; Werner *et al.*, 1994). Outside Iberia, reefs also may show peculiarities during this time interval of the late Early Kimmeridgian (Hypselocyclum–Divisum chrons). Coeval coral–microbolite reefs at La Rochelle, which recently were studied by Taylor and Palmer (1994), Schmid (1996), and Werner (personal communication), grew in slightly deeper water, as evidenced by the microencruster association. The microbolite development is unusually thick and largely substitutes coral growth in the late stages of reef growth, which together with a high individual number of nutrient-loving epifauna (serpulids, large oysters) hints at fairly elevated nutrient availability, although dysaerobic pulses are not directly documented here (Fig. 17b). In southwestern Germany, the same transgression gave rise to the partial interruption of terrigeneous sedimentation and to the development of peculiar small sponge–brachiopod reefs known as "lacunosa-Stotzen." Hexactinellid dominance and microencrusters indicate that these reefs certainly were positioned in waters not shallower than about 60 to 80 m, but nutrient increase might be indicated by the dominance of rhynchonellid brachiopods.

Since the bathymetry of these reefs can be assessed by the faunal characteristics of the oxic parts of reef development, the minimum position of a nutricline, which at least in Iberia was associated with an oxycline, can be roughly outlined (Fig. 18). It was strongly fluctuating but reached waters as shallow as about 30 m in central Portugal, 40 m in southwestern France, 60 m in southern Portugal, and possibly 80 m in southwestern Germany. However, such shallowness of the nutricline was largely restricted to the Hypselocyclum–Divisum transgression and is evidence of strong reduction in ocean water circulation which might be an effect of the climatic equilibration of a strongly rising sea level (Leinfelder, 1994a).

2.4.4e. Other Examples. *Reefs as indicators of basin structure and tectonic activity.* There are many other studied examples demonstrating the role of the general tectonic and paleogeographic setting on reef development. The Lusitanian Basin of Portugal provides many examples were reefs are indicative of basin configuration, tectonic setting, water circulation, and sea-level development. Among the examples are reef growth within siliciclastic fan deltas shed into strike–slip basins as indicators of tectonic quiescence, platform development as indicators of rising salt pillows, and patterns of contemporaneous reef and siliciclastic sedimentation indicative of longshore current systems (Leinfelder, 1994b, 1997). Examples from France, Switzerland, and Germany are in preparation for publication.

3. Intrajurassic Reef Development: Faunistic Evolution or Environmental Change?

Leinfelder (1994a) discussed the temporal evolution of reefs throughout the Jurassic period and stressed the fact that the abundance of reefs generally has increased with time, although there exist considerable differences in the trend between sponge and coral reefs and differences between the northern and southern Tethys shelf. Based on this compilation and some new data, a brief summary is given here.

3.1. Evolutionary Aspects of Reef Organisms

Siliceous sponge associations and siliceous sponge reefs are known from all three epochs of the Jurassic. On the northern Tethyan shelf, they strongly increase in abundance during the Mid- and Late Jurassic time, whereas on the southern Tethyan shelf, they occurred more widely during Early Jurassic time. With the exception of the lychniskid sponges, known since the later part of

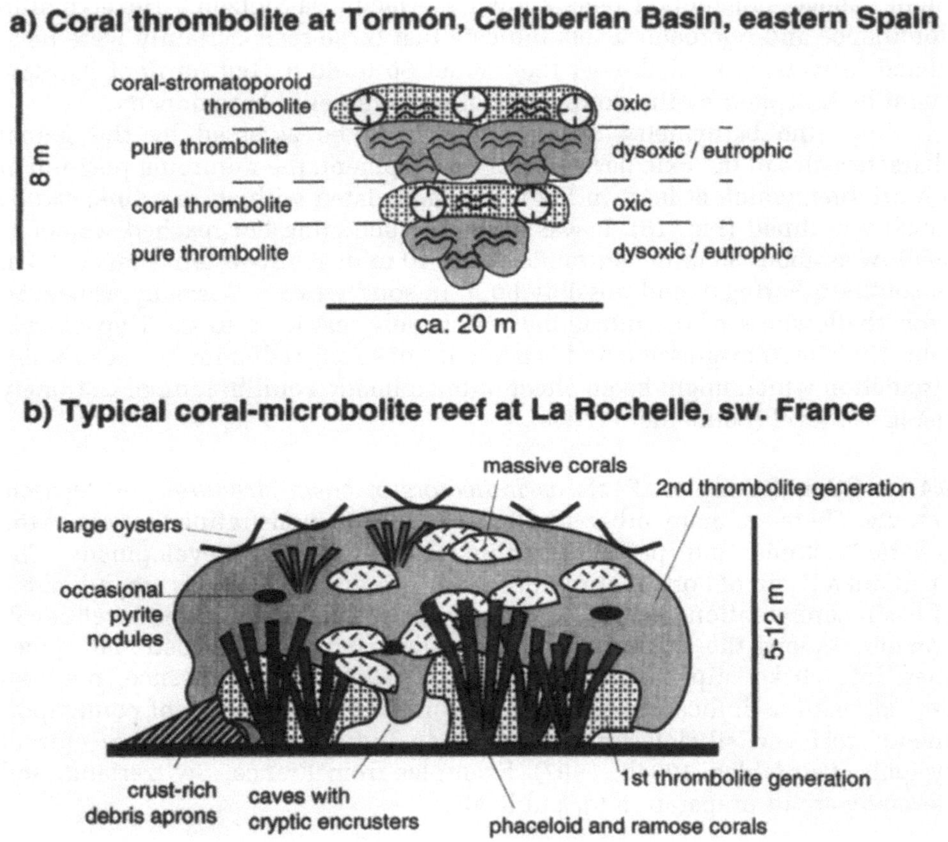

FIGURE 17. Examples of microbolite reefs from the late Early Jurassic (Hypselocyclum–Divisum chron) of southwestern Europe. (a) Repetitive succession of coral–microbolite to pure microbolite reef growth. (b) Coral reef with upward, outward, and downward increasing participation of microbolite crusts. (c) Lateral variability of a condensed level, partly with telescoped successions of dysaerobic and oxygenated reef facies. (d) Thick complex of largely pure microbolite. Siliceous sponges may occur but are refined to distinct levels.

Mid-Jurassic time, all other groups of these ultraconservative organisms existed prior to the Jurassic (Mehl, 1992). Although exact numbers of taxa are not available to date, the occurrence of lower Jurassic siliceous sponge mounds demonstrates that the general possibility of sponge reef formation existed throughout the entire Jurassic period. A certain evolutionary increase of sponge taxa therefore was caused by increasing availability of sponge reef habitats rather than vice versa.

Corals suffered a severe extinction at the Triassic–Jurassic boundary (see Chapter 7, this volume), and therefore are extremely rare in the early part of the Early Jurassic. Elmi (1987) mentions a small coral reef occurrence of

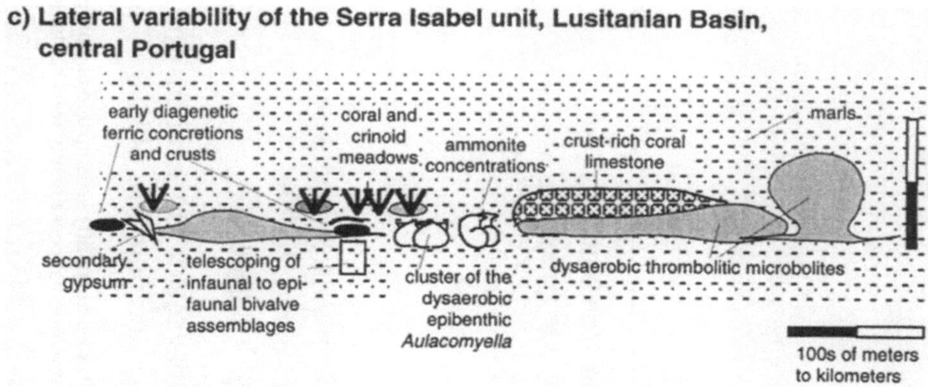

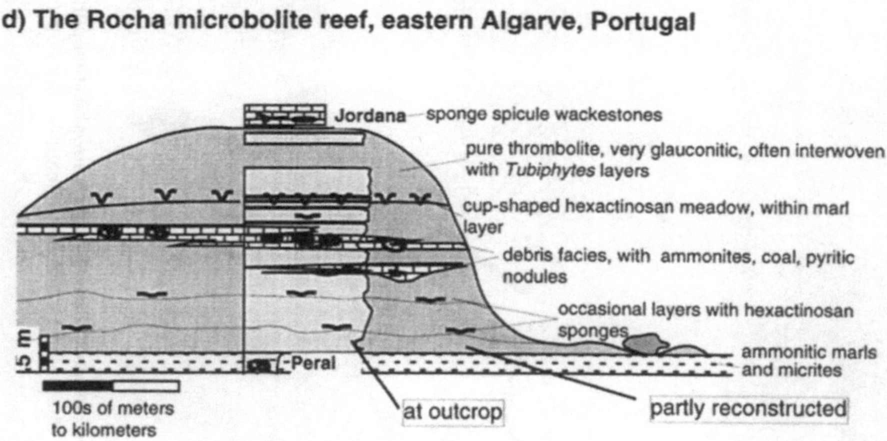

FIGURE 17. *Continued.*

Hettangian age in France, but it is still awaiting closer inspection. Besides this enigmatic Hettangian example, earliest coral reef occurrences are from the Sinemurian and are all situated on exotic terranes later accreted to the western margin of North America (Stanley and McRoberts, 1993), which was interpreted by Stanley and Beauvais (1994) and Stanley (1996) as an expression of long-term isolation of corals, surviving from Triassic time, while inhabiting exotic islands of the paleo-Pacific. From the Pliensbachian and particularly Toarcian onward, coral reefs became more frequent and it is since that time that available coral taxa were sufficient to form variable types of coral reefs. Numbers of generic taxa, as compiled from the literature (references in Leinfelder, 1994a), yield about 60 genera for the Early Jurassic with all important major groups still existing. About 100 genera were available during the Mid-Jurassic with about 130 genera in the Late Jurassic (data from various sources, compiled in Leinfelder, 1994a). Despite the fact that Jurassic corals

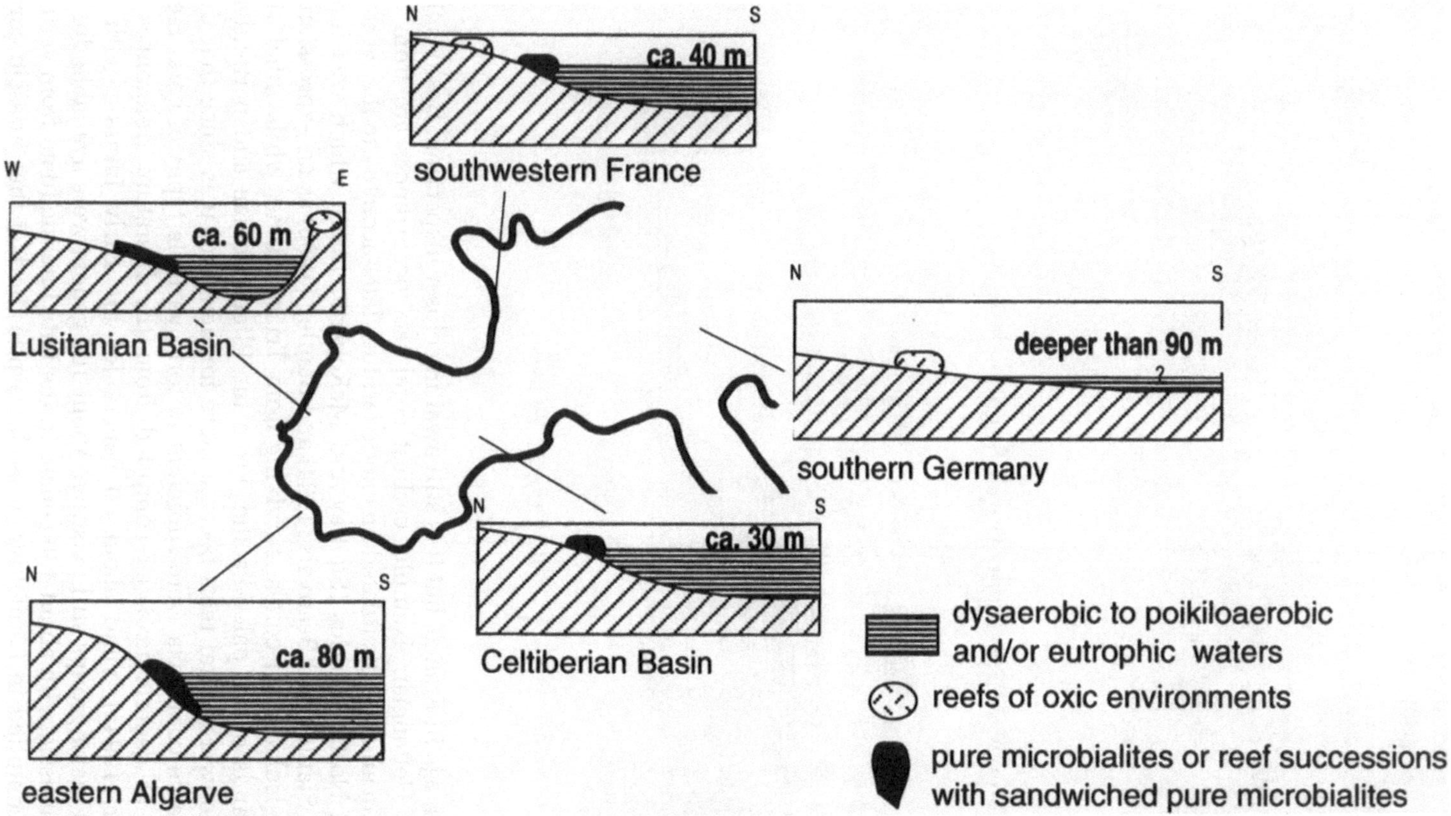

FIGURE 18. The contemporaneous reefs shown in Fig. 17 are interpreted as reflecting eutrophication pulses that were frequently accompanied by oxygen depletion. Aerobic parts of the reefs allowed for bathymetric interpretation and depth mapping of a nutricline–oxycline that occasionally were as shallow as 30 to 40 m in some examples.

urgently need taxonomic revision, it is obvious again that the increase in coral reef abundance need not necessarily be due to an adaptive radiation of corals. Nevertheless, the coral fauna diversified and the development of specialized taxa such as very morphovariable species or brackish water specialists allowed for the conquest of previously hostile environments, and this was a positive evolutionary feedback on the size of the "reef window" (see Section 4).

3.2. Sea-Level Development

The expansion of sponge reefs and coral reefs on the northern Tethys shelf correlates considerably well with general sea-level rise (Fig. 19). During Early Jurassic time there was considerable terrigeneous influx of clastics, permitting reef growth only in more or less protected, wide platform areas such as central Portugal (new discovery of sponge facies: Duarte and Krautter, 1998) and southern France (coral reefs). Southern Germany still was dominated by terrigeneous sediments derived from the Vindelician basement uplift and this prevented reef growth. Sea level kept rising well into Late Jurassic time, increasingly improving the environmental conditions suitable for the growth of both coral and sponge reefs by further reducing terrigeneous influx. Only during Late Jurassic time were the shelf areas flooded widely enough to provide suitable carbonate-dominated outer ramp settings on a large scale, where siliceous sponge mounds would develop in a more than 7000 km wide belt extending from Rumania to Texas. Productivity of these carbonate shelves was higher than the increase of accommodation by the general sea-level rise so that a generally shallowing succession developed. The peak of reef formation during the Oxfordian did not coincide with the peak in sea-level rise that occurred later, at the Kimmeridgian–Tithonian boundary (Haq *et al.*, 1988). The relative retreat of reefs already during the Kimmeridgian was caused by tectonic reactivation. Rift tectonics in the northern Atlantic graben systems also resulted in renewed terrigeneous sedimentation and uplifts elsewhere. However, this tectonically and possibly climatically induced general retreat of reefs was punctuated by flooding events of higher order. In many areas, these short-termed events brought back episodical reef growth, particularly during the Hypselocyclum–Divisum, Eudoxus, and Beckeri chrons.

3.3. Tectonic Control

Reef development along the northern Tethys shelf was strongly contrasted by the pattern of reef growth on the southern Tethys shelf, which cannot be correlated with the general sea-level development. Reefs are much rarer, which is particularly true of the Late Jurassic, at least for the western part of the Tethys. During the Early Jurassic, ramps or enechelon grabens still existed yet were rare (Stanley, 1988). They provided environments for coral and sponge reef development on the shallow and deeper shelf. Tectonic but not

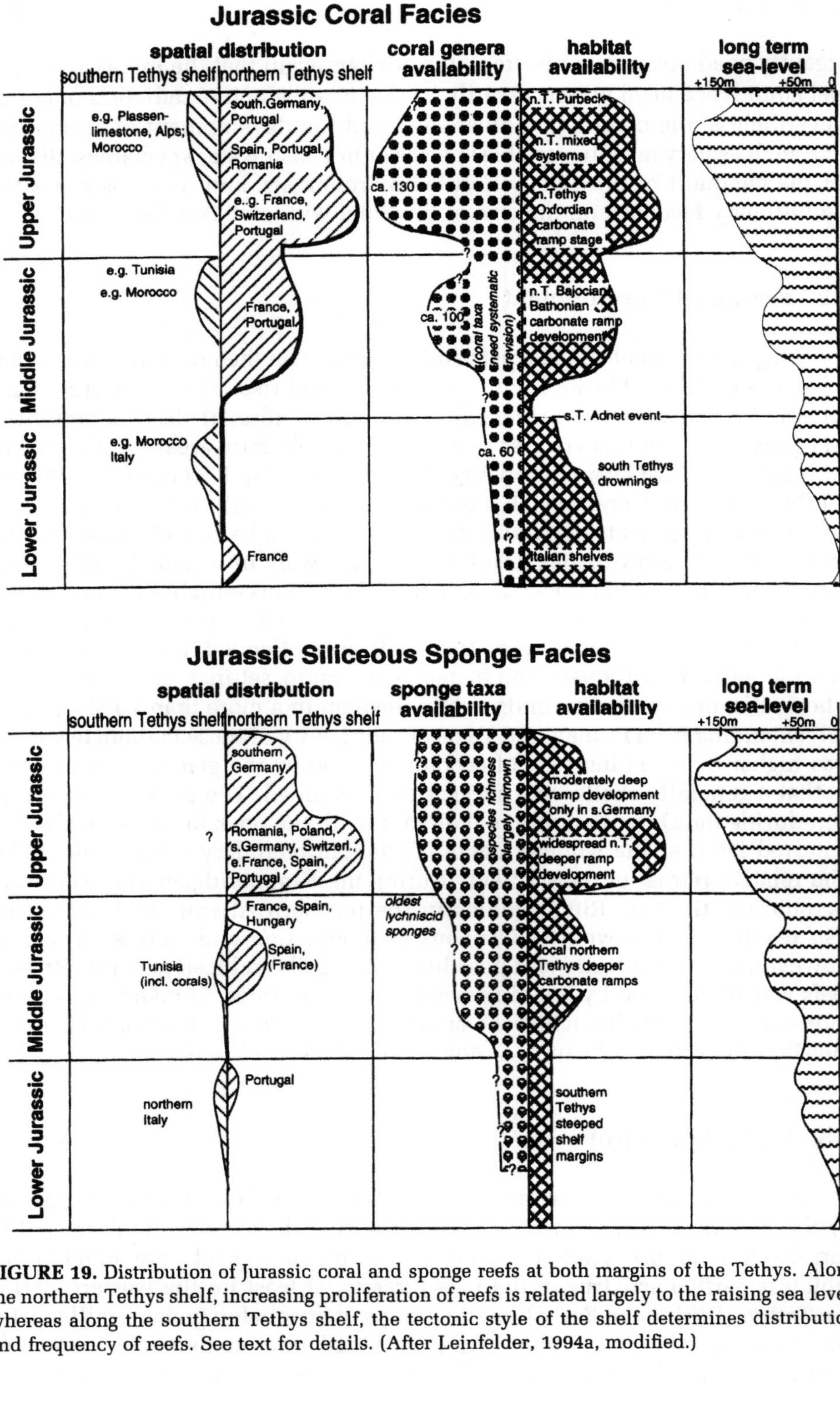

FIGURE 19. Distribution of Jurassic coral and sponge reefs at both margins of the Tethys. Along the northern Tethys shelf, increasing proliferation of reefs is related largely to the raising sea level, whereas along the southern Tethys shelf, the tectonic style of the shelf determines distribution and frequency of reefs. See text for details. (After Leinfelder, 1994a, modified.)

contemporaneous drowning of many of these areas made these potential reef habitats disappear during the Early and Mid-Jurassic. Only at the margins of rare, steeply bound, probably transpressional horst structures could coral reefs occasionally still grow during the Late Jurassic (e.g., Plassen limestone and Italian examples; for references see Section 2.1). Hence, tectonic influence on reef development and reef distribution is particularly obvious on the southern Tethys shelf but also in the Lusitanian Basin of central Portugal which is an Atlantic rift basin (Wilson *et al.*, 1989). Another interesting feature is that Upper Jurassic coral reefs of the southern Tethys shelf appear to exhibit much less microbolite crusts and only moderate to low diversities, although detailed studies are necessary to prove this qualitative impression. If true, this might be due to strongly oligotrophic conditions, which during the Jurassic did not yet allow maximum diversities (see Section 2.3.1).

4. Conclusions

Starting from a limited stock of surviving reef fauna after the Triassic–Jurassic mass extinction, Jurassic reefs became increasingly numerous during the course of the Jurassic, and reefs from the Late Jurassic represent one of the major peaks in reef evolution in Earth history. Jurassic reefs were not yet as specialized as modern reefs for highly oligotrophic settings but instead covered a broad array of reefal environments, ranging from shallow, high-energy down to several hundred meters depth, from zero sedimentation to considerably elevated sedimentation rates, from fully marine to brackish, and from mildly oligotrophic to eutrophic and even dysaerobic. However, these very different settings were not occupied by one single reef type, but were characterized by very different reef varieties, with end-member groups consisting of coral reefs, siliceous sponge reefs, and pure microbolite reefs. The many reef types and subtypes not only were statically related to the existence of distinct environments but also could respond and to some degree adapt themselves to environmental change. Besides sedimentological analysis, it is specifically the interpretation of physiological abilities of Jurassic reef organisms by constructional morphological analysis and semiquantitative to quantitative assessments of the trophic and diversity structure of reef associations that allows such conclusions.

This makes Jurassic reefs a most valuable tool for paleoenvironmental, including paleoclimatic, paleooceanographic, sea-level as well as paleostructural analysis. A particular feature in Jurassic reef analysis is the possibility of detecting eutrophic and dysaerobic episodes from development of pure microbolites in regimes of strongly reduced sedimentation, which normally prevents development of black shales. Lateral arrangement of coeval but different reef types allows for interpretation of general shelf type as well as small-scale differences of submarine morphology. The development of classical sponge mounds is particularly characteristic for homoclinal outer ramps, whereas coral–siliceous sponge–microbolite reefs may be characteristic for steepened slopes. Reefs of the coral–debris type are largely restricted to flat ramps or at the shallow shelf break of mildly inclined depositional margins. These debris-pile reefs express problems with the large amounts of calciclastic material produced by wave action and bioerosion (Leinfelder, 1992). This makes them perfect targets for oil exploration provided primary porosity is partially preserved or secondary vuggy leaching porosity developed. If developing along a shelf break in direct proximity of a steep bypass margin, high-energy coral–debris reefs can export large amounts of generated calciclastic material, which enables microbial mats to stabilize the remainder of the debris. This gives rise to easily recognizable coral–microbolite–debris reefs (Leinfelder, 1992).

An important theme is the restriction or proliferation of reefs during sea-level change. Major reef episodes are correlated with third-order rises and highstands, but individual reefs normally grew during forth, fifth, or even higher order. This is particularly true of coral reefs on terrigeneously influenced ramps. Sponge mounds may show composite, cluster-type architecture, with individual, stacked reef bodies also being related to fourth- and fifth-order floodings or early highstands.

In comparing the structure of modern and Jurassic reefs, similarities and differences exist. The calcareous siliceous sponge–microbolite mudmounds of Jurassic time to the present knowledge are unparalleled by modern analogues, although some siliciclastic cold-water sponge–microbial mats or mounds and some tropical deep-water lithistid associations share certain similarities. Within the Jurassic sponge facies, a wide variety of types and settings occurred. Environmental parameters for these reefs are resolved by a combined approach of paleoecological and sedimentological analysis. The controversial bathymetry of Jurassic siliceous sponge reefs is partly caused by the lack of differentiating sponge reef types. Lithistid-dominated reefs were generally but not necessarily more shallow than hexactinellid-dominated reefs. Siliceous sponge reefs can be transitional with the lower part of the coral reef zone, but are normally below the coral reef zone and possibly deeper than about 50 m. The lower distribution boundary of hexactinosan reefs might well be down to more than 100 m, with the lower limit probably defined by the existence of oxygen-depleted bottom waters or sediment accumulations. The wide distribution of Upper Jurassic sponge reefs especially along the northern Tethys shelf seas is a reflection of the high sea level. Modern shelves are rugged in the preferred depth zone of Jurassic sponge reef growth, due to glacial sea-level fluctuations that make the environment too variable for the development of sponge mounds. Also, the low sea-level of today causes stronger terrigeneous influx, which may bypass shallow-water coral reefs but may suppress reef growth in the deeper water.

Coral reefs and pure microbolite reefs existed both during the Jurassic and today. Just like their Jurassic counterparts, modern microbolite reefs are restricted to extreme settings, such as hypersaline, intertidal, or freshwater ponds or highly abrasive settings. However, the eutrophic–dysaerobic pure

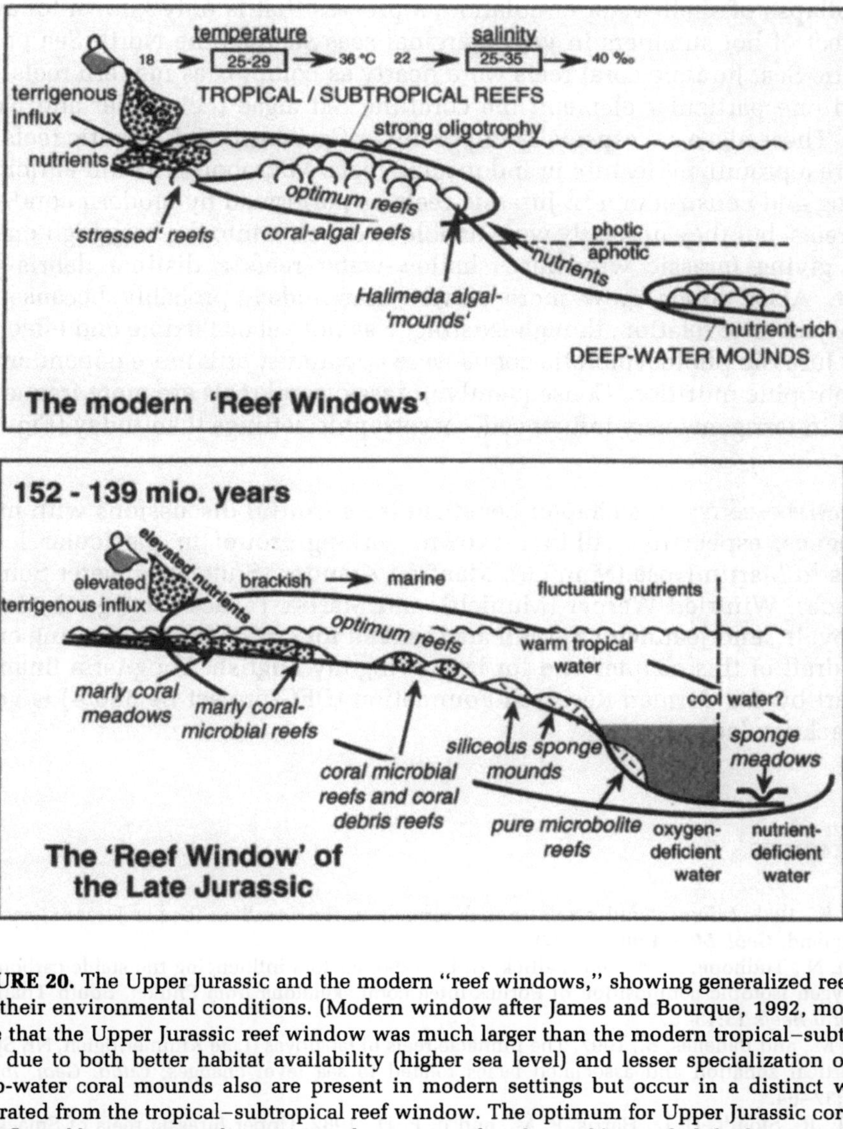

FIGURE 20. The Upper Jurassic and the modern "reef windows," showing generalized reef types and their environmental conditions. (Modern window after James and Bourque, 1992, modified). Note that the Upper Jurassic reef window was much larger than the modern tropical–subtropical one, due to both better habitat availability (higher sea level) and lesser specialization or reefs. Deep-water coral mounds also are present in modern settings but occur in a distinct window separated from the tropical–subtropical reef window. The optimum for Upper Jurassic coral reefs (as reflected by maximum diversities and pronounced reef growth) was slightly deeper and closer to coastlines than for modern coral reefs. (After Leinfelder and Nose, 1999.)

microbolite type of Early and Late Jurassic of southwest Europe and the Atlantic has no direct modern counterpart, save for the microbolite reefs of some hyperalkaline lakes with anoxic bottom waters such as the Satonda Lake (Kempe and Kazmierczak, 1993). The fairly frequent occurrence of pure microbolite reefs, particularly during some time episodes of the Late Jurassic, is an expression of the greenhouse-type climate of that time, causing temporary collapse of shelf water circulation, a process that is only known today as an effect of hot summers in very marginal seas such as the North Sea or the Adriatic Sea. Jurassic coral reefs were nearly as complex as modern reefs, but lacked one particular element: the coralline red algae (Leinfelder and Nose, 1999). These algae are especially important in Caribbean and Atlantic reefs but also are a prominent feature in Indopacific reefs. Microbolites could anticipate binding and construction in Jurassic reefs as performed by modern coral–red algal reefs, but they normally were unable to develop into the very high-energy zone, giving Jurassic windward shallow-water reefs a distinct debris pile aspect. Also, corals grew more slowly than today, probably because the photosymbiotic relation, though existing, was not yet as flexible and effective. Many Jurassic photosymbiotic corals were apparently still more dependent on heterotrophic nutrition. Consequently, Jurassic coral reefs are more frequently found in terrigeneously influenced, mesotrophic settings than today (Fig. 20).

ACKNOWLEDGMENTS: This chapter benefited from fruitful discussions with many colleagues, especially within my own working group: In particular I owe thanks to Martin Nose (Munich), Manfred Krautter (Stuttgart), Dieter Schmid (Munich), Winfried Werner (Munich), and Martin Takacs (Stuttgart). George Stanley, Jr., and Jeannette Yarnell are thanked for critically commenting on an early draft of this chapter and for improving my English. Long-term financial support by the German Research Foundation (DFG-project Le 580/4) is gratefully acknowledged.

References

Ali, O. E., 1983, Microsolenid corals as rock-formers in the Corallian (Upper Jurassic) rocks of England, *Geol. Mag.* **120**:375–380.

Allison, N., Tudhope, A. W., and Fallick, A. E., 1996, Factors influencing the stable carbon and oxygen isotopic composition of Porites lutea coral skeletons from Phuket, South Thailand, *Coral Reefs* **15**:43–57.

Aurell, M., and Bádenas, B., 1997, The pinnacle reefs of Jabaloyas (Late Kimmeridgian, NE Spain, Vertical zonation and associated facies related to sea level changes, *Cuad. Geol. Ibérica* **22**:37–64.

Baria, L. R., Stoudt, D. L., Harris, P. M., and d, P. D., 1982, Upper Jurassic reefs of Smackover Formation, United States Gulf Coast, *Am. Assoc. Petrol. Geol. Bull.* **66**:1449–1482.

Baumeister, J., 1997, Funktionsmorphologie und Paläoökologie regulärer jurassischer Echinoiden des nordwestlichen Tethys-Schelfs, unpublished dissertation (Dr. rer. nat.), Faculty of Geo- and Biosciences, University of Stuttgart.

Baumeister, J. G., and Leinfelder, R. R., 1998, Constructional morphology of three Upper Jurassic echinoids, *Palaeontology* **42**(2):203–219.

Baumgärtner, M., and Reyle, M., 1995, Oberjurassische Rampenentwicklung in der Region von Jabaloyas und Arroyo Cerezo (Keltiberikum; Spanien), *Profil* **8**:339–361.

Bertling, M., 1993, Ecology and distribution of the Late Jurassic Scleractinian *Thamnasteria concinna* (Goldfuss) in Europe, *Palaeogeogr. Palaeoclimatol. Palaeoecol.* **105**:311–335.

Bertling, M., 1997a, Bioerosion of Late Jurassic reef corals—Implications for reef evolution, *Proc. 8th Int. Coral Reef Sym.* **2**:1663–1668.

Bertling, M., 1997b, Structure and function of coral associations under extreme siltation stress—A case study from the northern German Upper Jurassic, *Proc. 8th Int. Coral Reef Sym.* **2**:1749–1754.

Brachert, T., 1992, Sequence stratigraphy and paleo-oceanography of an open-marine mixed carbonate/siliciclastic succession, Late Jurassic; South Germany, *Facies* **27**:179–216.

Buddemeier, R. W., and Kinzie, R. A., 1976, Coral growth, *Oceanogr. Mar. Biol. Annu. Rev.* **14**:183–225.

Coates, G. A., and Jackson, J. B. C., 1987, Clonal growth, algal symbiosis, and reef formation by corals, *Paleobiology* **13**: 363–378.

Conway, K. W., and Barrie, J. V., 1997, Modern Hexactinellid sponge reefs on the western Canadian continental margin, 18th IAS Reg. Europ. Meeting, Heidelberg, 1997, Abstracts, *Gaea heidelbergensis* **3**:105.

Conway, K. W., Barrie, J. V., Austin, W. C., and Luternauer, J. L., 1991, Holocene sponge bioherms on the western Canadian continental shelf, *Continental Shelf Res.* **11**: 771–790.

Crevello, P., and Harris, P., 1984, Depositional models for Jurassic reefal buildups, in: *The Jurassic of the Gulf Rim* (W. P. S. Ventress, D. G. Bebout, B. F. Perkins, and C. H. Moore, eds.), SEPM Gulf Coast Section, Proceedings of the 3rd Annual Research Conference, Tulsa, OK, pp. 57–102.

Dayton, P. K., Mordida, B. J., and Bacon, F., 1994, Polar marine communities, *Am. Zool.* **34**:90–99.

Deusch, M., Friebe, A., and Krautter, M. 1991, Spongiolithic facies in the Middle and Upper Jurassic of Spain, in: *Fossil and Recent Sponges* (J. Reitner and H. Keupp, eds.), Springer, New York, pp. 498–505.

Dietl, G., Dietl, O., Kapitzke, M., Rieter, M., Schweigert, G., Ilg, A., and Hugger, R., 1996, Der Nusplinger Plattenkalk (Weißer Jura)—Grabungskampagne 1996, *Jh. Ges. Naturk. Württemberg* **153**:185–203.

Helm, C., 1997, Faunistische Untersuchungen an einem Fleckenriff des Oberjura (florigemma-Bank; Süntel), unpublished diploma Thesis, Fachbereich Geowissenschaften, University of Hannover.

Draganescu, A., 1976, Constructional to corpuscular sponge–algal, algal and coralalgal facies in the Upper Jurassic carbonate formation of central Dobrogea (the Casimcea Formation), in: *Carbonate Rocks and Evaporites* (D. Patrullius *et al.*, eds.), Internat. Coll. Carb. Rocks/Evapor. Roumania, Guidebook Series, Vol. 15, pp. 3–43.

Dromart, G., Gaillard, C., and Jansa, L. F., 1994, Deep-marine microbial structures in the Upper Jurassic of Western Tethys, in: *Phanerozoic Stromatolites II* (J. Bertrand-Sarfati and C. Monty, eds.), Kluwer, Dordrecht, pp. 295–318.

Duarte, L. V., and Krautter, M., 1998, Siliceous sponge mud mounds from the Toarcian of the Lusitanian Basin, Portugal: Facies, stratigraphy and sequential evolution, Abstr. Jurassic Symposium, Vancouver, 1998.

Ellis, P. M., Wilson, R. C. L., and Leinfelder, R. R., 1990, Controls on Upper Jurassic carbonate buildup development in the Lusitanian Basin, Portugal, in: *Carbonate Platforms. Facies, Sequences and Evolution* (M. E. Tucker, J. L. Wilson, P. D. Crevello, J. R. Sarg, and J. F. Read, eds.), *Int. Assoc. Sediment. Spec. Publ.* **9**:169–202.

Elmi, S., 1987, Le Jurassique inférieur du Bas Vivarais (sud-est) de la France (avec la collaboration de R. Mouterde, C. Ruget, Y. Almeras, and G. Naud), 2ième Colloqu. Centr. Internat. Etudes du Lias (C.I.E.L.), Lyon, 27–30 Mai 1986, *Cah. Inst. Cathol. Lyon Sér. Sci.* **1**:63–189.

Errenst, C., 1990a, Das korallenführende Kimmeridgium der nordwestlichen Iberischen Ketten und angrenzender Gebiete (Fazies, Paläogeographie und Beschreibung der Korallenfauna), Teil 1, *Palaeontographica Abt. A* **214**:121–207.

Errenst, C., 1990b, Das korallenführende Kimmeridgium der nordwestlichen Iberischen Ketten und angrenzender Gebiete (Fazies, Paläogeographie und Beschreibung der Korallenfauna), Teil 2, *Palaeontographica Abt. A* **215**:1–42.

Fagerstrom, J. A., 1987, *The Evolution of Reef Communities*, Wiley, New York.

Fezer, R., 1988, Die oberjurassische karbonatische Regressionsfazies im südwestlichen Keltiberikum zwischen Griegos und Aras de Alpuente (Prov. Teruel, Cuenca, Valencia; Spanien), *Arb. Inst. Geol. Paläont. Univ. Stuttgart N.F.* **84**:1–119.

Flügel, E., Alt, T., Joachimski, M., Riemann, V., and Scheller, J., 1993, Korallen-Kalke im oberen Malm (Unter-Tithon) der Südlichen Frankenalb (Laisacker, Marching, Mikrofazies-Merkmale und Fazies-Interpretation, *Geo. Bl. NO-Bayern* **43**:33–56.

Fürsich, F. T., 1981, Salinity-controlled benthic associations from the Upper Jurassic of Portugal, *Lethaia* **14**:203–223.

Fürsich, F. T., and Werner, W., 1986, Benthic associations and their environmental significance in the Lusitanian Basin (Upper Jurassic, Portugal), *N. Jb. Geol. Paläont. Abh.* **172**:271–329.

Fürsich, F. T., and Werner, W., 1991, Palaeoecology of coralline sponge–coral meadows from the Upper Jurassic of Portugal, *Paläont. Z.* **65**:35–69.

Gaillard, C., 1983, Les biohermes à spongiaires et leur environnement dans l'Oxfordien du Jura méridional, *Docum. Lab. Géol. Lyon* **90**:1–515.

Geister, J., 1983, Holozäne westindische Korallenriffe: Geomorphologie, Ökologie und Fazies, *Facies* **9**:173–284.

Geister, J., 1992, Modern reef development and Cenozoic evolution of an oceanic island/reef complex: Isla de Providencia (Western Carribean Sea, Colombia), *Facies* **27**:1–70.

Geister, J., and Lathuilière, B., 1991, Jurassic coral reefs of the northeastern Paris basin (Luxembourg and Lorraine), Excursion A 3, Excursion-Guidebook, VI. Int. Sympos. on Fossil Cnidaria, Münster.

Ginsburg, R. N., 1997, Coral reefs are cities under the sea. Transparencies and teachers notes to introduce coral reefs to students and adults, (manuscript), Project Miami R.A.R.E., The Hachette Filipacchi Foundation.

Greb, L., Saric, B., Seyfried, H., Broszonn, T., Brauch, S., Gugau, G., Wiltschko, C., and Leinfelder, R., 1996, Ökologie und Sedimentologie eines rezenten Rampensystems an der Karibikküste von Panamá, *Profil* **10**:1–168.

Gwinner, M. P., 1976, Origin of the Upper Jurassic limestones of the Swabian Alb (Southwest Germany), *Contr. Sedimentol.* **5**:1–75.

Gygi, R. A., and Persoz, F. 1986, Mineralostratigraphy, litho- and biostratigraphy combined in correlation of the Oxfordian of the Swiss Jura range, *Eclogae geol. Helv.* **79**:455–491.

Hantzpergue, P., 1988, *Les Plates-Formes Recifales du Jurassique Nord-Aquitain*, Livret-Guide Excursion GFEJ, Poitiers.

Haq, B. U., Hardenbol, J., and Vail, P. R., 1988, Mesozoic and Cenozoic chronostratigraphy and cycles of sea level change, *SEPM, Spec. Publ.* **42**:71–108.

Henrich, R., Hartmann, M., Reitner, J., Schäfer, P., Freiwald, A., Steinmetz, S., Dietrich, P., and Thiede, J., 1992, Facies belts and communities of the Arctic Vesterisbanken Seamount (Central Greenland Sea), *Facies* **27**:71–104.

Henrich, R., Freiwald, A., Wehrmann, A., Schäfer, P., Samtleben, C., and Zankl, H., 1996, Nordic cold-water carbonates: Occurrences and controls, in: *Global and Regional Controls on Biogenic Sedimentation. I. Reef Evolution* (J. Reitner, F. Neuweiler, and F. Gunkel, eds.), research reports. *Göttinger Arb. Geol. Paläont.* **Sb 2**: 35–52.

Herrmann, R., 1996, Entwicklung einer oberjurassischen Karbonatplattform: Biofazies, Riffe und Sedimentologie im Oxfordium der Zentralen Dobrogea (Ost-Rumänien), *Berliner geowiss. Abh. E* **19**:1–101.

Highsmith, R. C., 1979, Coral growth rates and the environmental control of density banding, *J. Exp. Mar. Biol. Ecol.* **37**:105–125.

Hobbie, J. E., Holm-Hansen, O., Packard, T. T., Pomeroy, L. R., Sheldon, R. W., Homas, J. P., and Wiebe, W. J., 1972, A study of the distribution and activity of microorganisms in the ocean water, *Limnol. Oceanogr.* **17**:544–555.

Hubbard, J. A., and Pocock, Y. P., 1972, Sediment rejection by scleractinian corals: a key to palaeo-environmental reconstruction, *Geol.Rundschau* **61**:598–626.

Insalaco, E., 1996a, Upper Jurassic microsolenid biostromes of northern and central Europe: Facies and depositional environment, *Palaeogeogr. Palaeoclimatol. Palaeoecol.* **121**:169–194.

Insalaco, E., 1996b, The use of Late Jurassic coral growth bands as palaeoenvironmental indicators, *Palaeontology* **39**:413–431.

Insalaco, E., Hallam, A., and Rosen, B., 1997, Oxfordian (Upper Jurassic) coral reefs in Western Europe: Reef types and conceptual depositional model, *Sedimentology* **44**:707–734.

James, N. P., 1983, Reef environments, in: *Carbonate Depositional Environments* (P.A. Scholle, D. G. Bebout, and C. H. Moore, eds.), *Am. Assoc. Petrol. Geol., Mem.* **33**:345–440.

James, N. P., and Bourque, P., 1992, Reefs and mounds, in: *Facies Models. Response to Sea Level Change* (R. G. Walker and N. P. James, eds.), Geological Association of Canada, Waterloo, pp. 323–347.

Jansa, L. F., Pratt, B. R., and Dromart, G., 1989, Deep water thrombolite mounds from the Upper Jurassic of offshore Nova Scotia, in: *Reefs, Canada and Adjacent Areas* (H. H. J. Geldsetzer, N. P. James, and G. E. Tebbutt, eds.), *Can. Soc. Petroleum Geol. Mem.* **13**:725–735.

Kauffman, E. G., and Scott, R. W., 1976, Basic concepts of community ecology and paleoecology, in: *Structure and Classification of Paleocommunities* (R. W. Scott and R. R. West, eds.), Dowden, Hutchinson, and Ross, Stroudsburg, pp. 1–28.

Kempe, S., 1990, Alkalinity: The link between anaerobic basins and shallow water carbonates? *Naturwissenschaften* **77**:426–427.

Kempe, S., and Kazmierczak, J., 1993, Satonda Crater Lake, Indonesia: Hydrogeochemsistry and biocarbonates, *Facies* **29**:1–32.

Kempe, S., and Kazmierczak, J., 1994, The role of alkalinity in the evolution of ocean chemistry, organisation of living systems and biocalcification processes, in: *Past and Present Biomineralization Processes. Considerations about the Carbonate Cycle* (F. Doumenge, ed.), IUCN-COE Workshop, Monaco, 1993, *Bull. Inst. Océanogr. Monaco No. Spéc.* **13**:61–117.

Kempe, S., Kazmierczak, J., Reimer, A., Landmann, G., and Reitner, J., 1996, Microbialites and hydrochemistry of the Crater Lake of Satonda—a status report, in: *Global and Regional Controls on Biogenic Sedimentation. I. Reef Evolution. Research Reports* (J. Reitner, F. Neuweiler, and F. Gunkel, eds.), *Göttinger Arb. Geol. Paläont.* **Sb2**:59–63.

Keupp, H., Brugger, H., Galling, U., Hefter, J., Herrmann, R., Jenisch, A., Kempe, S., Michaelis, W., Seifert, R., and Thiel, V., 1996, Paleobiological controls of Jurassic spongiolites, in: *Global and Regional Controls on Biogenic Sedimentation. I. Reef Evolution. Research Reports* (J. Reitner, F. Neuweiler, and F. Gunkel, eds.), *Göttinger Arb. Geol. Paläont.* **Sb2**:209–214.

Koch, R., 1996, Paleogeography, microfacies and diagenesis of Upper Jurassic (Malm) "Reef"-limestones in the Geislingen-Eybtal area (Swabian Alb), in: *Global and Regional Controls on Biogenic Sedimentation. I. Reef Evolution. Research Reports* (J. Reitner, F. Neuweiler, and F. Gunkel, eds.), *Göttinger Arb. Geol. Paläont.* **Sb2**:215–220.

Koch, R., Senowbari-Daryan, B., and Strauss, H., 1994, The Late Jurassic "Massenkalk Fazies" of Southern Germany: Calcareous sand piles rather than organic reefs, *Facies* **31**:179–208.

Krautter, M., 1994, Observations on *Eudea clavata* Lamouroux (Calcarea) from the Upper Jurassic of Portugal, in: *Proceedings of the 4th International Porifera Congress "Sponges in Time and Space" Amsterdam 1993* (R. W. M. van Soest, T. M. G. van Kempen, and J. C. Braekman, eds.), Balkema, Amsterdam, pp. 29–34.

Krautter, M., 1995, Kieselschwämme als potentielle Indikatoren für Sedimentationsrate und Nährstoffangebot am Beispiel der Oxford-Schwammkalke von Spanien, *Profil* **8**:281–304.

Krautter, M., 1996, Kieselschwämme aus dem unterjurassischen Misonekalk der Trento-Plattform (Südalpen, Taxonomie und phylogenetische Relevanz, *Paläont. Z.* **70**:301–313.

Krautter, M., 1997, Aspekte zur Paläökologie postpaläozoischer Kieselschwämme, *Profil* **11**:199–324.

Laternser, R., 2000, Oberjurassische Korallenriffe von Nordostfrankreich (Lothringen) und Südwestdeutschland—unpublished dissertation, University of Stuttgart.

Leão, Z. M. A. N., and Ginsburg, R. N., 1997, Living reefs surrounded by siliciclastic sediments: The Abrolhos coastal reefs, Bahia, Brazil, *Proc. 8th Int. Coral Reef Sym.* **2**:1767–1772.

Legarreta, L., 1991, Evolution of a Callovian–Oxfordian carbonate margin in the Neuquén Basin of west-central Argentina: Facies, architecture, depositional sequences and global sea-level changes, *Sediment. Geol.* **70**:209–240.

Leinfelder, R. R., 1986, Facies, stratigraphy and paleogeographic analysis of Upper? Kimmeridgian to Upper Portlandian sediments in the environs of Arruda dos Vinhos, Estremadura, Portugal, *Münchn. Geowiss. Abh.* **A7**:1–215.

Leinfelder, R. R., 1992, A modern-type Kimmeridgian reef (Ota Limestone, Portugal, Implications for Jurassic reef models, *Facies* **26**:11–34.

Leinfelder, R. R., 1993a, Upper Jurassic reef types and controlling factors.—A preliminary report, *Profil* **5**:1–45.

Leinfelder, R. R., 1993b, A sequence stratigraphic approach to the Upper Jurassic mixed carbonate—siliciclastic succession of the central Lusitanian Basin, Portugal, *Profil* **5**:119–140.

Leinfelder, R. R., 1994a, Distribution of Jurassic reef types: A mirror of structural and environmental changes during breakup of Pangea, in: *Pangea: Global Environments and Resources* (B. Beauchamp, A. F. Embry, and D. Glass, eds.), *Can. Soc. Petrol. Geol. Mem.* **17**:677–700.

Leinfelder, R. R., 1994b, Karbonatplattformen und Korallenriffe innerhalb siliziklastischer Sedimentationsbereiche (Oberjura, Lusitanisches Becken, Portugal), *Profil* **6**:1–207.

Leinfelder, R. R., 1997, Coral reefs and carbonate platforms within a siliciclastic setting: General aspects and examples from the Late Jurassic of Portugal: *Proc. 8th Int. Coral Reef Sym*, Panama City, V.Z., p. 1737–1742, Smithsonian Tropical Research Institute, Balboa, Panama.

Leinfelder, R. R., and Ginsburg, R. N., 1998, Städte unter Wasser—Gibt es so etwas? in: *Riffe—Ein faszinierendes Thema für den Schulunterricht. Materialien für die Fächer Biologie, Erdkunde und Geologie* (R. R. Leinfelder, U. Kull, and F. Brümmer, eds.), *Profil* **13**:105–116.

Leinfelder, R. R., and Keupp, H., 1995, Upper Jurassic mud mounds: Allochthonous sedimentation versus autochthonous carbonate production, in: *Mud Mounds: A Polygenetic Spectrum of Fine-grained Carbonate Buildups* (J. Reitner and F. Neuweiler, coord.), *Facies* **32**:17–26.

Leinfelder, R. R., and Nose, M., 1999, Increasing complexity—decreasing flexibility. A different perspective of reef evolution through time, *Profil* **17**:135–147.

Leinfelder, R. R., and Nose, M., submitted for publication, Physical and geological controls of Upper Jurassic reef growth, in: *Reef Systems in Earth History* (E. Flügel and H. Zankl, eds.), Springer, Heidelberg–Berlin.

Leinfelder, R. R., and Schmid, D. U., 2000, Mesozoic reefal thrombolites and other microbolites, in: *Microbial Sediments* (R. Riding, ed.), Springer, Berlin, pp. 289–294.

Leinfelder, R. R., and Werner, W., 1993, The systematic position and palaeoecology of the alga *Marinella lugeoni* Pfender, *Zitteliana* **20**:105–122.

Leinfelder, R. R., and Wilson, R. C. L., 1989, Seismic and sedimentologic features of Oxfordian–Kimmeridgian syn-rift sediments on the eastern margin of the Lusitanian Basin, Portugal, *Geol. Rundschau* **78**:81–104.

Leinfelder, R. R., and Wilson, R. C. L., 1998, Third-order sequences in an Upper Jurassic rift-related second-order sequence, central Lusitanian Basin, Portugal, in: *Mesozoic and Cenozoic Sequence Stratigraphy of European Basins* (P.-C. de Graciansky, J. Hardenbol, T. Jacquin, and P. R. Vail, eds.), SEPM Spec. Publ. Vol. 60, pp. 507–525.

Leinfelder, R. R., Krautter, M., Nose, M., Ramalho, M. M., and Werner, W., 1993a, Siliceous sponge facies from the Upper Jurassic of Portugal, *N. Jb. Geol. Paläont. Abh.* **189**:199–254.

Leinfelder, R. R., Nose, M., Schmid, D. U., and Werner, W., 1993b, Microbial crusts of the Late Jurassic: Composition, palaeoecological significance and importance in reef construction, *Facies* **29**:195–230.

Leinfelder, R. R., Krautter, M., Laternser, R., Nose, M., Schmid, D. U., Schweigert, G., Werner, W., Keupp, H., Brugger, H., Herrmann, R., Rehfeld-Kiefer, U., Schroeder, J. H., Reinhold, C., Koch, R., Zeiss, A., Schweizer, V., Christmann, H., Menges, G., and Luterbacher, H., 1994, The origin of Jurassic reefs: Current research developments and results, *Facies* **31**:1–56.

Leinfelder, R. R., Werner, W., Nose, M., Schmid, D. U., Krautter, M., Laternser, R., Takacs, M., and Hartmann, D., 1996, Palaeoecology, growth parameters and dynamics of coral, sponge and microbolite reefs from the Late Jurassic, in: *Global and Regional Controls on Biogenic Sedimentation. I. Reef Evolution. Research Reports* (J. Reitner, F. Neuweiler, and F. Gunkel, eds.), *Göttinger Arb. Geol. Paläont.* **Sb2**:227–248.

Lévi, C., and Lévi, P., 1988, Nouveaux spongiaire Lithistides bathyaux à affinités crétacées de la Nouvelle Calédonie, *Bull. Mus. Nat. d'Histoire Naturelle Paris* **10**(A2):241–263.

Mackie, G. O., Lawn, I. D., and Pavans de Ceccatty, M., 1983, Studies on hexactinellid sponges. II. Excitability, conduction and coordination of responses in *Rhabdocalyptus dawsoni* (Lambe, 1873), *Phil. Trans. Royal Soc. London B Biol. Sci.* **301**:401–418.

Matheos, S. D., and Morsch, S. M., 1990, Geochemistry and paleoecological aspects of coral-bearing limestones from the Late Jurassic at the southern end of the Sierra de la Vaca Muerta, Neuquén Basin, Argentina, *N.Jb. Geol. Paläont. Abh.* **181**:159–169.

Matyszkiewicz, J., 1996, The significance of *Saccocoma*-calciturbidites for the analysis of the Polish epicontinental Late Jurassic basin: an example from the southern Cracow–Wielun Upland, Poland), *Facies* **34**:23–40.

Matyszkiewicz, J., 1997, Microfacies, sedimentation and some aspects of diagenesis of Upper Jurassic sediments from the elevated part of the Northern peri-Tethyan shelf: A comparative study on the Lochen area (Schwäbische Alb) and the Cracow area (Cracow–Wielun Upland, Polen), *Berlin. Geowiss. Abh. (E)* **21**:1–111.

Mehl, D., 1992, Die Entwicklung der Hexactinellida seit dem Mesozoikum—Paläobiologie, Phylogenie und Evolutionsökologie, *Berliner Geowiss. Abh. (E)* **2**:1–164.

Meyer, R. K. F., and Schmidt-Kaler, H., 1989, Paläogeographischer Atlas des süddeutschen Oberjura (Malm), *Geol. Jb. A* **115**:3–77.

Montaggioni, L. F., and Camoin, G. F., 1993, Stromatolites associated with coralgal communities in the Holocene high-energy reefs, *Geology* **21**:149–152.

Montgomery, S. L., 1993, Cotton Valley Lime of East Texas: New pinnacle reef play in the Late Jurassic, in: *Petroleum Frontiers (Petroleum Information Corp.)*, Vol. 10, pp. 1–55.

Montgomery, S. L., 1996, Cotton Valley Lime pinnacle reef play: Branton Field, AAPG *Bull.* **80**:617–629.

Morsch, S. M., 1989, Scleractinian corals from the Oxfordian La Manga Formation in the Neuquén Basin, Argentina, *Mem. Assoc. Australas. Palaeontols.* **8**:303–306.

Nose, M., 1995, Vergleichende Faziesanalyse und Palökologie korallenreicher Verflachungsabfolgen des iberischen Oberjura, *Profil* **8**:1–237.

Nose, M., submitted for publication, The modern (sub)tropical "reef window," in: *Reef Systems in Earth History* (E. Flügel and H. Zankl, eds.), Springer, Heidelberg–Berlin.

Nose, M., and Leinfelder, R., 1997, Upper Jurassic coral communities within siliciclastic settings (Lusitanian Basin, Portugal, Implications for symbiotic and nutrient strategies), *Proc. 8th Int. Coral Reef Symp.* **2**:1755–1760.

Nose, M., and Leinfelder, R. R., submitted for publication, The Upper Jurassic (sub)tropical "reef window," in: *Reef Systems in Earth History* (E. Flügel and H. Zankl, eds.), Springer, Heidelberg–Berlin.

Odum, E. P., 1983, *Grundlagen der Ökologie*, Thieme, Stuttgart.

Palmer, T. J., and Fürsich, F. T., 1981, Ecology of sponge reefs from the middle Jurassic of Normandy, *Palaeontology* **24**:1–23.

Paulsen, S., 1964, Aufbau und Petrographie des Riffkomplexes von Arnegg im höheren Weißen Jura der Schwäbischen Alb (Württemberg), *Arbeiten aus dem Institut für Geologie und Paläontologie der Universität Stuttgart, N.F.* **42**:1–98.

Pisera, A., 1997, Upper Jurassic siliceous sponges from the Swabian Alb: Taxonomy and Paleoecology, *Palaeontol. Polonica* **57**:1–216.

Pittet, B., Strasser, A., and Dupraz, C., 1995, Palaeoecology, palaeoclimatology and cyclostratigraphy of shallow-water carbonate–siliciclastic transitions in the Oxfordian of the Swiss Jura. IAS-16th Regional Meeting of Sedimentology—5ème Congrès Français de Sédimentologie, *ASF, Field Trip Guide Book, Publication ASF, Paris* **23**:225–254.

Ponsot, C. M., and Vail, P. R., 1991, Sequence stratigraphy of the Jurassic: New data from the Paris-London Basin, EUG VI-Congress, Strasbourg, *Terra Cognita* **28**:48.

Pümpin, V. F., 1965, Riffsedimentologische Untersuchungen im Rauracien von St. Ursanne und Umgebung (Zentraler Schweizer Jura), *Eclogae geol. Helvetiae* **58**:799–876.

Ramalho, M., 1988, Sur la découverte de biohermes stromatolithiques à spongiaires siliceux dans le Kimméridgien de l'Algarve (Portugal), *Comun. Serv. Geol. Portugal* **74**:41–55.

Reid, R. E. H., 1968, Bathymetric distribution of Calcarea and Hexactinellida in the present and the past, *Geol. Mag.* **105**:546–559.

Reiswig, H. M., and Mackie, G. O. 1983, Studies on Hexactinellid Sponges. III. The taxonomic status of Hexactinellida within the Porifera, *Phil. Trans. Royal Soc. London B Biol. Sci.* **301**:419–428.

Reitner, J., 1993, Modern cryptic microbialite/metazoan macies from Lizard Island (Great Barrier Reef, Australia)—Formation and concepts, *Facies* **29**:3–40.

Reitner, J., and Keupp, H., 1991, The fossil record of the haploslerid excavating sponge Aka de Laubenfels, in: *Fossil and Recent Sponges* (J. Reitner and H. Keupp, eds.), Springer, Berlin, pp. 102–120.

Reitner, J., Wörheide, G., Thiel, V., and Gautret, P., 1996, Reef caves and cryptic habitats of Indo-Pacific reefs—Distribution patterns of coralline sponges and microbialites, in: *Global and Regional Controls on Biogenic Sedimentation. I. Reef Evolution. Research Reports* (J. Reitner, F. Neuweiler, and F. Gunkel, eds.), *Göttinger Arb. Geol. Paläont.* **Sb2**:91–100.

Renesto, S., and Viohl, G., 1997, A sphenodontid (Reptilia, Diapsida) from the Late Kimmeridgian of Scharnhaupten (southern Franconian Alb, Bavaria, Germany), *Archaeopteryx* **15**:27–46.

Rheinheimer, G., 1980, *Aquatic Microbiology*, 2 ed., Wiley and Sons, New York.

Riegl. B., 1995, Effects of sand deposition on scleractinian and alcyonacean corals, *Mar. Biol.* **121**:517–526.

Rogers, C. S., 1990, Response of coral reefs and reef organisms to sedimentation, *Mar. Ecol. Prog. Ser.* **62**:185–202.

Röper, M., Rothgaenger, M., and Rothgaenger, K., 1996, *Die Plattenkalke von Brunn (Landkreis Regensburg). Sensationelle Fossilien aus dem Oberpfälzer Jura*, Eichendorf Verlag, Eichendorf.

Rosendahl, S., 1985, Die oberjurassische Korallenfazies von Algarve (Südportugal), *Arb. Inst. Geol. Paläont. Univ. Stuttgart N.F.* **82**:1–125.

Roux, M., Bouchet, P., Bourseau J. P., Gaillard, C., Grandperrin, R., Guille, A. Q., Laurin, B., Monniot, C., Richer de Forges, B., Rio, M., Segonzac, M., Vacelet, J., and Zibrorius, H., 1991, L'étagement du benthos bathyal observe à l'aide de la soucoupe CYANA, in: *L'Environements Carbonate Bathyal en Nouvelle-Calédonie* (B. Lambert and M. Roux, eds.), *Doc. Travaux Inst. Géol. Albert-de-Lapparent* **15**:151–165.

Sartorio, D., 1989, Reef and open episodes on a carbonate platform margin from Malm to Cenomanian: The Cansiglio example (Southern Alps), *Mem. Soc. Geol. Ital.* **40**:91–97.

Schlichter, D., 1992, A perforated gastrovascular cavity in the symbiotic deep water coral *Leptoseris fragilis:* A new strategy to optimize heterotrophic nutrition, *Helgoländer Wissenschaftliche Meeresuntersuchungen* **45**:423–443.

Schmid, D. U., 1995, *"Tubiphytes" morronensis*—Eine fakultativ inkrustierende Foraminifere mit endosymbiontischen Algen, *Profil* **8**:305–317.

Schmid, D. U., 1996, Marine Mikrobolithe und Mikroinkrustierer aus dem Oberjura, *Profil* **9**:101–251.

Schmid, D. U., and Jonischkeit, A., 1995, The Upper Jurassic São Romão limestone (Algarve, Portugal, An isolated carbonate ramp, *Profil* **8**:319–337.

Schmid, D. U., and Leinfelder, R. R., 1996, The Jurassic *Lithocodium aggregatum–Troglotella incrustans* foraminiferal consortium, *Palaeontology* **39**:21–52.

Schuhmacher, H., 1976, *Korallenriffe. Verbreitung, Tierwelt, Ökologie*, BLV Verlagsgesellschaft, München.

Scotese, C. R., Walsh, D. B., Kraus, J. U., and Bocharova, N. Y., 1993, Plate evolution of Pangea from the Late Paleozoic to the Jurassic. Pangea Conference (Can. Soc. Petrol. Geol.), August 15–19, 1993, Calgary, book of abstracts, p. 274, and PALEOMAP Project, International Lithosphere Program,

Scott, R. W., 1988, Evolution of Late Jurassic and Early Cretaceous reef biotas, *Palaios* **3**:184–193.

Stanley, G. D. Jr., 1988, The history of Early Mesozoic reef communities: A three step process, *Palaios* **3**:170–183.

Stanley, G. D. Jr., 1996, Recovery of reef communities after the Triassic mass extinction, *Paleont. Soc. Sp. Publ.* **8**:370.

Stanley, G. D., Jr., and Beauvais, L., 1994, Corals from an Early Jurassic coral reef in British Columbia: Refuge on an oceanic island reef, *Lethaia* **27**:35–47.

Stanley, G. D., Jr., and McRoberts, C. A., 1993, A coral reef in the Telkwa Range, British Columbia: the Earliest Jurasic example, *Can. J. Earth Sci.* **30**:819–831.

Stanley, G. D., Jr., and Swart, P., 1995, Evolution of the coral–zooxanthellae symbiosis during the Triassic: A geochemical approach, *Paleobiology* **21**:179–199.

Steiger, T., 1981, Kalkturbidite im Oberjura der Nördlichen Kalkalpen (Barmsteinkalke, Salzburg, Österreich), *Facies* **4**:215–348.

Steiger, T., and Jansa, L. F., 1984, Jurassic limestones of the seaward edge of the Mazagan Carbonate Platform, Northwest African continental margin, Morocco, *Initial Reports, DSDP* **79**:449–491.

Steiger, T., and Wurm, D., 1980, Faziesmuster oberjurassischer Plattformkarbonate (Plassenkalke, Steiermark), *Facies* **2**:241–284.

Takacs, M., in prep., Vergleichende Analyse zeitgleicher oberjurassischer Korallen- und Kieselschwammriffe aus dem Schweizer Jura, unpublished dissertation, University of Stuttgart.

Taylor, P. D., and Palmer, T. J., 1994, Submarine caves in a Jurassic reef (La Rochelle, France) and the evolution of cave biotas, *Naturwissenschaften* **81**:357–360.

Trammer, J., 1988, Ecologic history of the Oxfordian sponge assemblage in the Polish Jura Chain, *Berliner Geowiss. Abh. A* **100**:44–45.

Turnšek, D., Buser, S., and Ogorelev, B., 1981, An Upper Jurassic reef complex from Slovenia, Yugoslavia, in: *European Fossil Reef Models* (D. F. Toomey, ed.), *Soc. Econ. Paleont. Mineral. Sp. P.* **30**:361–369.

Wagenplast, P., 1972, Ökologische Untersuchung der Fauna aus Bank- Und Schwammfazies des Weißen Jura der Schwäbischen Alb, *Arb. Inst. Geol. Paläont. Univ. Stuttgart N.F.* **67**:1–99.

Walker, K. R., and Alberstadt, L. P., 1975, Ecological succession as an aspect of structure in fossil communities, *Paleobiology* **1**:238–257.

Warme, J. E., Burke, R. B., Crevello, P. D., Halliwell-Hazlett, B., and Letsch, D. K., 1988, Evolution of the Jurassic High Atlas Rift, Morocco: Transtension, structural and eustatic controls on carbonate facies: *Am. Assoc. Petrol Geol. Mediterranean Basins Conf. Nice, Sept. 1988, Fieldtrip 9, Guidebook.*

Werner, W., 1986, Palökologische und biofazielle Analyse des Kimmeridge (Oberjura) von Consolação, Mittelportugal, *Zitteliana* **13**:1–109.

Werner, W., Leinfelder, R. R., Fürsich, F. T., and Krautter, M., 1994, Comparative palaeoecology of marly coralline sponge-bearing reefal associations from the Kimmeridgian (Upper Jurassic) of Portugal and Southwestern Germany, *Cour. Forsch. Inst. Senckenb.* **172**:381–397.

Wiedenmayer, F., 1980, Spicules and sponges in the lower Jurassic of the western Tethys, *Sedimenta* **8**:135–145.

Wilkinson, C. R., and Evans, E., 1989, Sponge distribution across Davies Reef, Great Barrier Reef, relative to location, depth, and water movement, *Coral Reefs* **8**:1–7.

Wilkinson, C. R., and Trott, L. A., 1985, Light as a factor determining the distribution of sponges across the central Great Barrier Reef, *Proc. 5th Int. Coral Reef Congr. Tahiti* **5**:125–130.

Wilson, R. C. L., Hiscott, R. N., Willis, M. G., and Gradstein, F. M., 1989, The Lusitanian Basin of west central Portugal: Mesozoic and Tertiary tectonic, stratigraphy and subsidence history, in: *Extensional Tectonics and Stratigraphy of the North Atlantic Margins* (A. J. Tankard, and H. Balkwill, eds.), *AAPG Mem.* **46**:341–361.

Chapter 9

Cretaceous Evolution of Reef Ecosystems

A Regional Synthesis of the Caribbean Tropics

CLAUDIA C. JOHNSON and ERLE G. KAUFFMAN

1. Introduction

Our modern world has a tropical region that contains reefs, rain forests, and the highest biodiversity on the globe. From the hot tropics to the cold polar regions, a series of convection cells drive today's atmospheric circulation. In the oceans, high-density water masses from the polar regions source the thermohaline, subsurface circulation of our modern seas. In stark contrast to today's relatively cold, interglacial, "icehouse world" is the warm Cretaceous or "greenhouse world." Atmospheric carbon dioxide levels of the Cretaceous were two to ten times more than present-day levels (Berner, 1994), resulting

CLAUDIA C. JOHNSON and ERLE G. KAUFFMAN • Department of Geological Sciences, Indiana University, Bloomington, Indiana 47405.

The History and Sedimentology of Ancient Reef Systems, edited by George D. Stanley Jr., Kluwer Academic/Plenum Publishers, New York, 2001.

in a world considerably warmer than today's. But this warmth was not distributed evenly through the Cretaceous and across all latitudes, as noted by Huber *et al.* (1995) and more recently by Frakes (1999), who identified the Late Cretaceous, (early) Turonian stage as the warmest interval of the Cretaceous period.

As a scientific community, we use the ocean record to look for evidence of paleoclimatic–oceanographic conditions expressing the warmth of the Cretaceous and its temporal distribution. Hay (1995) reported that most of the Early Cretaceous did not have permanent ice at the poles as it does today but had only seasonal sea ice in the polar regions. He also postulated that subsurface circulation initiated from high latitudes during much of the Early Cretaceous, as it does in the modern world. In contrast, the mid- and Late Cretaceous polar regions were ice-free and relatively warm, the oceans may have been driven primarily by halothermal circulation (Hay, 1988, 1995), and sea levels were elevated by 200–300 m at their highest levels compared to present-day. In the halothermal scenario, ocean water sank in the low latitudes due to density differences driven not by temperature but by evaporative processes that resulted in high salinity waters. These dense, highly saline water masses may have flowed in the subsurface from the low to higher latitudes, resulting in ocean circulation the opposite of today's (Brass *et al.*, 1982; Hay, 1988), warming the globe by meridional transfer of heat. The amount of heat transferred by the oceans and by the atmosphere is not yet resolved, but modeling experiments are investigating this partitioning (Barron *et al.*, 1993, 1995). If the oceans took on a greater role than the atmosphere in the transfer of heat during greenhouse times (Barron *et al.*, 1995) and if this major heat transfer is initiated in the marine tropics, how were Cretaceous reefs affected?

Not only ocean–air temperature but also the chemical constraints of the Cretaceous seas were different than those of today. To drive subsurface circulation from the low latitudes, salinity is predicted to be higher in the Cretaceous tropics. If so, did the tropics experience both higher temperatures and higher salinity and supertropical–supertethyan conditions relative to those of today, and did these conditions fluctuate temporally and geographically (Kauffman and Johnson, 1988; Johnson *et al.*, 1996)? Norris and Wilson (1998) investigated this issue from the Blake Plateau and found mid-Cretaceous evidence suggestive of higher temperatures but not of higher salinity. Does this interpretation suggest that we should turn our attention to restricted seas in search of evidence leading to sourcing of low latitude, subsurface, mid- and Late Cretaceous water masses?

Further complicating the issue of saline conditions is the existence of Cretaceous oceanic anoxic water masses. Oceanic anoxic events (OAEs) correspond closely to transgressions (Jenkyns, 1980) and eustatic highstands (Arthur *et al.*, 1987). OAEs are recorded globally, but the formation, lateral extent, depth, and composition of these water masses and their relationship to low-latitude, saline water masses are under investigation.

If relatively high temperatures and ocean chemistry expressed in the saline and/or anoxic nature of water masses occurred in the Cretaceous tropics, would we expect the same marine organisms to inhabit the low latitudes of the Cretaceous and of today? In other words, would the same organisms be expected to exist under two different sets of thermal and chemical conditions, such as those of the Cretaceous and today? Scleractinian corals occur during both time periods. Rudist bivalves do not. Rudists existed during the Late Jurassic and Cretaceous, went extinct near the Cretaceous–Tertiary boundary, and left no modern relatives. The temporal and geographic details of the coexistence of scleractinian corals and rudists and their respective roles in ecosystem development have been worked out locally and regionally but not globally. We are far from a stage-by-stage global synthesis of the geographic and ecological patterns of tropical marine ecosystem development contributed by these biotas and many questions remain unanswered.

Warm conditions that exceeded modern tropical temperatures existed at mid-Cretaceous low latitudes (Norris and Wilson, 1998), and cool-temperate conditions were suggested for Cretaceous high latitudes (Kauffman, 1975). Does this mean that Cretaceous climate zones, including the tropics and subtropics, were expanded geographically, and if so, how do we document this expansion? Is it correct to interpret the extinct rudists as reef builders and reefs as indicators of the tropics and to document the geographic expansion and contraction of rudist-dominated reefs through Cretaceous stages (Johnson, 1993)? Is it further legitimate to compare these empirical results to those obtained from model simulations that used different data as input parameters (Johnson *et al.*, 1996)? Or do rudists not build structures that can be called reefs like those we refer to in the modern tropics (Kauffman and Sohl, 1974)? Are rudists only sediment dwellers (Gili *et al.*, 1995) and as such cannot be used to interpret ancient reefal conditions? Many workers have compared rudist reefs to reefs of today and the geologic past, noting both similarities and differences (the first was Nelson, 1959), and his chapter is filled with citations of additional workers who used the term "reef." Hofling (1997) defined criteria for the identification of reefs and illustrated examples of Phanerozoic reef mounds and skeletal and nonskeletal bioherms and biostromes. Comprehensive reviews of the term "reef" for strata of many ages can also be found in Longman (1981) and Hussner (1994) and for the Holocene in Hubbard (1998). In concept, we agree with Hofling (1997) and feel there is enough variability in rudist morphology, paleoecology, and ecosystem development such that some rudists can be considered reef builders and others cannot. We base these conclusions on Cretaceous rocks from the Caribbean and circum-Caribbean region and define rudist reefs as topographically elevated, biogenic structures, including bafflestone mounds, but we focus on the concept of a reef as part of an ecosystem, influenced by biological, chemical and physical processes.

We focus our analyses of Cretaceous tropical marine biota on the Caribbean and circum-Caribbean. This large region has been affected by geologic and oceanographic changes from its origin in the Middle to Late Jurassic through the first part of the Cenozoic. During this time, the area has undergone shifts in the positions of land masses, island–arc systems and ocean basins, in sea level, and in the character and distribution of water masses. Only from the middle Cenozoic to the Recent has the area stabilized sufficiently to support the highly diverse, modern marine biotas that comprise the complex, shallow-water ecosystems for which it is noted today.

The history of the Caribbean is complex and involves deriving a plate from the eastern Pacific Ocean and moving it through the newly rifted Caribbean Sea to its present position (Pindell and Barrett, 1990). As would be expected, the sedimentological and stratigraphic expressions of these movements are difficult to unravel and detailed chronostratigraphic histories need to be worked out for much of the Caribbean proper. Volcanic rocks are numerous in the region, and volcaniclastic, siliciclastic, and carbonate rocks comprise the bulk of the sedimentary packages. During volcanically quiescent intervals, islands had carbonate perimeters colonized by platform biota. During volcanically active times, carbonates decreased, volcanic material increased, and biota fluctuated in response to these environmental changes. Local and regional sea-level histories in zones of active plate spreading or volcanism would be expected to vary from the global syntheses. In contrast, for the tectonically more passive regions of the circum-Caribbean region, such as northeastern Mexico and Texas, carbonate platforms were widespread, facies fully developed, and depositional histories were divided into sequence stratigraphic packages (e.g., Scott *et al.*, 1994).

In this chapter, our intent is to orient topics toward the paleobiological and sedimentological components of the Cretaceous history in the Caribbean and circum-Caribbean region. We proceed by presenting a summary of the geologic history, focusing on plate tectonic models, mid-Cretaceous carbonate platform development (Fig. 1), petroleum resources, and sea-level fluctuations. This is followed by a history of Caribbean ecosystems in which we present hypotheses regarding early coral–rudist relationships and patterns of Caribbean rudist bivalve species' changes across paleolatitudes. We then present a synthesis of Caribbean ecosystem development and conclude the chapter with a brief discussion of the extinction of rudists in the region. We do not propose to present definitive answers to the complexity of paleobiological and paleoenvironmental issues in such a large region, but instead we present patterns and associations and highlight the numerous "unknowns" in Cretaceous tropical marine paleontology.

We use the term "tropics" because we investigate patterns in low paleolatitudes and because the role of the tropics is essential to our understanding of the Cretaceous world. Many investigations have centered on temperate regions of North America, western Europe, South America, and elsewhere on the globe, but tropical temperatures, ocean circulation, and patterns of evolutionary diversification are less known and need to be tied into these better-known temperate areas.

FIGURE 1. On the horizon, an aerial view of two reef tracts, Lower Cretaceous, northernmost Mexico.

2. Caribbean Geologic History

The most recent compendium on general Caribbean geology is Volume H of the *Decade of North American Geology Series* entitled *The Caribbean Region* (Dengo and Case, 1990). This volume includes chapters on regional geology and geophysics, magmatic processes, energy and metallic resources, marine geology, and tectonic processes and evolution, as well as a comprehensive post-1960 bibliography on these subjects. Although paleontological and stratigraphic information can be found in the proceedings volumes of Caribbean geological conferences, paleontology and stratigraphy are underrepresented in Caribbean geology relative to other fields and underutilized in interpreting the complex geologic history of the Caribbean region. This is not true, however, for the well-documented stratigraphic sections surrounding the Gulf Coast region of the United States and Mexico and for the passive margin of northern South America. In these areas stratigraphic and paleontologic publications are numerous and local to regional correlations based on ammonite and foraminiferal biostratigraphic zonations are well established and integrated into global correlations.

2.1. Plate Tectonic Models

The Caribbean region is defined geologically by its plate boundaries (Fig. 2A), and the areas surrounding the Caribbean plate are informally termed circum-Caribbean (Fig. 2B). The first comprehensive model for the evolution of the Caribbean was published by Woodring (1928), but this model predated the universal acceptance of the theory of plate tectonics. Morris *et al.* (1990) summarized the major differences among published models as: the Caribbean plate predated Jurassic sea-floor spreading; the Caribbean plate formed *in situ* and was of Tethyan origin; the Caribbean plate was inserted as a unit from the Pacific; and the Caribbean plate was inserted from the Pacific but was affected by a spreading center in the Venezuelan Basin at 80 Ma, which was overridden by South America. Pindell and Barrett (1990) integrated Caribbean plate tectonic elements into a plate kinematic framework to produce a Pacific origin model, and this model appears to be the one most widely accepted by workers in the region.

Present motion, boundary configurations, and magnetic anomalies of the plate, subduction-related magmatism, and tectonic contacts between rock suites were investigated before Pindell and Barrett (1990) assigned a Pacific provenance for the plate along the southwestern margin of modern-day Mexico. Five phases of Caribbean plate evolution were cited and eight plate boundary maps, three pertaining to the Cretaceous, were produced. The 130-Ma map represents the Early Cretaceous Valanginian stage (Fig. 2B) and illustrates the rifting proto-Caribbean Sea, the locations of Jamaica and the proto-Greater Antilles along the southwestern margin of Mexico, the positions of South America and Africa, and the mid-Atlantic ridge. Earliest Cretaceous geologic histories are best known for the US Gulf Coastal region, many parts of Mexico, and northern South America, whereas publications are fewer for localities on the incipient Caribbean plate.

The 95-Ma map representing the mid-Cretaceous Cenomanian stage (Fig. 2C) illustrates the expanded proto-Caribbean seaway and the increased geographic distance between the Americas relative to the 130-Ma time frame. The

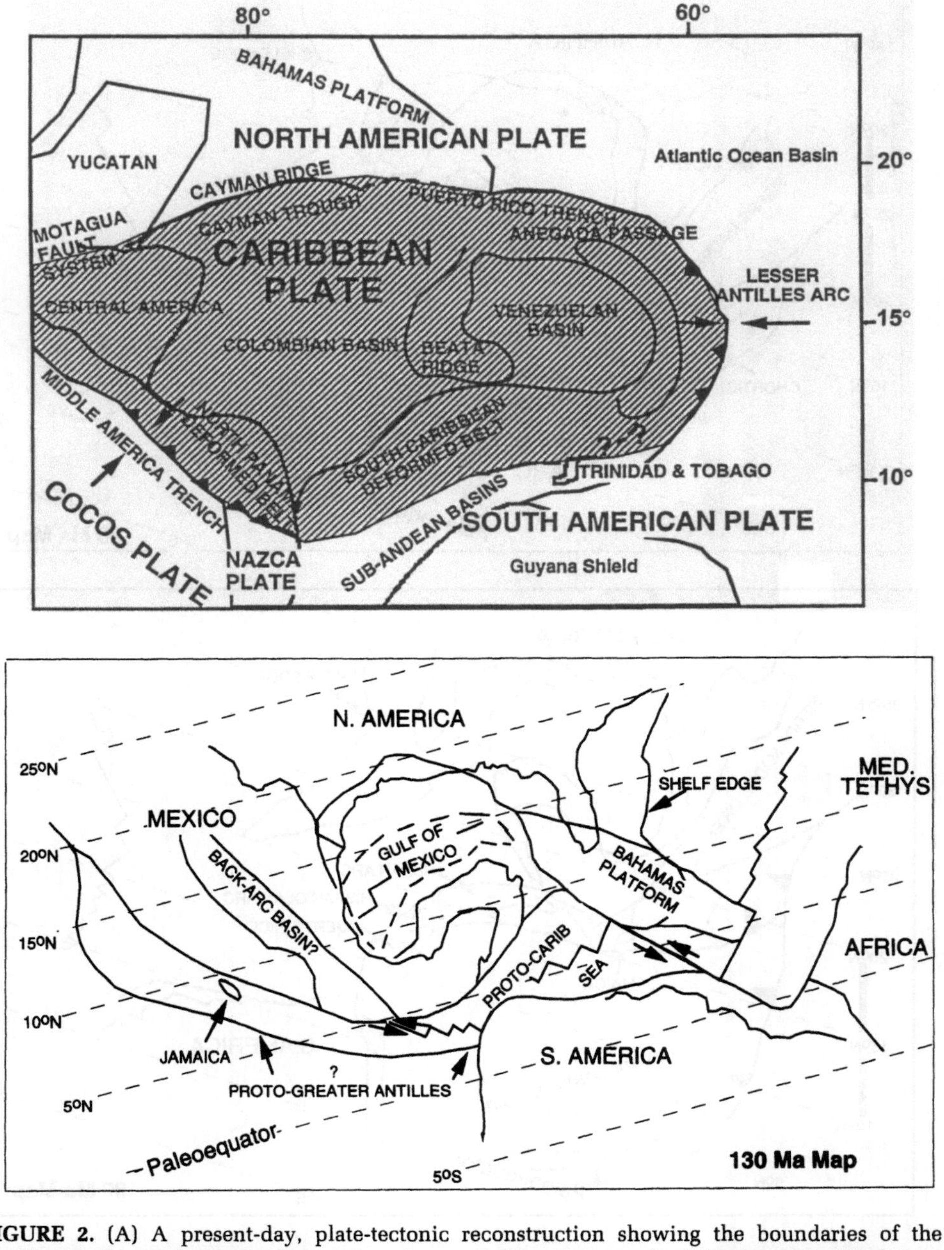

FIGURE 2. (A) A present-day, plate-tectonic reconstruction showing the boundaries of the Caribbean plate and surrounding region. Arrows indicate zones of subduction. (Modified from Case *et al.*, 1990), and Mann *et al.*, 1990.) (B) Valanginian-age distribution of landmasses and ocean basins at 130 Ma (after Pindell and Barrett, 1990). Note the size and position of the proto-Caribbean plate with the localities of Jamaica and the proto-Greater Antilles.

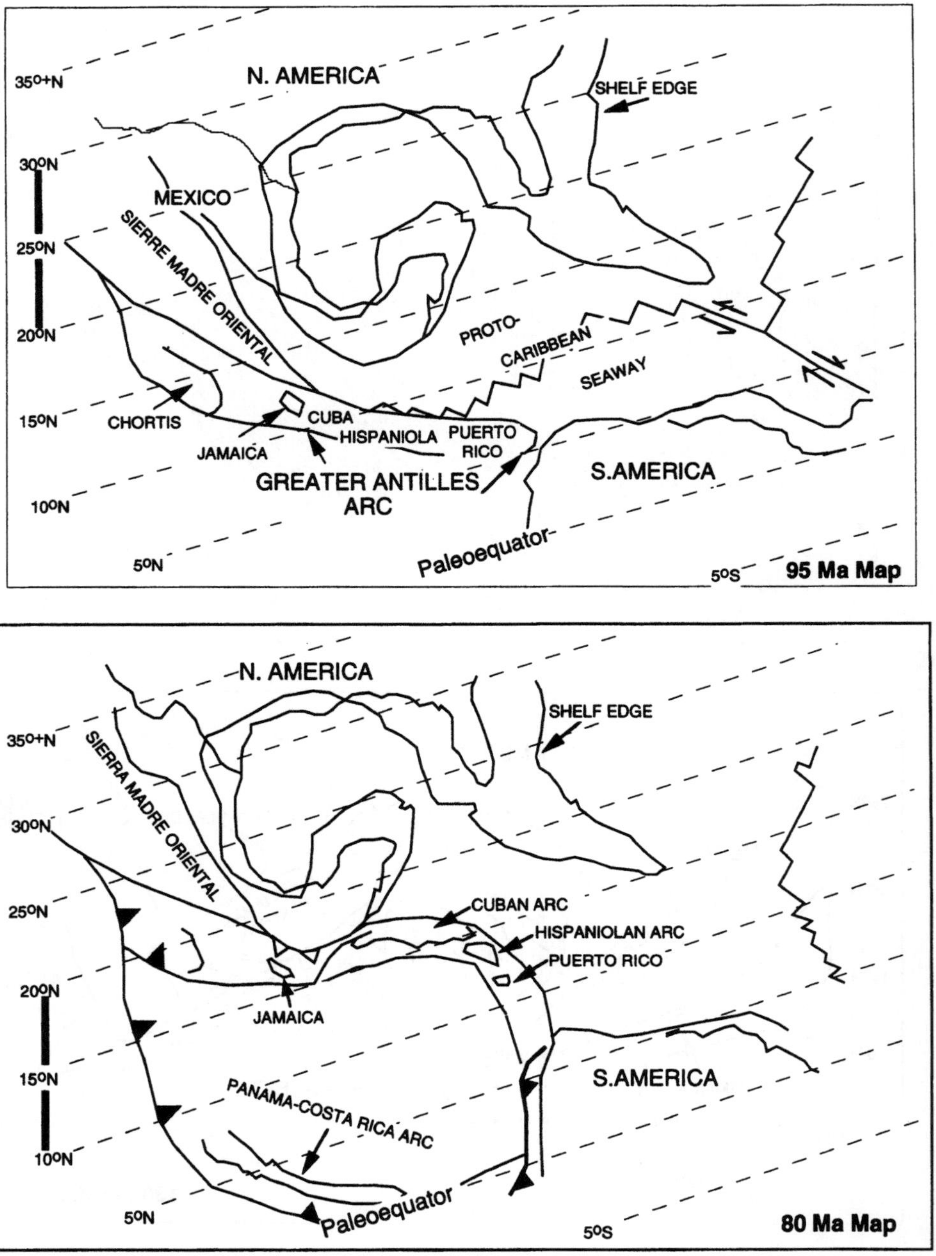

FIGURE 2. *Continued.* (C) Relative positions of upper Cenomanian landmasses, size of the proto-Caribbean Seaway at 95 Ma, and position of the proto-Caribbean plate (after Pindell and Barrett, 1990). Thick vertical line indicates paleolatitudes with highest mid-Cretaceous rudist species diversity. (D) Lower Campanian reconstructon at 80 Ma (after Pindell and Barrett, 1990), showing movement of the Caribbean plate into the expanded seaway between North and South America. Thick vertical line indicates paleolatitudes with highest Late Cretaceous rudist species diversity.

Caribbean plate is outlined to the southwest of Mexico. According to Pindell and Barrett (1990), equatorial spreading had begun in the Atlantic, and South America and Africa are farther apart than in the preceding time period. North America and Europe were also farther apart due to the spread of the North Atlantic Basin. The larger size of the Atlantic ocean basin implies that larval drift distances between the Caribbean and Mediterranean regions increased since Valanginian time. The extended shelf edge of northern South America and the expansive shelf edge along the US Atlantic coast and the Gulf of Mexico were potential sites for the development of shallow-water reefs. Pindell and Barrett (1990) noted that by the Campanian (Fig. 2D), the Caribbean plate migrated northeastward into the gap between the Americas, led by the Greater Antilles and the Aves Ridge arc systems. To the southwest, the Panama–Costa Rica arc developed, but the western boundary of the Caribbean plate is defined less clearly (Pindell and Barrett, 1990). The northern margin of South America is still passive, and the shelf edges along the southeastern United States and Gulf of Mexico are similar to those on the Cenomanian, 95-Ma map.

In viewing the tectonic maps from a biogeographic perspective, Pacific marine biota weere brought into the Caribbean during the plate's initial entry in the Cretaceous, and biota migrated across the Atlantic from the Mediterranean. Both routes were documented for Jurassic and Cretaceous biota by Imlay (1939) and have been updated by more recent studies cited later in this text. In viewing the plate tectonic reconstructions, it seems that ultimately a continuous, intra-Caribbean larval distribution route among localities in South America, the island arcs, and North America could have been established for shallow-water marine biota (Johnson, 1999). Alternatively, isolation of faunas and endemism could have occurred as terranes moved apart, as platforms were tectonically elevated, or as barriers were established (Johnson, 1993). On a larger scale, as the North Atlantic widened, trans-Atlantic larval flow may have been reduced and Caribbean endemism would have developed (Johnson and Kauffman, 1990). In consideration of forces destructive to marine biota, volcanic events would have decreased the amount of carbonate or siliciclastic, shallow-water marine habitats. Many of these biogeographic possibilities, namely dispersal, vicariance, and interchange, remain to be tested against faunal, sedimentological, and tectonic databases.

2.2. Carbonate Platforms and Petroleum Reservoirs

Wilson (1975) was the first to composite and illustrate the distribution of carbonates for the Early and mid-Cretaceous around the Gulf of Mexico. During the last decade, many of these platforms were reanalyzed, examined within depositional and sequence stratigraphic frameworks, and compiled into a global compendium entitled "Cretaceous Carbonate Platforms" (Simo *et al.*, 1993). In contrast to the well-studied region around the Gulf of Mexico, sites for carbonate deposition and biotic accumulation in the Caribbean proper, represented by the Proto-Caribbean Seaway and the developing plate, are more difficult to discern on the reconstructions and have fewer representations in the published literature. Regional plate motions, local tectonism expressed as fault blocks or horst and graben features, volcanism, sea-level histories, surface current patterns, and biotic buildups are some of the major factors that influenced the region's carbonate depositional environments. As end-members along a spectrum of depositional possibilities, the extended shelf regions around the Gulf of Mexico and Venezuela will be discussed separately from the smaller, more restricted carbonates that developed within the proto-Caribbean plate.

Texas, Mexico, and Venezuela may contain the richest history of mid-Cretaceous carbonate depositional environments in the region, most notably because these areas have the largest petroleum resources, many of which come from rudist reefs (Fig. 3A,B). The best-known carbonate units are described from mid-Cretaceous lithofacies representing distinct basin, basin–margin, and platform deposits that have been studied for decades and most thoroughly described by Enos (1974). In the Tampico Embayment, major oil accumulations are known from the Ebano–Panuco fields representing basinal facies, the Golden Lane fields representing platform facies, and the Poza Rica fields that represent basin–margin facies (Enos, 1974). Lithological and paleobiological descriptions of these facies are described below. The Poza Rica fields have cumulative production estimated at 2.5 billion barrels of oil (Wilson and Ward, 1993) and mid-Cretaceous carbonates may have provided 90% of Mexico's oil from the Tampico embayment (Enos, 1974).

Enos (1974) described the basinal facies as thin- to massive-bedded, dark gray mudstones and wackestones containing pelagic microfossils such as calcispheres, globigerinids, and tintinnids. Lenses and beds of cherts are common. The facies is recrystallized and dolomitized.

The platform facies is described as thick-bedded, light gray, skeletal grainstones to mudstones, consisting of a reef and back-reef facies. The general description of the reef facies provided by Enos (1974) notes that lithologically, coarse-grained skeletal grainstones are dominant, whereas biologically the rudists dominate, although corals, algae, gastropods, and stromotoporoids are minor components. Collins (1988) studied the reef core of the Valles Platform, the temporal equivalent of the Golden Lane, and did not discern a pattern of biotic succession of the macrofauna. Instead, Collins determined that the control on organisms was physical, with tall, upright rudists associated with higher mud content and less shell debris and recumbent rudists associated with more skeletal sand. The distribution of rudist biofacies across the El Abra Formation of the Valles platform was noted by Johnson *et al.* (1988). Across the platform, however, rudist reefs are discontinuous and aligned along the eastern edge (Enos, 1974). Elsewhere, for example, on the El Doctor platform, the rudists form a nearly continuous marginal zone a few hundred meters wide, with skeletal grainstones forming behind the rudist concentrations at the platform edge or pisolites and oolites forming locally near the edge of the various platforms. Wilson (1975) used the term "knoll-reef ramps" for this

FIGURE 3. (A) Mid-Cretaceous El Abra Formation, Valles Platform, Mexico, showing caprinids stained with oil (dark gray).

type of rudist reef at shelf margins and interpreted them as ecological, knoll-shaped reefs forming in linear belts on gentle slopes at the outer edges of shelf margins. The back-reef facies consists of thin- to thick-bedded grainstones to wackestones, with miliolid foraminifera, pellets and lumps of pellets, and rudists from the family Requieniidae. Stromatolites, anhydrites, and dolomitic intervals are present and evidence of currents is rare.

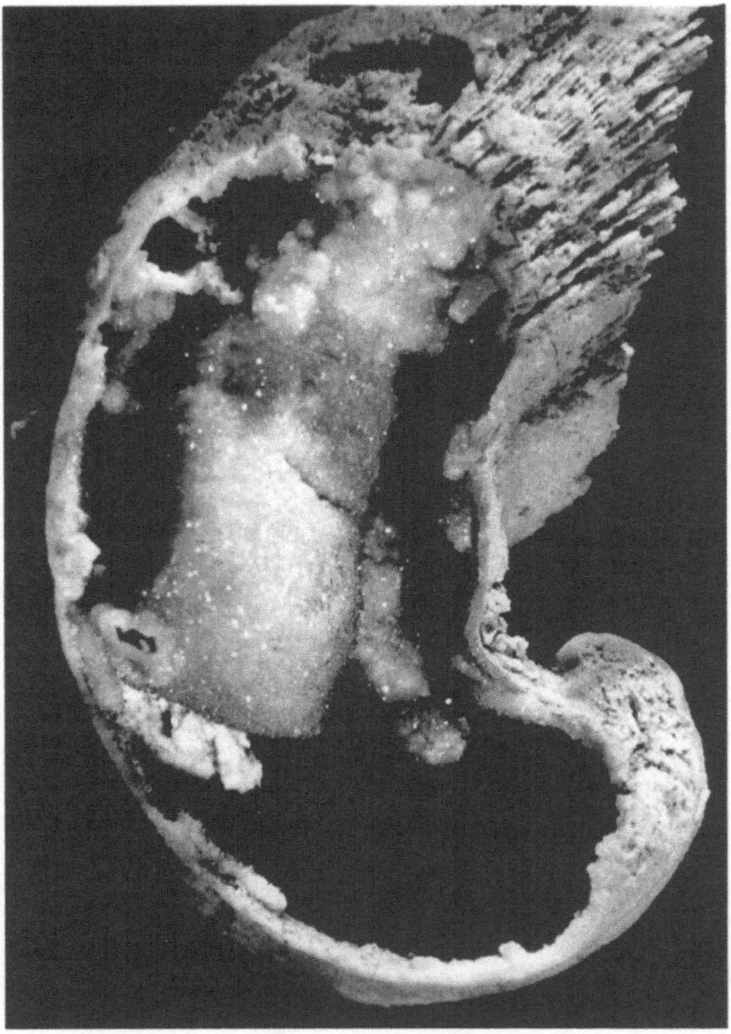

FIGURE 3. *Continued.* (B) Lower valve of a caprinid showing secondary porosity. Walls of body cavity (large cylinder in center) and those of accessory cavity (to right of center) have been replaced by silica.

The basin–margin facies is composed of three subunits: pelagic wackestones to mudstones with chert lenses; breccias with a carbonate mudstone or dolomite matrix and intraclasts of basinal mudstones, cherts, or platform fossil fragments; and detrital limestones composed of rounded, sand- to pebble-sized fragments of rudists and other platform fossils.

Structural contour maps and seismic reflection profiles indicate that some mid-Cretaceous platforms, such as the Golden Lane, had steep edges with about 1000 m of vertical relief. The relief was interpreted to result from vertical accretion of the platform–margin facies (Enos, 1974), resulting from subsidence and/or sea-level rise that occurred during the Albian (Wilson and Ward, 1993). Like many platforms in the region, the Golden Lane initiated on a Permian–Triassic(?) horst block that formed during the initial rifting of the Gulf of Mexico in the Late Jurassic; basement lithology of these horst blocks is varied and ranges across a variety of igneous and metamorphic rocks (Wilson and Ward, 1993). Other platforms, such as the El Doctor and Valles, contain evidence of faulting, but appear to reflect original depositional topography. These steep-sided, high-relief platforms of mid-Cretaceous age in northeastern and east–central Mexico were similar to the Capitan reefs and Bahamas Banks in terms of scale and depositional relief (Enos, 1974).

The robust descriptions of extensive carbonate units from the north in the Gulf Coast region and to the south in northwestern Venezuela (Vahrenkamp *et al.*, 1993) do not apply to rocks of most of the Caribbean plate. Many localities on the Caribbean plate have carbonate rocks interbedded with volcanic or volcaniclastic sediments, or more typically the rocks range from pure carbonate to pure volcaniclastic and grade through every combination of carbonate–volcaniclastic content possible. The carbonate-dominated rocks contain rudist bivalves as their numerically and ecologically dominant macrobiotic constituent, and arthropods, foraminifera, other bivalves, corals, and gastropods are less common in most mid- to Late Cretaceous situations; locally, these groups can occur in great numbers, e.g., the Albian of Puerto Rico where gastropod abundance is extremely high.

An example of the lithologic complexity expressed on Caribbean plate localities comes from the Lower Cretaceous Benbow Inlier in the northeastern part of Jamaica. Zans *et al.* (1962), Burke *et al.* (1968), Wright (1974), and Robinson and Lewis (1987) provided synopses of the geology and stratigraphy of this inlier and Chubb (1967, 1971) dated the beds based on rudistid bivalves in the carbonate units. The Benbow Inlier comprises two formations of Early Cretaceous age: the Devils Racecourse Formation and the overlying Rio Nuevo Formation (Robinson and Jackson, 1976). The Devils Racecourse Formation consists of over 1000 m of platform and slope facies that are composed of a series of lavas, volcaniclastic rocks, and laharic breccias intercalated with thin, hard, bluish gray, micritic, calcarenitic, and calciruditic limestones of shallow-platform origin. These limestones span Hauterivian through Late Aptian time (Chubb, 1967, 1971). The limestones represent short-lived development of small carbonate platforms or lenticular depositional areas during relative stabilization of the Jamaican platform between major volcanic eruptions from nearby arc–island sources.

The Albian Rio Nuevo Formation consists of a thick series of volcaniclastic sandstones, siltstones, and clays, and in the lower one third the formation contains a regionally lenticular, fossiliferous limestone termed the Seafield Limestone of probable middle Albian age. The Rio Nuevo Formation represents platform and slope facies during a major interval of arc–island volcanism, with the Seafield Limestone representing stabilization of the platform during a period of relatively low volcanic activity. The Rio Nuevo Formation is overlain by the Tiber Formation, a thick series of Albian or younger volcanic conglomerates, sands, and lava flows.

Another example of Caribbean plate geologic and lithologic complexity comes from the mid-Cretaceous of Puerto Rico. On this island, Briggs (1969) defined the type section of the Aguas Buenas Limestone Member of the Torecilla Breccia, a member first described by Semmes (1919). The member contains 30 m of dark gray, thick-bedded, finely crystalline limestone with an extensive rudist and gastropod fauna. The top of the member is faulted at the quarry, but is present nearby below thin-bedded, tuffaceous sandstone. The base of the member is poorly exposed and rests on hydrothermally altered volcanic rock. In total, the member is less than 60 m thick and is a lenticular, reefal limestone, probably Albian in age (Briggs, 1969). The abundance of volcaniclastic grains, algal-coated grains, and mollusk fragments in wackestones and the interbedded nature of these limestones with volcanic shale, shale, and volcanic mudstone to siltstone suggests a shallow-water, inner-platform setting for the lower two thirds of the section. A slight deepening trend occurs up-section, as represented by wackestones and packstones interbedded with thick- to massive-bedded, fossiliferous limestones such as a biolithite and requieniid rudists in growth position. The section is karstified at the top and faulted (Briggs, 1969).

Sea level is another variable to consider when unraveling the geologic history of localities on the Caribbean plate because it combines local effects such as island–arc activity with global effects related to the building and collapse of midocean ridges. Local sea-level curves derived from sedimentological, stratigraphic, and paleobiological indicators from Jamaica and Puerto Rico are compared to the global curve of Haq *et al.* (1987) and show major differences (Fig. 4). Antillean sections representing the lower Albian and the basal part of the middle Albian are similar to the global curve, but the same is not true for the upper Albian. In addition, the entire curve for Cenomanian sections on the islands is the opposite of that illustrated by the global curve; a period of transgression on the global chart is expressed as regression in Jamaica and Puerto Rico due to localized tectonism–volcanism. A middle Santonian regression is interpreted for both islands, although this may be the slightly offset version of the upper part of the middle and the late Santonian lowering of sea level shown globally, but correlations are poor. The next major lowering of relative sea level comes in the upper part of the middle and the early part of the late Campanian in both Jamaica and Puerto Rico, but this is expressed as transgressive on the global chart. Correlations are poor across the late part of the upper Campanian and across the Campanian–Maastrichtian boundary zone. The last major discrepancy between local and global sea level occurs in the lower Maastrichtian and the early part of the middle Maastrichtian, reflecting the independent history of the leading edge of the Caribbean plate.

In contrast to the generalized sea-level history compiled from localities on the Caribbean plate, nine Albian sequences were noted in Texas sections by

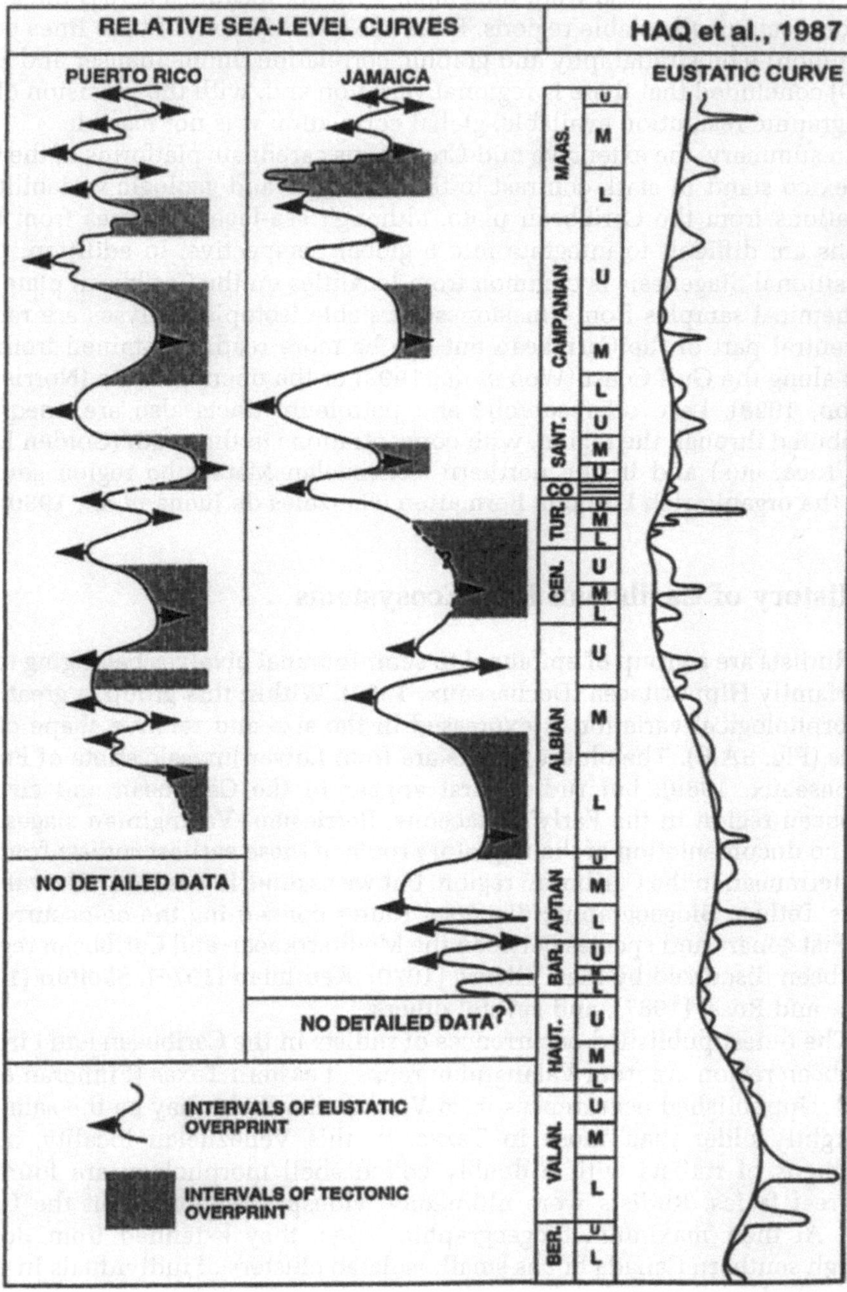

FIGURE 4. Global sea-level curve (right column), compared to islands at or near the leading edge of the Caribbean plate (middle and left columns), showing intervals of eustatic and tectonic (gray) influences. Modified from Johnson and Kauffman, 1996.

Scott *et al.* (1994). These were integrated with six localities across the globe, all from tectonically stable regions. From the development of time lines based on ammonite biostratigraphy and graphic correlation, Immenhauser and Scott (1999) concluded that there is regional variation and, with the precision of the stratigraphic resolution available, global correlation was not feasible.

In summary, the extensive mid-Cretaceous carbonate platforms of the Gulf of Mexico stand in stark contrast to the lithologic and geologic variability of formations from the Caribbean plate, although sea-level histories from both regions are difficult to integrate into a global perspective. In addition, post-depositional diagenesis is common from localities on the Caribbean plate and geochemical samples from limestones for stable isotopic analyses are rare in the central part of the Caribbean but can be more readily obtained from the areas along the Gulf Coast (Woo *et al.*, 1992) or the open Atlantic (Norris and Wilson, 1998). Last, oil reservoirs and petroleum tracts also are unequally distributed through the region, with concentrations in the north (Golden Lane, Poza Rica, etc.) and in the northern Venezuelan Maracaibo region sourced from the organic-rich La Luna Formation (Gonzales de Juana *et al.*, 1980).

3. History of Caribbean Reef Ecosystems

Rudists are a group of epifaunal to semi-infaunal bivalves belonging to the superfamily Hippuritacea (Dechaseaux, 1969). Within this group, a great deal of morphological variation is expressed in the size and relative shape of the valves (Fig. 5A,B). The oldest rudists are from Lower Jurassic strata of France (Dechaseaux, 1969), but rudists first appear in the Caribbean and circum-Caribbean region in the Early Cretaceous, Berriasian–Valanginian stages. We have no documentation of the migratory route of these earliest rudists from the Mediterranean to the Caribbean region, but we assume it was due to larval drift across Tethys. Biogeographic dispersal routes concerning the co-occurrences of rudist genera and species between the Mediterranean and Caribbean regions have been discussed by MacGillavry (1970), Kauffman (1975), Skelton (1982), Masse and Rossi (1987), and several others.

The oldest published occurrences of rudists in the Caribbean and circum-Caribbean region are from Valanginian rocks of eastern Texas (Finneran *et al.*, 1984). Unpublished occurrences from Venezuelan rocks may be the same age or slightly older than those in Texas. In this Venezuelan locality, a few specimens of rudists with a doubly coiled shell morphology are found in back-reef facies. Rudists were ultimately widespread throughout the Caribbean. At their maximum biogeographic range, they extended from Bolivia through southern Canada but as small, isolated clusters of individuals in these nontropical occurrences.

In this section, we first present hypotheses for the compositional changes in Cretaceous reef ecosystems. To put it simply, this involves a change from corals to rudists. Although some data used for interpretations on taxonomic changes come from the Caribbean and surrounding region, not all of it does,

FIGURE 5. (A) *Titanosarcolites* from the Maastrichtian Guinea Corn Formation, showing prone, doubly coiled form (uncoiled, nearly 5 times the hammer length; hammer is 30.5 cm). *Titanosarcolites* probably fed passively, with the valves closed, through a network of channels that were open to the outside.

and the hypotheses reflect a global database. For the Caribbean and circum-Caribbean region, we then present the temporal and geographic patterns of rudist species' occurrences before embarking on a temporal analysis of regional ecosystem changes. For this chapter, we compiled information from both published and unpublished resources. Unpublished sources are the US Geological Survey (USGS) open file reports, examination and report files, our field notes, and those of USGS workers involved in Caribbean research initiating in the 1960s. We also had limited access to unpublished oil company information or reports.

3.1. Hypotheses Proposed for the Change in Reef Composition from Coral to Rudist Domination

Late Jurassic and earliest Cretaceous reefs were composed of demosponge–algal and stromatoporoid–coral–algal paleocommunities, but soon thereafter coral–stromatoporoid–rudist, coral–algal–rudist, and then rudist paleocommunities dominated reefs for the remainder of the Cretaceous (Scott, 1988). In general, the transition from coral–algal–rudist to rudist reefs occurred earlier in the Gulf of Mexico (Aptian to middle Albian) (Scott, 1984b) than

FIGURE 5. *Continued.* (B) *Antillocaprina*, which fed by opening the valves like a normal bivalve; commissure is to the left at the junction of the ornament typical of left and right valves. Maastrichtian Guinea Corn Formation, central Jamaica.

it did in the Mediterranean (Cenomanian and Turonian) (Philip, 1980). We do not yet have sufficient data from the Pacific guyots to make an age assignment for the biotic turnover for that region of the world if it did occur.

There are numerous hypotheses regarding the Cretaceous compositional change in shallow-water inhabitants, and Scott (1984b, 1988, 1995) reduced these to two categories: environmental changes and biotic interactions. Scott suggested that environmental factors such as sea-level rises and productivity cycles, as well as the relatively shallower-water substrates occupied by rudists and relatively deeper-water substrates occupied by corals, were the reasons for the change in dominance. Kauffman and Johnson (1988) suggested that biotic influences, such as convergent evolution of morphology and competitive ecological advancements of rudists over corals, in combination with environmental changes such as increased salinity and temperature, were the reasons for the change in dominance. But these are only some of the more recent summaries; workers have been trying to determine the reasons for the change in dominance of rudists over corals for decades. For example, Beauvais and Beauvais (1974) analyzed the distribution of scleractinian corals and noted the decline of corals after the Turonian. They surmised that seas had increased turbidity that was unfavorable to the scleractinians in the Late Cretaceous. Lowering of sea level during the Santonian also was hypothesized to have stressed the corals more than the rudists (Philip, 1980) and thereafter the rudistids were dominant. Most recently, Stanley and Hardie (1998) proposed that the rudists displaced aragonitic hermatypic corals during a calcite II phase of seawater chemistry, a phase defined as a low Mg/Ca ratio. During this phase, the aragonitic corals declined as a result of the decreased Mg/Ca ratio of the seawater, but rudists may not have been affected by the shift in mineralogical composition and thereafter dominated by default. Conversely, rudists may have responded to the change from aragonite to calcite seas, but after a delayed period of time (Scott, 1988). Almost every environmental "stress," from warming trends to productivity cycles, has been discussed as a possible causal mechanism or at least a link to the change in biotic composition (Scott, 1988, 1995).

It is important to note the time periods of the Cretaceous for which these hypotheses are proposed. Scott (1984b) noted the Aptian to Middle Albian; Kauffman and Johnson (1988) identified the Albian through Turonian; Philip (1980) noted the Cenomanian and Turonian and the Santonian; Coates (1977) identified the Late Cretaceous; and Stanley and Hardie (1998) identified the mid- and Late Cretaceous as important turning points.

It appears there is no consensus on the timing or causal mechanisms of the biotic changeover. Large-scale, regional differences abound, especially in the timing of events, and we still lack sufficiently detailed observations from many places in the world, including many localities on the proto-Caribbean plate. To date, these hypotheses on Cretaceous biotic changes of one group by another, be they in the form of replacement or displacement, remain unresolved.

3.2. Temporal and Geographic Patterns of Rudist Evolution

Even though we cannot be sure of the reasons for the biotic changeover that occurred in the Caribbean and surrounding region, we do know that rudists dominated these ecosystems by the mid-Cretaceous and we can proceed with a regional analysis. Although this rudist database alone cannot attempt to address the issue of faunal changeover, it can illuminate other aspects of Caribbean paleontology and geology. Here we present the patterns of species' occurrences of rudist bivalves in the Caribbean and circum-Caribbean region. These patterns provide us with the temporal and geographic distribution of the group and show increases and decreases in numbers of species through time. The collection and normalization of the observed data set by statistical methods is the most refined yet compiled for Cretaceous rudist bivalves or for any set of Cretaceous paleotropical data. However, patterns presented here may be better used in the future if the analyses were calculated per million years or per depositional sequence instead of per Cretaceous stage, for Cretaceous stages are of varying durations and the results reflect these temporal biases. At present, however, stage-level temporal resolution is the most precise possible for this region. How these rudist patterns compare to other biotic groups, e.g., corals, algae, and so forth, remains to be seen in future studies.

The paleontological database consists of the geographic locations of 904 occurrences of rudist bivalves in the Caribbean and circum-Caribbean region. Per Cretaceous stage, all rudistid species occurrences were plotted on present-day mercator projection maps. Occurrences were assigned locality numbers and localities were transferred to the 130, 95, and 80 million year plate reconstructions of Pindell and Barrett (1990). The localities were placed within 5° paleolatitude increments on the reconstructions. Paleolatitude information was compiled from Smith *et al.* (1973), Sclater *et al.* (1977), Ross and Scotese (1988), and Pindell *et al.* (1988).

The total occurrences per 5° paleolatitude were tallied, as were the number of species and the number of 5° paleolatitude increments recorded per Cretaceous stage on the reconstructions (Table 1). These data comprised the input parameters for two equations designed to calculate the estimated expected distribution of species' ranges across levels or across paleolatitudes in this case (equations 7 and 8 of Koch and Morgan, 1988) (Table 2). The output parameters, consisting of the total number of estimated expected species per 5° paleolatitude, were then tallied and graphed per Cretaceous stage.

The position of the reef line was taken from Johnson (1993) and published in Johnson *et al.* (1996). Reef lines for the Cretaceous in Mexico and Texas were defined initially by Young (1983) for the Neocomian through Turonian and for the Campanian and Maastrichtian stages, but he noted the absence of reefs for the Coniacian and Santonian. Alencaster (1984) plotted and described the location and faunal composition of rudistid banks or rudistid-bearing limestones in the southern half of Mexico. The works of Young and Alencaster, together with field and literature data, were transferred as reef lines that were corrected for paleolatitudinal postions. Reef lines were drawn prior to and independent of the statistical analyses of species diversity.

Although the statistical method of Koch and Morgan (1988) was designed for stratigraphic levels, it is equally applicable to biogeographic studies or to

TABLE 1. Occurrences of Rudist Species (Observed dta) Per 5° paleolatitude Increment per Cretaceous Stage[a]

	A	B	C	D	E	F	G	H	I
35+			1		1		2	11	1
30–35			12	2	0	4	3	3	1
25–30			72	3	0	2	7	10	11
20–25			53	32	15	3	1	1	1
15–20	5	5	18	8	29	1	10	88	298
10–15	1	4	13	0	2	0	1	30	88
5–10	4	4	6	1	1	0	3	2	2
0–5N	3	8	4	1	0	2	0	0	0
0–5S	1				2	2	2	2	1

[a](A) Barremian: N = 14; s = 12; N ÷ s = 1.2; 6my = 2spp/my; (B) Aptian: N = 21; s = 17; N ÷ s = 1.2; 6my = 2.83spp/my; (C) Albian: N = 179; s = 76; N ÷ s = 2.4; 14.83my = 5.13spp/my; (D) Cenomanian: N = 47; s = 34; N ÷ s = 1.4; 3.77my = 9.02spp/my; (E) Turonian: N = 50; s = 17; N ÷ s = 2.9; 4.7my = 3.62spp/my; (F) Coniacian: N = 14; s = 13; N ÷ s = 1.1; 2.08my = 6.25spp/my; (G) Santonian: N = 29; s = 23; N ÷ s = 1.3; 3.12my = 7.37spp/my; (H) Campanian: N = 147; s = 53; N ÷ s = 2.8; 12.16my = 4.39spp/my; (I) Maastrichtian: N = 403; s = 116; N ÷ s = 3.5; 5.98my = 19.40spp/my; *N* = number of occurrences; s = number of species; N ÷ s = number of occurrences per species; my = million years; duration of the stage from Kauffman *et al.* (1993); spp/my = number of species per million years.

paleolatitudes as used herein. Koch and Morgan (1988) developed the equations to show the influence of sample size on observed species' ranges. Conclusions derived from their studies show that species richness patterns are affected by the distribution of sample size. They suggested that researchers test for sample size effects before interpreting biological or physical causes from observed databases.

The estimated expected diversity values of rudists range from 0 to 106 and are shown per stage in chronological order (Fig. 6). From the patterns, we can summarize that estimated, expected diversity values were low for the Barrem-

TABLE 2. Equations of Koch and Morgan (1988)[a]

$$E(N_{k,k+a}) = \sum_{i=1}^{s} \frac{(1-\hat{p}_i)^N[1-(1-\hat{p}_i)^{nk}]\,[1-(1-\hat{p}_i)^{nk+a}]}{[1-(1-\hat{p}_i)^N]\prod_{j=k}^{k+a}(1-\hat{p}_i)^{nj})}$$

$$E(N_{k,k}) = \sum_{i=1}^{s} \frac{(1-\hat{p}_i)^N[1-(1-\hat{p}_i)^{nk}}{[1-(1-\hat{p}_i)^N](1-\hat{p}_i)^{nk}}$$

$E(N_{k,k+a})$ = estimated expected number of species ranging from latitude 1(k) to latitude $k+a$

[a]s = observed species diversity; $\hat{p}_i = \dfrac{\text{number of occurrences of species } i}{\text{total number of occurrence}}$; n_j = number of collections at level$_j$; n_{ij} = number of occurrences of species i among the n_j collections.

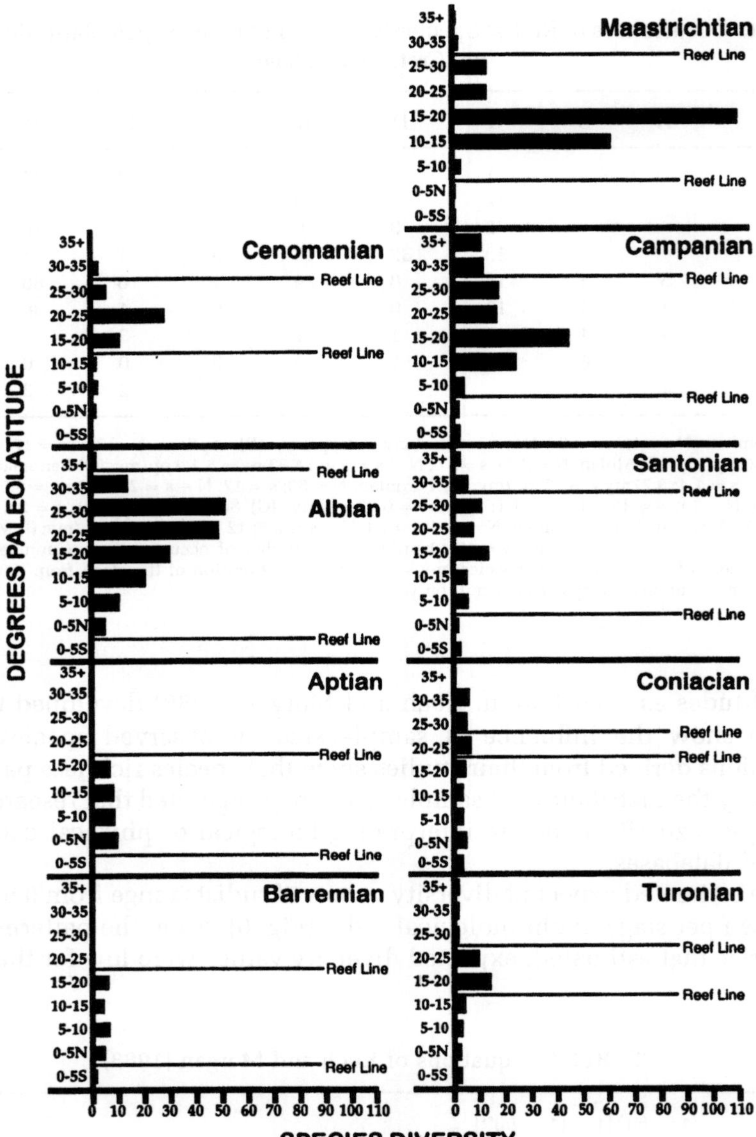

FIGURE 6. Barremian through Maastrichtian reef history and paleolatitudinal diversity trends of rudist bivalves through time (statistical results are graphed). Note fluctuations in diversity across paleolatitudes through time.

ian and Aptian, increased in the mid-Cretaceous Albian and Cenomanian stages, were low during the Turonian, Coniacian, and Santonian stages, and increased through the Campanian and Maastrichtian. If the peaks are viewed geographically, they are represented by the paleolatitudes encompassing northeastern Mexico and Texas for the Albian and Cenomanian and for the

Greater Antillean islands of Puerto Rico and Jamaica for the Campanian and Maastrichtian (Fig. 2C,D).

Two distinct patterns of rudistid diversity can be seen in the expected species diversity graphs (Fig. 6). The first pattern is characterized by a relatively uniform distribution of species across paleolatitudes. The Barremian, Aptian, Coniacian, and perhaps Santonian display this pattern, with low diversity within as well as outside of the reef line. For these stages, observed values are relatively even across all paleolatitudes, and an even distribution of expected diversity was predicted by examples illustrated by Koch and Morgan (1988).

The second pattern is characterized by a diversity increase roughly in the middle of the geographic area. In all stages displaying this pattern—the Albian, Cenomanian, Turonian, Campanian, and Maastrichtian—observed values were also relatively high (or peaked, to use the terminology of Koch and Morgan, 1988) about midpoint across paleolatitudes. This, too, was predicted by examples illustrated by Koch and Morgan (1988). In most of these stages, peaks are roughly in the center of the reef line, whether the reef line is expanded (e.g., Albian) or contracted (e.g., Cenomanian). A diversity peak occurring in the center of the geographic spread of the biota appears to be a typical tropical pattern for modern-day biota (Stehli *et al.*, 1967).

In summary, data from the reef line and diversity plots suggest the following:

1. The reef line and the nature of diversity changes exhibit both temporal and spatial fluctuations through the Cretaceous. When comparing the observed with the statistically derived data, the patterns illustrated are those predicted by Koch and Morgan (1988), who noted the influence of sample size on observed species' ranges. We may be able to attribute diversity patterns to a monographic effect, but we may also have to consider the coincidence of these diversity patterns to biologic and geologic events. The mid- and Late Cretaceous was a period of morphological diversification in shell shape, size, and ornamentation, experimentation in uncoiling and orientation of the valves relative to the substrate, and in ecological organization such as clustering of individuals. In addition, influences of plate tectonic movements, temporal and geographic changes in the ocean–climate system, surface current circulation patterns, and fluctuating sedimentological processes certainly affected the distribution of the group. We suggest that the paleotropics in this region of the world were characterized by dynamic rather than stable conditions.

2. All known Cretaceous rudists reefs were constructed wholly north of the paleoequator, and the southernmost extent of the reef line was either at or north of the paleoequator throughout the entire Cretaceous. These data indicate environmental asymmetry in this region of the globe.

3. The northern hemisphere location of Cretaceous rudist bivalve diversity peaks is more similar to that of Holocene hermatypic corals than to bivalves. Holocene corals attain their maximum diversity north of the equator in the Atlantic Ocean (Stehli and Wells, 1971), whereas Holocene gastropod

and lamellibranch diversity peak at the equator and decrease northward in the Atlantic Ocean (Abbott, 1954). Holocene bivalve data from Stehli *et al.* (1967) show Atlantic diversity peaks for species, genera, and families that coincide with the geographic equator. We are presently examining coral and nonrudist bivalve diversity patterns, integrated with geologic data, with the anticipation of gaining insights into the environmental conditions in this region.

3.3. Ecosystem Development

Three major evolutionary stages of Caribbean, shallow water, Mesozoic reefs can be distinguished. These are based on taxonomic composition, diversity, ecological complexity, framework structure and size: (1) a Late Jurassic–earliest Cretaceous (Oxfordian–Hauterivian: 155–124 Ma) stage dominated by small, low- to moderate-relief frameworks with low-diversity reef-building assemblages composed of skeletal algae (e.g., *Tubiphytes*), stromatoporoids, scleractinian corals, and calcareous sponges along open oceanic platform margins, and of coral–siliceous and calcareous sponge–stromatolite–calcareous algae assemblages in shallow platform and lagoonal environments (Crevello and Harris, 1984). Early diagenetic marine cements primarily bound and stabilized these reefs. Many species were conspecific or closely related to coeval Mediterranean Province taxa (e.g., Beauvais and Stump, 1976). Rudistid bivalves were almost wholly limited to shallow platform and lagoonal environments in this stage. (2) A late Early Cretaceous (Barremian–Albian: 124–97.2 Ma) stage was associated with maximum carbonate platform development in the history of the region. This stage was characterized by diverse reef types, including a barrier reef system with varied platform–lagoon frameworks around most of the Gulf of Mexico and on large, tectonically elevated platforms (Scott 1984a,b). Volcanic arc–islands and microcontinents of the Antilles contained fringing, pinnacle, and patch reefs (Kauffman and Sohl, 1974). Reefs and bafflestone frameworks were composed of coral–algal-dominated communities, mixed coral–rudistid communities, and rudistid bivalve-dominated communities. Rudistid-dominated reefs characterized inner platform and platform basin settings (Scott, 1984a), especially in mid-Tethyan, hypersaline and somewhat warmer water masses comprising the Supertethyan ocean–climate zone of Kauffman and Johnson (1988). Mixed coral–rudist communities characterized shelf margin reefs, and coral–algal-dominated frameworks characterized some fore-reef slope settings (Scott 1984a,b). Physical disturbances of the platform facies, identified sedimentologically, are numerous. (3) During a Late Cretaceous stage of reef development (Cenomanian–Maastrichtian: 97.2–67 Ma) virtually all reefs and frameworks within and marginal to the Caribbean Tethys were numerically and ecologically dominated by rudistid bivalves. Corals, bryozoans, stromatoporoids, and skeletal algae, though still locally diverse, comprise a small percentage of the reef biomass, most commonly as epibionts on rudistids, living between rudist reefs–frameworks or as part of pioneer stabilizing communities on which rudist-dominated reefs were built (Kauffman and Sohl, 1974). Binding of rudist reefs by other organisms was rare in stages 2 and 3.

The dramatic change in reef composition, from dominantly coral–algal communities to rudistid bivalve-dominated communities and from clonal to aclonal bioconstructors, coincided with increasing intensity of global climatic warming, a major eustatic rise in sea level, latitudinal expansion and contraction of the tropics and subtropics, episodic development of a superheated, possibly hypersaline ocean climate zone within Tethys, and rapid evolution of adaptive form or forms among various rudistid lineages. Thus, the major Cretaceous changes in reef composition may be the result of both environmental and biotic influences.

3.3.1. The Ecological Evolution of Stage 1: Late Jurassic–Earliest Cretaceous Caribbean Reefs

Immigrants from Mediterranean reef-building communities formed reefs in the Caribbean Province in the Oxfordian (155 Ma). These were small to moderate size (to 10 m in height), paucispecific, isolated frameworks, with inequitable communities of hermatypic corals, digitate and branching stromatolitic green algae, and siliceous sponges on carbonate ramps (Baria *et al.*, 1982; Scott, 1984a,b; Crevello and Harris, 1984). They also were represented by coral–calcisponge–calcareous algae-dominated frameworks to 40 m in height in shallower, higher energy settings (Scott, 1988). Most reefs were dominated by one or two framework-building species, without recorded succession. Beauvais and Stump (1976) reported only seven coral species, four of which were massive to digitate, framework-forming taxa in small Late Oxfordian–Kimmeridgian reefs from Mexico. Most species were conspecific with long-ranging Mediterranean species. A tally of Late Jurassic reefs of the Caribbean Province totaled overall alpha diversity at less than 25 species and coral diversity at less than 10 species; these numbers were much lower than those for coeval Mediterranean reefs (100–135 total species; 35 coral species). Therefore, despite high diversity and ecological complexity in Mediterranean reef source areas, complex reef ecosystems apparently could not be transplanted across the Tethys, a distance estimated at 1000–3000 km. It thus appears that relatively few members of European reef communities initially migrated to the Caribbean. These migrants had to re-evolve ecological relationships with other taxa before even the most basic reeflike structures could be constructed. These restrictions affected the early evolution of Caribbean Cretaceous ecosystems.

The second major development of stage 1 carbonate platforms in the Caribbean region initiated in the Kimmeridgian and terminated in the Valanginian (150–130 Ma). These were disjunct platforms scattered around the Caribbean–Gulf of Mexico basins (Knowles and Chinameca limestones: Scott, 1984a, 1988; Finneran *et al.*, 1984; Crevello and Harris, 1984). Individual reefs were larger and more diverse than those of the Oxfordian, but still were widely scattered. Cregg and Ahr (1984) described a large, wave-resistant Knowles limestone patch reef that was composed of algal boundstones with stromatoporoids and corals as major bioconstructors. This reef shows both vertical and lateral growth and zonation and is the oldest ecologically complex reef described from the Caribbean region. Finneran *et al.* (1984) reported reef succession in ecological time, with algal-coated, vase-shaped corals as a basal community, algal-encrusted laminar corals as a middle community, and large massive head corals at the top of the reef. Thus, a major increase in diversity and ecological complexity marked the Jurassic–Cretaceous boundary interval in the late part of stage 1. These Caribbean frameworks were evolutionarily catching up with those of the Mediterranean. The rare appearance of rudist bivalves within lagoons of these Caribbean reef facies marks an important event in which rudists begin to coexist with coral–algal–stromatoporoid-dominated communities that they would soon replace in primary reef ecospace during stage 2.

3.3.2. The Ecological Evolution of Stage 2: Lower Cretaceous Barremian–Albian Caribbean Reefs

The most extensive episode of carbonate platform and associated reef development in the Caribbean and surrounding region occurred during stage 2 (Barremian–Albian, or 124–97 Ma). Parts of this interval were characterized by: (1) accelerated rates of tectonoeustatic sea-level rise, possibly reflecting increased spreading rates and emplacement of the Pacific superplume (Larson, 1991); (2) widespread flooding of coastal areas, oceanic platforms, and epicontinental seas by warm, Tethyan water masses; (3) increased global warming and initiation of poleward, meridional heat flow from Tethys; (4) initial development of low-oxygen, oceanic water masses (OAEs) (Bralower *et al.*, 1994); and (5) during the Albian, by the possible establishment of a hypersaline, superheated water mass within the core of the Caribbean Tethys. These ocean–climate changes were accompanied by an expansion of tropical seas north of the Cretaceous paleoequator and by a major increase in reef development among both coral–algal- and rudistid-dominated communities.

Two extensive carbonate platforms developed around the margins of the Caribbean (Scott, 1984a,b, 1988): (1) a Barremian–Early Aptian platform represented by the Cupido and Sligo formations (to 600-m thick) rimmed the western and northern Gulf of Mexico. These platforms were characterized by diverse, shallow-water, tropical environments preserved as shelf-margin reef, shoal, lagoonal, and intertidal facies. Reefs are primarily known from shelf and ramp margins and were composed of mixed coral–algal and caprinid rudist frameworks and bioherms numerically dominated by caprinid rudists. (2) An even more extensive and environmentally diverse Albian (especially middle–late Albian) carbonate platform rimmed much of the Gulf of Mexico region and was associated with the first widespread development of relatively smaller, fringing and patch reefs on the Antillean Islands. Shelf-margin reefs around the Gulf of Mexico were composed of mixed coral–algal–rudist paleocommunities, and inner shelf environments and intrashelf basins supported rudist-dominated reefs. Thick sequences of carbonate platform facies suggest long-term stability of environments favorable to reef growth.

Lower Cretaceous, Neocomian through Aptian deposits of southwestern Mexico yielded 25-m thick biostromes from which rudists were collected at several stratigraphic levels (Alencaster and Pantoja-Alor, 1996). Lithologies varied from packstone–wackestone rudist beds, to *in situ*, tightly packed rudists (boundstone), to broken specimens representing storm deposits. Laterally, but from the same formation, biostromes yielded corals, rudists, and gastropods from a packstone–wackestone matrix. Boundstones higher in the section yielded rudists, corals, sponges, and some gastropods.

Detailed mapping of several Barremian–middle Aptian, Cupido limestone frameworks by the authors revealed dominance of loosely packed rudists (caprinids, requieniids) and oysterlike chondrodontid bivalves (Fig. 7). Chondrodont bivalves were overlain by stacked biostromes of recumbent requieniid rudists, which in turn were overlain by suberect caprinid or locally, early radiolitid rudists mixed with requieniids. Laterally, loose caprinid frameworks with small coral or stromatoporoid heads dominated the fore reef, tightly coiled requieniid and chondrodont biostromes dominated the proximal back reef, tightly coiled requieniids, such as *Requienia*, in biostromes dominated the mid-back reef, and loosley coiled, upward-growing, semi-infaunal *Requienia* dominated the distal lagoonal facies.

The regional Aptian extinction event did not greatly affect the rudist composition of reefs (Johnson and Kauffman, 1990). However, this extinction event does mark the acceleration of replacement–displacement of coral–algal communities by rudist-dominated communities in Caribbean reef ecosystems (Fig. 8).

For Albian strata (Fig. 9) Perkins (1974) modeled the distribution of reef-building taxa from 55 localities along the Glen Rose reef line in Texas and noted onshore ecological zonation from elevated bioherms composed of densely clustered, caprinid rudists with requieniid–chondrodontid bivalve communities at the base, to small, shallow-water, monopleurid biostromes and mounds, and to chondrodontid bivalve mats near the strand line, without significant corals.

The late Albian marked an expansion of Mesozoic Caribbean Province reefs associated with accelerated sea-level rise, flooding of shallow platforms and epicontinental basins by warm seas, and an increase in the size of the marine tropics, as defined by the limits of reef development, from the paleoequator to 35°N paleolatitude. A rudist-dominated barrier reef trend occurred along the Gulf of Mexico, as noted earlier for the Valles platform, behind which were diverse platform and lagoonal environments supporting a variety of smaller rudist-dominated reefs. At the northern margin of the late Albian Tethys (southern Arizona and New Mexico), mixed coral–algal reefs and other biogenic frameworks continued to flourish and developed complex ecological zonations (e.g., Scott and Brenckle, 1977; Campbell, 1988). In these buildups, substrates were first stabilized by crustose algae and encrusting

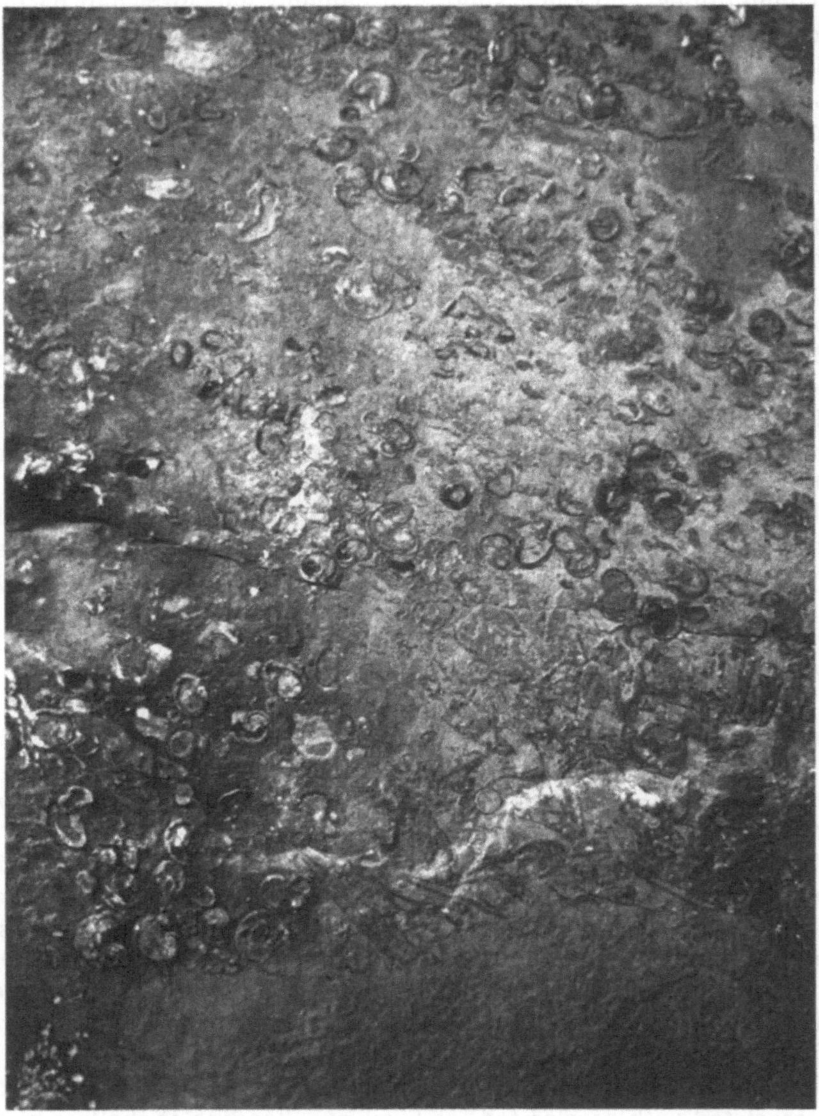

FIGURE 7. Cupido Formation, between Saltillo and Monterey, northeastern Mexico, showing lateral zonation and basic reeflike succession.

foraminifera and were overgrown by platy, microsolenid corals. These were overgrown by more massive corals, algae, and bushlike microsolenids, and capped by caprinid rudist biostromes. From reef-front to lagoon, communities changed from coral–algal–rudist barrier frameworks, to caprinid–radiolitid mounds and biostromes, to requieniid biostromes of several types, with or without chondrodont or oyster biostromes.

FIGURE 8. Minas Viejas, northeastern Mexico, showing moderately spaced Caprinidae. Coin diameter is about 3 cm.

Elsewhere in the Caribbean region, from southern Texas across Mexico to Guatemala, Belize, and Venezuela, late Albian rudists dominated all types of biogenic frameworks across the carbonate platforms. Small coral colonies are sparse in some of these rudist reefs. Thus in stage 2, both rudistid-dominated and coral–algal-dominated reefs existed, but rudistid-dominated reefs were most common toward the geographic center of the Caribbean Tethys, and significant coral–algal reefs or mixed coral–algal–rudist reefs mainly occurred near the northern margin of the region. This general biogeographic zonation may reflect the first stages of environmental changes in the core Tethys.

3.3.3. The Ecological Evolution of Stage 3: Upper Cretaceous Caribbean Reefs

The Upper Cretaceous history of Caribbean reefs reflects changes in the ocean–climate system and the effects of mass extinctions near the Cenomanian–Turonian and Cretaceous–Tertiary boundaries that decimated reef ecosystems worldwide. In each case, the reef ecosystems collapsed prior to the major extinction.

Johnson *et al.* (1996) reviewed the Upper Cretaceous history of the Caribbean region, noting frequent fluctuations in sea level, the effects on Late Cretaceous Caribbean reef ecosystems of alternating periods of (1) tropical warming or superheating associated with high evaporation rates and episodic development of hypersaline belts through the middle of the Caribbean Tethys;

FIGURE 9. Early Albian, Pipe Creek locality, Texas, showing packing density of individuals of the rudist Family Caprinidae. Reefs such as this have yielded petroleum reservoirs in the subsurface of Mexico; i.e., the Golden Lane fields. Hammer for scale to right of center.

and (2) periods of net heat export from Tethys via both surface and subsurface water masses, resulting in tropical cooling, constriction of the northern and southern reef lines, and both global and regional mass extinctions of reef ecosystems in the Cenomanian–Coniacian stages.

Coincident with these environmental changes, the widespread, ecologically complex, rudist-dominated reef ecosystems of the Late Albian collapsed during the Cenomanian; most reefs had disappeared by middle or late Cenomanian time. Nevertheless, those that mark this Cenomanian interval, e.g., the Colima rudist frameworks of western Mexico, persisted. The Colima buildups grew on mixed volcaniclastic–bioclastic sand that was locally stabilized by crustose algae. Large, recumbent, horn-shaped, caprinidlike rudists, such as *Immanites*, dominated the buildups and were mixed with smaller, cylindrical *Mexicaprina* and scattered, small radiolitids. *Immanites* were interpreted as "floating" in the sediment by Gili *et al.* (1995).

The primitive reef-building rudists were almost decimated by the Cenomanian extinction event. Most surviving groups did not form the main components of later Cretaceous reefs, but shifted ecological roles to become subdominant members of radiolitid- and hippuritid-dominated rudist reefs and/or to form small frameworks in protected tropical platform environments and in warm-temperate settings. The post-Cenomanian recovery of Caribbean frameworks was greatly delayed by continued heat export, a major sea-level fall in the middle Turonian, and a regional extinction event in the Coniacian. The tropics expanded again in the Santonian, spanning 5 to 30°N paleolatitude, associated with the second largest rise in Cretaceous sea level (Coniacian through middle Santonian). But reef development was reduced by high sea level, by continued heat export from the tropics, and by the still depauperate nature of the reef fauna as a result of slow recovery after the Cenomanian and Coniacian extinction events. Most reefs in the Coniacian and Santonian were relatively small, localized around topographic highs, and dominated by large, barrel-shaped radiolitid rudists such as *Durania*.

The Campanian and Maastrichtian marked the final and largest radiation of rudists and the development of rudist-dominated reefs of a very different character (Kauffman and Sohl, 1974). Reef-adapted, conical to cylindrical, rapidly growing, vertically oriented, possibly photosymbiotic radiolitid and hippuritid rudists numerically dominated the communities of these Late Cretaceous strata (Kauffman and Sohl, 1974). These were closely packed, mutually supportive (Fig. 10), and allowed development of elevated, resistant, reefal structures that are primarily documented in the Greater Antilles. Kauffman and Sohl (1974) interpreted these reefs in terms of ecological succession, with primitive caprinid rudists as the pioneer communities that acted to stabilize the substrate and provided protected habitats for subsequent communities of recumbent to erect radiolitids and hippuritids. Corals and algae occur in these reefs as small epibiont colonies and/or between rudists in the pioneer communities. In more protected carbonate platform settings, rudists dominated a diversity of more loosely constructed biogenic frameworks, most of which are preserved as bafflestone frameworks today. Cretaceous reefs and most reef-forming taxa underwent a dramatic, widespread extinction within carbonate platform settings in the middle Maastrichtian, 1.5–3 Ma prior to the Cretaceous–Tertiary boundary in the Caribbean region.

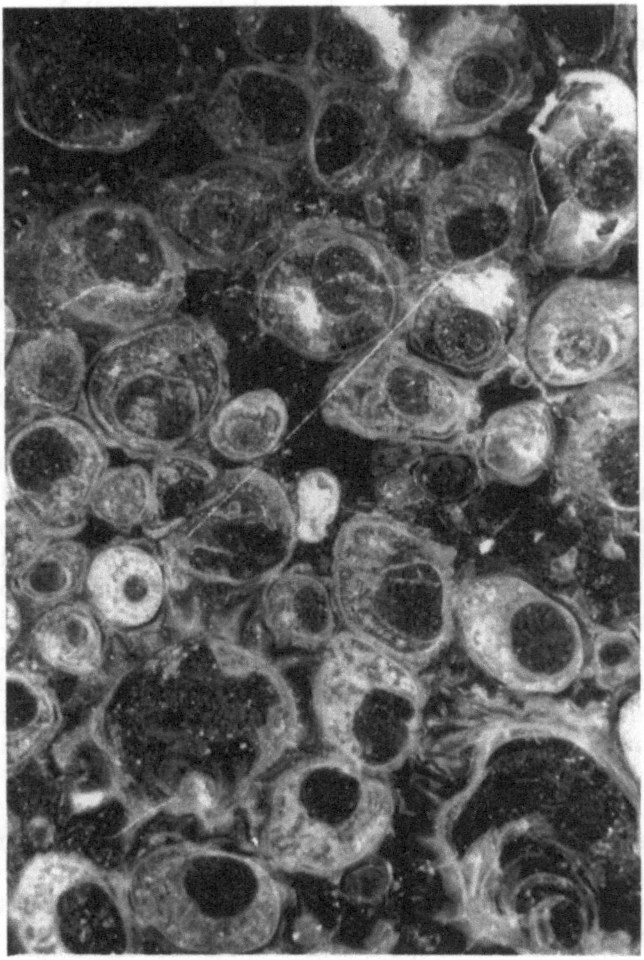

FIGURE 10. Cut and polished slab of cross-sections of densely packed, Upper Cretaceous rudists of the family Radiolitidae. Compare the packing density of individuals in Figs. 9 and 10 with any modern or ancient reef slab.

3.4. Extinctions of Rudists and Reef Ecosystems

The mid-Maastrichtian extinction event experienced by the rudists coincides approximately with extinction among other Cretaceous molluscan groups such as the inoceramid bivalves (MacLeod, 1994) and tropical gastropods. The most diverse and ecologically complex rudist reefs of the Caribbean region, documented in Mexico, Jamaica, and Puerto Rico, were abruptly terminated within carbonate platform facies (Johnson and Kauffman, 1996). They are immediately overlain by thin, low diversity, coral–algal biostromes, and throughout the Caribbean these were overlain by thick biostromes and low mounds of large, thick-shelled marine oysters. Survivors of this reef extinction event were few, mainly more primitive rudistid lineages with broad biogeographic ranges including temperate marine settings. No significant frameworks were formed after this extinction event. The last known rudist taxa of the Caribbean and surrounding region are isolated, small radiolitids and large caprinids, such as *Titanosarcolites*; these last occur within 2–3 m of the Cretaceous–Tertiary boundary in complete sections.

So why did the rudists go extinct? The answer has yet to be known definitively for rudists of the Caribbean region and for the global rudist biota. Johnson *et al.* (1996) attributed the reefal demise to increased rates of second-order sea-level fall associated with decreased pCO_2 (Berner, 1994) and climatic cooling (Barrera, 1994), but not to ocean heat transport, as incorrectly stated in Frank and Arthur (1999). MacLeod and Keller (1996) presented progress reports on the stratigraphic record of extinctions of many biotic groups and the environmental interpretations drawn from these analyses, but no comprehensive review of the extinction event was provided because of a lack of coverage of all fossil groups.

In the Caribbean region proximal to the Chixculub impact crater, data is constantly being reevaluated. The impact-tsunami origin of a clastic deposit in Texas was first published by Bourgeois *et al.* (1988). Elsewhere in the region, however, clastic deposits in Alabama and Georgia were originally interpreted as incised valley deposits from sea-level lowstands (e.g., Donovan *et al.*, 1988, and others). These same deposits were reinterpreted as impact-tsunami events by Hildebrand and Boynton (1990) and Smit *et al.* (1994). Stinnesbeck *et al.* (1996) reevaluated and collectively reinterpreted these regional clastic deposits as regressive deposits.

Newly discovered data are adding to the complexity. The Maastrichtian is being divided into finer scales of temporal resolution, with continued refinement of data and interpretations. For example, the addition of a benthic foraminiferal biostratigraphic zone, the *Plummerita hantkeninoides* zone, was placed above the *A. mayorensis* planktic zone for the last 170–200 ky of the Maastrichtian. Associated interpretations noted that a major regression occurred approximately 300 ky prior to the K/T boundary, followed by a warming and sea-level rise 50–100 ky prior to the K/T boundary (Pardo *et al.*, 1996). Similar interpretations based on stable isotope analyses yielded warming then cooling during the last 100 ky of the Maastrichtian (Li and Keller, 1998). Caribbean biostratigraphic data are not yet correlated into these newly refined data bases from Spain and elsewhere.

4. Conclusions

From a review of the literature and our own extensive experience throughout the Caribbean and circum-Caribbean region we conclude the following:

- The oldest occurrences of rudists are in Lower Cretaceous Berriasian–Valanginian rocks in Venezuela and Texas.

- Rudists were not major components of ecosystems until mid-Cretaceous time. We do not yet fully understand the change in taxonomic composition from coral–algal- to rudist-dominated ecosystems, but we attribute the change to both biologic and environmental (or geologic) processes.
- Rudists numerically and ecologically dominated mid- and Late Cretaceous ecosystems until their extinction.
- Lower and mid-Cretaceous rudists were predominantly associated with carbonate substrates. However, in Upper Cretaceous localities on the Caribbean plate and elsewhere, rudists are associated predominantly with carbonates, but they also are associated with volcaniclastic–siliciclastic sediments and all ranges in between. There is abundant sedimentological evidence to indicate that rudist reefs have been disturbed numerous times during their vertical accretion.
- Rudists are variable in their shell morphology, paleoecology, and roles in ecosystem development. We do not consider that every occurrence of rudists constitutes a reef, but we consider a rudist reef to be a topographically elevated, biogenic structure, including bafflestone mounds, composed predominantly of rudists. The reef ecosystem results from biological, chemical, and physical processes.
- Peaks in rudist species diversity occur between 20 and 30°N paleolatitude in the mid-Cretaceous, whereas peaks occur between 10 and 20°N paleolatitude in the Late Cretaceous. Environmental factors such as plate tectonic movements, ocean–climate interactions, surface-current circulation patterns, changes in sea level and possible changes in ocean temperature, chemistry, and nutrient cycles, as well as sedimentological processes in part can account for these latitudinal shifts in species diversity.
- As analyzed per stage, speciation was lowest in the Barremian, 2 spp/My over a period of 6 My, and highest in the Maastrichtian, 19.40 sp/My over a 5.98-My period.

The information presented in this chapter covers many issues pertinent to Caribbean paleobiology and sedimentology, but it does not cover the entire spectrum of related topics. However, a number of questions and hypotheses may be addressed from knowledge of the Mesozoic evolution and ecological development of Caribbean reef ecosystems, e.g., were the tropics a stable and equable environment over long time periods, as commonly assumed, and were the complex ecosystems that characterize reefs and other tropical communities the result of progressive evolution in a time-stable environment? From viewing hundreds of outcrops in the Caribbean and circum-Caribbean region, we suggest that neither is the case.

Acknowledgments: We acknowledge the USGS workers who shared their unpublished reports with us through the years. Johnson acknowledges funds from National Science Foundation grants EAR 9418081 and 0074603, and

Kauffman acknowledges EAR 93 04659. We thank P. J. Harries and R. Hofling for helpful reviews of the manuscript.

References

Abbott, P. T., 1954, *American Seashells*, Van Nostrand, New York.

Alencaster, G., 1984, Late Jurassic-Cretaceous molluscan paleogeography of the southern half of Mexico, in *Jurassic–Cretaceous Biochronology and Paleogeography of North America* (G. E. G. Westermann, ed.), Geological Association of Canada, Ontario, Canada, pp. 77–88.

Alencaster, G., and Pantoja-Alor, J., 1996, The rudist *Amphitriscoelus* (Bivalvia–Hippuritacea) in the lower Cretaceous of southwestern Mexico, *J. Paleontol.* **70**:399–407.

Arthur, M. A., Schlanger, S. O., and Jenkyns, H. C., 1987, The Cenomanian–Turonian oceanic anoxic event II: Palaeoceanographic controls on organic-matter production and preservation, in: *Marine Petroleum Source Rocks* (J. Brooks, and A. J. Fleet, eds.), Geological Society of London Special Publication No. 26, Geological Society of London, London, United Kingdom, pp. 401–420.

Baria, L. R., Stoudt, D. L., Harris, P. M., and Crevello, P. M., 1982, Upper Jurassic reefs of Smackover Formation, United States Gulf Coast, *Am. Assoc. Petrol. Geol. Bull.* **66**:1449–1482.

Barrera, E., 1994, Global environmental changes preceding the Cretaceous–Tertiary boundary: Early–late Maastrichtian, *Geology* **22**:877–880.

Barron, E. J., Peterson, W. H., Pollard, D., and Thompson, S., 1993, Past climate and the role of ocean heat transport: Model simulations for the Cretaceous, *Paleoceanography* **8**:785–798.

Barron, E. J., Fawcett, P. J., Peterson, W. H., Pollard, D., and Thompson, S. L., 1995, A "simulation" of mid-Cretaceous climate, *Paleoceanography* **10**:953–962.

Beauvias, L., and Beauvais, M., 1974, Studies on the world distribution of the upper Cretaceous Corals, Proceedings of the Second International Coral Reef Symposium, 1, Great Barrier Reef committee, *Brisbane* **2**:639–652.

Beauvais, L., and Stump, T. E., 1976, Corals, molluscs, and paleography of Lower Jurassic strata of the Cerro Poza Serna, Sonora, Mexico, *Palaeogeogr. Palaeoclimatol. Palaeoecol.* **19**:275–301.

Berner, R. A., 1994, Geocarb II: A revised model of atmospheric CO_2 over Phanerozoic time, *Am. J. Sci.* **294**:56–91.

Bourgeois, I., Hansen, T. A., Wiberg, P. L., and Kauffman, E. G., 1988, A tsunami deposit at the Cretaceous–Tertiary boundary in Texas, *Science* **241**:567–570.

Bralower, T. J., Arthur, M. A., Leckie, R. M., Sliter, W. V., Allard, D. J., and Schlanger, S. O., 1994, Timing and paleoceanography of oceanic dysoxia/anoxia in the late Barremian to early Aptian (Early Cretaceous), *Palaios* **9**:335–369.

Brass, G., Saltzman, E., Sloan, J., Southam, J., Hay, W., Holzer, W., and Peterson, W., 1982, Ocean circulation, plate tectonics and climate, in *Climate in Earth History, Studies in Geophysics*, National Academy Press, Washington, DC, pp. 83–89.

Briggs, R. P., 1969, Changes in stratigraphic nomenclature in the Cretaceous system East–Central Puerto Rico, Vol. B1274-0, U.S. Geol. Survey Bull., U.S. Geological Survey, Reston, VA, pp. 1–31.

Burke, K., Coates, A. G., and Robinson, E., 1968, Geology of the Benbow Inlier and surrounding areas, Jamaica, in: *Transactions of the Fourth Caribbean Geological Conference*, Port-of-Spain, 1965, Queens College Press, Flushing, NY, pp. 299–307.

Campbell, G. T., 1988, Depositional environments, biofacies and paleoecology of the Howell's Ridge Member of the Aptian–Albian u-bar formation, Little Hatchet Mountains, New Mexico, unpublished master of science thesis, University of Colorado, Boulder.

Case, J. E., MacDonald, W. D., and Fox, P. J., 1990, Caribbean crustal provinces: Seismic and gravity evidence, in: *The Caribbean Region* (G. Dengo and J. E. Case, eds.), The Geological Society of America, Boulder, CO, pp. 15–36.

Chubb, L. J., 1967, New rudist species from the Cretaceous rocks of Jamaica, *J. Geol. Soc. Jamaica* **IX:**24–31.

Chubb, L. J., 1971, Rudists of Jamaica, *Palaeontogr. Am.* **VII:**161–257.

Coates, A. G., 1977, Jamaican Cretaceous coral assemblages and their relationship to rudist frameworks, *Mem. Bureau Recherches Geol. Minieres* **89:**336–341.

Collins, L., 1988, The faunal structure of a mid-Cretaceous rudist reef core, *Lethaia* **21:**271–280.

Cregg, A. K., and Ahr, W. M., 1984, Paleoenvironment of an upper Cotton Valley (Knowles Limestone) patch reef, Milan County, Texas, in: *The Jurassic of the Gulf Rim* (W. P. S. Ventress, D. G. A. Bebout, B. F. Perkins, and C. H. Moore, eds.), Proceedings of the Third Annual Research Conference, Gulf Coast Section, Society of Economic Paleontologists and Mineralogists Foundation, Tulsa, OK, pp. 41–56.

Crevello, P. D., and Harris, P. M., 1984, Depositional models for Jurassic reefal buildups, in *The Jurassic of the Gulf Rim* (W. P. S. Ventress, D. G. A. Bebout, B. F. Perkins, and C. H. Moore, eds.), Proceedings of the Third Annual Research Conference, Gulf Coast Section, Society of Economic Paleontologists and Mineralogists Foundation, Tulsa, OK, pp. 57–102.

Dechaseaux, C., 1969, Origin and extinction, in *Treatise on Invertebrate Paleontology, Part N, Mollusca 6, Bivalvia* (R. C. Moore, ed.), Geological Society of America and University of Kansas Press, Lawrence, KS, p. N765.

Dengo, G., and Case, J. E. (eds.), 1990, *The Caribbean Region*, The Geological Society of America, Boulder, CO.

Donovan, A. D., Baum, G. R., Blechschmidt, G. L., Loutit, T. S., Pflum, C. E., and Vail, P. R., 1988, Sequence stratigraphic setting of the Cretaceous/Tertiary boundary in central Alabama, in: *Sea-level Changes: An Integrated Approach* (C. K. Wilgus, B. S. Hastings, C. G. Kendall, H. W. Posamentier, H. W. Ros, and J. C. VanWagoner, eds.), Special Publication Society of Economic Paleontologists and Mineralogists Special Publication No. 42, SEPM, Tulsa, OK, pp. 300–307.

Enos, P., 1974, Reefs, platforms, and basins of Middle Cretaceous in Northeast Mexico, *Am. Assoc. Petrol. Geol. Bull.* **58:**800–809.

Finneran, J. M., Scott, R. W., Taylor, G. A., and Anderson, G. H., 1984, Lowermost Cretaceous ramp reefs: "Knowles Limestone," SW flank of the east Texas basin, in: *The Jurassic of the Gulf Rim* (W. P. S. Ventress, D. G. Bebout, and C. H. Moore, eds.), Proceedings of the Third Annual Research Conference, Gulf Coast Section, Society of Economic Paleontologists and Mineralogists Foundation, Tulsa, OK, pp. 125–134.

Frakes, L. A., 1999, Estimating the global thermal state from Cretaceous sea surface on continental temperature data, in: *Evolution of the Cretaceous Ocean–Climate System* (E. Barrera and C. C. Johnson, eds.), Geological Society of America Special Paper No. 332, The Geological Society of America, Inc., Boulder, CO, pp. 49–58.

Frank, T. D., and Arthur, M. A., 1999, Tectonic forcings of Maastrichtian ocean–climate evolution, *Paleoceanography* **14:**103–117.

Gili, E., Masse, J. P., and Skelton, P. W., 1995, Rudists as gregarious sediment-dwellers, not reef-builders, on Cretaceous carbonate platforms, *Palaeogeogr. Palaeoclimatol. Palaeoecol.* **118:**245–267.

Gonzales de Juana, C., Iturralde de Arozena, J. M., and Picard Cadillat, X., 1980, *Geologia de Venezuela y de sus Cuencas Petroliferas: Ediciones Foninves*, Caracas, Venezuela.

Haq, B. V., Hardenbol, J., and Vail, P. R., 1987, Chronology of fluctuating sea levels since the Triassic, *Science* **235:**1156–1167.

Hay, W. W., 1988, Paleoceanography: A review for the GSA Centennial, *Geol. Soc. Am. Bull.* **100:**1934–1956.

Hay, W. W., 1995, Cretaceous paleocenography, *Geol Carpathica* **45:**257–266.

Hildebrand, A. R., and Boynton, W. V., 1990, Proximal Cretaceous/Tertiary boundary impact deposits in the Caribbean, *Science* **99:**263–275.

Hofling, R., 1997, *Eine erweiterte Riff-Typologie und ihre Anwendung auf kretazische Biokonstruktionen*, Bayerische Akademie der Wissenschaften, Mathematisch-Naturwissenschaftliche Klasse, Munich.

Hubbard, D. K., Burke, R. B., and Gill, I. P., 1998, Where's the reef: The role of framework in the Holocene, *Carbonates Evaporites* **13:**3–9.

Huber, B. T., Hodell, D. A., and Hamilton, C. P., 1995, Midde–Late Cretaceous climate of the southern high latitudes: Stable isotopic evidence for minimal equator-to-pole thermal gradients, *Geol. Soc. Am. Bull.* **7**:1164–1191.

Hussner, H., 1994, Reefs, an elementary principle with many complex realizations, *Beringeria Institut für Palaontolgieder Universität Wurzburg, Germany* **11**:1–99.

Imlay, R. W., 1939, Possible interoceanic connections across Mexico during the Jurassic and Cretaceous periods, in *Proceedings of the Sixth Pacific Science Congress*, Pacific Science Congress, USA, pp. 423–427.

Immenhauser, A., and Scott, R. W., 1999, Global correlation of middle Cretaceous sea-level events, *Geology* **27**:551–554.

Jenkyns, H. C., 1980, Cretaceous anoxic events: From continents to oceans, *Geol. Soc. London J.* **137**:171–188.

Johnson, C. C., 1999, Evolution of Cretaceous surface current circulation patterns, Caribbean and Gulf of Mexico, in: *Evolution of the Cretaceous Ocean-Climate System* (E. Barrera and C. C. Johnson, eds.), Geological Society of America Special Paper 332, Boulder, CO, p. 329–344.

Johnson, C. C., 1993, Cretaceous biogeography of the Caribbean region, Ph.D. thesis, University of Colorado, Boulder.

Johnson, C. C., and Kauffman, E. G., 1990, Originations, radiations and extinctions of Cretaceous rudistid bivalve species in the Caribbean Province, in: *Extinction Events in Earth History* (E. G. Kauffman and O. H. Walliser, eds.), Springer-Verlag, New York, pp. 305–324.

Johnson, C. C., and Kauffman, E. G., 1996, Maatrichtian extinction patterns of Caribbean province rudistids, in *Cretaceous-Tertiary Mass Extinctions: Biotic and Environmental Changes* (N. MacLeod and G. Keller, eds.), W.W. Norton, New York, pp. 321–373.

Johnson, C. C., Collins, L. S., and Kauffman, E. G., 1988, Rudistid biofacies across the El Abra Formation (Late Albian (?), Early–Middle Cenomanian), northeastern Mexico, in: *Transactions of the 11th Caribbean Geological Conference, Barbados*, Energy Division of the Ministry of Finance, Government of Barbados, Bridgetown, Barbados, pp. 1:1–1:12.

Johnson, C. C., Barron, E. J., Kauffman, E. G., Arthur, M. A., Fawcett, P. J., and Yasuda, M. K., 1996, Middle Cretaceous reef collapse linked to ocean heat transport, *Geology* **24**:376–380.

Kauffman, E. G., 1975, Dispersal and biostratigraphic potential of Cretaceous benthonic Bivalvia in the Western Interior, in: *The Cretaceous System in the Western Interior of North America* (W. G. E. Caldwell, ed.), The Geological Association of Canada Special Paper No. 13, Ottawa, Canada, Geological Association of Canada, St. John's, Newfoundland, Canada, pp. 163–194.

Kauffman, E. G., and Johnson, C. C., 1988, The morphological and ecological evolution of middle and upper Cretaceous reef-building rudistids, *Palaios* **3**:194–216.

Kauffman, E. G., and Sohl, N. F., 1974, Structure and evolution of Antillean Cretaceous rudist frameworks, *Verhandlungen Naturforschende Gessellschaft Basel* **84**:399–467.

Kauffman, E. G., Sageman, B. B., Kirkland, J. J., Elder, W. P., Harries, P. J., and Villamil, T., 1993, Molluscan biostratigraphy of the Cretaceous Western Interior Basin North America, in *Evolution of the Western Interior Basin* (W. G. E. Caldwell and E. G. Kauffman, eds.), Geological Association of Canada, Special Paper 39, Geological Association of Canada, St. John's, Newfoundland, Canada, pp. 397–434.

Koch, C. F., and Morgan, J. P., 1988, On the expected distribution of species' ranges, *Paleobiology* **142**:126–138.

Larson, R. L., 1991, Latest pulse of the Earth: Evidence for a mid-Cretaceous superplume, *Geology* **19**:547–550.

Li, L., and Keller, G., 1998, Abrupt deep-sea warming at the end of the Cretaceous, *Geology* **26**:995–998.

Longman, M. W., 1981, A process approach to recognizing facies of reef complexes, in: *European Fossil Reef Models* (D. F. Toomey, ed.), Society of Economic Paleontologists and Mineralogists Special Publication 30, SEPM, Tulsa, OK, pp. 9–40.

MacGillavry, H. J., 1970, Geological history of the Caribbean, *Koninkl Nederl Akad. Van Wetenschappen Proc. Series B* **73**:64–96.

MacLeod, K. G., 1994, Bioturbation, inoceramid extinction, and mid-Maastrichtian ecological change, *Geology* **22**:139–142.

MacLeod, N., and Keller, G., 1996, *Cretaceous–Tertiary Mass Extinctions: Biotic and Environmental Changes*, W. W. Norton, New York.

Mann, P., Schubert, C., and Burke, K., 1990, Plate 11: Neotectonic map of the Caribbean region, in: *The Caribbean Region* (G. Dengo and J. E. Case, eds.), The Geological Society of America, Boulder, CO.

Masse, J. P., and Rossi, T., 1987, Le Provincialisme Sud-caraibe a l'Aptien Inferieur. Sa signification dans le cadre de l'evolution geodynamique du domaine caraibe et de l'Stlantique Central, *Cretaceous Res.* **8**:349–363.

Morris, A. E. L., Taner, I., Meyerhoff, H. A., and Meyerhoff, A. A., 1990, Tectonic evolution of the Caribbean region: alternative hypothesis, in: *The Caribbean Region* (G. Dengo and J. E. Case, eds.), The Geological Society of America, Boulder, CO, pp. 433–457.

Nelson, H. F., 1959, Deposition and alteration of the Edwards Limestone, central Texas, in: *Symposium on Edwards Limestone in Central Texas*, Texas University Publication 5905, University of Texas at Austin, Bureau of Economic Geology, Austin, TX, pp. 21–95.

Norris, R. D., and Wilson, P. A., 1998, Low-latitude sea-surface temperatures for the mid-Cretaceous and the evolution of planktic foraminifera, *Geology* **26**:823–826.

Pardo, A., Ortiz, N., and Keller, G., 1996, Latest Maastrichtian and Cretaceous-Tertiary boundary foraminiferal turnover and environmental changes at Agost, Spain, in: *Cretaceous–Tertiary Mass Extinctions: Biotic and Environmental Changes* (N. MacLeod and G. Keller, eds.), W. W. Norton and Company, New York, pp. 139–171.

Perkins, B. F., 1974, Paleoecology of a rudist reef complex in the Comanche Cretaceous Glen Rose Limestone of Central Texas, *Geosci. Man* **8**:131–173.

Philip, J., 1980, Cretace superieur de Provence, *Geobios Mem. Spec.* **4**:99–109.

Pindell, J. L., and Barrett, S. F., 1990, Geological evolution of the Caribbean region: A plate-tectonic perspective, in: *The Caribbean Region* (G. Dengo, and J. E. Case, eds.), The Geological Society of America, Boulder, CO, pp. 405–432.

Pindell, J. L., Cande, S. C., Pitman III, W. C., Rowley, D. B., Dewey, J. F., LaBreque, J. L., and Haxby, W., 1988, A plate-kinematic framework for models of Caribbean evolution, *Tectonophysics* **155**:121–138.

Robinson, E., and Jackson, T. A., 1976, *Field Guide, The Benbow Inlier*, Geological Society of Jamaica, Kingston, Jamaica, pp. 1–4.

Robinson, E., and Lewis, J. F., 1987, *Field Guide to Aspects of the Geology of Jamaica*, Department of Geology, University of the West Indies, Mona, Jamaica.

Ross, M. I., and Scotese, C. R., 1988, A hierarchical tectonic model of the Gulf of Mexico and Caribbean region, *Tectonophysics* **155**:139–168.

Sclater, J. G., Hellinger, S., and Tapscott, C., 1977, The paleobathymetry of the Atlantic Ocean from the Jurassic to the present, *J. Geol.* **85**:1108–1127.

Scott, R. W., 1984a, Mesozoic biota and depositional systems of the Gulf of Mexico-Caribbean region, *Geol. Soc. Can. Spec. Paper* **27**:49–64.

Scott, R. W., 1984b, Evolution of Early Cretaceous reefs in the Gulf of Mexico, *Palaeontologr. Am.* **54**:406–412.

Scott, R. W., 1988, Evolution of Late Jurassic and Early Cretaceous reef biotas, *Palaios* **3**:184–193.

Scott, R. W., 1995, Global environmental controls on Cretaceous reefal ecosystems, *Palaeogr. Palaeoclimatol. Palaeoecol.* **119**:187–199.

Scott, R. W., and Brenckle, P. L., 1977, Biotic zonation of a Lower Cretaceous coral–algal–rudist reef, Arizona, in *Proceedings of the Third International Coral Reef Symposium* (R. N. Ginsburg, ed.), Rosensteil School of Marine and Atmospheric Science, University of Miami, Miami, FL, **2**:183–190.

Scott, R. W., Franks, P. C., Stein, J. A., Bergen, J. A., and Evetts, M. J., 1994, Graphic correlation tests the synchronous Mid-Cretaceous depositional cycles: Western Interior and Gulf Coast, in *Unconformity-Related Hydrocarbons in Sedimentary Sequences* (J. C. Dolson, M. L. Hendricks, and W. A. Westcott, eds.), Rocky Mountain Association of Geologists, Denver, pp. 89–98.

Semmes, D. R., 1919, *Geology of the San Juan District, Porto Rico. Scientific Survey of Porto Rico and the Virgin Islands*, New York Academy of Sciences, New York.

Simo, J. A. T., Scott, R. W., and Masse, J. P., 1993, Cretaceous Carbonate Platforms, *American Association of Petroleum Geologists Memoir 56*, AAPG Press, Tulsa, pp. 1–479.

Skelton, P. W., 1982, Aptian and Barremian rudist bivalves of the New World: Some Old World similarities, *Cretaceous Res.* **3**:145–153.

Smit, J., Roep, T. B., Alvarez, W., Claeys, P., and Montanari, A., 1994, Is there evidence for Cretaceous–Tertiary boundary-age deep-water deposits in the Caribbean and Gulf of Mexico? *Comment: Geol.* **22**:953–954.

Smith, A. G., Briden, J. C., and Drewry, G. E., 1973, Phanerozoic world maps, in: *Organisms and Continents Through Time* (N. F. Hughes, ed.), The Palaeontological Association, London, England, pp. 1–42.

Stanley, S. M., and Hardie, L. A., 1998, Secular oscillations in the carbonate mineralogy of reef-building and sediment producing organisms driven by tectonically forced shifts in seawater chemistry, *Palaeogeog. Palaeoclimatol. Palaeoecol.* **144**:3–19.

Stehli, F., and Wells, J. W., 1971, Diversity and age patterns in hermatypic corals, *Syst. Zool.* **20**:115–126.

Stehli, F. G., McAlester, A. L., and Helsley, C. E., 1967, Taxonomic diversity of Recent bivalves and some implications for Geology, *Geol. Soc. Am. Bull.* **78**:455–466.

Stinnesbeck, W., Keller, G., Adatte, T., Lopez-Oliva, J. G., and MacLeod, N., 1996, Cretaceous–Tertiary boundary clastic deposits in northeastern Mexico: Impact tsunami or sea-level lowstand? in: *Cretaceous–Tertiary Mass Extinctions: Biotic and Environmental Changes* (N. MacLeod and G. Keller, eds.), W. W. Norton, New York, pp. 471–517.

Vahrenkamp, V. C., Franssen, R. S. W. M., Grotsch, J., and Munoz, P. J., 1993, Maracaibo platform (Aptian–Albian), Northwestern Venezuela, in: *Cretaceous Carbonate Platforms, AAPG Memoir 56* (J. A. T. Simo, R. W. Scott, and J. P. Masse, eds.), AAPG Press, Tulsa, pp. 25–33.

Wilson, J. L., 1975, *Carbonate Facies in Geologic History*, Springer-Verlag, New York.

Wilson, J. L., and Ward, W. C., 1993, Early Cretaceous carbonate platforms of Northeastern and East-Central Mexico, in: *Cretaceous Carbonate Platforms, American Association of Petroleum Geologists Memoir 56* (J. A. T. Simo, R. W. Scott, and J. P. Masse, eds.), AAPG Press, Tulsa, pp. 35–50.

Woo, K. S., Anderson, T. F., Railsback, L. B., and Sandberg, P. A., 1992, Oxygen isotope evidence for high-salinity surface sea water in the mid-Cretaceous Gulf of Mexico: Implications for warm, saline deepwater formation, *Paleoceanography* **7**:673–685.

Woodring, W. P., 1928, Tectonic features of the Caribbean region, in: *Proceedings, 3rd Pan-Pacific Science Congress, Tokyo, 1926*, Vol. 1, Tokyo, Japan, pp. 401–431.

Wright, R. M., 1974, Field guide to selected Jamaican geological localities, in: *Mines and Geology Division*, Special Publication No. 1, Jamaican Geological Survey, Kingston, Jamaica, pp. 1–45.

Young, K., 1983, Mexico, in *The Phanerozoic Geology of the World II: The Mesozoic, B* (M. Moullade and A. E. M. Nairn, eds.), Elsevier, New York, pp. 61–88.

Zans, V. A., Chubb, L. J., Versey, H. R., Williams, J. B., Robinson, E., and Cooke, D. L., 1962, *Synopsis of the Geology of Jamaica*, Vol. 4, Geological Survey Department, Jamaica.

Chapter 10

The Role of Framework in Modern Reefs and Its Application to Ancient Systems

DENNIS K. HUBBARD, IVAN P. GILL and RANDOLPH B. BURKE

1. Introduction

Uniformitarianism—the principle that "the present is the key to the past"—forms the cornerstone of geologists' efforts to unravel the prehistory of our world. When comparing modern and fossil reefs, a fundamental problem

DENNIS K. HUBBARD • Department of Geology, Oberlin College, Oberlin, Ohio 44074. IVAN P. GILL • Department of Geology and Geophysics, University of New Orleans, New Orleans, LA. 70148 RANDOLPH B. BURKE • North Dakota Geological Survey, Bismarck, North Dakota 58505-0840.

The History and Sedimentology of Ancient Reef Systems, edited by George D. Stanley Jr., Kluwer Academic/Plenum Publishers, New York, 2001.

arises because the organisms that largely drive this important biological system have evolved over geological time. This has fueled a heated debate over the appropriateness of using modern reefs dominated by scleractinian corals as models for ancient ones built by stromatoporoids, sponges and a host of seemingly dissimilar organisms.

In siliciclastic systems, reconstructing the past is simplified by the dominance of processes that have remained relatively constant over the span of geologic time and are directly observable today. Moreover, the link between a particular process and the resulting sedimentary structures can be documented in modern systems and directly applied to ancient deposits, both in scale and timing of deposition. While a larger sedimentary sequence viewed in outcrop might represent decades, centuries, or eons, it can be broken down into individual beds representing specific events that occurred over time scales that can be studied in detail today.

By contrast, fossil reefs are a result of a myriad of processes that often are contemporaneous and occur over time scales that do not lend themselves to direct measurement. Moreover, many of these processes occur inside the reef and therefore, are difficult for the casual observer to study. Instead of seeing an orderly stack of short-term depositional cycles, the geologist studying fossil reefs often is faced with a single unit that has been formed by an amalgam of processes operating in concert over a very long period of time. This integration over a longer time scale makes direct observation and measurement of critical reef processes much more difficult.

On a larger scale, it is easier to make morphological comparisons between modern reefs (Fig. 1A) and their ancient counterparts (Fig. 1B). When swimming over a modern reef, however, one's eyes are drawn to the abundant and diverse organisms that form a veneer over the underlying carbonate surface (Fig. 2A). If this panorama is extrapolated to the interior of the reef, it seems logical to assume that modern reef fabrics are dominated by corals that are largely in life position. In marked contrast, many ancient reefs are not characterized by in-place framework. For example, stromatoporoid-rich deposits in the Devonian of Alberta, Canada (e.g., Mountjoy and Geldsetzer, 1981), are often a loose assemblage of calcifying organisms "floating" in a detrital matrix (Fig. 2B). While large-scale geomorphic differentiation of the "Devonian Great Barrier Reef" in the Canning Basin of Australia is undeniable, the reef itself contains as much (or perhaps more) encrusted and/or cemented substrate as it does primary calcifiers (Playford, 1980).

These visual differences between modern and many ancient reefs have led to discussions over whether modern reef models are critically undermined by the evolution of reef organisms through time. In this argument, the building blocks of the reef and the processes that have assembled them have changed so much over geologic time that modern reefs are fundamentally different from those generated eons ago. As such, modern reefs provide poor models for ancient ones, or, as Heinz Lowenstam once joked, "The present is the key to the Pleistocene—perhaps" (A. Conrad Neumann, personal communication).

Others (including the authors of this chapter) see changes in reefs through

FIGURE 1. (A) Aerial view of the atoll rim on Palau. Photo by John Ogden. (B) Aerial view of a Devonian reef margin, Canning Basin, Australia at a similar scale.

time in a different perspective. R. N. Ginsburg eloquently has compared reefs to a long-running play in which the actors change but the plot and the interplay among the main characters remain constant. The aim of this chapter is to show that the processes responsible for reef building have remained consistent over geologic time and provide a unifying perspective by which reefs can be compared. Important changes have occurred in the organisms that produce carbonate and those that reduce it to rubble. However, the resulting structure—the reef—has always reflected a dynamic balance between those

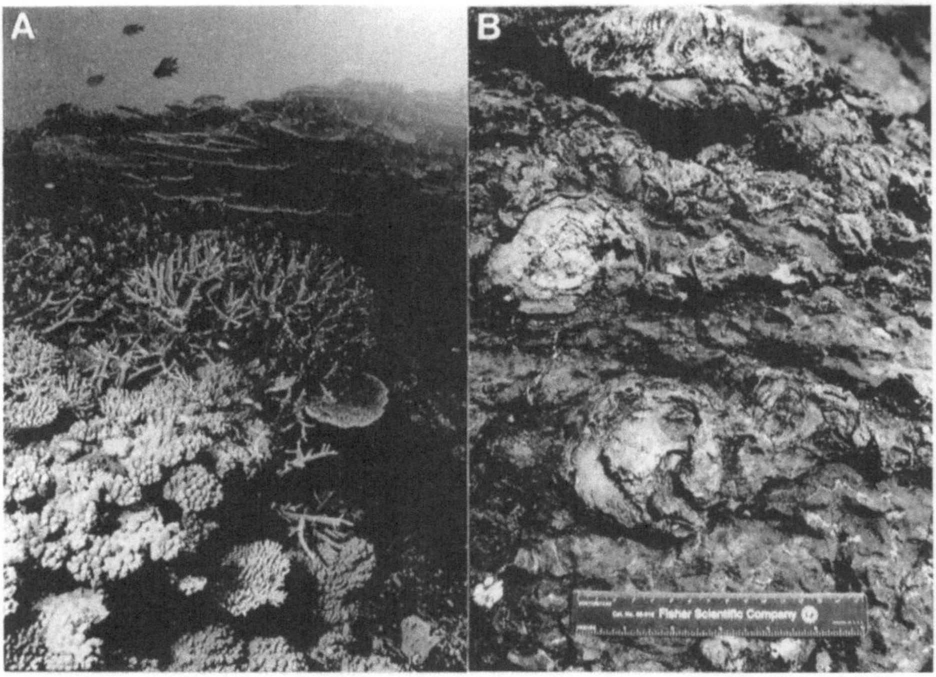

FIGURE 2. (A) Underwater photograph of the fore reef at Davies Reef, Great Barrier Reef, Australia. Water depth at the reef crest is approximately 5 m. (B) Stromatoporoids from the Devonian Grassi Lakes reef complex of Alberta, Canada. While locally abundant, these important calcifiers are often found "floating" in a finer matrix.

competing controls. Understanding reef development involves quantifying the roles that have been played by each group of organisms and trying to understand how they have responded to changes in environmental stresses (and to one another). Simply cataloging the individuals that are present or absent at a particular time in geologic history will not meet these needs. If we can transcend simple visual comparisons by focusing on how those images came about, then a clearer understanding should emerge of the link between the present and the past.

1.1. The Perspective of the Authors

No matter how objective we might try to be, no conclusion—and this includes those that are supported by rigorous statistical analysis—is made without bias. Therefore, it seems appropriate to spend a small amount of time considering the perspective from which this chapter has evolved. All three authors have studied modern coral reefs with a principal aim of understanding their internal character and the processes that are responsible. Our approach has been to examine a variety of reefs exposed to widely varying physical–oceanographic and biological processes. These studies have focused not only on the active and exposed reef crest but on the entire platform—from the beach to the shelf break. What has evolved is an appreciation for the variability in depositional styles that can be seen from site to site and across the shelf at a single locale.

Most of the information presented below is based on coring studies, using a rotary drilling system. As each new system was designed and built (we used two), the intent was to progressively reduce the shortcomings of their predecessors. It therefore is important for the reader to understand the limitations and biases of these sampling systems. First, each sample represents only a small portion of the reef (i.e., a cylinder 5–10 cm in diameter). The relationship between a drilled coral and its nearest neighbor is difficult to determine unless the core happens to penetrate the point of contact. Equally problematic, the coring process has the potential to bias what is sampled. Water circulating through the core bit can wash away sediment, rubble, and other important clues to what has accumulated between larger and more easily retained sections of the reef. If thin enough, platy corals can be ground up by the rotating bit, thereby skewing the coral census in deeper reefs where these growth forms are volumetrically important.

Our coring work has attempted to minimize these problems in two ways. First, a large number of cores were recovered within each environment all across the shelf. In many instances multiple cores were taken in close proximity to one another to examine the small-scale variability that might exist at a single site. Second, the Scarid system developed in 1988 (Fig. 3) was designed specifically to reduce core loss to the maximum extent possible in a rotary drilling system. The heavier weight of the drill (ca. 400 pounds) guarantees constant pressure on the bottom of the drill hole. This, plus the lower flow rate of water through the core barrel, greatly reduces the loss of detrital material from the core. In contrast to samples from drilling systems that are suspended from a tripod, cores taken with the Scarid drill often contain rubble intervals that are volumetrically more important than solid reef rock. Equally important, tactile feedback is provided through the drive chain and the handle used by the operator to raise and lower the drill. It is easy to discern between solid rock, sand, rubble, and open cavities encountered by the advancing drill bit. By carefully locating the transition from one substrate type to another, logging the relative abundance of solid reef rock, sediment, rubble, and void space is a simple matter.

Our application of this information is based on one simple premise. However elegant a model based on fossil examples might be, the actual structures that existed millions of years ago cannot be directly observed. Whether or not we would have called it a true reef if we had been able to dive on it 400 million years ago, we can only speculate. We do not intend to argue with specific models derived in this manner, as this would only add our opinions to those of the original researchers. What we do argue, however, is that modern reefs are the only directly observable features that we have

FIGURE 3. Underwater photographs of the Scarid drilling system used to recover cores from Buck Island, Lang Bank, and Puerto Rico. (A) The aluminum and stainless steel drill frame is supported by three adjustable legs. The drill motor (near the diver) travels along two stainless steel guides and is controlled by a hand crank on the opposite side of the drilling frame (arrow). Power and water are supplied through hoses from the surface. (B) Divers releasing the core barrel from the drill motor. Core is recovered in 5-feet sections. An extra inner core barrel sits on the bottom between the two divers. In addition to greatly increasing the control of the drill string, the system also provides tactile feedback to the diver/operator. As a result, changes between solid coral, rubble, sand, and void space can be easily detected and accurately logged.

available to study. They therefore represent the only logical place to begin.

All the cores described in this chapter come from features that many researchers from diverse backgrounds have called true reefs. Hopefully we provide a convincing argument that the structural integrity of these features is not based solely on the dominance of corals that are in growth position. If we can use these observations to agree that some modern reefs do not rely on in-place corals for their rigidity, then this criterion no longer can be considered as the yardstick against which their ancient counterparts should be measured. This clearly is not intended to imply that modern and ancient reefs dominated by in-place corals do not exist. The only point we make is that models of reefs based on the presence of a rigid "framework" in growth position must be expanded to include a host of other depositional fabrics and processes that may play a role as important as the initial biological generation of calcium carbonate.

1.2. Clarification of Terms Used in This Chapter

Many terms associated with reefs have evolved to the point that the original meaning is no longer retained. The result is often general confusion and arguments over a concept based more on semantics than substance. Before starting we, therefore, review two concepts that recur throughout this chapter. Two terms "*in-situ*" versus "in-place" need clarification. The former refers to any organism that was ultimately deposited in the general area where it lived (i.e., it remained within the same environment or reef zone). The latter is reserved for those organisms that are preserved in growth position and that retain their original point of attachment to the underlying substrate.

The second term that lies at the heart of the discussion is "framework." At one end of the spectrum, workers have reserved this term to describe only those internal fabrics in which (1) corals are largely in place and (2) the reef owes its rigidity to successive overgrowth of these and similar organisms in a largely interlocking network (e.g., Fagerstrom, 1987). This is referred to as in-place or primary framework. At the other end of the spectrum, framework has been used synonymously with the term "fabric" and includes anything found in a particular depositional sequence. In this chapter "framework" is intermediate between these two extremes. It encompasses both in-place skeletal elements (primary framework) and toppled or rolled corals, along with encrusting organisms and syndepositional cements that bind them together and add structural integrity to the reef (collectively termed secondary framework). "Framework" is placed in quotes in our historical discussion to avoid arguments over what a particular individual may have meant by the term.

1.3. Previous Reef Classification Schemes

Any discussion of the origin and evolution of reefs must start with an agreement over what we mean by the term. While this might seem simple and obvious, strongly divergent opinions remain among reef researchers as to what is a reef and what is not. In part, this arises from differences in perspective. To mariners, a reef is anything on the sea floor that might damage the keel of a vessel. As a result, coastal charts off the northeastern United States warn of "reefs" that owe their origin not to biological activity but to the most recent episode of continental glaciation (i.e., they are moraines and drumlins). Biologists tend to focus on the rich associations of calcifying organisms and high diversity/abundance associated with their "reefs," paying less attention to questions of relief or physical structure. In marked contrast, geologists place a primary emphasis on the three-dimensional feature that results. This chapter focuses on the last set of "reefs."

While it remains difficult to reconcile differences among biological, oceanographic, and geologic definitions of reefs, there are several attributes common to the spectrum of opinions expressed in the geological literature.

Inherent in most definitions is the idea that reefs are:

- Largely composed of and built by organisms that produced the calcium carbonate.
- Wave resistant to a significant degree.
- Elevated above their surroundings enough to exert at least localized control over oceanographic processes.

In his 1950 compendium on the Niagaran reefs of the Great Lakes region, Lowenstam (1950) introduced the term "framework" to describe the rigid, internal fabric of these topographically elevated structures. His criteria were largely limited to the three that are listed above and he made no apparent attempt to limit how these characteristics might be achieved. To the contrary, he advocated a process-based approach through which modern and ancient reefs might be compared. In Lowenstam's (1950) own words,

> If we formulate the principles of reef-building in terms of potential, we arrive at an ecological definition of reefs which should apply equally to reef communities of the past and those of the present, even though the reef-builders of different times were totally unrelated (p. 433).

In fact, the crux of his discussion of whether Niagaran structures were "true reefs" centered on the assumption that because they shed reef-derived detritus, they must have built to sea level and been rigid enough to withstand the vigorous wave action that occurred there.

His ideas on framework were echoed by Newell *et al.* (1953) in their classic work on the Capitan reef complex of west Texas and New Mexico. While we might argue how Newell and his colleagues intended for the term "framework" to be applied at the time, his later admonition should be kept at the forefront: ". . . a persistent misconception remains that organic reefs are mainly composed of a wave resistant framework of rigidly cemented *in-situ* skeletons of corals and algae" (Newell, 1971).

While it is difficult to accurately trace the transition from Lowenstam's initial discussions, it has been argued by some that framework is confined to only those fabrics created by interlocking organisms in growth position. As such, the rigidity of the reef is the result of this interlocking nature and anything that falls short is not a reef. In an excellent and comprehensive review of reefs through time, Fagerstrom (1987) describes true reefs—both modern and ancient—as those structures composed to a major extent of "large, colonial or gregarious, intergrown skeletal organisms in general growth position." Lesser amounts of "early, submarine, calcareous cement," also may be present. If there is any question of meaning, Fagerstrom further states that a feature is not a reef if there are "organisms not *in situ* and in growth position, and lack of colonial or gregarious growth habit (p. 12).

This insistence on in-place framework is not a new or unique position. Reef classification schemes from four seminal papers are compared in Fig. 4. In each case, the author highlights the role of organic growth in reef development. Dunham (1970) used the term "ecologic reefs" to describe buildups that

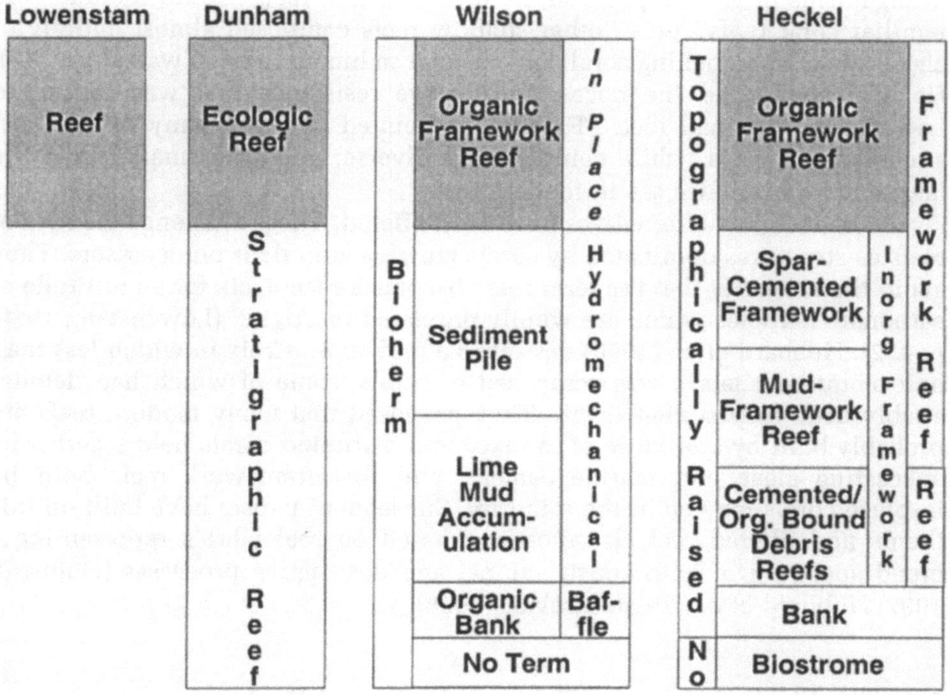

FIGURE 4. Summary of previous reef classifications. Dunham (1970) and Wilson (1975) restrict "true reefs" to those in which in-place and interlocking biological organisms represent the main framework element. Heckel (1974) also includes those features in which cementation and encrustation share the role in creating framework.

reflect the presence of significant topography at the time of deposition and owed that topography largely to organic binding." These are distinguished from "stratigraphic reefs," in which "organisms provided their skeletal debris, their bulk; but organisms did not provide their rigid framework" (p. 1932).

Wilson (1975) attributes any assemblage not dominated by in-place and interlocking organisms to hydromechanical deposition in which biological processes provide the initial building blocks but play little or no direct role at the time of final deposition. Heckel's (1974) treatment of reefs through time and space reflects a much more liberal view and the one closest to that offered in this chapter. While his "organic framework reefs" are limited to those dominated by calcifiers growing atop their predecessors, he allows for secondary cementation in creating the rigid framework of both modern and ancient reefs (i.e., his "inorganic framework reefs").

In perhaps the most complete summary of modern and fossil reefs to date, James (1983) expressed concerns with the severe limitations of reef models based largely on the presence of mostly in-place and interlocking calcifiers. In what probably represents the opposite extreme to requiring a preponderance of organisms in growth position, he proposed that reefs should include "the familiar coral reefs" plus "other shallow reefs composed almost entirely of algae, banks of branching coral, and skeletal sediment in deep water" (p. 347). He also challenged the necessity of wave resistance that was central to Lowenstam's original ideas. Finally, he pointed out that many of the most ancient fossil reefs, while complex and diverse, might be smaller than the largest coral heads we see in modern reefs.

More recently, alternatives have been offered. These challenge the idea of reefs as structures dominated by corals growing atop their predecessors. They avoid the necessity that the term reef "become a receptacle for a multitude of carbonate lenticles which are wholly unrelated in origin" (Lowenstam, 1950, p. 432). Hubbard *et al.* (1990) described a reef on St. Croix in which less than half of internal fabric was composed of corals, some of which had demonstrably been moved after death. They proposed that many modern reefs are probably built by a mixture of in-place and disrupted corals held together by encrusting algae and marine cement, and that framework reefs built by in-place corals may be in the minority. Subsequent papers have built on this theme and offered reef classifications based on reef fabrics representing a broad spectrum of both constructional and destructive processes (Hubbard, 1997; Hubbard *et al.*, 1998; Insalaco, 1998).

1.4. The Study of Modern Reefs: A Historical Review

As discussed above, the prevailing view of modern reefs often ignores the importance of processes going on out of sight and operating on temporal scales beyond a short visit to the reef. The abilities of slower-growing algae to contribute substantially to the reef (Adey, 1977; Adey *et. al.*, 1977a) and of grazing fish (Ogden, 1977), boring sponges (Moore and Shedd, 1977), and burrowing worms to break the reef down (Kiene and Hutchings, 1994; Scoffin and Macintyre, 1997; Glynn, 1997) are often overlooked as they seem small compared to the bulk of the corals on the surface of a living reef. What has emerged from this is a picture of modern reefs dominated by corals growing atop one another and primarily in very shallow water.

Until recently, direct observation of the modern reef interior was hampered by an inability to core through reefs at a reasonable cost. After World War II, long cores taken to document the impact of atomic bomb tests provided a glimpse of the interior of Enewitak Atoll (Ladd and Schlanger, 1960). Explosive exposures and dredging excavations for marine projects (Adey *et al.*, 1977b; Lighty *et al.*, 1982) provided valuable but limited information. In the 1970s, workers from the Smithsonian Institution began a program of systematic reef coring using small, portable coring equipment and a more sophisticated, submersible system described by Macintyre (1978). The introduction of these inexpensive systems enabled academic researchers to launch a revolution in reef studies based on a systematic examination of the interiors of modern reefs in a variety of settings. A flurry of activity over the following decades provided a detailed catalogue of the organisms that built modern reefs, while also addressing important questions about reef accretion under the influence of the rising Holocene sea level (e.g., Adey and Burke, 1976; Cabioch *et. al.*, 1995; Davies and Hopley, 1983; Montaggioni and Fuare, 1997). Generally, these studies focused on emergent reefs where it was presumed that the greatest accumulations of Holocene material would be found. Notable exceptions include Macintrye *et al.* (1982) and Hubbard *et al.* (1985, 1986, 1997).

Early coring reports stressed high rates of accretion and the dominance of coral in the depositional sequence. Adey and Burke (1977) stated that "bench reefs have a coral or coralline framework throughout" (p. 68), and "the 5- to 10-m thick cap of [bank barrier] reefs is structurally quite similar to that of the bench reefs" (p. 71). Simplified core logs often emphasized the corals that were present but failed to report their actual volumetric importance. For example, Davies and Hopley (1983) showed continuous intervals of their "branching coral facies" and "head coral facies," which often occupied nearly 100% of the cores that were illustrated. While the message of these papers may have been more complex, discussions of coral dominance and rapid accretion fueled the argument for modern reefs being built by in-place corals and coralline algae. With no detailed information on actual recovery, willing readers could easily form the impression of dominant in-place framework throughout. Peter Davies (personal communication) estimates average coral recovery in his cores from the Great Barrier Reef to be near 30%.

If a group of reef scientists were to swim over a present-day reef, there would be little disagreement over what to call it. Based on what they could see on the reef surface, many would probably envision a fabric of in-place corals that are disrupted by biological and physical degradation only to the extent that a spectrum of secondary fabrics might be superimposed on a dominantly in-place constructional framework. It is the apparent disparity between this perception of modern reef interiors and what is observed in many fossil reefs that has led to questions about the appropriateness of modern reefs as models for their ancient counterparts.

The primary aim of this chapter is to show that a careful look at the interior of many modern reefs will reveal a fabric that clearly is not dominated by in-place and interlocking organisms. Given the absence of in-place framework in these examples, it seems inappropriate to make framework the sole criterion by which ancient structures are judged. While in-place interlocking framework is one way to build rigid and elevated structures, it is by no means the only one.

2. Examples from Some Modern Caribbean Reefs

2.1. Methods

Over the past three decades, the authors have cored reefs from widely varying settings. Unlike many earlier studies, these investigations ventured

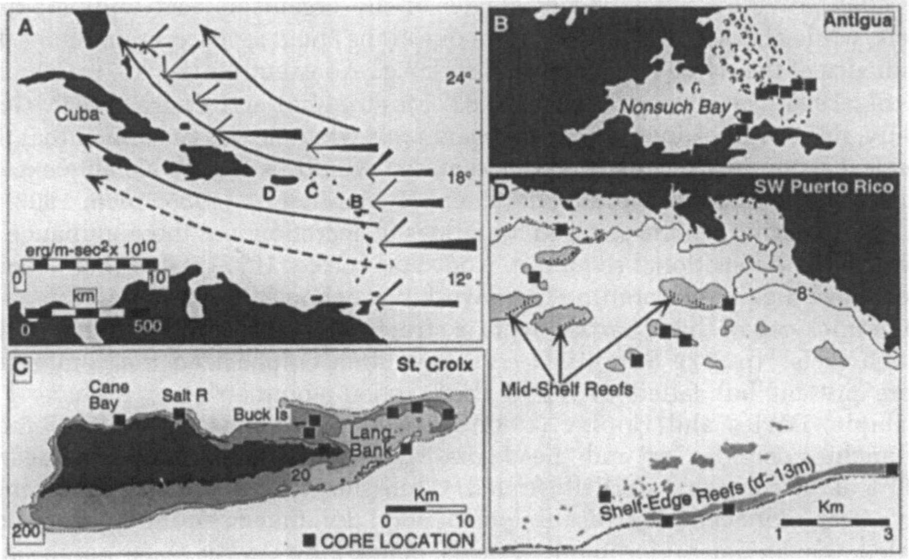

FIGURE 5. (A) Map of the Caribbean region showing locations discussed in the text and core sites shown in B through (D) Rose diagrams depict the strength of waves from each direction. Note that the strongest trade-wind influence and therefore the highest wave energy centers around 15°N latitude. Dominant hurricane and tropical-storm paths (Hubbard, 1989; National Climatic Center, 1981) are indicated by solid and dashed arrows, respectively. (B) Map of Nonsuch Bay in eastern (i.e., windward) Antigua showing core locations (closed squares). After Macintyre et al (1985). (C) Map of St. Croix showing core locations (closed squares). Several cores were typically recovered at each site. The carbonate budget discussed in this chapter was derived from cores and other measurements at Cane Bay. (D) Map of the shelf off southwestern Puerto Rico near La Parguera. The solid squares show core locations (multiple cores were taken at several sites).

into deeper water (up to 30 m) and recovered cores from shelf-wide transects. Among other findings, these cores demonstrate that thick reef accumulations need not be confined to shallow water or to areas where reefs have succeeded in reaching present-day sea level. In addition, such findings provide detailed information on the interiors of eastern Caribbean reefs that have developed under a wide variety of oceanographic conditions.

The following discussion is based on 74 cores (total length = 693 m) recovered from coral reefs in water depths up to 30 m around Antigua, St. Croix, and southwestern Puerto Rico (Fig. 5, Table 1). Core length averaged over 12 m, with the longest core exceeding 30 m in length. The time represented in the cores ranged from the present to ca. 11,000 ybp in the Holocene core interval. Over half the cores reached underlying Pleistocene strata and one passed through the entire Pleistocene record. The environments that are represented in our cores include reef crests, hardgrounds, patch reefs, fore-reef slopes, submerged reef ridges, shelf–bank margins, and the walls of one submarine canyon. Samples were recovered using two diver-operated

TABLE 1. Statistics for Cores Discussed in this Chapter. The Environment of Deposition, Water Depth at the Time of Deposition, Average and (Maximum) Core Length, and the Age Range Contained within the Cores Is Provided

Sites	No. of cores	Environment	Water depth (m)	Core length (m) avg. (max)	Age range (Ka)
Antigua	8	Patch reef,	0–1	9.5 (17.5)	Pres–7500
		reef crest,	1		
		black reef,	1–2		
		fore reef	5		
St. Croix:					
Buck Is.	8	Reef crest,	<1	12.6 (17.5)	Pres–7500 +
		lagoon pvmt.,	2–3		Pleist
		fore reef,	5–12		
		open shelf	5		
Cane Bay	7	Shelf edge	8–25	4.0 (7.0)	Pres–6400
Lang Bank	9	Shelf/bank edge	12–24	10.6 (20.0)	4900–9100 +
					Pleist
Salt R.	10	Canyon walls,	15–30	7.4 (24.8)	700–11100
		reef crest,	<1		
		shelf edge,	7		
Tague Reef	13	Reef/fore reef	0–4	7.7 (13.0)	Pres–7000
Puerto Rico	19	Reef/shoal crest,	<1	16.8 (30.0)	Pres–9900 +
		reef crest	1		Pleist
		Submerged ridge	5–7		
		Shelf edge	10–16		

drilling systems. An earlier version used a small drill motor suspended from a tripod and was similar to the system described by Macintyre (1978). A later version (Fig. 3) used a heavier and more rigid frame that provided greater stability and allowed for much more reliable core logging.

Cores were typically recovered in 5-feet sections. With the Scarid system, any change in drilling character was described and vertically referenced using a centimeter scale permanently mounted on the drill frame. As a result, logs kept on the surface accurately reflected the location of each sample as well as the extent of sand, rubble, and open voids in between. The earlier system shared several problems with all other rotary drills. Because the drill string was suspended from a tripod, accurately recording the position of voids and intervals of sediment and rubble was more difficult. Also, because of its lighter weight, the drill tended to bounce in the hole and made the potential for sand and rubble loss greater than with the Scarid system. The ideas presented below therefore rely more heavily on later cores collected with the Scarid drill. Nevertheless, it should be noted that the basic picture remains the same regardless of which cores are examined.

In the laboratory, samples were slabbed longitudinally and corals were identified to species level. The corals described in this chapter have been combined into two groups that have been shown to have large-scale environmental significance. The first group includes the branching corals (primarily *Acropora palmata* and *A. cervicornis*) that are more common in shallow water but can occur to depths greater than 10 m. The second group includes head corals (primarily *Montastraea* spp.) and platy corals that are more common at depths greater than 5 m. Information on species diversity and distribution is undoubtedly lost in this approach, but we assume that the vagaries of framework generation rely far more on larger-scale biological and environmental interactions than on the individual species that are present. Individual cores are described in greater detail in papers by the authors listed in the references.

Radiocarbon ages of unaltered coral were determined at commercial laboratories in which the within- and between-laboratory variations were shown to be small. Most dated samples were examined by X-ray diffraction to detect mineralogical changes after death and in thin section to find any secondary calcite or aragonite overgrowths that might induce errors in the absolute age. The measured ages were corrected for metabolic changes imparted by the corals and isotopic effects related to seawater chemistry using tree-ring calibrations that are widely accepted (Stuiver *et al.*, 1993; Talma and Vogel, 1993; Vogel *et al.*, 1993). Water depth at the time of deposition was determined by subtracting the depth of each sample below present sea level from the elevation of sea level at that time (using the Caribbean sea-level curve of Lighty *et al.*, 1982). Accretion rates were computed by dividing the vertical separation between samples by the span in their radiocarbon ages [results are reported in meters of accretion per thousand years (m/Ka)]. The rates reported below should be thought of as averages as they do not take into account the 50- to 100-year errors in individual radiocarbon dates.

The relative importance of coral, sand, rubble, and void are based on their linear abundance in a core or group of cores. This approach is equivalent to the point-intercept method commonly used to characterize the percent cover of organisms on modern reefs. Discussions continue on the best method of measuring reef cover, and their summary is beyond the scope of this chapter. However, the slight differences in results from one method to the next are trivial compared to the magnitude of the numbers reported below. It should be kept in mind that some of the corals that were encountered in the cores are demonstrably not in growth position. Therefore, the recovery values provided below for corals represent the absolute maximum for in-place primary framework.

2.2. Accretion and Framework Production

It has been argued that the presence of corals in life position provides the rigidity necessary to support the reef structure (Fagerstrom, 1987). We would counter that: while in-place calcifiers will certainly provide rigidity, this is not the only way to build wave-resistant structures. Recolonization of overturned corals, overgrowth by coralline algae, and stabilization by syndepositional cements can all contribute significantly to the structural integrity of reefs, both modern and ancient. Furthermore, these secondary processes are facilitated by

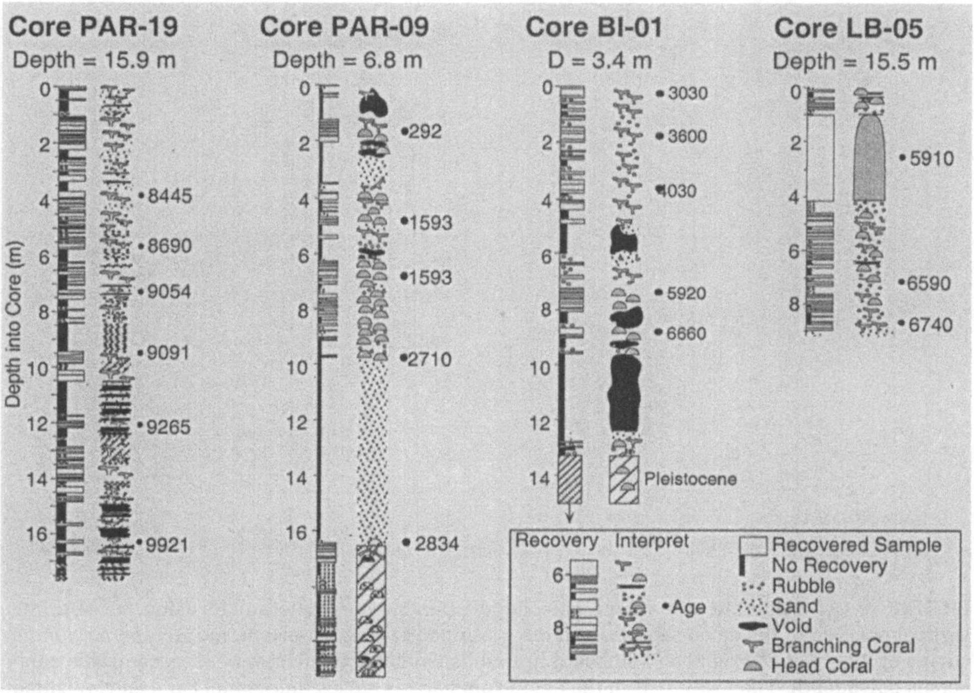

FIGURE 6. Representative core logs from Caribbean reefs discussed in this chapter. While in-place and interlocking framework is no doubt represented in sections of these cores, it is not overwhelmingly dominant. Secondary overgrowths by coralline algae and contemporaneous submarine cementation are probably as responsible for the rigidity of these reefs as is framework in the sense proposed by earlier investigators.

the close packing of disrupted biological debris, a process that by itself can substantially increase a reef's ability to withstand wave attack. The cores described below contain examples of all these accretionary styles and there is general acceptance that the structures from which they were recovered are excellent examples of modern coral reefs.

Representative core logs from several eastern Caribbean reefs are provided in Fig. 6. Clearly, much of the recovered coral material in the 74 cores is not in-place and interlocking framework. In some areas, recovery of coral (both in place and otherwise) was high. At Cane Bay, on the northwest coast of St. Croix (Hubbard *et al.*, 1985), recovery generally ranged from 11 to 52%. On Lang Bank, east of St. Croix, recovery from one deeper-water (ca. 10 m) reef dominated by the coral *A. palmata* averaged 61%, but the abundance of coral in eight other cores was much lower (Hubbard, unpublished data). Likewise, along the windward (east) side of Antigua, recovery exceeded 50% in one 8-m core through the reef crest, but average recovery was much lower in all other cores (Macintyre *et al.*, 1985). For example, one core through a patch reef penetrated 13 m of unconsolidated *A. cervicornis* coral debris. These examples

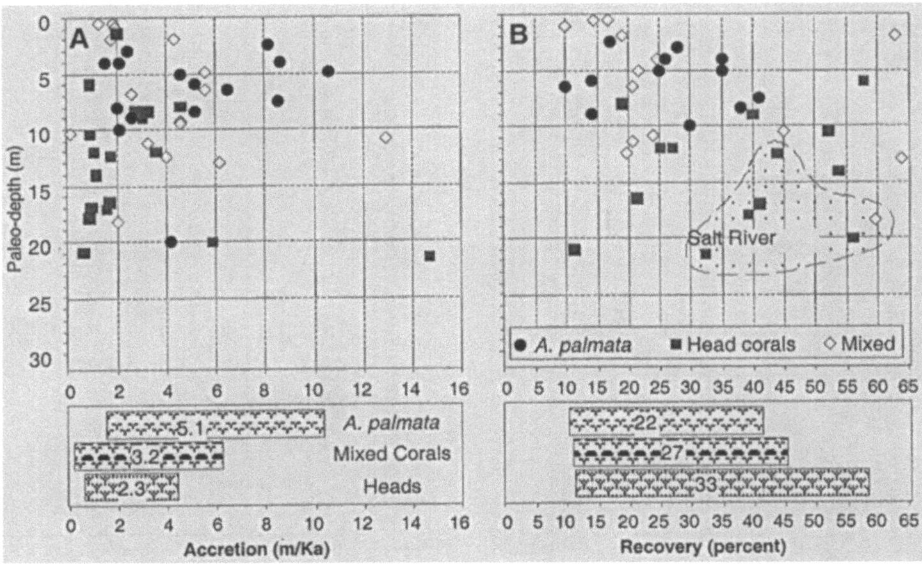

FIGURE 7. (A) Accretion rate versus paleodepth. Depths were determined using the absolute vertical position of the dated samples and the presumed sea level based on the *Acropora palmata* curves of Lighty *et al.* (1982) and Hubbard (unpublished data). (B) Recovery of recognizable coral versus paleodepth. The lower part of the figure summarizes ranges (horizontal bars) and weighted averages (numbers on bars) of recovery and accretion rate. After Hubbard et al (1998).

represent the upper limits of coral recovery. In general, coral recovery for the 74 cores averaged between 20 and 30%, with sand and recovered rubble accounting for a greater portion of the cores.

Coral recovery and reef accretion rates are summarized in Fig. 7. In general, branching corals and mixed-coral assemblages occurred in shallower depths than head corals. In core intervals dominated by branching *A. palmata*, accretion rate averaged 5.1 m/Ka (range = 1.5–10.5 m/Ka; Fig. 7A). In intervals dominated by head corals, accretion averaged 2.3 m/Ka (range = 0.8–6.0 m/Ka). Assemblages containing a mix of branching and head corals accreted at rates averaging 3.2 m/Ka. The recovery of coral (to be differentiated from total recovery, which included sand and rubble) ranged from an average of 22% in branching corals to 33% in head corals, with considerable overlap (Fig. 7B). Clearly, the average coral recovery of 27% for all 74 cores does not reflect a dominance by coral, either in place or otherwise, in the depositional fabric.

2.3. Criteria for In-Place versus Displaced Corals

While we feel that it is not necessary for corals to be in life position to make up reef framework, it is still worth considering which corals might be in place in any reef. The core slabs in Fig. 8 show the spectrum of preservation seen in our cores, ranging from fresh and largely unaltered material to heavily reworked substrate that is barely recognizable as coral. As a result, a clear delineation of in-place versus toppled and rolled corals can be difficult. Nevertheless, there are some criteria that might be useful in distinguishing assemblages of displaced coral from in-place colonies growing one atop the other.

2.3.1. Basal Contact with Other Corals

Whatever substrate a reef might start on, if it is built by interlocking corals in life position, then there should be some evidence of basal contacts between successive corals. Few of the recovered samples in the 74 cores show evidence for such a basal contact. More typically, coral clasts were rimmed by cemented sediments or coralline algae that bind adjacent fragments together. Basal contacts will be missed if the core barrel passes along the side of a colony. Nevertheless, if the corals that were encountered were part of a suite of largely in-place colonies, one would expect to find, in almost 2300 feet of core, some appreciable and unequivocal evidence of corals growing atop one another. What emerges from the cores as discussed here is a mixture of in-place corals and dislodged colonies, packed together either loosely or within a matrix of encrusting algae and varying amounts of cemented sediment. These are often separated by open cavities or intervals filled with sand or reef rubble. The presence of the open cavities implies that close packing of coral colonies and fragments, encrustation by algae, and binding by marine cement are sufficient to hold the surrounding reef fabric together in a rigid and resistant mass.

2.3.2. The Degree of Bioerosion or Encrustation

Once a coral dies, it is almost immediately subjected to processes that will systematically obliterate surface detail and excavate material from the exposed colony. It is this process of bioerosion (and secondarily, physical abrasion) that produces much of the sand- and mud-sized particles found within the reef. In addition, exposed substrates become quickly covered by coralline algae as well as a host of other epibionts and these develop in a reasonably predictable manner (Parsons, 1993). Figure 9 shows the broken edge of an in place *Acropora palmata* branch that had died as a result of widespread coral disease at most about 2 years earlier. The branch was already heavily bored by worms, sponges, and mollusks. In addition, encrustation by coralline algae and foraminifers had already begun. The cycle of boring, sedimentary infill, and cementation (Zankl and Multer, 1977) can completely destroy original coral texture in less than a decade (H. Zankl, personal communication).

The only way to prevent the eventual destruction of any reef organism by bioerosion is to bury it concurrent with or shortly after death (Parsons, 1993; Scoffin and Macintyre, 1997). Because of this, the potential for in-place coral preservation is greatly reduced. Corals broken off the reef surface collect along

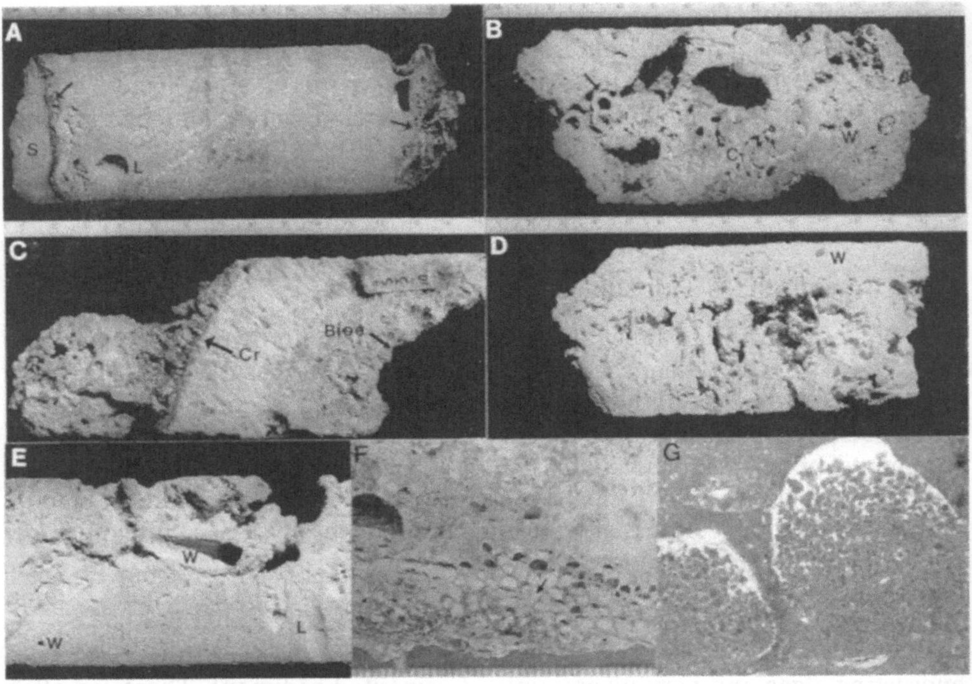

FIGURE 8. Core slabs and photomicrographs that illustrate the spectrum of preservation in the cores. Unless otherwise indicated, up is to the left and scale is in centimeters. (A) Relatively fresh *Acropora palmata* from the shelf break off Puerto Rico. The top (left) end of the sample is draped with cemented sediment (S) over a layer of coralline algae (arrow). The lower edge is bored and encrusted (arrow). Water depth at the site is 15.9 m. Sample PAR19-65 from 6.75 m below the reef surface. Approximate age = 9,000 ybp. (B) Bored and encrusted head coral (*Siderastrea* sp.) from the mid-shelf reefs off Puerto Rico. The coral has been mostly replaced through repetitive cycles of boring, sedimentary infill, and cementation. Serpulid tubes can be seen at the far left (arrow). Worm burrows (w) and sponge galleries (c) also are common. Water depth at the site is 6.8 m. Sample PAR9-36 from 7.5 m below the reef surface. Approximate age is 1400 ybp. (C) Bored and encrusted sample of *Montastraea* sp. From the west wall of Salt River submarine canyon, St. Croix, US Virgin Islands (USVI). The sample is from a horizontal core; the reef surface is to the left. The bioeroded colony has been covered by a sediment drape crust (Cr) and has been excavated from within the reef interior (Bioe). Water depth is 20 m. Sample SR3-10 from 1.8 m into the wall. Approximate age is 3300 ybp. (D) Heavily bored head coral from the eastern edge of Lang Bank, St. Croix, USVI. Most galleries have been excavated by clionid sponges, but polychaete worm tubes (w) are also present. Present water depth is 15.5 m. Sample LB5-38 from 5.3 m below the reef surface. Approx. age = 6400 ybp. (E) Partially bored sample of *A. palmata* from Buck Island lagoon, St. Croix, USVI. Both serpulid and polychaete worm tubes (w) are present. Present water depth is 3.4 m. Sample BI1-7 from 1.0 m below the lagoon floor. Approximate age is 4000 ybp. (F) Bored and infilled head coral from the west wall of Salt River submarine canyon, St. Croix, USVI. Sponge borings (arrow, probably *Cliona lampa*) have been infilled by cement that was indurated by submarine cement. Water depth is 20 m. Sample SR1-12 from 2 m into the canyon wall. Approximate age is 4800 ybp. Scale in milli m. (G) Photomicrograph of sediment infills in F.

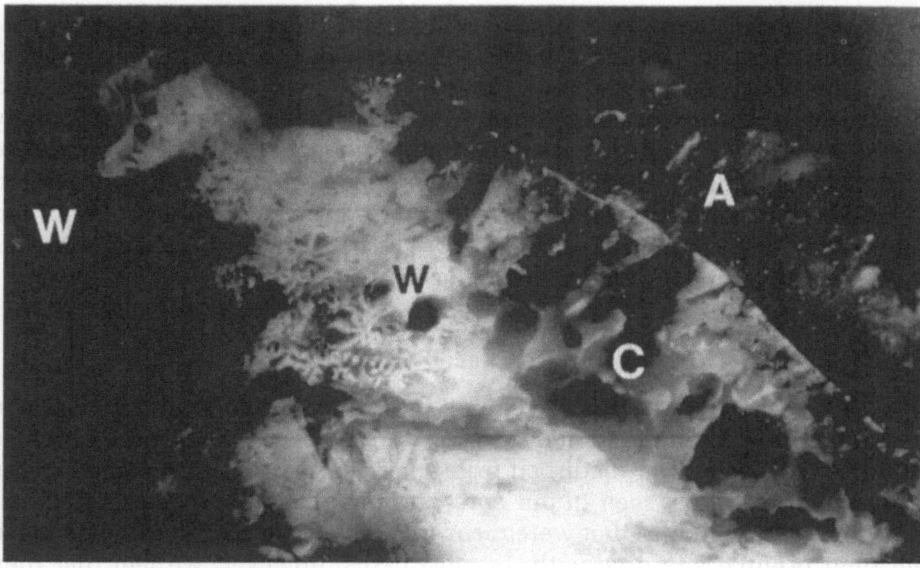

FIGURE 9. Close-up photograph of a broken branch tip of *Acropora palmata*. While dead for less than 2 years, the branch is already infested with boring sponges (c) and worms (w). The branch surface has also been encrusted by coralline algae and foraminifera. Algal turf (A) dominates the upper branch surface. Field of view ca. 15 cm across.

the backreef margin (Hubbard *et al.*, 1991) and on the fore reef (Harmelin-Vivien, 1985; Hubbard *et al.*, 1986) where they form collections of poorly sorted debris in all stages of degradation. Burial processes are accelerated during storms when thick accumulations of reef debris provide a substrate for renewed coral recruitment (Hubbard, 1993). During this process, large quantities of fresh coral detritus are removed from the zone of active encrustation and subsequently are cemented together or bound by microbial or algal crusts (Carter *et al.*, 1989). This forms the foundation on which the new reef builds, and the processes that are responsible are integral to reef accretion and aggradation. The resulting fabric of disrupted and broken colonies is as much a part of the reef as are the living corals found on the surface or in growth position within the interior.

2.3.3. Colony Orientation as an Indicator of In-Place Preservation

In cores, it is impossible to directly observe an entire coral colony. As a result, calical orientation, consistent with that of a live colony, has been used to infer in-place preservation. For example, Macintyre and Glynn (1976) concluded that *A. palmata* found in cores from Panama were largely in life position because the upper colony surface in the core was consistent with the top side of living colonies. As a test of this assumption, the orientation of 84 broken *A. palmata* branches was measured in Buck Island Underwater National Monument. Each counting transect started with an overturned branch. Despite this strong bias toward overturned colonies, only 18% of the measured branches were upside down. If the overturned branches that were artificially selected as starting points are removed from the data set, then the abundance of overturned branches drops below 3%. Therefore, while colony orientation can be useful as an indicator of corals that have been moved (i.e., overturned branches or those sitting at high angles), a predominance of upright branches is a less reliable indicator of in-place preservation.

2.3.4. Separating In-Place from Displaced Corals

While no single criterion is reliable by itself, the tools described above may be helpful, when combined, to discriminate between in-place colonies and those that have been disturbed from their original position. Inasmuch as these criteria are rarely seen in our cores, it seems reasonable to argue that at least some of the colonies that were recovered have been moved some distance from their original site of growth. In most cases, however, we feel that they probably remain somewhere within the reef zone where they grew.

All 74 cores came from rigid and topographically elevated reef structures. On the reef surface, coral growth was still contributing carbonate to the system. The average recovery rate (ca. 27%) of corals in our cores represents the upper threshold for in-place and interlocking framework in the reefs that were sampled (i.e., assuming that all corals were in place, which they were not). The rigidity of those reefs therefore must be derived from a combination of (1) in-place and interlocking coral growth, (2) close packing of disrupted colonies, and (3) the subsequent stabilization of both by encrusting organisms and submarine cement. The absolute percentage of each is of less importance than is the realization that the resulting structure, its rigidity, and the nature of its internal fabric are not the sole result of in-place and interlocking primary framework growth. They instead are the products of a complex interplay between constructional and degradational processes that reflect the physical and biological regimes in which they developed. It is argued that this also was the case in ancient times.

3. Where's the Reef?

If we are to use modern reefs as models for ancient systems, then those models must be based on comprehensive data from the reef interior, not extrapolations from the surface. In the reefs just described, coral comprises less that 30% of their internal fabrics. Of that, some portion is demonstrably out of place and it seems logical that the rigidity of these reefs cannot be explained by the presence of in-place coral "framework" alone.

Many of the reefs discussed here have been in some way recognized for their importance as "representative" or "particularly well-developed" examples of modern coral reefs and few would hesitate to characterize them as "true coral reefs." Yet, cores through these reefs reveal interiors dominated not by successive generations of corals growing one atop the other in an orderly fashion but instead by more variable assemblages of both in-place and dislodged coral, secondary encrusters, and submarine cement. This broad spectrum of reef fabrics provides the cornerstone of the reef classification proposed below.

At this point, three questions remain. First, how "typical" are these examples of Holocene reefs in general? Second, if they are in fact reasonably representative, then what processes are responsible for the fabrics that we see? And finally, how do we fit all of this into the context of modeling ancient reefs that have evolved throughout geologic time?

3.1. How Representative Are the Caribbean Reefs Described Here?

Adey and Burke (1977) and Geister (1977) noted that Caribbean reefs vary widely in response to known regional patterns of wave energy and water clarity. Based on historical data from the National Climatic Center, Hubbard (1989) proposed that hurricane frequency also played an important and predictable role in this pan-Caribbean spectrum of reef types. Adey and Burke (1977) argued that the sequences of corals that comprise those reefs therefore will reflect the control of those factors framed within the context of rising Holocene sea level.

The cores described in this chapter have been recovered from across the eastern Caribbean and the pattern that has been described is felt to be representative of that broad area. Coral recovery averaged 27%. Recovery in cores from the western Caribbean (i.e., Galeta Point, Panama: Macintyre and Glynn, 1976) averaged 23% (weighted average computed from data provided by Ian Macintyre (written communication). Recovery ranged from less than a few percent in peripheral areas to as high as 47% in a core through the main reef. Because these values include recovered rubble, the percentage of coral falls somewhere below these figures. Whichever value is used, recovery compares favorably with that from eastern Caribbean reefs.

The simplified logs of Davies and Hopley (1983) show that along the exposed windward margins of the northern and central Great Barrier Reef, cores reaching the underlying Pleistocene contained an average of 44% coral from their "branching" and "massive coral facies." Inasmuch as some portion of this material probably is not in growth position, this figure represents an absolute upper limit to in-place framework development. Peter Davies (Australian Bureau of Mineral Resources, personal communication) provided a rough estimate of 30% for the average recovery of solid material within the cores described in his paper. Again, the values fall near the recovery rates for eastern Caribbean reefs and a dominance of in-place and interlocking corals is difficult to defend.

3.2. How Do Modern Reefs Form?

In most modern reefs, corals provide the raw material from which the structure is built. Of these corals, some remain in place, some are toppled and remain close to where they grew and some are transported distances of a few meters to many tens of meters. The proponents of in-place framework reefs contend that undisturbed corals provide the bulk of the preserved fabric and that their interlocking nature and their tendency to overgrow one another provides the rigidity that allows the reef to withstand wave energy and rise above its surroundings. While the authors of this chapter recognize that in-place and interlocking framework does occur in certain circumstances, it is proposed that most reefs are not simply biologically produced structures that formed in response to local ecological factors. They can only be understood as the product of biological and sedimentological processes that are both constructional and degradational in nature.

On any modern reef, once a coral dies, it is immediately attacked by organisms seeking either food or shelter. Burrowing sponges, urchins, worms, and mollusks generate silt as they excavate complex and often extensive galleries in their quest for shelter within the reef interior (Figs. 8 and 9). At the same time, a variety of fishes graze on algal turfs and secondary epibionts. Some, like parrotfish with their highly adapted jaw structure, remove large chunks of carbonate, which is ground up during the digestive process and passed through as sand and mud. Similarly, urchins graze on bottom algae, producing substantial quantities of sand. Together, these animals reduce large volumes of solid substrate to sediment of all sizes.

The result is a mix of well-preserved corals, partially destroyed ones and intervals in which multiple episodes of boring, sedimentary infill, encrustation, and cementation have all but obliterated the original coral structure (Fig. 8). Within these, corals deposited as or buried by storm-generated detritus stand the best chance of preservation. In many of the reefs that were cored, the volume of sediment produced by bioerosion exceeds the amount of solid substrate that is left behind. Even after hurricanes and other storms have removed much of this excess (Hubbard, 1993), the remaining detritus can still be as volumetrically important as recognizable coral within the reef.

In a landmark paper, Land (1979) pointed out the need to quantify not only primary constructional processes of coral growth, but also the factors that degrade those fundamental building blocks and redistribute the resulting sedimentary debris. While bioerosion has been widely recognized over the past three decades (Neumann, 1966; Kiene and Hutchings, 1994; Scoffin and Macintyre, 1997), it has been thought of as a process that destroys framework rather than one that contributes to the primary fabric of the reef interior. Likewise, sedimentation has generally been thought of as a process that creates deposits that might be related to the reef but are not part of the reef itself. In fact, it is still widely believed that quantitatively understanding physical processes is a frustrating goal with little useful application to carbonate sedimentology. Recent studies have shown this to be untrue (Stearn and Scoffin, 1977; Hubbard *et al.*, 1981, 1990; Sadd, 1984).

Figure 10 summarizes the processes responsible for reef building along the northwest coast of St. Croix in the eastern Caribbean Sea (Hubbard *et al.*, 1990). Based on the total cover and production rates of the present benthic community, the annual rate of carbonate production along this shelf margin was computed at 1.21 kg/m^2 per year of reef surface. If this is representative of carbonate production over the past few thousand years, then the coral recovered in seven cores from the reef can account for only 41% (0.50 kg/m^2 per year) of the calcium carbonate that should be found within the reef interior. The remaining 59% was reduced to sediment and rubble over time. Some of this loose material (0.41 kg/m^2 per year) has been reincorporated into the reef as fill within open voids. If we allow for the 0.30 kg/m^2 per year of additional sediment that is removed from the reef, primarily by major storm waves (Hubbard, 1993), then the carbonate budget "balances."

Volumetrically, the sediment produced by bioerosion exceeded the amount of recognizable coral found in the Cane Bay cores. This pattern repeats itself in the remainder of the 74 cores discussed in this chapter. Given that over half the volume of the reef interior is occupied by sediment, rubble, and open voids, and much of the coral is demonstrably not in place, ancient reefs with little in-place framework should come as no surprise.

3.3. The Evolution of Reefs through Time

While the information provided above has hopefully brought modern and ancient reefs closer together, significant differences still remain. In addition to the evolution of reef dwellers, the relative importance of sedimentary infill, marine cement, and biological alterations also has undergone substantial changes through geologic time. Modifying Ginsburg's analogy of a long-running play, we envision something more akin to a never-ending prizefight in which two well-matched combatants effectively counter strategic moves that momentarily give one of them a competitive advantage over the other. On one side are the calcifiers that provide the raw material for reef building and on the other are the myriad processes that physically and biologically reduce solid carbonate to sediment.

The evolutionary patterns of the major reef taxa that build reefs and those that break them down are summarized in Fig. 11. The left-hand column shows the temporal distribution of reefs through geologic time (James, 1983). The taxa that are responsible for those buildups are derived from data by Fagerstrom (1987) and Wood (1993, 1995). The right-hand column tracks the evolution of bioeroders, in particular fish, urchins and boring mollusks.

The earliest carbonate buildups were stromatolites composed of microbes (cyanophytes) that trapped sediment. It is easy to imagine the problems that such organisms faced after the appearance of bony fish in the Ordovician and the progressive specialization of complex jaw structures over time. Are the decline of cyanophytes and the rise of calcareous stromatoporoids after this time perhaps responses to grazing pressure? The early Mesozoic era marked the first appearance of teleostean fish and the subsequent development of the

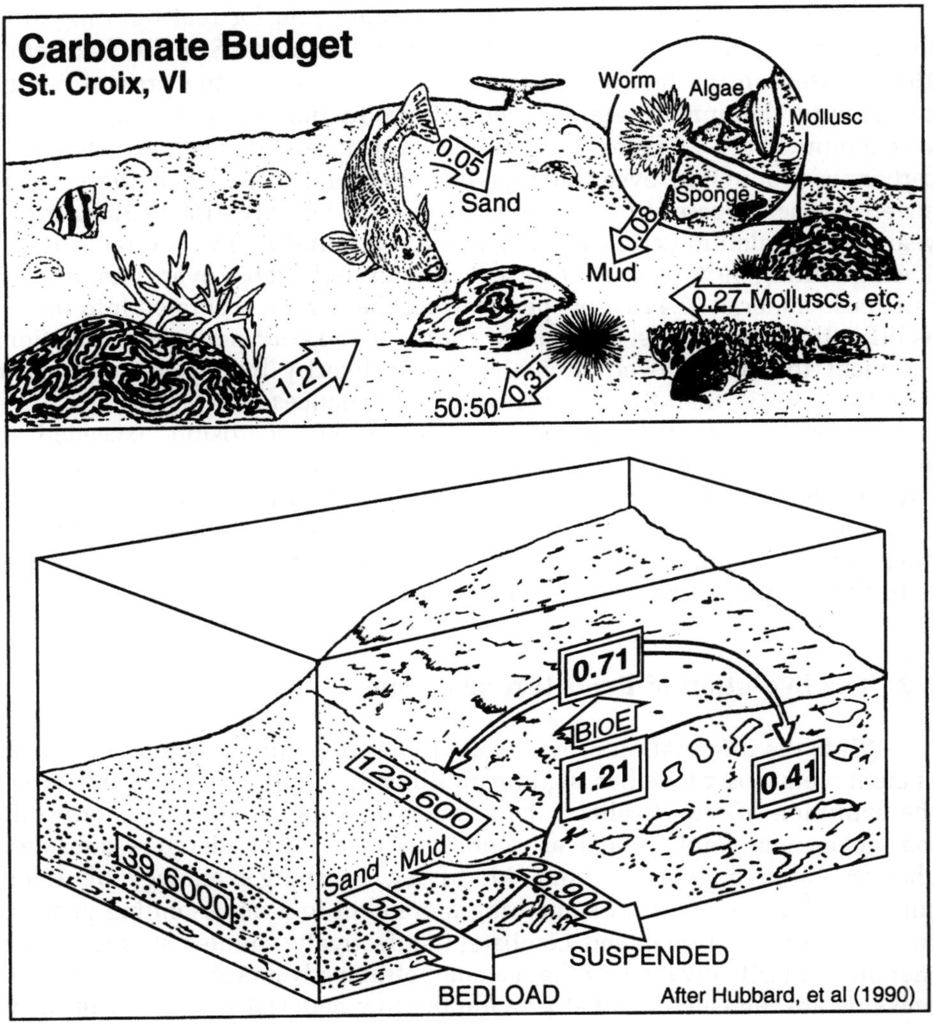

FIGURE 10. Diagram illustrating the relative importance of constructional and degradational processes within the reef at Cane Bay on the northwestern coast of St. Croix, USVI. The top panel illustrates the constructional and destructional contributors to reef fabric. Total production by corals and other calcifiers averages 1.21 kg of carbonate per square meter of reef surface each year. Bioerosion reduces 0.71 kg/m^2 of that to sediment annually, leaving only 0.50 kg/m^2 per year of recognizable coral and coralline algae to form primary framework. The lower panel describes the fate of the sediment produced by bioerosion. Of the 0.71 kg/m^2 per year of sediment produced, approximately 0.41 kg/m^2 is reincorporated within the reef interior. The remainder is stored in sand channels within the reef until storms can remove enough of it to prevent the reef from being buried in its own detritus. On Cane Bay's north shore, all but 39,600 kg of the 123,600 kg of sediment stored in the intervening channels is swept away. The result is a reef fabric that is comprised of more sediment, voids and encrusting algae than recognizable coral. (After Hubbard *et al.*, 1990.)

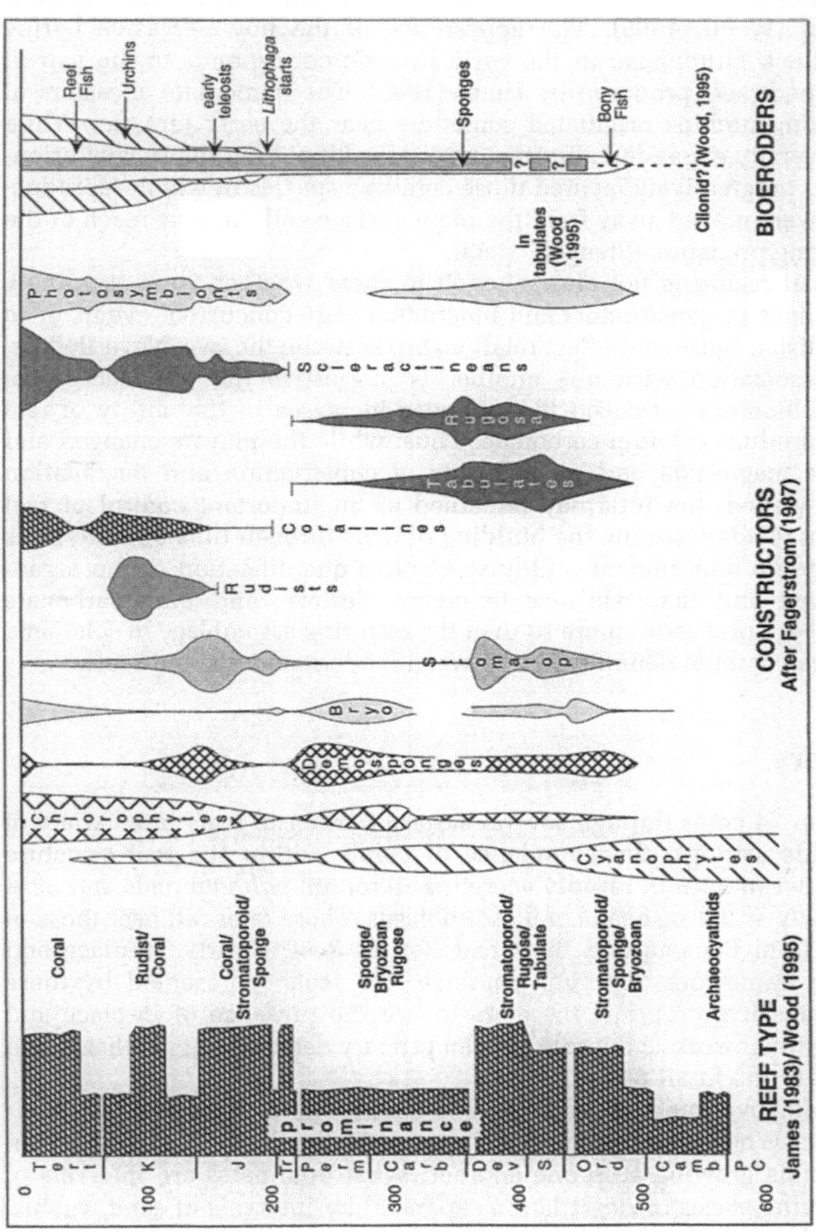

FIGURE 11. The evolution of framework constructors and bioeroders through geologic time. These are compared to the development of carbonate buildups according to James (1983). While the data are provided on a crude scale, there is a seeming pattern of alternating surges by each. It is proposed that this represents ongoing competition between organisms that create reef and those that destroy it. While the relative balance between construction and degradation has changed back and forth as evolutionary pressures have changed, the basic processes have remained in place. Comparing modern and ancient reefs should be rooted in these processes.

pharyngeal mill used today by parrotfish to grind up reef fragments. At about this time, both coralline algae and scleractinean corals began their rapid radiation. Direct evidence of boring clionid sponges dates back to Devonian time and galleries tentatively attributed to these organisms occur in even earlier rocks (Wood, 1995). The appearance of the now-pervasive boring bivalve mollusk *Lithophaga* in the early Jurassic corresponds to the gap in well-developed reefs proposed by James (1983). The immediate ancestors of today's grazing urchins originated sometime near the early Jurassic. While coralline diversity expanded after the onset of echinoid herbivory, evolutionary selection progressively favored those coralline species in which reproductive organs were moved away from the plant surface and the easy reach of the jaws of grazing predators (Steneck, 1983).

The fossil record is not clear enough to show whether these seemingly parallel changes in constructors and bioeroders were concurrent events or to firmly establish a cause-and-effect relationship between the two. Nevertheless, the close association with one another is suggestive that the successive radiation of bioeroders tracked the stepwise increases in the ability of reef builders to produce calcium carbonate. Thus, while the players changed and the absolute magnitude and mechanisms of construction and degradation waxed and waned, the interplay remained as an important control of reef development. Understanding the building of reefs through time and the links between modern and ancient buildups requires quantification of the organisms involved and their abilities to create, destroy, and bind carbonate substrate. These processes, more so than the resulting assemblage of calcifiers, represent the common denominator between modern and ancient reefs.

4. Summary

Based on 74 cores through several eastern Caribbean reefs in a variety of oceanographic settings, the abundance of corals within the reef structure averaged between 22% in rapidly accreting *Acropora palmata* reefs and 33% in more slowly accreting head coral assemblages. These rates fall near those of other reefs from Panama and the Great Barrier Reef. Clearly, in-place and interlocking framework does not dominate the reefs represented by these cores. It seems inappropriate therefore to use the presence of in-place and interlocking framework as the sole or even primary determinant of what is and is not a reef in the fossil record.

Swimming over modern reefs, there is little room for argument over their classification. When we examine their interiors's details, we find fewer than expected corals growing atop one another. What dominates are intervals of coral in varying stages of degradation separated by intervals of sand, rubble, or open voids. Corals and other large fragments are further bound together by overgrowths of coralline algae and to a lesser extent contemporaneous marine cements. In contrast to the ordered fabric envisioned in framework-based models, we find a mix of in-place and *in situ* coral, reef-derived sediment,

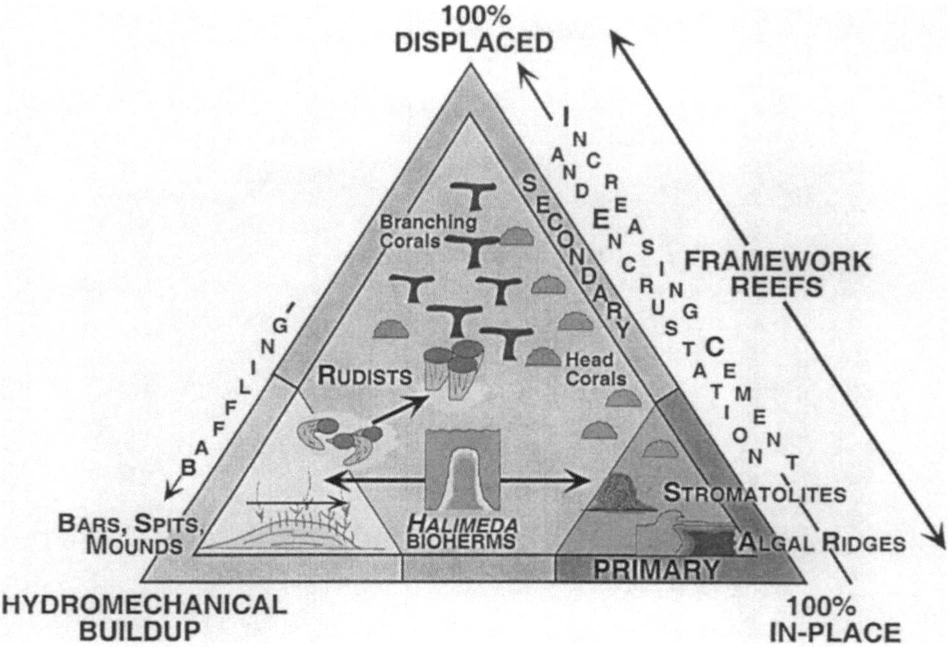

FIGURE 12. Ternary diagram showing the relative importance of in-place corals and other calcifiers ("primary framework"), material that has been dislodged but reincorporated into the reef structure by cementation and encrustation ("secondary framework"), and debris moved out of the reef and deposited by mostly hydromechanical processes. The position of intermediate examples on the diagram is generalized and is not intended to represent an absolute position for all buildups of the type depicted. Coral symbols are the same as for Fig. 6. After Hubbard et al (1998).

encrusting overgrowths, and synsedimentary cement. If the latter is the model that is compared to the fossil record, then the difference between the two narrows and a close look at modern reefs can shed considerable light on the processes that are responsible for the fabrics of their fossil counterparts.

The ternary diagram in Fig. 12 provides a starting point for classifying carbonate buildups. It is based on the relative importance of (1) in-place organisms, (2) disrupted organisms that remain close to where they grew and are bound within the reef by biological overgrowth or submarine cement, and (3) reef-derived material that is deposited in a largely physical regime far removed from the reef. To an extent, its end members are similar to a ternary diagram proposed by Gerhard and Burke (1991). While we disagree with some aspects of that classification, we agree wholeheartedly (e.g., Hubbard *et al.*, 1990) with their contention that carbonate production, destruction and transport create a boad spectrum of reef fabrics.

Primary framework reefs are those in which the vast majority of the reef is composed of calcifying organisms in growth position. The best modern examples include Caribbean (Fig. 13A) and Pacific algal ridges (Fig. 13B) that

FIGURE 13. (A) Algal ridge complex 400 m off the south coast of St. Croix, USVI. The reefs face the open Caribbean and dominant hurricane tracks to the south (see Fig. 5). (B) Algal ridge on the western shore of Tikehau in the Tuomotu Chain. Pacific swell exclude predators in the same way as storm waves and Caribbean swell do along the south coast of St. Croix.

form in high-energy environments where grazing predators are largely excluded. Some stromatolitic buildups that dominated early Paleozoic reefs (Fig. 14) probably share this lack of predators as a key to their success. Modern stromatolites thrive on Lee Stocking Island (Bahamas) because periodic burial by migrating sandbars discourage predators (Dill and Shinn, 1986).

Most modern and ancient reefs contain some number of organisms such as corals and stromatoporoids that have been dislodged and reincorporated

FIGURE 14. Stromatolite mound from the Canning Basin of Australia. These small patches are scattered in a depositional interval dominated by micrite and replacement cements.

into the reef structure through overgrowth by other organisms (e.g., coralline algae) and varying quantities of submarine cement. Most if not all of these reefs from which our cores were recovered are of this type and we classify them as secondary framework reefs. An excellent fossil example occurs in an exposed Holocene reef in the Dominican Republic (Mann *et al.*, 1984). Its internal structure is a mix of coral colonies that are clearly in place and those that have been rolled and redeposited. While the general zonation patterns are clear, the internal fabric is a mix of primary and secondary framework elements. Sherman *et al.* (1999) described a submerged Pleistocene reef off Oahu in Hawaii that consists of "*in-situ* coral and coralline algal framework along with coarse algal grainstone and rudstone that may fill voids in the framework as well as constitute entire core sections" (p. 1084). Fagerstrom and Weidlich (1999) described an abundance of upright and clearly in-place sponges occurring within the core of the upper Capitan reef complex mixed with colonies that have been overturned.

Following convention from ternary diagrams used to classify sedimentary rocks, we have used a cutoff of 70–80% in-place organisms as the transition from primary framework reefs to secondary framework reefs. Those structures are those in which more than 70–80% of their fabric is made up of fragments that (1) have been exported from the reef, (2) were deposited by purely mechanical means, and (3) do not serve as a foundation for further reef development. Such deposits have been classified as hydromechanical build-ups. These include sand shoals, mud mounds related to baffling by sea grasses, many debris aprons, and alochthonous reef blocks deposited in deeper water (e.g., the "satellite reefs" of Mikulic and Kluessendorf, 1999, in Thornton Quarry; the "cipit blocks" of Zankl *et al.*, 1987, in the Triassic Alps). Careful examination of a number of modern and ancient reefs may eventually refine these cutoffs in a way that is less arbitrary and reflects either (1) important changes in physical or biological factors, or (2) natural gaps in the spectrum between end-members.

Figure 15 is our attempt to relate the fabrics described in Fig. 12 to energy level (wind, waves, storm activity), grazing intensity (mostly by scarids and urchins), and nutrient levels. It draws heavily on discussions by Adey and Burke (1977), Hallock (1988) and Wood (1993). The greatest potential for primary framework reefs exists when grazers that disrupt primary framework are excluded either through deposition in low-energy settings or by organisms that are specially adapted to life in the surf zone. In modern reefs, the best examples are algal ridges, where the tight encrusting growth form provides structural resistance to wave attack. High wave energy excludes an active grazing community that ordinarily prevents thick accumulations of coralline algae in less exposed settings. Early stromatolites that developed before the evolution and radiation of grazers were able to build substantial structures in ancient reef settings. In these buildups, it is important to separate those depositional sequences that formed resistant structures from those that simply trapped sediment (i.e., bafflers).

The majority of the reefs will be dominated by secondary framework. Storm action will inevitably disrupt some portion of the live and dead reef cover. Nutrients also are important (see Chapter 11, this volume). As nutrient levels increase, a greater abundance of algal turfs will discourage larval recruitment by calcifiers (Pennings, 1997), while encouraging grazing. Increased nutrients also will favor the proliferation of infaunal borers, especially boring sponges (Moore and Shedd, 1977). As a result, reef substrates in areas of high nutrients will be less developed and weakened due to infaunal boring and primary calcifiers. Algal crusts will be more susceptible to grazing and cements perhaps will be discouraged by higher phosphate levels (Simskiss, 1964). Encrustation and cementation will be favored in higher-energy reef sites with lower levels of nutrients.

The classification scheme proposed above retains the three criteria listed in our introductory remarks. While not all the framework is in place, the main elements of the internal fabric are largely biologically derived. The structural rigidity of the reef is not maintained solely by corals in growth position but by

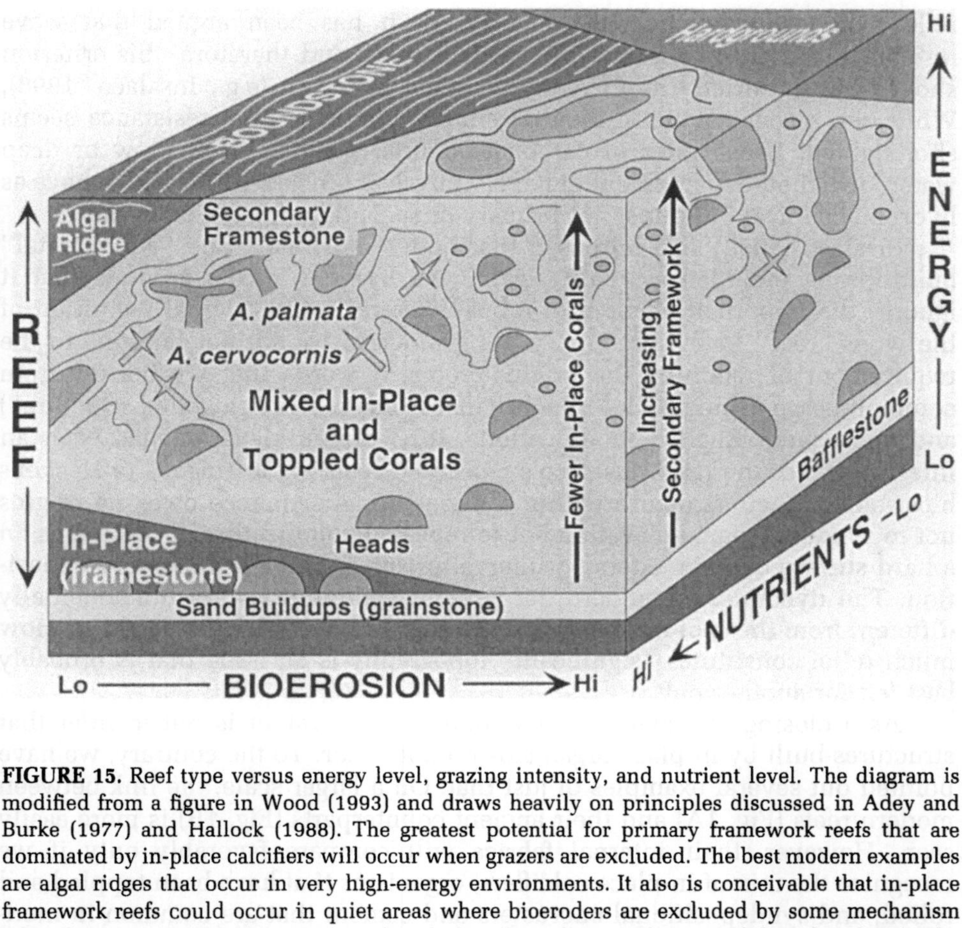

FIGURE 15. Reef type versus energy level, grazing intensity, and nutrient level. The diagram is modified from a figure in Wood (1993) and draws heavily on principles discussed in Adey and Burke (1977) and Hallock (1988). The greatest potential for primary framework reefs that are dominated by in-place calcifiers will occur when grazers are excluded. The best modern examples are algal ridges that occur in very high-energy environments. It also is conceivable that in-place framework reefs could occur in quiet areas where bioeroders are excluded by some mechanism other than wave energy (e.g., stromatolites before the radiation of grazing fish and urchins). The likelihood of toppled and rolled corals being part of the framework will increase with wave energy and to a lesser extent extent with bioerosion. As wave energy increases and progressively excludes bioeroders, both encrustation and cementation will increase.

a combination of in-place corals, those that have been toppled, and syndepositional encrustation and/or cementation. Finally, all the structures that have been studied display some degree of topographic relief to the extent that they impact local oceanographic processes. A structure need not reach sea level for it become a reef (e.g., the ideas of Lowenstam, 1950). Vagaries of sea-level rise and other physical constraints have resulted in topographically significant structures, primarily along shelf breaks, that display all the attributes of reefs as described above.

While the proposed classification broadens the definition of reef based solely on in-place framework, it does restrict reefs to those structures that reflect the characteristics just mentioned. It has been argued that wave resistance can only be inferred in ancient reefs, and therefore, this criterion should not be strictly applied to carbonate buildups (e.g., Insalaco, 1998). While we recognize this difficulty, simply ignoring wave resistance seems shortsighted. Loose piles of carbonate debris, whether in shallow or deep water, owe their development to largely physical processes and will behave as hydromechanical buildups, not primary or secondary framework reefs.

Insalaco (1998) also proposed that both "constratal" and "suprastratal" buildups be included as reefs. Again, we disagree on the grounds that it ignores the important element of relief inherent in the original definition of the word "reef" in the dictionary. As geologists, we are not immune to the requirement of retaining the original roots of words that we borrow from common usage. Going back to the original Norse (i.e., a "ridgelike structure") and subsequent English modifications, topographic elevation has been an integral part of the term. Insalaco's constratal reefs are analogous to Heckel's biostromes. Their structure within a depositional sequence owes its origins not to syndepositional elevation but to repeated colonization of organisms on a hard surface over an extended interval of surrounding sedimentary deposition. The dynamics of modern hardground communities are fundamentally different from those of reefs and the two should not be lumped together. How much relief constitutes "significant" topography is an issue that is probably best left for another day.

As a closing statement, we reiterate that our intent is not to infer that structures built by in-place organisms do not occur. To the contrary, we have pointed out several examples of just that. On a larger scale, the link between modern reefs (Fig. 1A) and their ancient counterparts (Fig. 1B) is more easily seen. However, their internal fabrics will compare favorably only if we recognize the mix of in-place calcifiers, organisms that have been toppled and rolled, and syndepositional encrusters and cement that are common in modern reefs. While much of the bioeroded debris is deposited in flanking beds at the base of the reef slope, much or it remains within the reef and constitutes an important part of the resulting fabric. Whether the reef core is made up of corals, stromatolites, or other carbonate producers, recognizing the balance between constructional and destructive processes is the key to comparing modern and ancient reefs.

If we allow for secondary framework in ancient reefs, then many examples of framework reefs can be found in the rock record. Jurassic and Triassic reefs of western Europe are built of in-place (Fig. 16A) and overturned phaceloid *Retiophyllia* sp. (Fig. 16B). The gross morphology of these scleractinians resembles that of *Porites porites* in modern Caribbean reefs (Fig. 16C). Well-preserved head corals comprise a large percentage of the internal fabric of a Pleistocene reef exposed in Windley Key quarry in Florida (Fig. 16D). Many of these appear to be in place. Whatever the relative abundance of in-place versus displaced corals, both are an important component of "true reefs."

Because of the confusion that it has caused, it is tempting to try and purge the word "framework" from our vocabulary and use terms such as "fabric" or

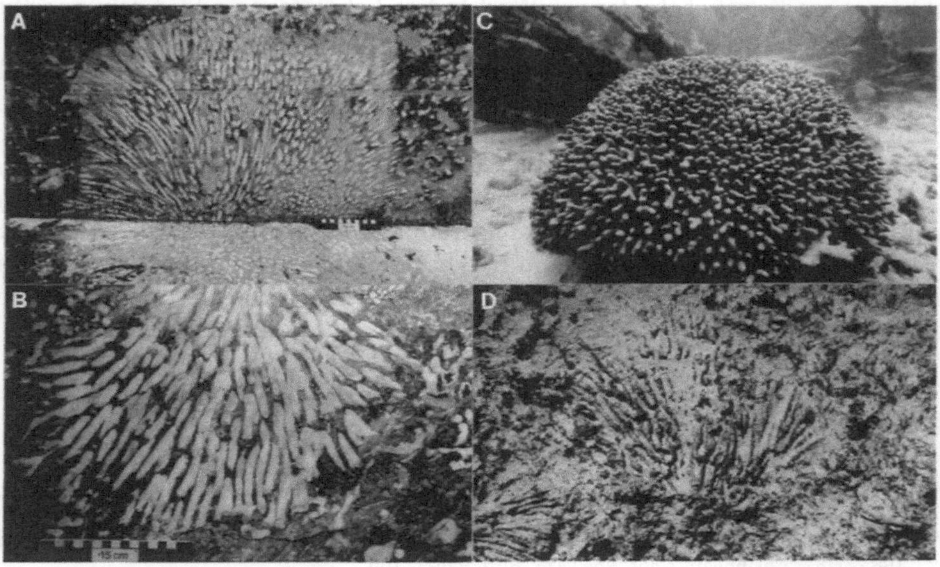

FIGURE 16. (A) In-place colony of Triassic *Retiophyllia* sp. from the Adnet Quarry in Austria. The surrounding groundmass is dominated by detrital sediments. The colony is 80 cm across and is exposed in both the quarry wall and floor (foreground). (B) Overturned colony of *Retiophyllia* sp. from the Adnet Quarry. Cement binds the fingerlike coral branches together. The fact that the otherwise fragile colony has remained intact may reflect encrustation or cementation prior to the colony being overturned. (C) Live *Porites porites* colony in the backreef of Tague Bay, St. Croix, USVI. The colony is approximately 1 m across. (D) Pleistocene *Diploria* sp. (plan view) from Windley Key quarry in southern Florida.

"texture" in their place. Unfortunately, the idea of "framework" is so ingrained in the literature that it is here to stay. Dunham (1970) noted,

> Some workers, including myself, have tried to resolve the problem of the reef concept by practically striking the word "reef" from their vocabularies, using "buildup," "margin," "mound," and such terms as substitutes. This is unsatisfactory evasion. We now recommend retaining the word, but with qualification as needed (p. 1932).

We offer the classification scheme in this chapter in the same spirit.

ACKNOWLEDGMENTS: The data presented represent nearly a dozen studies spanning two decades. It is impossible to adequately acknowledge everyone by name who has contributed to this effort. The work has been sponsored by several research grants from the National Undersea Research Program of National Oceanic and Atmospheric Administration (NOAA), the National Science Foundation, the National Institute for Global Environmental Change (NIGEC) The National Park Service, and NOAA Sea Grant. Both drilling systems were developed at West Indies Laboratory with funds provided by NOAA. Over the years, we have received enormous support from colleagues, students, and staff at the former West Indies Laboratory on St. Croix and the University of Puerto Rico's Marine Laboratory on Isla Magueyes. In particular, we wish to acknowledge Walter Adey, Robert Dill, Ian Macintyre, and Jack Morelock who have been with us in the field and have provided ideas and critical feedback throughout the careers of the authors. Ian kindly provided data on recovery from his Panama cores. Peter Davies allowed access to his cores from the Great Barrier Reef and provided considerable insight, along with David Hopley, on styles of reef accretion in the GBR. To these colleagues, plus Al Fagerstian and John Fandolfi, who provided valuable criticisms of the original manuscript only and many other valued associates, we offer our heartfelt thanks.

References

Adey, W. H., 1977, Shallow water Holocene bioherms of the Caribbean Sea and West Indies, *Proc. Third Int. Coral Reef Symp.* **2**:xxi–xxiv.

Adey, W. H., and Burke, R. B., 1976, Holocene bioherms (algal ridges and bank–barrier reefs) of the eastern Caribbean, *Geol. Soc. Am. Bull.* **87**:95–109.

Adey, W. H., and Burke, R. B., 1977, Holocene bioherms of Lesser Antilles-geologic control of development, in: *Reefs and Related Carbonates—Ecology and Sedimentology*, AAPG Studies in Geology, Vol. 4 (S. H. Frost, M. P. Weiss, and J. Saunders, eds.), AAPG, Tulsa, OK, pp. 67–81.

Adey, W. H., Adey, P. J., Burke, R. B., and Kaufman, L., 1977a, The Holocene reef systems of eastern Martinique, French West Indies, *Atoll Res. Bull.* **21B**:1–40.

Adey, W. H., Macintyre, I. G., Stuckenrath, R., and Dill, R. F., 1977b, Relict barrier reef system off St. Croix: Its implications with respect to late Cenezoic coral reef development in the western Atlantic, *Proc. Third Int. Coral Reef Symp.* **2**:15–21.

Cabioch, G., Faure, G., and Montaggioni, L. F., 1995, Holocene initiation and development of New Caledonian fringing reefs, SW Pacific, *Coral Reefs* **14**:131–140.

Carter, B., Simms, M., Moore, C., Roberts, H., and Lugo-Fernandez, A., 1989, Modern carbonate environments of St. Croix and the Caribbean: a general overview, in: *Terrestrial and Marine Geology of St. Croix, U.S. Virgin Islands* (D. K. Hubbard, ed.), West Indies Laboratory Spec. Pub. No 8, St. Croix, pp. 111–116.

Davies, P. J., and Hopley, D., 1983, Growth facies and growth rates of Holocene reefs in the Great Barrier Reef, *BMR J. Australian Geol. Geophys.* **8**:237–252.

Dill, R. F., and Shinn, E. A., 1986, Living lithified columnar stromatolites: Exuma Island, Bahamas, in: *Abstracts—SEPM Mid-year Mtg. Society of Economic Paleontologists and Mineralogists*, Tulsa, OK, V.3:28.

Dunham, R. J., 1970, Stratigraphic versus ecologic reefs. *AAPG Bull.* **54**:1931–1932.

Fagerstrom J. A., 1987, *The Evolution of Reef Communities*, John Wiley and Sons, New York.

Fagerstrom, J. A., and Weidlich, O., 1999, Origin of the upper Capitan-massive limestone (Permian), Guadalupe Mountains, New Mexico–Texas: Is it a reef? *GSA Bull.* **111**:159–176.

Geister, J., 1977, The influence of wave exposure on the ecology and zonation of Caribbean coral reefs, *Proc. Third Int. Coral Reef Symp.* **1**:23–29.

Gerhard, L. C., and Burke, R. B., 1991, *Reefs, Banks and Bioherms: A Genetic and Semantic Continuum*, Kansas Geological Survey Open-File Report, Lawrence, KS, 90-11.

Glynn, P., 1997, Bioerosion and coral growth: a dynamic balance, in: *Life and Death of Coral Reefs* (C. Birkeland, ed.), Chapman and Hall, New York, NY, pp. 68–94.

Hallock, P., 1988, The role of nutrient availability in bioerosion: Consequences to carbonate buildups, *Paleogeogr. Paleoclimatol. Paleoecol* **63**:275–291.

Harmelin-Vivien, M., 1985, Atoll de Tikehau, *Proc. Fifth Intl. Coral Reef Symp.* **1**:213–266.

Heckel, P. H., 1974, Carbonate buildups in the geologic record, in: *Reefs in Time and Space* (L.

F. LaPorte, ed.), SEPM Spec. Pub. 18, Society of Economic Paleontologists and Mineralogists, Tulsa, OK, pp. 90–154.

Hubbard, D. K., 1989, Terrestrial and Marine Geology of St. Croix, U.S. Virgin Island, West Indies Laboratory Spec. Pub. No. 8, St. Croix, pp. 85–94.

Hubbard, D. K., 1993, Hurricane-induced sediment transport in open-shelf tropical systems—An example from St. Croix, U.S. Virgin Islands, *J. Sedimentary Petrol.* **62**:946–960.

Hubbard, D. K., 1997, Dynamic processes of coral-reef development, in: *Life and Death of Coral Reefs* (C. Birkeland, ed.), Chapman and Hall Publishers, New York, NY, pp. 43–67.

Hubbard, D. K., Sadd, J. L., Miller, A. I., Gill, I. P., and Dill, R. F., 1981, *The Production, Transportation and Deposition of Carbonate Sediments on the Insular Shelf of St. Croix, U.S. Virgin Islands*, Tech. Report No MG-1, West Indies Laboratory, St. Croix.

Hubbard, D. K., Burke, R. B., and Gill, I. P., 1985, Accretion in shelf-edge reefs, St. Croix, U.S.V.I., in: *Deep-Water Carbonates* (P. D. Crevello and P. M. Harris, eds.), SEPM Core Workshop No. 6, Society of Economic Paleontologists and Mineralogists, Tulsa, OK, pp. 491–527.

Hubbard, D. K., Burke, R. B., and Gill, I. P., 1986, Styles of reef accretion along a steep, shelf-edge reef, St. Croix, U.S. Virgin Islands, *J. Sediment. Petrol.* **56**: 848–861.

Hubbard, D. K., Burke, R. B., and Gill, I. P., 1998, Where's the reef: The role of framework in the Holocene, *Carbonates Evaporites* **13**:3–9.

Hubbard, D. K., Parsons, K. M., Bythell, J. C., and Walker, N. D., 1991, The effects of Hurricane Hugo on the reefs and associated environments of St. Croix, U.S. Virgin Islands—a preliminary assessment: *J. Coastal Res.* **7**:33–48.

Hubbard, D. K., Gill, I. P., Burke, R. B., and Morelock, J., 1997, Holocene reef backstepping—southwestern Puerto Rico shelf, *Proc. 8th Int. Coral Reef Symp.* **2**:1779–1784.

Hubbard, D. K., Miller, A. I., and Scaturo, D., 1990, Production and cycling of calcium carbonate in a shelf-edge reef system (St. Croix, U.S. Virgin Islands): Applications to the nature of reef systems in the fossil record, *J. Sediment. Petrol.* **60**:335–360.

Insalaco, E., 1998, The descriptive nomenclature and classification of growth fabrics in fossil scleractinian reefs, *Sediment. Geol.* **118**:159–186.

James, N. P., 1983, Reef environment, in: *Carbonate Depositional Environments* (P. A. Scholle, D. G. Bebout, and C. H. Moore, eds.), APG Memoir, 33, American Association of Petroleum Geologists, Tulsa, OK, pp. 345–440.

Kiene, W. E., and Hutchings, P. A., 1994, Bioerosion experiments at Lizard Island, Great Barrier Reef, *Coral Reefs* **13**:91–98.

Ladd, H. S., and Schlanger, S. O., 1960, Drilling operations on Eniwetok Atoll, *USGS Prof. Paper* 1044–9612.

Land, L. S., 1979, The fate of reef-derived sediment on the north Jamaican island slope, *Marine Geol.* **29**:55–71.

Lighty, R. G., Macintyre, I. G., and Stuckenrath, R., 1982, *Acropora palmata* reef framework: A reliable indicator of sea level in the western Atlantic for the past 10,000 years, *Coral Reefs* **1**:125–130.

Lowenstam, H. A., 1950, Niagaran reefs of the Great Lakes area, *Geol. Soc. Am. Mem.* **67** (2):215–248.

Macintyre, I. G., 1975, A diver operated hydraulic drill for corins submerged substrates, *Atoll Res. Bull.* **336**:1–7.

Macintyre, I. G., and Glynn, P. W., 1976, Evolution of a modern Caribbean fringing reef, Galeta Point, Panama, *AAPG Bull.* **60**:1054–1072.

Macintyre, I. G., Burke, R. B., and Stuckenrath, R., 1982, Core holes in the outer fore reef off Carrie Bow Cay, Belize: A key to the Holocene history of the Belizean barrier reef complex, *Proc. Fourth Int. Coral Reef Symp.* **2**:567–574.

Macintyre, I. G., Multer, H. G., Zankl, H. L., Hubbard, D. K., Weiss, M. P., and Stuckenrath, R., 1985, Growth and depositional facies of a windward reef complex, Nonsuch Bay, Antigua, W.I., *Proc. Fifth Int. Coral Reef Symp.* **6**:605–610.

Mann, P., Burke, K., Kulstad, R., and Taylor, F. W., 1984, Subaerially exposed Holocene coral reef, Enriquillo Valley, Dominican Republic, *GSA Bull.* **95**:1084–1092.

Mikulic, D. G., and Kleussendorf, J., 1999, *The Classic Silurian Reefs of the Chicago Area, Ill*, State Geological Survey Guidebook 29, Illinois State Geological Survey, Champaign, IL.

Montaggioni, L. F., and Faure, G., 1997, Response of reef coral communities to sea-level rise: A Holocene model from Mauritius (western Indian Ocean), *Sedimentology* **44**:1053–1070.

Moore, C. H., and Shedd, W. W., 1977, Effective rates of sponge bioerosion as a function of carbonate production, *Proc. Third Int. Coral Reef Symp.* **2**:499–506.

Mountjoy, E. W., and Geldsetzer, H. H. J., 1981, Devonian stratigraphy and sedimentation, southern Rocky Mountains, in: *Field Guides to Geology and Mineral Deposits* (R. I. Thompson and D. G. Cook, ed.), Geol. Assoc. Canada Annual Meeting, Geological Association of Canada, St. Johns, NF, Canada, pp. 195–224.

National Climatic Center, 1981, *Tropical Cyclones on the North Atlantic Ocean, 1871–1980* (with annual updates), National Climatic Center, Ashville, NC.

Neumann, A. C., 1966, Observations on coastal erosion in Bermuda and measurements of the boring rate of the sponge, *Cliona lampa, Limnol. Oceanogr.* **11**:92–108.

Newell, N. D., 1971, An outline history of tropical organic reefs. in: *American Museum Novitates,* No. 2465, New York Museum of Natural History.

Newell, N. D., Rigby, J. K., Fischer, A. G., Whiteman, A. J., Hickox, J. E., and Bradley, J. S., 1953, *The Permian Reef Complex of the Guadalupe Mountains Region, Texas and New Mexico*, W. H. Freeman, San Francisco.

Ogden, J. C., 1977, Carbonate sedimentation production by parrotfish and sea urchins on Caribbean reefs, in: *Reefs and Related Carbonates—Ecology and Sedimentology* (S. H. Frost, M. O. Weiss, and J. B. Saunders, ed.), AAPG Studies in Geology 4, pp. 281–288.

Parsons, K. M., 1993, Taphonomic attributes of mollusks as predictors of environment of deposition in modern carbonate systems: Northeastern Caribbean, Ph.D. dissertation, Department of Geological Sciences, University of Rochester, Rochester, NY.

Pennings, S. C., 1997, Indirect interactions on coral reefs, in: *Life and Death of Coral Reefs* (C. Birkeland, ed.), Chapman and Hall Publishers, New York, NY, pp. 249–272

Playford, P., 1980, Devonian "Great Barrier Reef" of the Canning Basin, Western Australia, *AAPG Bull.* **64**:814–840.

Sadd, J. L., 1984, Sediment transport and $CaCO_3$ budget on a fringing reef, Cane Bay, St. Croix, U.S. Virgin Islands, *Bull. Mar Sci.* **35**:221–238.

Scoffin, T. P., and Macintyre, I. G., 1997, Taphonomy of coral reefs a review, *Coral Reefs* **11**:57–77.

Sherman, C. E., Fletcher, C. H., and Rubin, K. H., 1999, Marine and meteoric diagenesis of Pleistocene carbonates from a nearshore submarine terrace, Oahu, Hawaii, *J. Sediment. Res.* **69**:1083–1097.

Simskiss, L., 1964, Phosphates as crystal poisons of calcification, *Biol. Rev.* **39**:487–505.

Steneck, R. S, 1983, Escalating herbivory and resulting adaptive trends in calcareous algal crusts, *Paleobiology* **9**:44–61.

Stern, C. W., and Scoffin, T. P., 1977, Carbonate budget of a fringing ref, Barbados, *Proc. Third Int. Coral Reef Symp.* **2**:471–477.

Stuiver, M., Long, A., Kra, R. S., and Devine, J. M., 1993, Calibration—1993, *Radiocarbon* **35**:1–214.

Talma, A. S., and Vogel, J. C., 1993, A simplified approach to calibrating C14 dates, *Radiocarbon* **35**:317–322.

Vogel, J. C., Fuls, A., Visser, E., and Becker, B., 1993, Pretoria calibration curve for short lived samples, *Radiocarbon* **35**(1):73–86.

Wilson, J. L., 1975, *Carbonate Facies in Geologic Time*, Springer-Verlag, New York.

Wood, R., 1993, Nutrients, predation and the history of reef-building, *Palaios* **8**(6):526–543.

Wood, R., 1995, The changing biology of reef building, *Palaios* **1**(6):517–529.

Zankl, H., and Multer, H. G., 1977, Origin of some internal fabrics in Holocene reef rocks, St. Croix, U.S. Virgin Islands, *Proc. Third Int. Coral Reef Symp.* **2**:127–133.

Zankl, H., Wilson, J. L., and Bosellini, A., 1987, *Outcrop Models for Seismic Stratigraphy:* Field Guide for ERICO Field Trip through the Triassic Alps of Southern Europe, University of Marburg, Marburg, Germany.

Chapter 11

Coral Reefs, Carbonate Sediments, Nutrients, and Global Change

PAMELA HALLOCK

PAMELA HALLOCK • Department of Marine Sciences, University of South Florida, St. Petersburg, Florida 33701.

The History and Sedimentology of Ancient Reef Systems, edited by George D. Stanley Jr., Kluwer Academic/Plenum Publishers, New York, 2001.

1. Introduction

As the 21st century begins, studies of coral reefs, carbonate sediments, and limestones will continue to be fundamental to understanding the past, present, and future of marine ecosystems and global climate. An intellectually challenging aspect of carbonate research is the plethora of paradoxes associated with the biology of carbonate-secreting organisms, carbonate geochemistry, and carbonate depositional ecosystems. Discovering new paradoxes, deciphering existing ones, and deepening understanding of old ones undoubtedly will continue to engage carbonate researchers well into the new century.

The responses of coral reef communities to environmental changes are of interest to governmental agencies, environmental organizations, and the reef-using public, as well as to carbonate sedimentologists and paleontologists. Ever increasing human populations have profoundly altered global biogeochemical cycles through activities that will soon double the concentration of atmospheric CO_2 over preanthropogenic levels and that already have doubled the rate of nitrogen input to terrestrial ecosystems over preanthropogenic rates (e.g., Vitousek *et al.*, 1997). The major goals of this chapter are (1) to summarize basic concepts relating to coral reefs and carbonate sedimentation, (2) to update models relating to shallow-water carbonate production and sedimentation in response primarily to nutrient supply, and (3) to discuss possible consequences of human-influenced global change on reef communities.

2. Coral Reefs and Carbonate Sediments: The Basics

For biologists, coral reefs are marine communities characterized by abundant corals. For geologists, those corals need to be constructing a biogenic reef, which is a limestone structure or buildup produced by biological as well as geological processes. Biogenic reefs can be constructed predominantly by corals, coralline algae, or even oysters or serpulid worms. Ideally, a biogenic reef is a significant, rigid skeletal framework that influences deposition of sediments in its vicinity and is topographically higher than surrounding sediments. A coral reef is such a rigid skeletal structure in which stony corals are major framework constituents.

The major chemical constituent of carbonate sediments and limestones is calcium carbonate ($CaCO_3$). Organisms secrete $CaCO_3$ either as calcite or aragonite. The mineralogical difference is in crystal structures; calcite forms rhombohedral crystals while aragonite forms orthorhombic crystals. Another difference is in the chemical stability of the minerals at temperatures and pressures found on land and in the oceans. Aragonite more readily precipitates in warm seawaters that are supersaturated with $CaCO_3$, but it is less stable in cooler seawaters and in freshwater. Thus, the energy required to precipitate and maintain an aragonite or calcite shell differs depending on the carbonate saturation state of the waters in which an organism lives. In addition, aragonite is structurally stronger than calcite, a characteristic of particular importance for organisms with erect growth forms, including branching corals and some large bryozoans. Energetic and structural aspects of biomineralization have likely played roles in the adaptation of shelled organisms to their environments.

2.1. Environmental Requirements for Coral Reef Growth

Mention of coral reefs leads many persons, whether scientist or not, to envision a tropical island surrounded by warm, crystal-clear seawater, beneath the surface of which lies a profusion of corals and colorful fish (Fig. 1). Such pictures are no accident, for they illustrate the basic requirements for reefs that are constructed by zooxanthellate corals (i.e., those hosting dinoflagellate algal symbionts) and associated communities (Table 1).

Coral reefs require subtropical to tropical temperatures year round: seasonal temperature ranges may be from occasional winter lows of about 14°C to summer highs of about 30°C, with optimum summer temperatures of 23–29°C. Prolonged exposure to temperatures even a degree or two above normal

FIGURE 1. Oceanic-influenced Pacific coral reef showing dominance by framework-building corals. (Photograph by J. C. Halas; reprinted from Hallock *et al.*, 1993.)

TABLE 1. Summary of Environmental Factors Influencing Coral Reefs

- Optimum temperature: 23–29°C, winter mean >18°C
  - Problems: many species killed at <14°C, >31°C
- Optimum salinity: "normal marine," 33–37°
  - Problems: prolonged exposure to <30° and >40°
- Optimum solar radiation: to at least 30°N/S latitude
  - Problems: diminished water transparency (algal blooms, sediments); too much sunlight in very calm weather
- Water motion: some is essential for metabolic exchange and uptake of food
  - Problems: very calm weather; major storms
- Terrigenous sediments: inhibit reef growth
  - Optimum: Oceanic banks and islands and off passive margins and deserts
  - Problems: rivers and streams, mountainous coastlines (active margins); muddy sediments worse than sands
- Nutrients: fixed nitrogen, phosphorus, and trace nutrients
  - Optimum: minimal; oceanic waters of subtropical gyres
  - Problems: upwelling zones, river plumes, local sources
- Oxygen, carbon dioxide, and pH
  - Optimum: oxygen concentrations near saturation; pH, 8.1–8.4
  - Problems: limited water motion, excessive respiration
- Combined stresses:
  - Hot, still weather: heat stress, more UVB, oxygen stress
  - River runoff: lower salinity, more sediments, nutrients, turbidity
  - Upwelling: lower temperature, more nutrients
  - Sewage: more nutrients, higher biological oxygen demand

summer temperatures commonly promotes coral bleaching, which is the expulsion of the zooxanthellae. Winter cold fronts also can damage or kill corals, though on the southern Great Barrier Reef of Australia and in the Persian Gulf some species can tolerate exposure to temperatures below 10°C.

Corals also require seawater of normal marine salinities, typically in the 33 to 37° range. Though many corals can tolerate limited exposure to rainwater or freshwater runoff, reefs are generally not found immediately offshore from even small streams. At the other extreme, in the Persian Gulf some coral species have adapted to salinities in excess of 40°.

Pictures of beautiful reefs in crystal-clear waters also are no accident. Corals, with their zooxanthellae, require abundant sunlight to grow and calcify. Particles in the water that reduce water transparency, such as sediments or abundant phytoplankton, are detrimental to reef growth. Even the island setting is typical, for while there are (or, in many cases, were) fringing reefs on continental shorelines around the Caribbean, East Africa, the Middle East, and Southeast Asia, the most spectacular coral reefs thrive in settings removed from significant terrestrial runoff, which includes freshwater, sediment, and nutrients. At the same time, coral reefs require a shallow substrate on which to grow; islands or shoals on continental shelves and volcanic structures in the open ocean provide that necessary substrate.

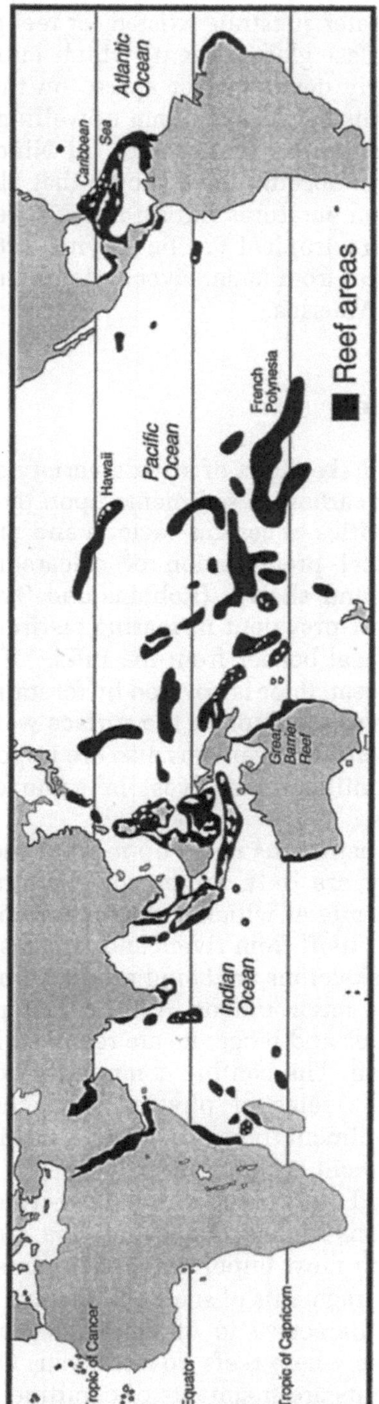

FIGURE 2. Global distribution of coral reefs.

As a result, coral reefs are found throughout the tropical regions (Fig. 2) where suitable shallow-water substrate existed for reef growth to occur as sea level rose following the last glacial event, which culminated about 18,000 years ago. Reefs are poorly developed or sparse on the eastern sides of the Atlantic and Pacific Oceans because regional upwelling produces cool, nutrient-laden currents that suppress reef growth. El Niño/Southern oscillation events during the past two decades have shown that bleaching and mortality in response to elevated temperatures during El Niño years further limits reef development in the eastern tropical Pacific (Glynn, 1988). Reef development also is suppressed by runoff from major river systems such as the Amazon and Orinoco Rivers of South America.

2.2. Carbonate Sediments

Calcareous shells and skeletons of a wide variety of protists, plants, and animals become biogenic carbonate sediments upon the death of those organisms. The metabolic activities of certain bacteria and microalgae also contribute to the biogeochemical precipitation of calcareous muds in seawater overlying shallow banks and shelves (Robbins and Blackwelder, 1992). Carbonate sediments are most prevalent in marine environments that are separated by distance or physical barrier from the influx of sediments from land. Nearly half the modern ocean floor is covered by foraminiferal ooze, the empty shells of protists that live as plankton in the surface waters of the open ocean. Shells and skeletons of benthic organisms also are important sediment constituents, especially on continental shelves, in some coastal areas, and on oceanic banks and shoals.

Whether biogenic constituents make up most of the bottom sediments in an area or whether they are only minor contributors depends on several factors. One factor is the rate at which sediments from land are entering the marine environment via runoff from rivers and streams. Another factor is the rate at which shells and skeletons are being produced by the biotic communities living in the marine environment. A third factor is the rate at which sediments, both terrigenous and biogenic, are removed from that environment by transport or dissolution. The benthic community not only produces sediments, it also affects the rates of physical and chemical breakdown of sediments and rates of sediment transport. Thus, benthic communities strongly influence rates of sediment accumulation.

Lees (1975) recognized three classes of shallow-water carbonate sediments based on their major constituents. He called the simplest group *foramol sediments* after two of the most important constituents: benthic foraminifers and mollusks, especially fragments of snail and bivalve shells. Lees noted that foramol sediments are characteristic of temperate shelves, but sometimes dominate in tropical areas where reefs do not occur. Other important constituents of foramol sediments are fragments of coralline algae, echinoid spines and plates, bryozoans, barnacles, worm tubes, and azooxanthellate corals.

James (1997), in his comprehensive review of cool-water carbonates, suggested the term "heterozoan association" to describe this assemblage, because constituents other than foraminifers and mollusks, particularly bryozoan and azooxanthellate coral debris, often dominate such sediments.

Lees (1975) recognized two major classes of shallow-water carbonate sediments that are more prevalent in subtropical and tropical environments. *Chloralgal sediments* are dominated by the remains of calcareous green algae such as *Halimeda*; foramol constituents, particularly molluscan debris and foraminiferal shells, are typical secondary components. Chloralgal sediments are prevalent in expansive shallows like Florida Bay and the Bahama Banks and in deep-euphotic settings including the lagoon of the Australian Great Barrier Reef. *Chlorozoan sediments* are found around coral reefs and have coral and calcareous algal remains as the dominant constituents. Coralline algae, foraminifers, and mollusk and echinoid fragments are important secondary components. Bryozoan, barnacle, worm–shell, and azooxanthellate–coral debris are usually uncommon in chlorozoan sediments because these organisms thrive best in waters with richer food supplies than do corals. James (1997) recommended combining the chloralgal and chlorozoan sediment classes into one, the "photozoan association," in recognition of the role that photosynthesis plays in the calcification of zooxanthellate corals and calcareous green algae, and included the products of active biogeochemical and geochemical precipitation, specifically lime muds, ooids, and carbonate cements. However, as will become evident later in the chapter, there is a solid ecological basis for distinguishing between chlorozoan and chloralgal sediments.

2.3. Organism–Sediment Interactions

Carbonate sediments produced by benthic communities are primarily found on continental shelves, oceanic banks and atolls, and nearshore environments where terrigenous input is limited. Whether sediments accumulate in place or whether they are transported away depends both on physical factors, such as the strength of waves and currents, and on the ability of the benthic community to hold sediments in place. Organisms that project upward from the sediment, slowing water motion and providing quieter places for sediments to settle, are called *bafflers*. Organisms that live in or directly on the sediment, holding or encrusting it in place, can be considered *binders* (Fagerstrom, 1987).

Microalgae and bacteria grow and create mats directly on sediments that accumulate where wave and current motion is limited or intermittent. Bacterial filaments bind these mats, which can resist as much as 10 times more wave or current energy than is required to move similar unbound sediments (Grant and Gust, 1987). Stromatolites are biogenic reefs constructed by this process. Modern stromatolites are found in Shark's Bay, Australia (Logan *et al.*, 1974) and at several localities on the Bahama Banks (Reid and Browne, 1991).

In current-swept environments, specialized sponges may live in and on the surface layers of skeletal sediments, binding them in place. Coralline red algae can colonize the surface of sediment-filled sponges, forming solid substrate on which other organisms settle and grow. Such communities produce sponge–algal bioherms along the margins of some western Caribbean banks (Hallock *et al.*, 1988).

A variety of elongate, upward-projecting plants and animals baffle water motion and trap sediments. On modern shallow shelves, seagrass beds effectively stabilize sediment over vast areas. The blades and fronds of seagrasses and associated macroalgae slow water flow, allowing suspended sediments to settle out. Sediments are then held in place by extensive seagrass root and rhizome systems, as well as by the holdfasts of the algae. Other common bafflers and binders include sponges and octocorals such as sea whips and sea fans.

Stony corals, which include both scleractinian and milleporian corals, are the major biogenic framework builders of modern reefs (Fagerstrom, 1987). These organisms grow upward or outward in branching, massive or platy morphologies; secreting substantial quantities of calcium carbonate, while trapping even greater quantities of sediment within and in the lee of the reef framework. Crustose coralline algae encrust the reef framework and enclose sediments into the massive, wave-resistant structures we recognize as coral reefs. Geochemically precipitated aragonite or high-magnesium calcite cements also play a major role in creating the massive, wave-resistant structures that coral reef communities have constructed.

The three-dimensional topography of the reef provides abundant habitats for the diverse array of species that dwell within the reef structure. All contribute to the reef community in some way, many to the reef structure itself and all to energy flow within the community. Some of these species are encrusters, some are sediment producers, and some are wholly soft-bodied and have little direct influence on the reef structure. Many species even contribute to the breakdown of the reef structure by boring into it or scraping away at it as they graze. Such organisms are known collectively as bioeroders (Neumann, 1966).

On a healthy, actively accreting reef, bioerosion contributes to the diversity of habitats within the massive reef structure (Hutchings, 1986). Organisms that scrape away reef limestone as they graze algae include sea urchins, chitons, and some snails. Many reef fish feed by breaking or scraping off bits of coral or coralline algae. Organisms that bore or etch their way into the reef include bacteria, fungi, several varieties each of sponges, worms, clams, and sea urchins. However, if reef growth slows in response to natural or anthropogenic environmental stresses, the rates of destruction can exceed rates of accretion and the reef may cease to exist (Glynn, 1988; Hallock, 1988).

3. The Nutrient Paradox

A key paradox of coral reef systems for much of the 20th century was how such diverse and "productive" communities could thrive in the clearest, most nutrient-deficient (oligotrophic) surface waters in the oceans (e.g., Odum and Odum, 1955; Wells, 1957). Hallock and Schlager (1986) demystified this "nutrient paradox" by explaining why reef-building corals are highly adapted to clear, nutrient-poor waters. They provided several arguments for why excess nutrients are detrimental to reef accretion, including reduced water transparency, phosphate inhibition of $CaCO_3$ crystal formation, community change, and increased rates of bioerosion. Reef communities respond to changes in nutrient supply on both local and regional scales (Birkeland, 1997).

The nutrient paradox can be more readily understood by distinguishing different aspects of biological productivity (Hallock and Schlager, 1986). The term "nutrients" refers primarily to fixed nitrogen (NH_4^+, NO_2^-, NO_3^-) and phosphate ions (PO_4^{3-}) required by all organisms to synthesize proteins and nuclear material for cell maintenance, growth, and reproduction.

Primary productivity is the rate at which energy is fixed into organic carbon (i.e., simple sugars) during photosynthesis. Gross primary productivity (GPP) is the total amount of energy fixed to organic carbon per unit time, regardless of whether the organic matter is used directly for respiration or further synthesized into more complex compounds for metabolism and growth. GPP is controlled primarily by light availability, so in shallow, clear, reef waters, GPP can be extremely high, up to 18 g C m^{-2}/day (Odum, 1959).

The nutrient paradox arose because many researchers failed to recognize that net primary productivity (NPP) is quantitatively quite different from GPP. NPP is the amount of energy fixed in photosynthesis minus the energy lost during respiration, that is, the amount of energy in the form of organic matter available for growth of organisms. NPP, which largely determines "harvestable production," is primarily a function of nutrient supply and respiration rates. Since nutrient supply is low in clear oceanic waters and since respiration rates increase with temperature, NPP on reefs tends to be very low. Thus, on coral reefs, "high" productivity is deceptive: photosynthetic rates are high but so are respiration rates; nutrient supplies are so low that net (harvestable) production may be negligible.

4. Advantages of Algal Symbiosis

Coral reefs are special because they are characterized by mixotrophic organisms, with autotrophic and heterotrophic organisms playing secondary roles (Hallock and Schlager, 1986). Autotrophic (meaning self-feeding) organisms fix their own organic carbon using sunlight and dissolved inorganic nutrients but do not feed on organic matter. Typical autotrophs (i.e., "plants") include phytoplankton, benthic algae, and seagrasses. Heterotrophic (meaning feeding on others) organisms must consume organic matter to live and grow. Heterotrophs include many kinds of bacteria and protists and most animals. Mixotrophic organisms can both feed and photosynthesize, often because they are symbiotic associations in which viable algal cells are maintained within the cells or tissues of protozoan or animal hosts. Although zooxanthellate

FIGURE 3. Foraminifera with algal endosymbionts, including (a) *Amphistegina*, (b) *Laevipeneroplis*, and (c) *Peneroplis* (scale bar = 1 mm).

corals are the best known mixotrophs, larger foraminifers (Fig. 3), giant clams (*Tridacna* spp.), sea anemones, soft corals, and some sponges and ascidians also commonly host symbiotic algae or cyanobacteria (i.e., photosynthetic bacteria).

The coral polyp (Fig. 4), with its tentacles armed with stinging nematocysts, is an effective predatory animal. A biological paradox is that corals have a very high prey–capture surface to body-volume ratio and in that respect are one of the most highly specialized predators in the animal kingdom (Yonge, 1930), yet food capture provides only about 10% of their daily energy needs (Falkowski *et al.*, 1993). Within the coral's tissue lie thousands of live algal cells known as zooxanthellae that can harvest solar energy through photosynthesis. About 90% of the coral's energy supply comes from photosynthetic products produced by their zooxanthellae. The solution to this paradox lies in recognizing that corals capture organic matter required for growth and reproduction, rather than to meet their energy needs.

4.1. Energy from Photosynthesis

Hallock (1981) mathematically explored the hypothesis that algal symbiosis provides host–symbiont units with substantial energetic advantages over similar organisms that lack algal symbionts (Fig. 5a). Basic assumptions of the

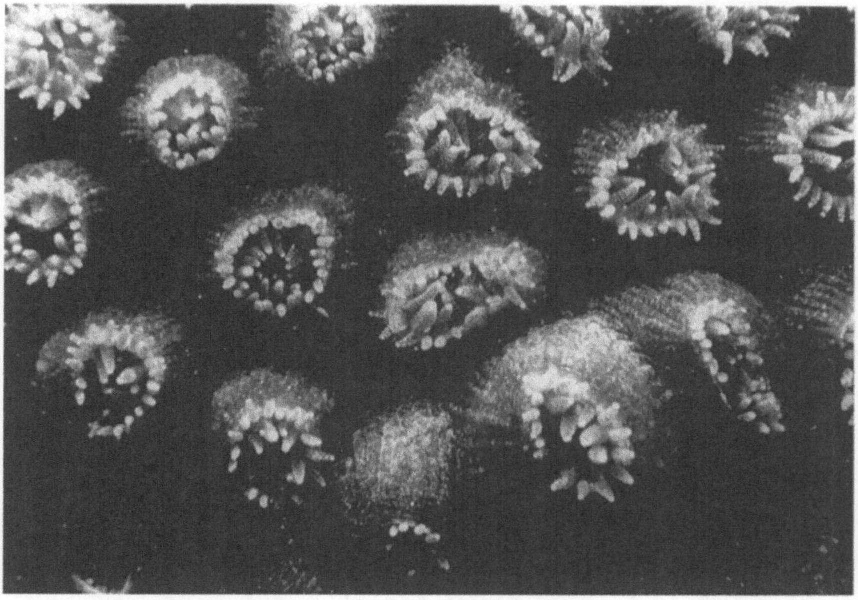

FIGURE 4. Polyps of *Montastrea annularis* showing tentacles and central mouth. (Photo by S. Bowser.)

model were (1) that nutrients are required for growth in proportion to their ratios in organic matter, (2) that a mixotrophic organism could uptake nutrients either as dissolved inorganic nutrients utilized by the zooxanthellae or as organic carbon ingested by the host, and (3) that application of the model was limited to the upper euphotic zone where there is ample light for photosynthesis. Hallock (1981) postulated that in truly nutrient-deficient environments the most significant concentrations of essential nutrients are in organic matter and that algal symbiosis enables the symbiotic unit to function as a primary producer that obtains nutrients by feeding. The model results indicated that algal symbiosis provides the symbiotic unit with several orders of magnitude energetic advantage over nonsymbiotic competitors in nutrient-deficient environments. Thus, once established, algal symbiosis provides significant energetic advantage for specialization of the host to symbiosis, providing considerable insight into repeated episodes of specialization of probable symbiont-bearing organisms in the fossil record (e.g., Hallock, 1982; Cowen, 1983).

Subsequent research (e.g., Falkowski *et al.*, 1993; Steven and Broadbent, 1997) indicates that it is essential for the host to maintain control of the supply of fixed nitrogen. The photosynthetic process is limited by the availability of solar energy and CO_2, and is relatively independent of nutrient supply. Symbionts that are nitrogen-limited fix one to two orders of magnitude more carbon than is needed to maintain stable cell densities within the coral host.

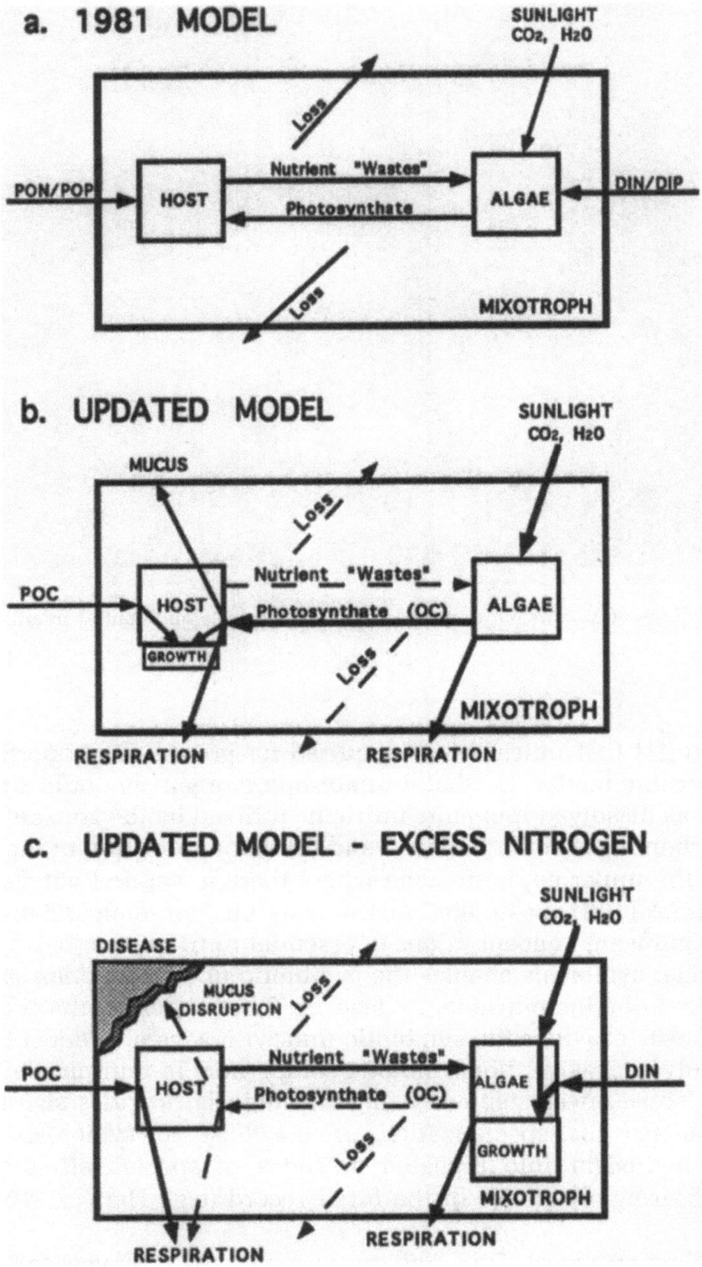

FIGURE 5. (a) Model of algal symbiosis of Hallock (1981), based on assumptions that essential nutrients were limiting and could be taken up by either the host as particulate organic carbon containing particulate organic nitrogen (PON) and phosphorus (POP) or by the symbionts as dissolved inorganic nitrogen (DIN) or phosphorus (DIP). (b) Updated model of algal symbiosis based on subsequent studies indicating that it is essential for the host to maintain control of the

About 90–99% of the carbon fixed by the zooxanthellae, consisting mostly of carbohydrates and lipids and often referred to as photosynthate, is translocated to the host (Fig. 5b), where it is used as an energy source for respiration and for synthesis of mucus. With its metabolic energy requirements met by the photosynthate from its symbionts, the host can use most of the proteins and phosphatic acids ingested in captured food for its own growth and reproduction.

However, if exposed to chronically elevated sources of inorganic fixed nitrogen, the zooxanthellae retain their photosynthate to synthesize proteins for their own growth and reproduction. The consequences are an increase in algal densities within the host and a decrease in translocation of photosynthate to the host. Moreover, photosynthesis per unit area of coral does not increase, even if zooxanthellae densities double or triple, because photosynthesis is limited by availability of light and CO_2. Thus, coral symbiotic associations require low ambient nutrient concentrations (Falkowski *et al.*, 1993). The biological consequences of chronic nutrification, including changes in mucus production and zooxanthellae densities, may contribute to disease and bleaching in corals.

Corals produce mucus to shed sediments (Rogers, 1990) and to fend off bacterial attack (Santavy and Peters, 1997). If mucus is produced in excess, bacteria can bloom in the mucus and kill the corals by oxygen depletion, by accumulation of sulfide poisons at the coral surface below the mucus layer and by predation by bacteria on weakened coral polyps (Garrett and Ducklow, 1975; Ducklow and Mitchell, 1979). If insufficient mucus is produced, bacteria can attack directly. The updated conceptual model of symbiosis (Fig. 5b), modified to include chronic nutrification (Fig. 5c), represents possible consequences of reduced photosynthate translocation to the coral host, and therefore reduced mucus production. New diseases are being reported regularly, particularly in western Atlantic corals (e.g., Santavy and Peters, 1997). Black-band disease (Fig. 6), which is caused by pathogenic outbreaks of a naturally occurring assemblage of microorganisms (Richardson *et al.*, 1997), is now reported from both the western Atlantic and Indo-Pacific, particularly in areas where the reefs are otherwise stressed (Peters, 1997). Santavy and Peters (1997) suggested that the disruption in mucus production associated with chronic anthropogenic stresses could be a factor in the proliferation of previously uncommon diseases.

supply of fixed nitrogen reaching the symbionts. When severely nutrient limited but in ample sunlight the symbionts photosynthesize excess organic carbon (OC), which is released to the host for use in respiration and for carbohydrate-rich compounds such as coral mucus. Thus, protein-rich food (i.e., particulate organic carbon, POC) captured by the host can be used by the host for growth. (c) Updated model indicating a possible consequence of excess fixed nitrogen, resulting in breakdown of the symbiosis: When dissolved inorganic nitrogen (DIN) is readily available to the symbiotic algae, they retain their photosynthate for growth. This reduces the supply of photosynthate to the host for respiration and mucus production, forcing the host to utilize its food for respiration and metabolism, and increasing the host's susceptibility to disease.

FIGURE 6. Black band disease (*Phormidium coralyticum*) on a *Colpophyllia natans* coral in Key Largo National Marine Sanctuary: (a) live coral, (b) disease line, and (c) dead coral. (Photograph by J. C. Halas; reprinted from Hallock *et al.*, 1993.)

Zooxanthellae–cell numbers per unit area of coral host can increase up to threefold in corals exposed to elevated inorganic nitrogen levels (Falkowski *et al.*, 1993). Thus, the coral has more competition for oxygen at night, when photosynthetic oxygen is not directly available, and greater potential for oxygen toxicity during the day when photosynthesis is taking place (e.g., Shick *et al.*, 1996). Furthermore, increased algal cell densities result in darker coral tissue. All these factors potentially make zooxanthellate corals more vulnerable to stress in summer, especially during episodes of hot, still weather, the times when coral bleaching events are most likely to occur (Glynn, 1996).

This newer information strongly supports earlier predictions about the advantages of algal symbiosis in tropical, nutrient-deficient environments and the potential of symbiosis as an adaptive mechanism to explain specialization of host organisms. The more recent studies also explain why these symbioses fail to compete effectively when nutrient supply increases, with possible geological consequences ranging from facies changes (e.g., Hallock, 1988) to mass extinctions (e.g., Hallock, 1987).

4.2. Algal Symbiosis and Calcification

Because scleractinian corals and foraminifers with algal symbionts are such prolific producers of calcium carbonate skeletons, conventional wisdom has long held that photosynthesis by the symbionts promotes calcification by splitting of bicarbonate (e.g., ter Kuile, 1991) and removal of CO_2 (Eq. 1):

$$Ca^{2+} + 2HCO_3^- \Leftrightarrow CO_2 \text{ (to photosynthesis)} + CaCO_3 \text{ (calcification)} + H_20 \quad (1)$$

McConnaughey (1989) and McConnaughey and Whelan (1997) have proposed the reverse interpretation. They postulate that lack of CO_2 limits photosynthesis in warm, shallow, alkaline aquatic environments and that calcification provides protons that make CO_2 readily available from the much more abundant bicarbonate ions (Eq. 2). That is, calcification:

$$Ca^{2+} + HCO_3^- \Leftrightarrow CaCO_3 + H^+ \quad (2)$$

promotes photosynthesis:

$$HCO_3^- + H^+ \Leftrightarrow CH_20 + O_2 \quad (3)$$

According to this hypothesis, the electron capture phase of photosynthesis provides ATP for active transport of Ca^{2+} and H^+ ions, promoting calcification and making bicarbonate ion a viable source of CO_2 for the organic carbon synthesis phase of photosynthesis. Erez (1983) found essentially normal calcification rates in symbiont-bearing foraminifers that had been treated with an herbicide that blocks photosystem II, the carbon fixation step in photosynthesis, but not photosystem I, the initial step in which solar energy is captured and fixed into ATP. ATP also may provide the energy for removal of ammonium, phosphate, and magnesium ions that inhibit calcification (ter Kuile, 1991). If calcifying animals with algal endosymbionts can use sunlight rather than food resources as the source of energy for ATP production, that would account for the significant enhancement of calcification by light that has long been recognized in corals (Goreau, 1959) and foraminifers (Duguay and Taylor, 1978; Erez, 1983).

4.3. Significance of Algal Symbiosis to Ecosystems

Besides the benefits to individual organisms, the advantages of algal symbioses to ecosystems in nutrient-deficient environments are tremendous. In ecosystems dominated by primarily autotrophic and heterotrophic nutrition, food webs are based on autotrophic primary producers. There typically is a 90% or greater loss of energy and biomass to respiration with each heterotrophic link in the food chain (e.g., Odum, 1959). That is, less than 10% of the biomass at one trophic level is transferred to the next level.

Mixotrophic nutrition in contrast provides essentially "free links" in the food web. That is, when a mixotrophic organism such as a zooxanthellate coral captures a unit of prey, the majority of the digestible organic matter in the prey can be used by the host for growth and reproduction. Photosynthesis by the symbionts provides energy for the host's respiratory needs (e.g., Falkowski *et al.*, 1993). Thus, the biomass transfer from the prey to the mixotroph potentially can be far greater than the typical 10%. For this reason, reef-building corals can accumulate organic matter that was formerly dispersed in great volumes of seawater into reef biomass. Mixotrophic nutrition is a fundamental trophic strategy of critical importance to ecosystems occupying nutrient-deficient euphotic environments. Furthermore, mixotrophic organisms that calcify can provide substratum for whole ecosystems to develop and flourish.

5. $CaCO_3$ Production and Nutrient Gradients

Rates of nutrient supply are a major controlling mechanism for benthic communities in shallow, tropical environments (Hallock and Schlager, 1986; Birkeland, 1987). Characteristics of the oceanic waters within a region provide "background" conditions, i.e., the clearest waters and minimum nutrient supply. Local upwelling or terrestrial runoff add nutrients to that "background" supply. Where clear, nutrient-deficient (oligotrophic), tropical–subtropical oceanic waters interact with benthic communities, reef-building corals dominate (Fig. 7). Where modest nutrient flux from terrestrial or upwelled sources affect benthic communities, either locally or regionally, coral cover diminishes in favor of macroalgae and sponges. Where sufficient runoff or upwelling elevate nutrient flux to support significant phytoplankton densities in the water column, benthic communities are dominated by heterotrophic organisms.

Carbonate sedimentation and reef development also can be characterized along nutrient gradients (e.g., Hallock, 1988; Triffleman *et al.*, 1992). In geologic terms, "oligotrophic" includes the range of conditions that support true reef framework (Fig. 7). Rimmed coral reef development occurs where sufficient coral framework can grow and accrete, encrusted by coralline algae, providing a wave-resistant rim to hold the sediment production by other members of the reef community, particularly calcareous algae and larger foraminifers. Moreover, the oligotrophic range on the gradient includes both oceanic reef systems, such as those of volcanic islands and atolls in the Pacific, and continentally influenced reef systems, which include the those of southeast Asia and the Caribbean (Birkeland, 1997).

Mesotrophic refers to truly intermediate conditions where light penetration is sufficient to support prolific calcareous algal production and where dissolved and particulate nutrient supplies are sufficiently high to favor macroalgae and sponge domination of the benthos, including abundant bioeroding faunas (Fig. 8). Molluscan and echinoid sediment contributions, particularly by grazing taxa, also can be significant. The result is a benthos that

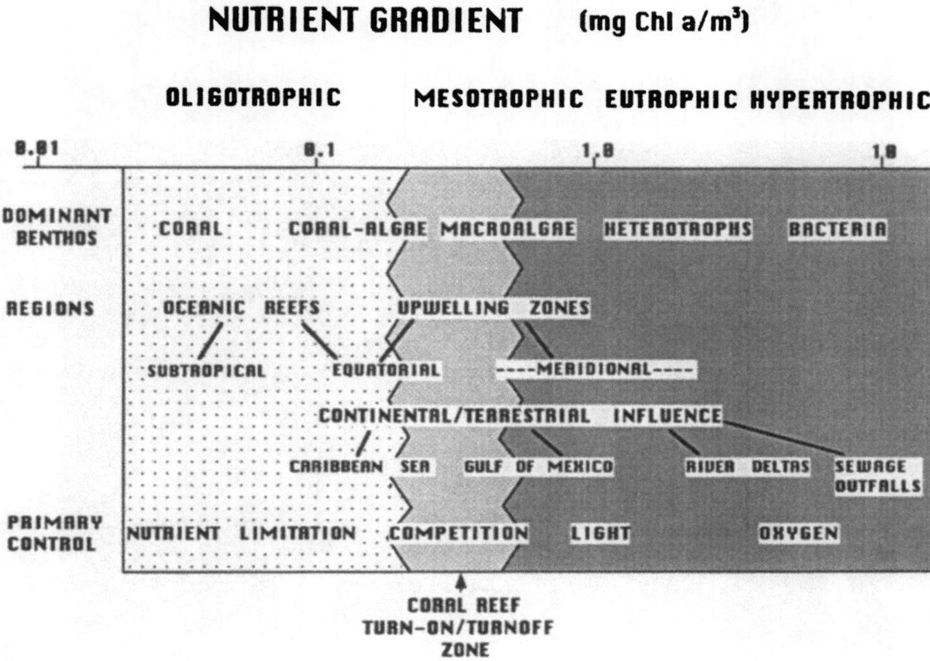

FIGURE 7. The nutrient gradient in low-latitude, euphotic marine waters, presented in milligrams of chlorophyll per cubic meter of seawater. Nutrient flux and benthic communities: "Primary control" refers to major factors controlling benthic community structure, i.e., nutrient limitation favors mixotrophic organisms (i.e., corals), "biotic factors" such as competition for space favors fast-growing macroalgae at intermediate rates of nutrient supply, light limitation of photosynthetic benthos occurs as phytoplankton densities increase in response to relatively high rates of nutrient flux, and oxygen limitation occurs when the supply of organic carbon exceeds the rate of mixing of oxygen to the benthos. "Dominant benthos" refers to visually dominant benthic organisms under that nutrient flux. Nutrient flux regions contrasts oceanic and upwelling zones with regions influenced by terrestrial runoff. The "coral reef turn-on/turnoff zone" refers to the range of nutrient flux near the limits for reef development, under which changes in local physical or biotic factors can turn on or turn off reef growth.

can produce carbonate sediments at high rates. But without net framework accretion, physical processes can move sediment offshelf. In geologic terms, this is the realm of chloralgal bioherms or of hardgrounds, depending on physical processes.

From a geologic context, eutrophic can be used to denote the region of the gradient where nutrient flux is sufficient to limit light penetration, and therefore calcareous algal production. Even so, if terrigenous or pelagic sedimentation is limited, benthic carbonate production may continue to dominate, produced by members of the heterozoan association (James, 1997), especially oysters or other bivalves, predatory and detritus-feeding gastropods, grazing and burrowing echinoids, bryozoans, and azooxanthellate corals.

FIGURE 8. Example of an upwelling-influenced benthic community dominated by macroalgae and sponges on Seranilla Bank, Nicaraguan Rise, Caribbean Sea. Sea fans are common, while reef-building corals are minor components of the community. (Photograph by W. C. Jaap; reprinted from Hallock *et al.*, 1993.)

Hypertrophic environments occur where nutrient flux to surface waters is so high as to support dense stands of phytoplankton. The supply of organic matter frequently outpaces oxygen supply to benthic communities, resulting in episodic hypoxia or anoxia in the benthos. Characteristic sediments in areas of hypertrophication include diatom-rich, organic-rich, or phosphatic sediments. Because coral reefs are typically relatively shallow and exposed to wave and current action, hypertrophication is not likely to occur on a coral reef (Fig. 7) without human intervention such as installation of a sewage outfall or serious runoff from heavily fertilized fields. Even in the well-studied case of sewage eutrophication of Kaneohe Bay, Hawaii, organic matter did not outpace oxygen supply (Smith *et al.*, 1981).

One natural mechanism for short-term eutrophication of an oceanic reef community might be a submarine volcanic megaplume as suggested by Vogt (1989). Nutrient concentrations are hundreds of times higher in deep seawater than in subtropical surface waters. Therefore, if a large submarine volcanic eruption suddenly heated a volume of deep water enough that it rose to the ocean surface, that plume would stimulate a major plankton bloom. If an oceanic island was downcurrent from where the megaplume surfaced, the reef community would be temporarily damaged. Another commonly invoked hypothesis—that of rising sea level and flooding of the shelves by anoxic water (e.g., Arthur and Schlanger, 1979; Copper, 1994)—could certainly increase nutrient flux and profoundly change shallow-water communities, but probably would not directly eliminate them by oxygen stress. Surface-mixing processes on open shelf environments typically oxygenate waters as deeply as light can penetrate. Furthermore, even very rapid sea-level rise occurs on time scales of at least millenia (e.g., Blanchon and Shaw, 1995), while community change in response to nutrients occurs on the order of years to decades (e.g., Smith *et al.*, 1981; Cockey *et al.*, 1996).

5.1. Ecological Basis for Gradient Subdivisions

The ecological parameters influencing each community transition (Figs. 7, 9a,b) are quite different. Benthic community shift from coral-dominated, to coral/algae-dominated, to macroalgae/sponge-dominated appear to be primarily controlled by nutrient supply and by biotic factors such as competition for space, bioerosion, and disease.

Because zooxanthellate corals are mixotrophs, i.e., they both feed and photosynthesize and thereby can recycle scarce nutrients, they are most successful and competitive for space under the very nutrient-poor conditions, such as subtropical Pacific atolls (Figs. 7, 10a). Using sea-surface chlorophyll estimates as an indicator of nutrient flux (e.g., Laws and Redalje, 1979), Hallock *et al.* (1993) defined oligotrophic waters as those with chlorophyll concentrations consistently less than 0.1 mg/m^{-3} (Fig. 7). Under these conditions, coral growth rates may be somewhat suppressed (Fig. 9a), but calcification appears to be largely a function of gross photosynthesis, which is a function of light. Thus, corals can actively accrete framework, at rates of up to 10 kg $CaCO_3$ m^{-2}/year (e.g., Kinsey, 1985) under very low nutrient conditions (Fig. 9a). Coralline algae play an indispensable role as encrusters, stabilizing the reef framework. Under these same conditions, larger foraminifers, which also have algal symbionts and are physiologically analogous to corals in many respects (Lee and Anderson 1991), often are major sediment producers. But sediment production over the whole reef system is nutrient limited (Fig. 7), so the most active accretion is on the reef rim, with lower rates of accretion in back reef environments, resulting in Pacific-type oceanic reefs (Fig. 10a).

As nutrient flux slightly increases, either from terrestrial runoff or subsurface sources such as equatorial, coastal, or topographic upwelling, to consistently support sea surface chlorophyll concentrations of about 0.1–0.2 mg/m^{-3}, the carbonate machine approaches its most productive mode. Calcareous green algae are the major beneficiaries of the increased nutrients, and thereby are principally responsible for the increase in carbonate production (Figs. 7, 9b). Individual coral growth rates may increase as food supplies increase (Fig. 9a). But increasing densities of phytoplankton in the water column provide nonsymbiotic animals (aerobic heterotrophs: Fig. 7) with critical food sources. Perhaps even more importantly, many coral competitors and predators

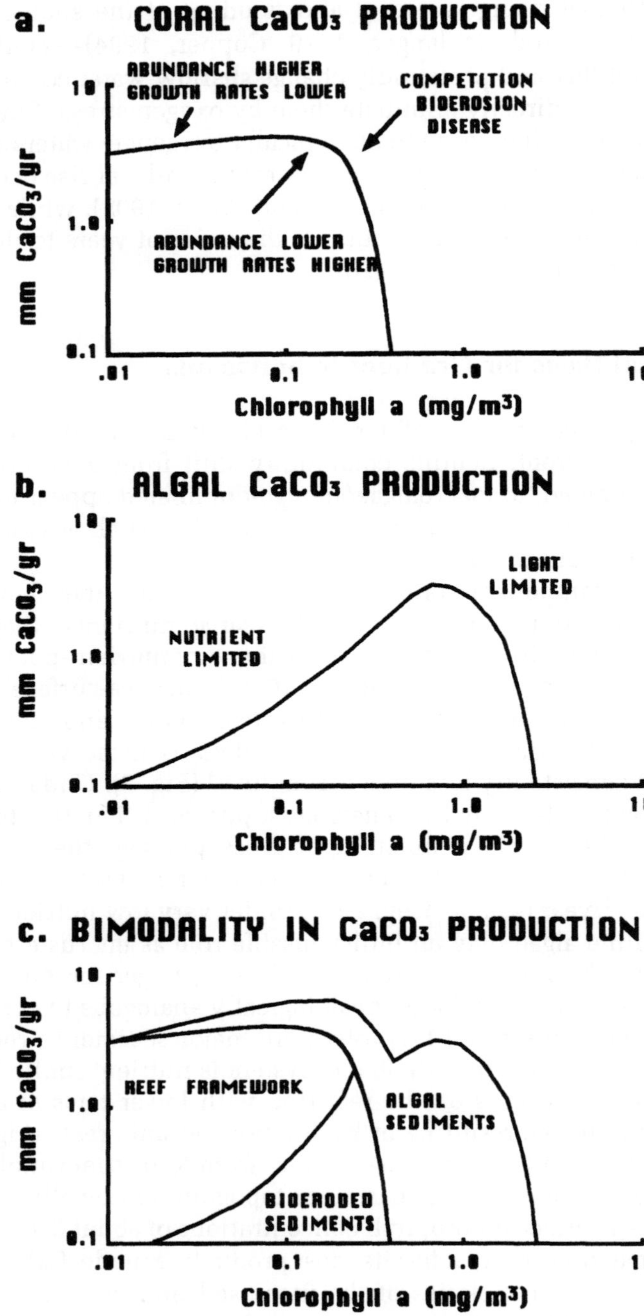

FIGURE 9. (a) Calcium carbonate production (mm $CaCO_3$/yr) by corals relative to the nutrient gradient presented in Fig. 7. In very nutrient-limited waters, growth rates of individual corals may be nutrient limited. However, because food resources are so limited, mixotrophic corals are abundant relative to competitors. As nutrient and food supplies increase, growth rates of individual corals may increase, but so do growth rates of algae and heterotrophic organisms that

have planktotrophic larvae (i.e., larvae that feed on plankton), so their survival is dependent on plankton densities. Coral larvae, on the other hand, do not feed but are provided zooxanthellae or an energy supply by the parent. Coral larvae are most competitive when there is little else in the water either to eat them or to compete with them for space on which to recruit (Birkeland, 1977).

Among the animals benefiting from increased planktonic food supplies are the bioeroders: lithophagid clams, clionid and other boring sponges, and a variety of boring worms. Similarly, the bioeroding surface grazers, including echinoids, gastropods, and fish, benefit not only from increased larval survival, but also from increased growth rates of filamentous algae. As a result, bioerosion rates increase with increasing nutrient flux (Fig. 9c), as does production of molluscan and echinoid skeletal debris.

Growth rates of adult corals often remain high or even increase somewhat as nutrient flux increases (Fig 9a), but net framework production declines as coral cover declines, and especially as bioerosion rates increase and convert framework to sediments. At the same time, total carbonate production increases, as algal, molluscan, and echinoid production increases. The result is Caribbean/Western Atlantic-style carbonate banks (Figs. 7, 10).

The shift from chlorozoan sedimentation (coral reefs) to chloralgal sedimentation (dominated by calcareous green algal production) appears to result from community response to increasing nutrient supplies (Fig. 9c). On the other hand, the transition from chloralgal sedimentation to eutrophic heterozoan sedimentation, at least in low-latitude environments, is largely a response to light limitation. As nutrient flux increases, plankton densities increase and water transparency declines. At chlorophyll concentrations of 1 mg/m^{-3} or more, light penetration is restricted to a few meters at most. With most primary production occurring within the water column, there are abundant food supplies for filter- and detritus-feeding benthos, including bivalves, bryozoans, and burrowing echinoids. Bioerosion can be very active; Kinsey (1985) reported bioerosion rates of 6.5 kg $CaCO_3$ m^{-2}/year under eutrophic conditions in Kanoehe Bay, Hawaii.

In the somewhat analogous subtropical to temperate transition from chloralgal to heterozoan sedimentation, geochemical factors likely reduce the energetic advantage of photosynthetic production of aragonite by calcareous algae, so that even within the euphotic zone, calcareous organisms in the temperate benthic community are dominated by coralline red algae, which produce calcite skeletons, and by animals without zooxanthellae (James, 1997).

---

can outcompete the corals for recruitment space and that bioerode coral skeletons. (b) Calcareous algal production generally increases with nutrient supply until increasing phytoplankton densities, which are also increasing relative to nutrient supply, severely restrict light reaching the benthic community. (c) Combining the coral and algal production curves indicates that carbonate production may be bimodal, with maximal reef framework accretion in very oligotrophic environments; some framework accretion and maximum carbonate production occur in marginally oligotrophic environments and algal carbonate production is maximum in mesotrophic environments. Note $\log_{10}$ scales on both axes.

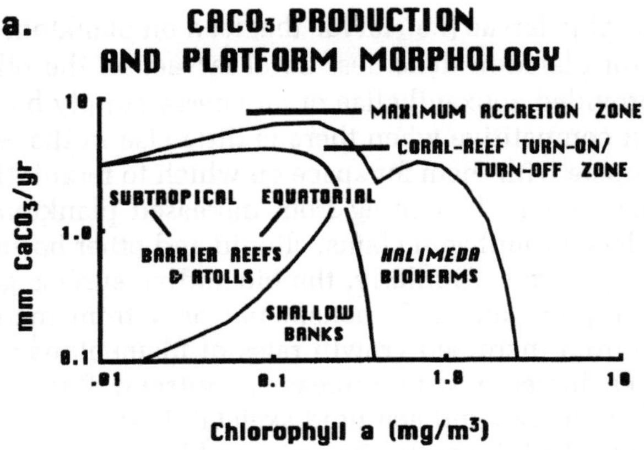

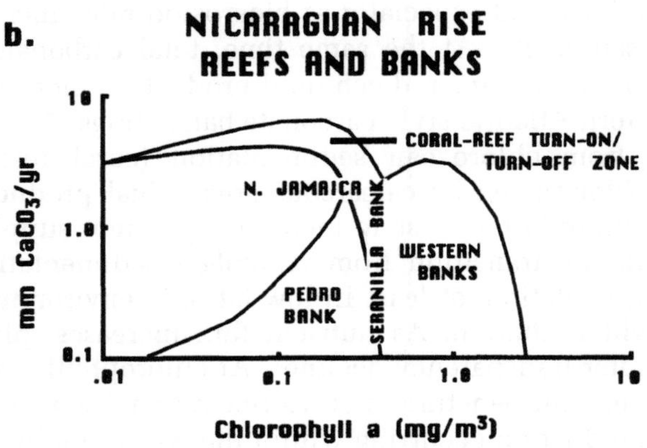

FIGURE 10. (a) Predictions of platform morphology relative to nutrient supply and carbonate production. Note that the zone of maximum potential for accretion overlaps with the coral reef turn-on/turnoff zone (see also Fig. 7). (b) The Nicaraguan Rise of the western Caribbean (see Fig. 11) crosses the coral reef turn-on/turnoff zone. Historically, reefs of northern Jamaica were coral–algal reefs (since 1982 they have shifted to algal dominated according to Hughes 1994). Pedro Bank has some coral reef development, prolific algal production, and appears to be in "catch-up" mode. Seranilla Bank has no modern coral reef development, the windward margin is dominated by a macroalgae/sponge-dominated community, and appears to be in "turnoff" mode. Deep (30–60 m) *Halimeda* bioherms line the margins of the western banks of the Nicaraguan Rise; bank-top sediments are dominated by calcareous algal and molluscan debris.

The final transition on the nutrient gradient to "hypertrophic" conditions (Fig. 7) is controlled by oxygen supply to the benthos. This transition therefore is strongly influenced by primary production of organic matter in the overlying surface waters and by physical and chemical factors that influence the input and solubility of oxygen, such as mixing processes driven by waves and currents, water temperature, and salinity. Thus, a 2-m deep dead-end canal in the Florida Keys would be much more prone to hypertrophication than a 2-m deep location offshore the same island, even if the nutrient flux to the canal was somewhat less than that to the open shelf site.

5.2. The "Maximum Accretion to Turnoff" Paradox

The importance of reef framework to reef accretion and its sensitivity to increasing nutrient flux provides insight into why drowned reefs can occur in close proximity to seemingly thriving catch-up or keep-up reefs (terminology of Neumann and Macintyre, 1985) or why even keep-up reefs may abruptly cease accretion, becoming give-up or drowned reefs. Shelves and carbonate banks in the western Caribbean (Fig. 10) and Gulf of Mexico provide excellent examples of the relationship between nutrient flux and reef accretion. The Alacran Reef on the Yucatan peninsula has some of the highest recorded accretion rates of the Holocene (Macintyre *et al.*, 1977), yet is surrounded by drowned reefs. Coastal zone color scanner (satellite) imagery of the western Caribbean, especially the Yucatan peninsula, show chlorophyll plumes that indicate the borderline mesotrophic conditions of this region (e.g., Garrett, 1985). Both Roberts and Murray (1983) and Hine *et al.* (1987) reported prolific calcareous algal production on shelves of the Nicaraguan–Honduran continental shelf and Nicaraguan Rise (Fig. 11).

A coral reef turn-on/turnoff zone (terminology of Buddemeier and Hopley, 1988) presently occurs on the Nicaraguan Rise between Pedro and Seranilla Banks (Triffleman *et al.*, 1992), corresponding to increased nutrient flux (Figs. 10b, 11). As the Caribbean current accelerates to the west, topographically induced turbulent mixing of uppermost nutricline waters into surface waters results in significantly higher chlorophyll concentrations, with no detectable cooling of surface waters (Hallock and Elrod, 1988). Pedro Bank has well-developed coral reefs along its windward margin. The benthos on Seranilla and Rosalind Banks, near the middle of the Nicaraguan Rise, are dominated by calcareous and fleshy algae and sponges. However, islands and shoals on the windward sides of these banks indicate Pleistocene reef development, suggesting that the coral reef turn-on/turnoff point was shifted westward during the last interglacial episode before the present one (Triffleman *et al.*, 1992).

5.3. *Halimeda* Bioherms

Along the western banks of the Nicaraguan Rise (Figs. 10b, 11), subsurface nutrients and strong currents promote prolific chloralgal sediment production, which is stabilized by sponges and coralline algae, resulting in deep (30–60 m) *Halimeda* bioherms (Fig. 12) (Hine *et al.*, 1987). Similar *Halimeda* bioherms, with accumulation rates of 5.9 m/1000 years, occur in the Makassar Straits of Indonesia and are associated with monsoon-forced upwelling (Roberts *et al.*, 1988). Carbonate accumulation rates of 1.75 m/1000 years have

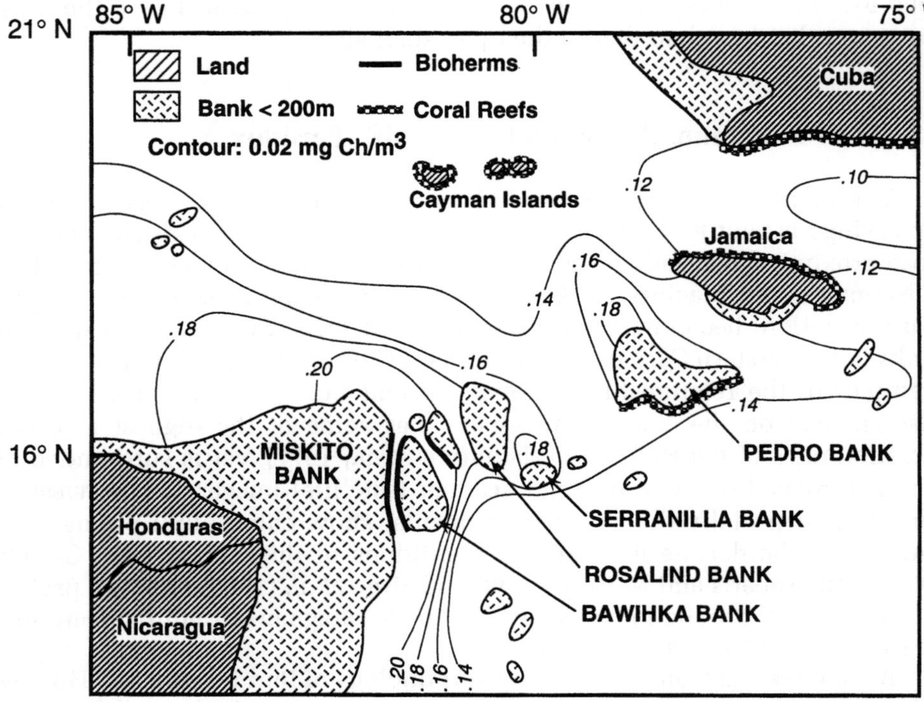

FIGURE 11. The Nicaraguan Rise region of the western Caribbean Sea. Coral reefs occur along the southern margin of Pedro Bank and as recently as the early 1980s along the north coast of Jamaica. Major banks to the west of Pedro Bank, i.e., Seranilla, Rosalind, Bawihka, and Miskito Banks, have chloralgal sedimentation but lack coral reefs. *Halimeda* bioherms line the current-swept facing margins of Miskito and Bawihka Banks. Sea-surface chlorophyll concentrations west of the coral reef turnoff zone between Pedro and Seranilla Banks exceed 0.16 mg/m^{-3}. Broken arrows indicate general direction of flow of the Caribbean Current. (Adapted from Molinari et al., 1981).

been estimated for *Halimeda* buildups at similar depths in the northern lagoon of the Great Barrier Reef (Davies and Marshall, 1985), attributed to tidal pumping of nutrient-rich uppermost thermocline water into the deep lagoon (Wolanski *et al.*, 1988).

What all three of these locations have in common are:

1. Geographic settings in warm, tropical waters.
2. Relatively clear surface waters that permit sufficient light to penetrate to 30–60 m to support prolific *Halimeda* photosynthesis and growth.
3. A physical oceanographic mechanism that periodically forces uppermost nutricline waters over the substratum where the *Halimeda* are growing.
4. Temperatures in uppermost nutricline waters that are only a degree or two cooler than surface waters, as maintenance of relatively warm temperatures over the banks is crucial to both rapid algal growth rates

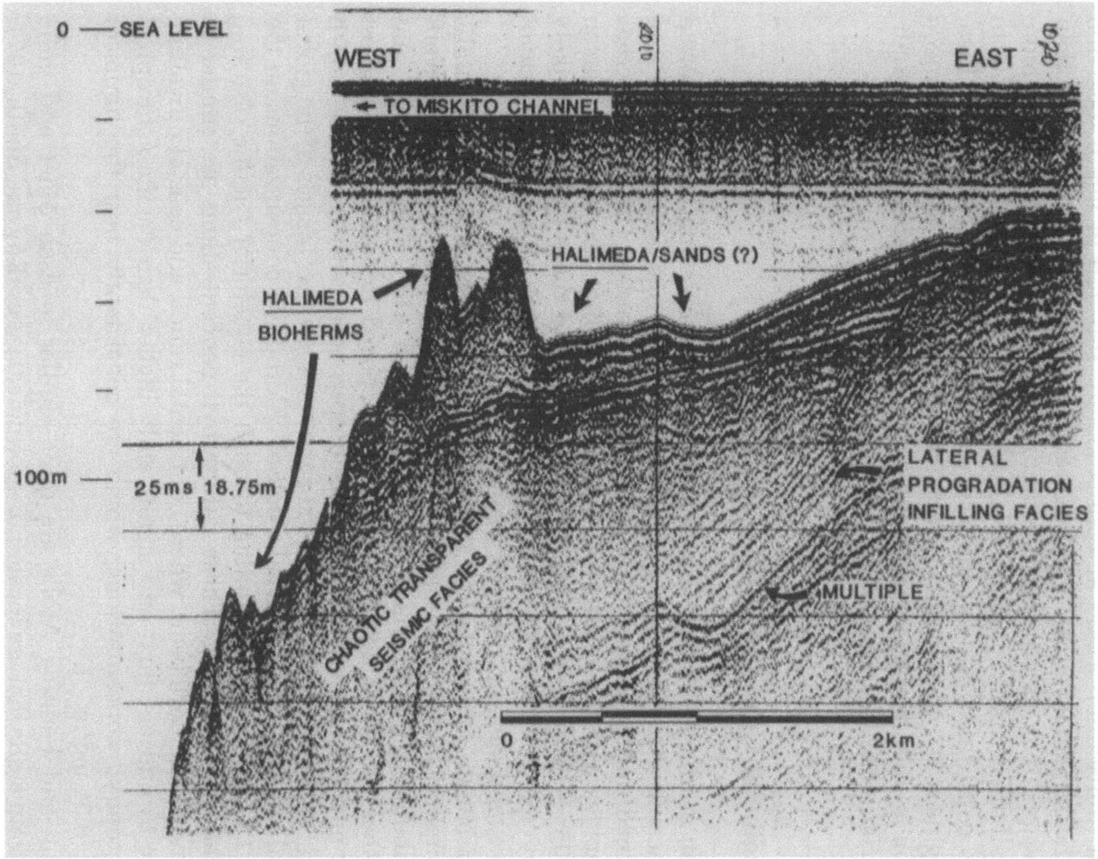

FIGURE 12. Seismic profile (1 cu. in. air gun) of *Halimeda* bioherms along the western margin of Bawihka Bank on the Nicaraguan Rise, western Caribbean Sea. Deeper bioherms are inactive; the progression of bioherms from deepest to the west and shallowest to the east probably represents backstepping as sea level rose during the past 15,000 years. (Modified from Hine *et al.*, 1987.)

and geochemical factors that promote aragonite precipitation by photosynthetic mechanisms.

In the case of the Nicaraguan Rise, sea surface temperatures average 27–29°C. In the open Caribbean south of the rise, the subsurface chlorophyll maximum (nutricline) occurs at approximately 100–120 m depth, near the base of the mixed layer, at temperatures approximately 1.5°C cooler than surface waters. The chlorophyll maximum carries three to five times more chlorophyll than surface waters (Hallock *et al.*, 1991). As the Caribbean current is constricted through the seaways of the western Nicaraguan Rise, bank margins at 30–60 m depth are fertilized by these richer waters. *Halimeda* production is stabilized by sponges and coralline algae, enabling the accumulation of *Halimeda* bioherms in areas of intense currents.

5.4. Equatorial Upwelling and Guyots

Larsen *et al.* (1995), Ogg *et al.* (1995), and Wilson *et al.* (1998) reported that Cretaceous through Eocene limestone-capped guyots drilled on ODP Leg 144 drowned during passage through equatorial latitudes. The latitude crisis zone coincides with the equatorial upwelling belt of high productivity (Larsen *et al.*, 1995; Wilson *et al.*, 1998). Final drowning of these banks was likely caused by the combination of decelerated carbonate accumulation, guyot subsidence, and eustatic sea-level rise.

Do carbonate sedimentation patterns on modern equatorial Pacific atolls provide support for the Larsen *et al.* (1995) hypothesis that equatorial upwelling contributed to the drowning of guyots? Canton Atoll, Fanning Atoll, and several others in the equatorial Pacific are exposed to some of the highest inorganic nutrient concentrations found in open ocean surface waters (Smith and Jokiel, 1975a,b). Composite coastal zone color scanner images of the global ocean (Hallock *et al.*, 1993) reveal a belt of chlorophyll concentrations on the order of 0.1–0.25 mg/m^{-3} running through the equatorial Pacific, resulting from equatorial upwelling. These chlorophyll concentrations are similar to those found in the Caribbean, where continental runoff and coastal upwelling contribute to nutrient flux, and significantly higher than Pacific subtropical gyre chlorophyll concentrations, which are consistently near the limits of detection (0.04 mg/m^{-3}).

Canton and Fanning Atolls are quite similar to each other and quite different from atolls that are located in subtropical gyres (Roy and Smith, 1971; Smith and Jokiel, 1975a,b). The equatorial atolls have relatively shallow, turbid lagoons with relatively high organic productivity. Both lagoons are nearly landlocked, with very limited passes. Sources of the fine sediments in Fanning Lagoon appeared to be primarily molluscan and bioeroded coral, while foraminiferal faunas were dominated by eutrophication-tolerant *Ammonia*, rather than by larger foraminifers with algal symbionts. Menard (1982) also contrasted "healthy" atoll reef-systems in nonequatorial regions with "nonhealthy" systems in equatorial regions, though he attributed the differences to rainfall.

In updated models of carbonate sedimentation with increasing nutrient flux (Figs. 7, 10), modern oceanic coral reefs within the equatorial upwelling zone and Caribbean reefs are exposed to nutrient fluxes that approach mesotrophic, resulting in conditions approaching or within the turn-on/turnoff zone (Figs. 7, 10a). Can the argument be developed that Cretaceous through Eocene equatorial upwelling might have pushed carbonate buildups through the accelerated accretion zone and into accretion failure? Taxonomic differences in carbonate-producing communities may strengthen the argument, since according to Ogg *et al.* (1995) the bank-top biotas were not framework builders, but rather maintained upward growth of the platform by overproduction of bioclastic debris. The argument that upwelling regions were more nutrient rich and that subtropical gyre regions were more nutrient poor in greenhouse oceans was put forth by Fischer and Arthur (1977) and was further developed by Hallock and Schlager (1986) and Hallock *et al.* (1991). On the modern Nicaraguan Rise, the transition from reef accretion to turn-off occurs where sea surface chlorophyll concentrations exceed about 0.16 mg/m^{-3} (Fig. 11), demonstrating that quite subtle increases in nutrient flux can suppress accretion on carbonate banks. Turnover rates of greenhouse oceans were estimated to have been one to two orders of magnitude slower than modern rates (e.g., Bralower *et al.*, 1994). Thus, subsurface waters of the uppermost nutricline, which are the sources of waters brought to the surface in upwelling zones, were predictably somewhat more nutrient rich than they are today. Widespread phosphatic deposition in coastal upwelling zones of Cretaceous age (Cook and McElhinney, 1979) support this argument.

6. Coral Reefs and Global Change

Rapidly increasing human populations are altering the Earth's environments at rates approaching catastrophic. Local threats to coral reefs and other coastal ecosystems are increasing in relative proportion to increasing human populations in upland and coastal watersheds (e.g., Bryant *et al.*, 1998). The single most predictable correlation is increasing sediment and nutrient flux to coastal waters with increasing human populations in the watershed (e.g., Walsh, 1984). At the global scale, anthropogenic alteration of the chemistry of the Earth's atmosphere and oceans will have consequences for coral reef ecosystems in the next few centuries (Fig. 13). Humans are currently performing "experiments" on the biosphere that should enhance scientific understanding of crisis events in the fossil record. At the same time, the history of reef systems may provide insights into how modern carbonate depositional systems will respond to global change.

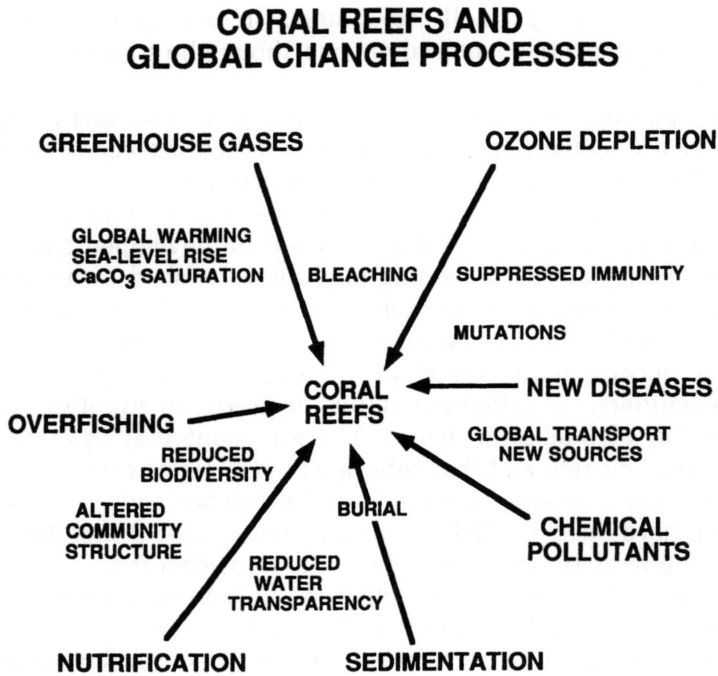

FIGURE 13. Diagram illustrating selected global change factors (bold, also see text) that influence coral reef growth and development. Some examples of how those environmental changes influence reefs are also shown.

6.1. Coastal Sedimentation

Coastal sedimentation is considered by many reef researchers to be the single greatest threat to fringing reef systems (Hatcher *et al.*, 1989). Vast areas of fringing coral reefs, particularly in the Caribbean, Asia, and Africa, are threatened or have been lost to increasing input of terrestrial sediments to coastal zones. Upland deforestation, agriculture practices, and road construction that promote soil erosion, coastal clearing, and desertification are all contributing factors. Volume and composition of the sediments, coupled with changes in water chemistry (especially addition of nutrients, pesticides, and herbicides), are factors dictating the rate and magnitude of loss of nearshore reefs. For example, slash-and-burn or clearcut deforestation of high-relief tropical uplands results in very rapid increase in the delivery of terrigenous sediments to the coastal zone. Any reefs within the rapidly extending sediment plume likely will be irrevocably lost to burial. If the predominant sediments from the deforested area are muds, the range of destruction will be greater than if the sediments are predominantly sands. Slash-and-burn deforestation also mobilizes nutrients, resulting in nutrient plumes that extend far beyond the mud plumes (Hallock *et al.*, 1993). The unique aspect of increased coastal sedimentation is the finality of the loss of fringing reefs because they are buried. Species unique to these environments may be threatened with extinction.

6.2. Anthropogenic Nutrient Flux to Coastal Systems

Ever increasing human populations, necessitating land clearing and increasingly intensive use of fertilizers, have doubled the annual flux of nitrogen to the Earth's ecosystems (Vitousek *et al.*, 1997). In coastal waters, the results are a spectrum of effects, from nutrification, which elicits shifts in benthic community structure (Cockey *et al.*, 1996), to hypertrophication, which creates or expands dead zones in bays, estuaries, and deltas (Turner and Rabalais, 1994).

Sources of waterborne anthropogenic nutrients are numerous and range from fertilizers to sewage to nutrients mobilized by deforestation and soil erosion (Vitousek *et al.*, 1997). Satellite images of chlorophyll pigments in Amazon, Orinoco, and Mississippi River plumes show that even remote reefs of the western Atlantic and Caribbean are within the range of influence of anthropogenic nutrients (Hallock *et al.*, 1993). Nutrients also are airborne; nitrous oxides from burning of fossil fuels influence ocean chemistry far downwind of major cities (Fanning, 1989). Windborne nutrients from areas of desertification or dryland agriculture settle thousands of miles from their sources of origin (Swap *et al.*, 1992).

Nutrification is a serious threat to communities dominated by algal symbiont-bearing organisms. Virtually all field, model, and physiological studies indicate that algal symbiosis is most advantageous in nutrient-deficient environments (e.g., Hallock, 1981; Smith *et al.*, 1981; Falkowski *et al.*, 1993). The consequences of nutrification range from subtle increase to complete takeover of the benthic community by fleshy macroalgae, sponges, and other non–reef-building organisms, accompanied by increased rates of bioerosion of existing skeletal substrate.

6.3. Overfishing

Reefs adjacent to or near heavily populated areas, especially in the Caribbean and Southeast Asia, also are suffering from what has been called Malthusian overfishing (Munro, 1983), in which every fish over a few centimeters in length that is not toxic is being harvested by human populations desperate for protein. Worldwide, even on remote reefs, fishing, often using illegal and damaging techniques, has removed most of the larger more valuable vertebrate and invertebrate species.

Among the most predictable results of overfishing is algal overgrowth of coral reefs. The documented change in Jamaican reefs over the past 30 years, from 60–80% coral cover in the 1970s to less than 10% coral cover in the

1990s (Hughes, 1994) has been attributed to the combined effects of hurricane disturbance and the mass dieoff of the spiny sea urchin, *Diadema antillarum*, in 1983 (Lessios *et al.*, 1984). Without this important invertebrate grazer and with much of the coastline subjected to Malthusian overfishing (Munro, 1983), coupled with increasing nutrient input to the coastal zone by rapidly increasing human populations (Lapointe *et al.*, 1997), Jamaica's once famous coral reefs are now algal-dominated hardbottom communities (Hughes, 1994).

Wood (1993) speculated that the evolution of specialized grazers and predators were key events in the evolution of reef ecosystems. Modern overexploitation of such organisms is providing an ongoing, unintentional experiment to test her hypothesis.

6.4. Other Anthropogenic Chemicals

Other chemicals, both man-made and naturally occurring, are entering marine systems at ever increasing rates. The kinds of chemicals include hydrocarbons, heavy metals, and medical and agricultural products. The effects of hydrocarbons and heavy metals can be locally devastating (Glynn *et al.*, 1989; Jackson *et al.*, 1989) but are not truly "global" problems. The effects of rapidly increasing input to aquatic systems of pesticides and herbicides, many with neurological and estrogenic effects that can be concentrated up food chains, are mostly unknown. But that does not mean they are insignificant. Certain pesticides can significantly reduce recruitment of coral larvae to affected substrata (Richmond, 1993). Pollutants also can impair the defense mechanisms of reef organisms (Santavy and Peters, 1997).

6.5. Ozone Depletion and Ultraviolet Radiation

Many researchers and scientific agencies do not consider ozone depletion to be a problem for tropical marine communities because (1) the catalytic effect of stratospheric ice crystals and sunlight, which enable chlorofluorocarbon (CFC) molecules to attack ozone molecules, results in the most extreme ozone depletion at high latitudes (e.g., Cutchis, 1982); (2) international treaties have resulted in reduction of use of CFCs so that the rate of ozone destruction has declined and net repair is expected to begin early in the 21st century (World Meterological Association 1995); (3) intensities of ultraviolet (UV) radiation are naturally high and seasonally variable at lower latitudes (Frederick *et al.*, 1989), so tropical biotas are believed to be adapted to relatively high intensities of UV (e.g., Glynn, 1996; Shick *et al.*, 1996); (4) and conventional wisdom holds that biologically damaging wavelengths of ultraviolet (UVB) radiation (UVB) only penetrate a few meters into seawater.

Unfortunately, scientists may be overlooking evidence of UV stress, particularly in reef communities near the higher latitudinal limits for reef growth. Stratospheric ozone depletion has progressed to the extent that pre-1960s summer solstice intensities of biologically damaging UVB are now experienced throughout the summer months at subtropical latitudes (Shick *et al.*, 1996). Recent major volcanic events, such as the El Chichon eruption in Mexico in 1982 and the Mt. Pinatubo eruption in the Philippines in 1991, have further contributed to ozone depletion. Very explosive eruptions inject SO_2 molecules into the stratosphere that provide substrate for CFCs to attack ozone molecules (Vogelmann and Ackerman, 1993). The El Chichon eruption accelerated ozone depletion over northern mid-latitudes by up to 10% (Hoffmann and Solomon, 1989) and Mt. Pinatubo increased ozone depletion at tropical latitudes by about 4% (Randel *et al.*, 1995). Low latitude ozone depletion between 1969 and 1990 was approximately 10%, representing up to a 25% increase in UVB radiation reaching the sea surface (Shick *et al.*, 1996). Following the Mt. Pinatubo eruption, UVB intensities were about 35% higher than intensities in the 1960s.

The idea that UVB does not penetrate to any significant depths in marine waters is only partly correct. Indeed, UVB is rapidly absorbed by waters with significant densities of phytoplankton or concentrations of dissolved organic matter, especially refractory organic compounds (tannins) of terrestrial origin (Bricaud *et al.*, 1981). Clear, nutrient-deficient tropical waters generally do not contain significant concentrations of either phytoplankton or tannins, and as a result potentially damaging doses of UVB can penetrate tens of meters (Gleason and Wellington, 1993). Furthermore, there is some evidence that algae (including zooxanthellae) that are adapted to very low light intensities may capture high-energy UV wavelengths and fluoresce them to visible wavelengths, which then can be used for photosynthesis (Falkowski *et al.*, 1990). These organisms may be sensitive to slight increases in UVB. The general rule that organisms thrive near their upper physiological limit for an environmental parameter (e.g., Glynn, 1996), which has been used to explain coral bleaching in response to 1–2°C increases in sea surface temperature, should not be ignored for UVB radiation.

Reef-dwelling foraminiferal populations, particularly *Amphistegina* spp., appear to be suffering damage from increasing UVB (Talge *et al.*, 1997). Beginning in summer 1991, following the Mt. Pinatubo eruption, *Amphistegina* populations in the Florida Keys began to exhibit visible loss of algal endosymbionts. Sampling at Heron Island, Australia, and Montego Bay, Jamaica, in 1993 indicated that this previously unknown malady was widespread in *Amphistegina* populations. Monthly monitoring of populations in the Florida Keys between 1992 and 1996 revealed that symbiont loss accelerated each March, to summer maxima in June or early July, with partial population recovery by late summer each year. This pattern indicates a relationship to the solar light cycle, which peaks with the summer solstice in the Florida Keys, not the seasonal temperature cycle, which peaks in August or September (Talge *et al.*, 1997). The kinds of damage exhibited by the afflicted foraminifers also implicate UVB damage, which typically affects photosynthesis, protein synthesis, DNA, and behavior in protists (e.g., Hadar and Worrest, 1991).

The emergence of coral bleaching as a problem over the past two decades may involve sublethal stress to corals by ultraviolet radiation. The first global coral bleaching event occurred during the 1982–1983 El Niño/Southern Oscillation Event (e.g., Williams and Bunckley-Williams, 1990), which also coincided with the significant ozone depletion event following the El Chichon eruption (Vogelmann and Ackerman, 1993). Corals stressed by elevated water temperature have diminished ability to synthesize mycosporinelike amino acids, which are UV-absorbing materials believed to act as sunscreens (Lesser, 1996). As a result, the combined effects of elevated temperature and UVB are particularly damaging to zooxanthellate corals (Glynn *et al.*, 1993).

Cyanobacteria play a prominent role on declining reefs. Black-band disease in corals (Fig. 6) is an assemblage of microbes, including cyanobacteria (e.g., Richardson *et al.*, 1997). Secondary infestations of cyanobacteria are commonly observed in foraminifers that exhibit symbiont loss (Hallock *et al.*, 1995). Nuisance "algal" outbreaks often are profuse growths of cyanobacteria (e.g., Butler *et al.*, 1995). Many species of cyanobacteria live in very shallow water or intertidally, where they are naturally exposed to high intensities of UVB. Since cyanobacteria evolved prior to the formation of the Earth's ozone layer, a hypothesis worth testing is that elevated intensities of UVB may actually favor the proliferation of certain cyanobacteria.

6.6. New Diseases

Epidemics of previously rare or unknown diseases are ravaging corals and other reef organisms of the Caribbean and elsewhere (e.g., Santavy and Peters, 1997; Richardson *et al.*, 1998). The whole array of global change factors may be contributing to the proliferation of new diseases (Fig. 13). The most obvious problem is global transport of microbes. The disease that decimated the spiny sea urchin, *Diadema antillarum*, is suspected of having entered the Caribbean in ballast water of ships passing through the Panama Canal (Lessios, 1988).

Another potential source of new diseases involves the ever increasing volumes of urban and agricultural wastewater entering marine systems. Though most terrestrial and freshwater microbes are killed by seawater, there are undoubtedly survivors. Furthermore, mutagenic chemical pollutants and UVB may be producing new microbe genotypes. Thus, human activities are adding new microbes to coastal systems, changing the selective pressures on existing microbe assemblages, and increasing mutation rates. Those same human activities are increasing the stress levels on larger protistan, invertebrate, and vertebrate populations of reefs and other marine systems (Fig. 13). It should not be surprising if microbial communities once again take over shallow tropical environments, as they did after several mass extinction events in the geologic record (e.g., Fagerstrom, 1987; Copper, 1994).

6.7. Atmospheric CO_2 and Global Climate Change

Human activities, including fossil fuel burning, agricultural and industrial practices, and changes in terrestrial vegetation, are increasing the concentrations of greenhouse gases in the atmosphere at unprecedented rates (e.g., Watson *et al.*, 1990; Vitousek *et al.*, 1997). The most significant greenhouse gases are water vapor (H_2O), carbon dioxide (CO_2), methane (CH_4), and nitrous oxide (NO_2). Greenhouse gases are so-called because they absorb infrared radiation reflected from the Earth's surface that would otherwise escape into space. As the concentrations of greenhouse gases increase in the Earth's atmosphere, more solar energy is held within the atmosphere and the average temperature of the Earth increases.

The most predictable effect of global warming is expansion of the subtropical and boreal regions of the Earth and warming of the polar regions (e.g., Barron *et al.*, 1995). During the Cretaceous period, when atmospheric CO_2 concentrations were several times higher than today (e.g., Berner, 1994), oxygen isotopic evidence indicates that polar temperatures were on the order of 10–14°C (e.g., Huber *et al.*, 1995). In the Quaternary world, warming of polar regions melts high latitude glaciers, causing sea-level rise (e.g., Blanchon and Shaw, 1995). On human time scales, sea-level rise may be the least significant CO_2-related factor for coral reefs, though possibly the most important for humans, especially those living on low-lying islands or coastal regions.

Since the establishment of the symbiosis between scleractinian corals and zooxanthellae during the Late Triassic (Stanley, 1981), times of elevated concentrations of atmosphere CO_2, specifically the Cretaceous and Eocene, were times of widespread coral distribution and diversity, but not of coral reefs (e.g., Hallock, 1997). The major shallow-benthic carbonate producers of these episodes were the rudistid bivalves in the Cretaceous and the coralline algae and larger foraminifers during the Paleocene–Eocene. Concentrations of atmospheric CO_2, which were two to ten times higher during the Cretaceous through Eocene than during the preanthropogenic Holocene (Berner, 1994), probably had two major effects that are detrimental to coral reef development: temperatures outside the optimum range for reef development and significantly reduced carbonate saturation levels in surface waters.

The effects of greenhouse gas-induced global change on tropical sea surface temperatures are controversial (e.g., Huber *et al.*, 1995; D'Hondt and Arthur, 1996). Kauffman and Johnson (1988) postulated a Cretaceous "super-Tethys" tropical region characterized by elevated sea-surface temperatures and salinities. They suggested that advantages provided by bivalve physiology, particularly the ability to actively move water over their gills, might have enabled rudistid dominance over suspension-feeding corals in the warm, saline shallow shelf and bank-top environments. The Kauffman/Johnson hypothesis also was based on the well-known sensitivity of modern corals to temperatures above 30°C, above which they are prone to bleaching and mortality (e.g., Glynn, 1996). Skelton *et al.* (1997) also concluded that rudistids and corals had quite different ecological requirements. The rudists, which likely fed on suspended phytoplankton, proliferated on shallow, protected platform tops where sedimentation rates were relatively high and where temperature and salinity were variable. The corals they examined probably hosted zooxanthellae, and therefore required clear water. These corals predominated in slightly deeper, open water settings, where temperature and salinity were less variable.

Early global climate models of "greenhouse" conditions predicted elevated tropical temperatures (e.g., Barron, 1989), while later models (Barron *et al.*, 1995; Johnson *et al.*, 1996) incorporated mechanisms of more effective heat transport into higher latitude, resulting in lower tropical sea surface temperatures. Oxygen isotopic evidence (D'Hondt and Arthur, 1996) indicates tropical sea surface temperatures during the Paleogene in the low 20s (°C). Adams *et al.* (1990) argued that larger foraminiferal faunas in the Eocene required tropical temperatures similar to modern conditions. However, Eocenelike carbonate facies, dominated by coralline algae, bryozoans, and larger foraminifera, at water temperatures of 20–24°C on the southwest Australian shelf (James *et al.*, submitted), indicates that lower tropical temperatures could have been real.

Neither scenario is encouraging for the future of coral reefs. Higher tropical sea-surface temperatures will increase temperature stress and probability of bleaching events. Cooler tropical sea surface temperatures will reduce carbonate saturation levels, making the precipitation of aragonite increasingly difficult. On the other hand, atmospheric global change may be relatively advantageous for coralline algae and larger foraminifers, regardless of its influence on tropical sea-surface temperatures. Greenhouse episodes in the geologic past expanded habitats for larger foraminifers as subtropical belts moved into higher latitudes and as sea level rose and flooded continental shelves. Bleaching in larger foraminifers shows no correlation to summer temperatures (Hallock *et al.*, 1995). Even the loss of the reef-building corals may not be important to larger foraminifers, because times of proliferation and dominance of carbonate shelves by larger foraminifers in the geologic record were generally times of suppressed reef building by corals (Hallock, 1997).

6.8. Atmospheric CO_2 and Ocean Chemistry

The effects of increasing CO_2 on ocean chemistry is an even more intriguing issue, because of potential effects on photosynthesis and calcification. It is a paradox of carbonate geochemistry that lower concentrations of atmospheric CO_2 actually promote supersaturation of seawater with respect to $CaCO_3$, thereby promoting calcification. At the same time, the availability of CO_2 for photosynthesis is reduced, making the McConnaughey (1989) model of calcification as a source of CO_2 for photosynthesis appear more advantageous for photosynthetic organisms, both calcareous autotrophs and algal symbionts (Hallock, 1997). Thus, lower atmospheric CO_2 concentrations strongly favor coral reef development. In contrast, higher CO_2 concentrations may render calcification less important as a source of CO_2 for photosynthesis in aquatic systems according to the McConnaughey model.

In oceanic waters, calcite and aragonite become progressively easier to precipitate, and thereby energetically more advantageous to biomineralize, as the saturation of seawater with respect to $CaCO_3$ increases. Most temperate to tropical surface waters are supersaturated with respect to $CaCO_3$ (Morse and Mackenzie, 1990). However, as the degree of supersaturation increases, Mg^{2+} interferes with the formation of the calcite crystal structure. Apparently organisms can deal with this in several ways. One way is to use energy to actively exclude the Mg^{2+} to maintain the integrity and strength of low-Mg calcite shell. Larger rotaliid foraminifers appear to utilize this mechanism. Other organisms, such as coralline algae and echinoderms, secrete their calcite shells or skeletons close to equilibrium with seawater. As the $CaCO_3$ saturation level increases, the shell incorporates more Mg^{2+} in the crystal structure, up to 18–20% in very warm, alkaline seawater (Morse and Mackenzie, 1990). However, incorporation of Mg^{2+} weakens the crystal structure, so that high-Mg calcite shells are weaker than those of low-Mg calcite. The biomineralization strategy, found in scleractinian corals and calcareous green algae, is to secrete aragonite. Aragonite is much stronger than calcite and Mg^{2+} does not interfere with biomineralization. However, aragonite is energetically expensive to secrete and maintain at reduced saturation levels.

Shell mineralogies have played a role in the history of shallow-water carbonate sedimentation through geologic history (Hallock, 1997). As atmospheric CO_2 concentrations rose (Berner, 1994), aragonite-secreting corals, though highly diverse, lost dominance on Cretaceous reefs and carbonate platforms to the mixed calcite/aragonite-secreting rudistid bivalves (Kauffman and Johnson, 1988). Following the Cretaceous–Tertiary boundary extinctions, which included the rudists, the variable-Mg calcite coralline algae predominated the few Paleocene reefs and were joined by low-Mg calcite larger rotaliid foraminifers by the Late Paleocene (Hallock *et al.*, 1991). By the Eocene, when atmospheric CO_2 was about two to three times preanthropogenic Holocene levels (Berner, 1994), aragonite-secreting corals again were diverse and common but still were not building reefs (Frost, 1977). It was not until the Oligocene, when CO_2 levels were declining to near modern levels, that corals reemerged as the major reef builders. Using the Eocene as a predictor, doubling to tripling of current CO_2 levels, as are expected in the 21st century (Watson *et al.*, 1990), may lower saturation levels and thereby favor calcitic foraminifers and coralline algae over aragonitic corals and calcareous green algae. This is an ongoing anthropogenic experiment for which young scientists today may witness the outcome.

7. The Future of Coral Reefs

Rapidly increasing human populations are threatening coral reefs on multiple fronts. Nutrification, sedimentation, chemical pollution, and overfishing are significant and often interrelated local threats of global extent. Ozone depletion has increased biologically damaging UVB to the point where pre-1960s summer solstice intensities are now experienced throughout the summer months at subtropical latitudes. Increasing concentrations of atmospheric CO_2 threatens to destabilize climate and induce global warming. Corals under temperature stress lose the ability to synthesize protective sunscreens. Thus, the combination of ozone depletion and global warming may be particularly nefarious. Biologically damaging UV also reduces fitness, rendering reef-building organisms more susceptible to emerging diseases. Chemical pollutants and nutrification also may provide stress that increases the probability of pathogenic infections. Perhaps globally the progressive reduction of $CaCO_3$ saturation of sea-surface waters in response to increasing atmospheric CO_2 is most serious.

The fossil record of biogenic reefs and carbonate-producing communities has much to contribute to our understanding of living species and to predictions of how communities may respond to anthropogenically induced environmental stress. The prognosis is not bright. Carbonate-producing communities are geologically both fragile and ephemeral, flourishing under favorable conditions but suffering the most from widespread extinctions under regional or global environmental perturbations. Ironically, the volcanic islands and atolls of the vast subtropical Pacific Ocean, which should otherwise provide refugia for many reef-dwelling species, are in exceptionally clear oceanic waters and therefore may be most susceptible to biologically damaging UV radiation.

References

Adams, C. G., Lee, D. E., and Rosen, B. R., 1990, Conflicting isotopic and biotic evidence for tropical sea-surface temperatures during the Tertiary, *Palaeogeogr. Palaeoclimatol. Palaeoecol.* **77**:289–313.

Arthur, M. A., and Schlanger, S. O., 1979, Cretaceous "oceanic anoxic events" as causal factors in development of reef-reservoired giant oil fields, *Am. Assoc. Petrol. Geol. Bull.* **63**:870–885.

Barron, E. J., 1989, Studies of Cretaceous climate, in: *Understanding Climate Change*, America Geophysical Union Monograph 52 (W. A. Berger, ed.), American Geophysical Union, Washington, DC, pp. 149–157.

Barron, E. J., Fawcett, P. J., Peterson, W. H., Pollard, D., and Thompson, S. L., 1995, A "simulation" of mid-Cretaceous climate, *Paleoceanography* **10**:953–962.

Berner, R. A., 1994, Geocard II: A revised model for atmospheric CO_2 over Phanerozoic time, *Am. J. Sci.* **294**:56–91.

Birkeland, C., 1977, The importance of the rate of biomass accumulation in early successional stages of benthic communities to the survival of coral recruits, *Proc. 3rd Int. Coral Reef Symp.* **1**:16–21.

Birkeland, C., 1987, Nutrient availability as a major determinant of differences among coastal hard-substratum communities in different regions of the tropics, in: *Differences between Atlantic and Pacific Tropical Marine Coastal Systems: Community Structure, Ecological Processes, and Productivity* (C. Birkeland, ed.), UNESCO, Paris, pp. 45–90.

Birkeland, C., 1997, Geographic differences in ecological processes on coral reefs, in: *Life and Death of Coral Reefs* (C. Birkeland, ed.), Chapman and Hall, New York, pp. 273–297.

Blanchon, P., and Shaw, J., 1995, Reef drowning during the last deglaciation: Evidence for catastrophic sea-level rise and ice-sheet collapse, *Geology* **23**:4–8.

Bralower, T. J., Arthur, M. A., Leckie, R. M., Sliter, W. V., Allard, D. J., and Schlanger, S. O., 1994, Timing and paleoceanography of oceanic dysoxia/anoxia in the Late Barremian to Early Aptian (Early Cretaceous), *Palaios* **9**:335–369.

Bricaud, A., Morel, A., and Prieur, L., 1981, Absorption by dissolved organic matter of the sea (yellow substance) in the UV and visible domains, *Limnol. Oceanogr.* **26**:43– 53.

Bryant, D., Burke, L., McManus, J., and Spalding, M., 1998, *Reefs at Risk: A Map-Based Indicator of Potential Threats to the World's Coral Reefs*, World Resources Institute, Washington, DC.

Buddemeier, R. W., and Hopley, D., 1988, Turn-ons and turn-offs: Causes and mechanisms of the initiation and termination of coral reef growth, *Proc. 6th Int. Coral Reef Symp. Australia* **1**:253–261.

Butler, M. J. IV, Hunt, J. H., Herrnkind, W. F., Childress, M. J., Bertelsen, R., Sharp, W., Matthews, T., Field, J. M., and Marshall, H. G., 1995, Cascading disturbances in Florida Bay, USA: Cyanobacterial blooms, sponge mortality, and implications for juvenile spiny lobsters, *Panuirus argus, Mar. Ecol. Progr. Ser.* **129**:119–125.

Cockey, E. M., Hallock, P., and Lidz, B. H., 1996, Decadal-scale changes in benthic foraminiferal assemblages off Key Largo, Florida, *Coral Reefs* **15**:237–248.

Cook, P. J., and McElhinney, M. W., 1979, A reevaluation of the spatial and temporal distribution of sedimentary phosphate deposits in the light of plate tectonics, *Econ. Geol.* **74**:315–330.

Copper, P., 1994, Ancient reef ecosystem expansion and collapse, *Coral Reefs* **13**:3– 11.

Cowen, R., 1983, Algal symbiosis and its recognition in the fossil record, in: *Biotic Interactions in Recent and Fossil Benthic Communities* (J. J. S. Tevesz and P. L. McCall, eds.), Plenum Press, New York, pp. 431–479.

Cutchis, P., 1982, A formula for comparing annual damaging ultraviolet (DUV) radiation doses at tropical and mid-latitude sites, in: *The Role of Solar Ultraviolet Radiation in Marine Ecosystems* (J. Calkins, ed.), Plenum Press, New York, pp. 213–228.

Davies, P. S., and Marshall, J. F., 1985, *Halimeda* bioherms—low energy reefs, northern Great Barrier Reef, *Proc. 5th Int. Coral Reef Cong.* **5**:1–7.

D'Hondt, S., and Arthur, M. A., 1996, Late Cretaceous oceans and the cool tropic paradox, *Science* **271**:1838–1841.

Ducklow, H. W., and Mitchell, R., 1979, Observations on naturally and artificially diseased tropical corals: a scanning electron microscope study, *Microb. Ecol.* **5**:215–223.

Duguay, L. E., and Taylor, D. L., 1978, Primary production and calcification by the soritid foraminifera *Archaias angulatus* (Fichtel and Moll), *J. Protozool.* **25**:356–361.

Erez, J., 1983, Calcification rates, photosynthesis and light in planktonic foraminifera, in: *Biomineralization and Biological Metal Accumulation* (P. Westbroek and E. J. de Jong eds.), Reidel, Dordrecht, pp. 307–312.

Fagerstrom, J. A., 1987, *The Evolution of Reef Communities*, Wiley, New York.

Falkowski, P. G., Jokiel, P. L., and Kinzie, R. A. III, 1990, Irradiance and corals, in: *Coral Reefs—Ecosystems of the World 25, Coral Reefs* (Z. Dubinsky, ed.), Elsevier, New York, pp. 89–108.

Falkowski, P. G., Dubinsky, Z., Muscatine, L., and McCloskey, L., 1993, Population control in symbiotic corals, *BioScience* **43**:606–611.

Fanning, K. A., 1989, Influence of atmospheric pollution on nutrient limitation in the ocean, *Nature* **339**:460–463.

Fischer, A. G., and Arthur, M. A, 1977, Secular variations in the pelagic realm, *Soc. Econ. Paleontol. Mineral. Spec. Publ.* **25**:19–50.

Frederick, J. E., Snell, H. E., and Haywood, E. K., 1989, Solar ultraviolet radiation at the earth's surface, *Photochem. Photobiol.* **50**:443–450.

Frost, S. H., 1977, Cenozoic reef systems of the Caribbean—prospects for paleoecologic synthesis, in: *Reefs and Related Carbonates—Ecology and Sedimentology* (S. H. Frost, M. P. Weiss, and J. B. Saunders, eds.), Stud. Geol. No. 4, American Association of Petroleum Geologists, Tulsa, OK, pp. 93–110.

Garrett, P., and Ducklow, H. W., 1975, Coral diseases in Bermuda, *Nature* **253**:349–350.

Garrett, W. E. (ed.), 1985, *Atlas of North America: A Space Age Portrait of a Continent*, National Geographic Society, Washington, DC.

Gleason, D., and Wellington, G., 1993, Ultraviolet radiation and coral bleaching, *Nature* **365**:836–838.

Glynn, P. W., 1988, El Niño warming, coral mortality and reef framework destruction by echinoid bioerosion in the eastern Pacific, *Galaxea* **7**:129–160.

Glynn, P. W., 1996, Coral reef bleaching: facts, hypotheses and implications, *Global Change Biol.* **2**:495–510.

Glynn, P. W., Szmant, A. M., Corcoran, E. F., and Cofer-Shabica, S. V., 1989, Condition of coral reef cnidarians from the northern Florida reef tract: Pesticides, heavy metals, and histopathologic examination, *Mar. Poll. Bull.* **20**:568–576.

Glynn, P. W., Imai, R., Sakai, K., Nakano, Y., and Yamazato, K., 1993, Experimental responses of Okinawan (Ryukyu Islands, Japan) reef corals to high sea temperature and UV radiation, *Proc. 7th Int. Coral Reef Symp.* **1**:27–37.

Goreau, T. F., 1959, The physiology of skeleton formation in corals. I. A method for measuring the rate of calcium deposition by corals under different conditions, *Biol. Bull.* **116**:59–75.

Grant, J., and Gust, G., 1987, Prediction of coastal sediment stability from photopigment content of mats of purple sulfur bacteria, *Nature* **330**:244–246.

Hadar, D.-P., and Worrest, R. C., 1991, Effects of enhanced solar ultraviolet radiation on aquatic ecosystems, *Photochem. Photobiol.* **53**:717–725.

Hallock, P., 1981, Algal symbiosis: A mathematical analysis, *Mar. Biol.* **62**:249–255.

Hallock, P., 1982, Evolution and extinction in larger foraminifera, *Proc. 3rd N. Am. Paleontol. Conv.* **1**:221–225.

Hallock, P., 1987, Fluctuations in the trophic resource continuum: A factor in global diversity cycles? *Paleoceanography* **2**:457–471.

Hallock, P., 1988, The role of nutrient availability in bioerosion: consequences to carbonate buildups, *Palaeogeogr. Palaeoclimatol. Palaeoecol.* **63**:275–291.

Hallock, P., 1997, Reefs and reef limestones in Earth history, in: *Life and Death of Coral Reefs* (C. Birkeland, ed.), Chapman and Hall, New York, pp. 13–42.

Hallock, P., and Elrod, J. A., 1988, Oceanic chlorophyll around carbonate platforms in the western Caribbean: Observations from CZCS data, *Proc. 6th Int. Coral Reef Symp.* **2**:449–454.

Hallock, P., and Schlager, W., 1986, Nutrient excess and the demise of coral reefs and carbonate platforms, *Palaios* **1**:389–398.

Hallock, P., Hine, A. C., Vargo, G. A., Elrod, J. A., and Jaap, W. C., 1988, Platforms of the Nicarguan Rise: Examples of the sensitivity of carbonate sedimentation to excess trophic resources, *Geology* **16**:1104–1107.

Hallock, P., Premoli Silva, I., and Boersma, A., 1991, Similarities between planktonic and larger foraminiferal evolutionary trends through Paleogene paleoceanographic changes, *Palaeogeogr. Palaeoclimat. Palaeoecol.* **83**:49–64.

Hallock, P., Muller-Karger, F. E., and Halas, J. C., 1993, Coral reef decline. *Res. Explor.* **9**:358–378.

Hallock, P., Talge, H. K., Cockey, E. M., and Muller, R. G., 1995, A new disease in reef-dwelling foraminifera: Implications for coastal sedimentation, *J. Foram. Res.* **25**:280–286.

Hatcher, B. G., Johannes, R. E., and Robertson, A. I., 1989, Review of research relevant to the conservation of shallow tropical marine systems, *Oceanogr. Mar. Biol. Annu. Rev.* **27**:337–414.

Hine, A. C., Hallock, P., Harris, M. W., Mullins, H. T., Belknap, D. F., and Jaap, W. C., 1987, *Halimeda* bioherms along an open seaway: Miskito Channel, Nicaraguan Rise, SW Caribbean Sea, *Coral Reefs* **6**:173–178.

Hoffmann, D. J., and Solomon, S., 1989, Ozone destruction through heterogenous chemistry following the eruption of El Chichón, *J. Geophys. Res.* **94**:5029–5041.

Huber, B. B. T., Hodell, D. A., and Hamilton, C. P., 1995, Middle-Late Cretaceous climate of the southern high latitudes—Stable isotope evidence for minimal equatorial-to-pole thermal gradients, *Geol. Soc. Am. Bull.* **107**:1164–1191.

Hughes, T. P., 1994, Catastrophes, phase shifts, and large-scale degradation of a Caribbean coral reef, *Science* **265**:1547–1551.

Hutchings, P. A., 1986, Biological destruction of coral reefs, *Coral Reefs* **4**:239–252.

Jackson, J. B. C., Cubit, J. D., Keller, B. D., Batista, V., Burns, K., Caffey, H. M., Caldwell, R. L., Garrity, S. D., Gelter, C. D., Gonzalez, C., Guzman, H. M., Kaufmann, K. W., Knap, A. H., Leving, S. C., Marshall, M. J., Sleger, R., Thampson, R. C., and Weit, E., 1989, Ecological effects of a major oil spill on Panamanian coastal marine communities, *Ecology* **243**:37–44.

James, N. P., 1997, The cool-water carbonate depositional realm, in: *Cool-Water Carbonates* (N. P. James and J. A. D. Clarke, eds.), SEPM (Society for Sedimentary Geology) Spec. Publ. No. 56, Tulsa, OK, pp. 1–20.

James, N. P., Collins, L., Bone, Y., and Hallock, P., 1999, Subtropical carbonates in a temperate realm: Modern sediments on the southwest Australian shelf, *J. Sed. Res.* **69**:1297–1321.

Johnson, C. C., Baron, E. J., Kauffman, E. G., Arthur, A. A., Fawcett, P. J., and Yasuda, M. K., 1996, Middle Cretaceous reef collapse linked to ocean heat transport, *Geology* **24**:376–380.

Kauffman, E. G., and Johnson, C. C., 1988, The morphological and ecological evolution of middle and upper Cretaceous reef building rudistids, *Palaios* **3**:194–126.

Kinsey, D. W., 1985, Metabolism, calcification and carbon production. I. Systems level studies, *Proc. 5th Int. Coral Reef Cong.* **4**:505–526.

Lapointe, B. E., Littler, M. M., and Littler, D. S., 1997, Macroalgal overgrowth of fringing coral reefs at Discovery Bay, Jamaica: Bottom-up versus top-down control, *Proc. 8th Int. Coral Reef Symp.* **1**:927–932.

Larsen, R. L., Erba, E., Nakanishi, M., Bergersen, D. D., and Lincoln, J. M., 1995, Stratigraphic, vertical subsidence, and paleolatitude histories of Leg 144 guyots, in: *Proceedings of ODP Scientific Results, 144* (J. A. Haggerty, I. Premoli SilvaF. Rack, and M. K. McNutt, eds.), College Station, TX (Ocean Drilling Program), pp. 915–933.

Laws, E. A., and Redalje, D. G., 1979, Effect of sewage enrichment on the phytoplankton populations of a tropical estuary, *Pac. Sci.* **33**:129–144.

Lee, J. J., and Anderson, O. R., 1991, Symbiosis in foraminifera, in: *Biology of Foraminifera* (J. J. Lee and O. R. Anderson, eds.), Academic Press, New York, pp. 157–220.

Lees, A., 1975, Possible influence of salinity and temperature on modern shelf carbonate sedimentation, *Mar. Geol.* **19**:159–198.

Lesser, M. P., 1996, Elevated temperatures and ultraviolet radiation cause oxidative stress and inhibit photosynthesis in symbiotic dinoflagellates, *Limnol. Oceanogr.* **41**:271–283.

Lessios, H. A., 1988, Mass mortality of *Diadema antillarum* in the Caribbean: What have we learned? *Annu. Rev. Ecol. Syst.* **19**:371–393.

Lessios, H. A., Robertson, D. R., and Cubit, J. D., 1984, Spread of *Diadema* mass mortality through the Caribbean, *Science* **226**:335–337.

Logan, B. W., Read, J. F., Hagan, G. M., Hoffman, P., Brown, R. G., Woods, P. J., and Gebelein, C. D., 1974, Evolution and diagenesis of Quaternary carbonate sequences, Shark Bay, Western Australia, *Am. Assoc. Petrol. Geol. Mem.* **22**:1–358.

McConnaughey, T. A., 1989, Biomineralization mechanisms, in: *Origin, Evolution, and Modern Aspects of Biomineralization in Plants and Animals* (R. E. Crick, ed.), Plenum Press, New York, pp. 57–73.

McConnaughey, T. A., and Whelan, J. F., 1997, Calcification generates protons for nutrient and bicarbonate uptake, *Earth-Sci. Rev.* **42**: 95–117.

Macintyre, I. G., Burke, R. G., and Stuckenrath, R., 1977, Thickest recorded Holocene reef section, Osla Perez core hole, Alacran Reef, Mexico, *Geology* **5**:749–754.

Menard, H. W., 1982, The influence of rainfall on the morphology and distribution of atolls, in: *The Ocean Floor: Bruce Heezen Commemorative Volume* (R. A. Scrutten and M. Talwani, eds.), Wiley, Chichester, pp. 305–311.

Molinari, R. L., Spillane, M., Brooks, I., Atwood, D., and Duckett, C., 1981, Surface currents in the Caribbean Sea as deduced from Lagrangian observations, *J. Geophys. Res.* **86**:6537–6542.

Morse, J. W., and Mackenzie, F. T., 1990, *Geochemistry of Sedimentary Carbonates.* Developments in Sedimentology 48, Elsevier, Amsterdam.

Munro, J. L. (ed.), 1983, Caribbean coral reef fishery resources, *ICLARM Stud. Rev.* **7**:1–276.

Neumann, A. C., 1966, Observations on coastal erosion in Bermuda and measurements of the boring rates of the sponge *Cliona lampa*, *Limnol. Oceanogr.* **11**:92–108.

Neumann, A. C., and Macintyre, I., 1985, Reef response to sea level rise: Keep-up, catch-up or give-up, *Proc. 5th Int. Coral Reef Congr.* **3**:105–110.

Odum, E. P., 1959, *Fundamentals of Ecology*, Saunders, Philadelphia.

Odum, H. T., and Odum, E. P., 1955, Trophic structure and productivity of a windward coral reef community on Eniwetok Atoll, *Ecol. Monogr.* **25**:291–320.

Ogg, J. G., Camoin, G. F., and Vanneau, A. A., 1995, Limalok Guyot: Depositional history of the carbonate platform from downhole logs at Site 871 (lagoon), in: *Proceedings of ODP Scientific Results, 144* (J. A. Haggerty, I. Premoli SilvaF. Rack, and M. K. McNutt, eds.), College Station, TX (Ocean Drilling Program), pp. 233–253.

Peters, E. C., 1997, Diseases of coral-reef organisms, in: *Life and Death of Coral Reefs* (C. Birkeland, ed.), Chapman and Hall, New York, pp. 114–139.

Randel, W. J., Wu, F., Russell, J. M. III, Waters, J. W., and Froidevaux, L., 1995, Ozone and temperature changes following the eruption of Mt. Pinatubo, *J. Geophys. Res.* **100**:16,753–16,764.

Reid, R. P., and Browne, K. M., 1991, Intertidal stromatolites in a fringing Holocene reef complex, Bahamas, *Geology* **19**:15–18.

Richardson, L. L., Kuta, K. G., Schnell, S., and Carlton, P. G., 1997, Ecology of the black band disease microbial consortium, *Proc. 8th Int. Coral Reef Symp.* **1**:597–600.

Richardson, L. L., Goldberg, W. M., Kuta, K. G., Aronson, R. B., Smith, G. W., Ritchie, K. B., Halas, J. C., Feingold, J. S., Miller, S., 1998, Florida's mystery coral-killer identified, *Nature* **392**:587–588.

Richmond, R. H., 1993, Coral reefs: Present problems and future concerns resulting from anthropogenic disturbance, *Am. Zool.* **33**:524–536.

Robbins, L. L., and Blackwelder, P. L., 1992, Biochemical and ultrastructural evidence for the origin of whitings: A biologically induced calcium carbonate precipitation mechanism, *Geology* **20**:464–468.

Roberts, H. H., and Murray, S. P., 1983, Controls on reef development and the terrigenous–carbonate interface on a shallow shelf, Nicaragua (Central America), *Coral Reefs* **2**:71–80.

Roberts, H. H., Aharon, P., and Phipps, C. V., 1988, Morphology and sedimentology of *Halimeda* bioherms from the eastern Java Sea (Indonesia), *Coral Reefs* **6**:161–172.

Rogers, C. S., 1990, Responses of coral reefs and reef organisms to sedimentation, *Mar. Ecol. Prog. Ser.* **62**:185–202.

Roy, K. J., and Smith, S. V., 1971, Sedimentation and coral reef development in turbid waters: Fanning Lagoon, *Pac. Sci.* **25**:234–248.

Santavy, D. L., and Peters, E. C., 1997, Microbial pests: Coral diseases in the Western Atlantic, *Proc. 8th Int. Coral Reef Symp.* **1**:607–612.

Shick, J. M., Lesser, M. P., and Jokiel, P. L., 1996, Effects of ultraviolet radiation on corals and other coral reef organisms, *Global Change Biol.* **2**:527–545.

Skelton, P. W., Gili, E., Rosen, B. R., and Valldeperas, F. X., 1997, Corals and rudists in the late Cretaceous: A critique of the hypothesis of competitive displacement, *Bol. R. Soc. Esp. Hist. Nat. (Sec. Geol.)* **92**:225–239.

Smith, S. V., and Jokiel, P. L., 1975a, Water composition and biogeochemical gradients in the Canton Atoll lagoon: 1. Lagoon description; design of system analysis; salt and water budget, *Mar. Sci. Comm.* **1**:75–100.

Smith, S. V., and Jokiel, P. L., 1975b, Water composition and biogeochemical gradients in the Canton Atoll lagoon: 2. Budgets of phosphorus, nitrogen, carbon dioxide, and particulate materials, *Mar. Sci. Comm.* **1**:165–207.

Smith, S. V., Kimmerer, W. J. Laws, E. A. Brock, R. E., and Walsh, T. W., 1981, Kaneohe Bay sewage diversion experiment: Perspectives on ecosystem responses to nutrient perturbation. *Pac. Sci.* **35**: 279–402.

Stanley, G. D., Jr., 1981, The early history of scleractinian corals and its geologic consequences, *Geology* **9**:507–511.

Steven, A. P. L., and Broadbent, A. D., 1997, Growth and metabolic responses of *Acropora palifera* to long-term nutrient enrichment, *Proc. 8th Int. Coral Reef Symp.* **1**:867–872.

Swap, R., Garstang, M. Greco, S., Talbot, R., and Kallberg, P., 1992, Saharan dust in the Amazon Basin, *Tellus Ser. B* **44:**133–149.

Talge, H. K., Williams, D. E., Hallock, P. and Harney, J. N., 1997, Symbiont loss in reef foraminifera: Consequences for affected populations, *Proc. 8th Int. Coral Reef Symp.* **1:**589–594.

ter Kuile, B., 1991, Mechanisms for calcification and carbon cycling in algal symbiont-bearing foraminifera, in: *Biology of Foraminifera* (J. J. Lee and O. R. Anderson eds.), Academic Press, New York, pp. 73–89.

Triffleman, N. J., Hallock, P., Hine, A. C., and Peebles, M. W., 1992, Morphology, sediments, and depositional environments of a small carbonate platform: Serranilla Bank, Nicaraguan Rise, Southwest Caribbean Sea, J. Sed. Petrol. 62:591–606.

Turner, R. E., and Rabalais, N. N., 1994, Coastal eutrophication near the Mississippi River delta, *Nature* **368:**619–621.

Vitousek, P. M., Mooney, H. A., Lubshenco, J., and Melillo, J. M., 1997, Humandomination of Earth's ecosystems, *Science* **277:**494–499.

Vogelmann, A. M., and Ackerman, T. P., and Turco, R. P., 1992. Enhancements in biologically effective ultraviolet radiation following volcanic eruption, *Eos* **359:**47–49.

Vogt, P. R., 1989, Valcanogenic upwelling of anoxic, nutrient-rich water: A possible factor in carbonate-bank/reef demise and benthic faunal extinctions? *Geol. Soc. Am. Bull.* **101**(10):1225–1245.

Walsh, J. J., 1984, The role of ocean biota in accelerated ecological cycles: A temporal view, *BioScience* **34:**499–507.

Watson, R. T., Rodhe, H., Oeschger, H., and Siegenthaler, U., 1990, Greenhouse gases and aerosols, in: *Climate Change. The IPCC Scientific Assessment* (J. T. Houghton, G. J. Jenkins, and J. J. Ephraums, eds), Cambridge University Press, Cambridge, pp. 1–40.

Wells, J. W., 1957, Coral reefs, *Geol. Soc. Am. Mem.* **67**(1):609–631.

Williams, E. H. Jr., and Bunckley-Williams, L., 1990, The worldwide coral reef bleaching cycle and related sources of coral mortality, *Atoll Res. Bull.* **335:**1–71.

Wilson, P. A., Jenkins, H. C., Elderfield, H., and Larson, R. L., 1998, The paradox of drowned carbonate platforms and the origin of Cretaceous Pacific guyots, *Nature* **392:**889–894.

Wolanski, E., Drew, A., Abel, K., and O'Brien, J. O., 1988, Tidal jets, nutrient upwelling, and their influence on the productivity of the alga *Halimeda* in the Ribbon Reefs, Great Barrier Reef, *Est. Coast. Shelf Sci.* **26:**169–201.

Wood, R., 1993, Nutrients, predation and the history of reef building, *Palaios* **8:**526–543.

World Meterological Association, 1995, *Scientific Assessment of Ozone Depletion: 1994*, World Meterological Organization Global Ozone Research and Monitoring Project, Report No. 37, World Meterological Organization, Geneva.

Yonge, C. M., 1930, Studies on the physiology of corals. I. Feeding mechanisms and food, Great Barrier Reef Expedition, 1928–29, *Sci. Rep.* **1:**15–57.

Glossary

Accretion The vertical aggradation or horizontal progradation of the reef over time. Typical rates of accretion range from 1 to 5 m per 1000 years. It is important to understand that this is not the same thing as coral growth, which is typically one to two orders of magnitude higher.

Activation energy The additional amount of energy required for a particle to go from one energy state to another, such as the transition from an ion in a fluid to an atom in a crystalline solid.

Active transport Cellular process, requiring energy expenditure whereby ions or molecules are actively transported against concentration gradients.

Actualistic reef pattern Global geographic distribution of reefs that matches the distribution of modern shallow-water coral reefs. Opposite is nonactualistic reef pattern.

Aeronian The middle of three substages in the early Silurian (Llandoverian stage).

Ahermatypic From "herma," a mound. (1) Said of corals that lack symbiotic zooxanthellae; (2) non-reef building (see *hermatypic*).

Algal ridge An accumulation of pure coralline algae, typically formed in areas of very high wave energy where grazers that might feed on these algae are excluded. Algal ridges are often the culmination of vertical accretion by corals to a depth shallow enough to allow algae to overgrow both in-place and toppled corals.

Algal symbiosis The close association (symbiosis), usually intracellular, between a protistan or invertebrate host and unicellular algae (see symbionts).

Alkaline Waters containing more than average amounts of dissolved carbonates.

Anabarites A tubular calcareous marine microfossil with three-radiate symmetry, typical of Late Vendian and Early Cambrian.

Anoxia [anoxic] A state in the watermass or atmosphere denoting a lack of oxygen [=anaerobic].

Anthropogenic Changes directly or indirectly related to humans and the growth of human populations.

Aragonite The orthorhombic mineral form of calcium carbonate ($CaCO_3$); aragonite precipitates more readily than calcite in waters highly saturated with respect to $CaCO_3$ and dissolves more readily than calcite in undersaturated waters.

Archaeans A kingdom of very simple, bacterialike procaryotic organisms (=Archaebacteria), some of which live under extreme temperature conditions but others under normal surface conditions (including also methanogens) or anaerobically.

Archaeocyatha A group of sessile marine Cambrian fossils possessing porous calcareous cuplike aspicular skeletons, possible relatives of demosponges.

Archaeocytes A cell type characteristic of the demosponges, which rather than having a specific task as do most other sponge cells, performs various tasks.

Architecture [as applied to reefs] The structure of a reef in terms of their shape, growth and relationship of frame-building skeletons, and space between these (e.g., reef flat, spur and groove, breaker zone, etc.).

Ascidian Common name for members of the phylum Urochordata; also called tunicates.

Atoll A ringlike structure of coral reef surrounding a lagoon, usually with one or more islands on the reef rim.

ATP Adenosine triphosphate; a phosphatic compound that supplies energy for many biochemical cellular processes by undergoing enzymatic hydrolysis to ADP (adenosine diphosphate).

Autochthonous Occurring in the place of origin, such as sediments that were deposited and preserved in the place of their formation.

Automicrites From autochthonous micrites, carbonate muds derived from processes or deposition within the reef, e.g., via organomineralizing factors, where organic matter from reef inhabitants is precipitated *in situ* through microbial or biomineralizing activity.

Autotrophic Self-feeding; refers to an organism that fixes its own organic carbon using sunlight and dissolved inorganic nutrients and does not feed on organic matter; e.g., phytoplankton, benthic algae, and seagrasses.

Autotrophs Organisms that acquire nourishment for their metabolic activities by using inorganic matter to synthesize food such as photosynthetically from sunlight.

Autotrophy An ecological strategy wherein an organism is capable of living by utilizing inorganic nutrients exclusively.

Azooxanthellate Organisms that lack dinoflagellate symbionts; usually used in reference to corals (synonym: nonzooxanthellate).

Backreef Landward side of a reef, including the area and sediments between the reef crest/algal ridge and the land; corresponds to reef flat and lagoon of barrier reef and platform margin reef systems.

Bacteria [=superkingdom Eubacteria] Single-celled prokaryotes, including photosynthetic, chemosynthetic, and heterotrophic forms commonly aggregated as microbial mats or "slime molds." May play a significant role in reef accretion, cementation, or recycling.

Baffler guild Reef builders that trap (fine) sediment by the baffling effect (reduction of current velocity). Erect growth and weak skeletonization predominant.

Bafflers Organisms that project upward from the sediment, slowing water motion and providing quieter places for sediments to settle.

Bafflestone mound A reef-type mound of sediment deposited around skeletal organisms such as rudist bivalves. Deposition occurred because of the organism's presence, which caused the sediment to drop out of the current and settle to the bottom of the sea floor.

Benthic Pertaining to bottom-dwelling marine organisms.

Benthos Bottom-dwelling microbes, plants or animals occupying the substrate [adjective benthic: bottom-dwelling life, resting or fixed to the substrate].

Binder guild One of the reef guild types. The binder guild is occupied by reef builders that bind sediment, encrust other reef builders, or produce sediment *in situ* by their activity.

Binders Organisms that live in or directly on sediment, holding or encrusting it in place.

Bioclastic Consisting of fragments of shells or organic structures that have been moved individually from their places of origin.

Bioclasts Sedimentary grains or fragments produced biologically by cementation or precipitation of skeletons.

Bioeroders Organisms that contribute to the breakdown of the reef structure by boring into it or scraping away at it as they graze (see *bioerosion*).

Bioerosion Erosion of a shell, rock, or reef substrate by biological action such as grazing by gastropods or echinoids; mechanical abrasion, e.g., by echinoid spines; boring and etching activities, e.g., by sponges, worms, bivalves, microalgae, fungi, and many other kinds of organisms and microbes; and any other biological process that mechanically or chemically breaks shell or rock down to finer sediments or to dissolved constituents in seawater.

Biofilm A typically thin (two to many micrometers) organic layer containing microbial and/or eukaryotic communities in mucilage and degrading organic matter.

Biogenic Processes or structures that are produced biologically, i.e., by the mediating activity of organisms (e.g., reefs, stromatolites).

Biogenic reef A limestone structure or buildup produced by biological as well as geological processes; a rigid skeletal structure that influences deposition of sediments in its vicinity and that is topographically higher than surrounding sediments; may be constructed predominantly by corals, coralline algae, or a variety of other kinds of organisms.

Bioherm A moundlike, massive carbonate structure raised above the seafloor and produced by biological processes. It is a biogenic feature enclosed in rocks of a different lithology. The structure is distinctly domelike with vertical dimensions roughly equal or exceeding those of the horizontal (see *biostrome*).

Biomass The living material in an ecosystem represented by the total living organic matter.

Biomineralization Mineral precipitation by organisms, including biogenic calcification.

Biostrome Distinctly bedded, lenticular, rather than dome-shaped carbonate deposits composed of fossil organisms or their skeletal remains. In stratigraphic profile, the horizontal dimensions exceed those of the vertical. In life a biostrome had little relief above the seafloor (see *bioherm*).

Black-band disease A common disease in head-forming scleractinian corals, characterized by a dark band of pathogen complex that advances on the living coral tissue, consuming it, and leaving dead coral skeleton behind the disease line; the disease consists of an assemblage of microbes including the cyanobacterium *Phormidium corallyticum* Ruetzler.

Bleaching When referring to zooxanthellate corals, the process of expulsion of the zooxanthellae symbionts, commonly in response to stress.

Bryozoans Members of the phylum Bryozoa, the so-called moss animals.

Buildup A generalized term for any carbonate mass that rises above the surrounding seafloor. A buildup may be a reef, a baffled structure (e.g., a mud mound built up around sea grass) or a purely hydromechanical buildup (e.g., a carbonate sandbar).

Calcareous algae Primarily refers to calcifying members of the algal division Chlorophyta, informally called green algae; modern calcareous algae that precipitate aragonite skeletons.

Calcification The process of precipitation of calcium carbonate ($CaCO_3$).

Calcifier Any organism that produces calcium carbonate. This may be in the form of an internal skeleton, an exoskeleton or spicules. While bioeroders produce carbonate sediment, they are not considered to be calcifiers unless they secrete carbonate as part of their body structure.

Calcimicrobes Microbes (Cyanobacteria and various single-celled green algae) calcified through impregnation within or encrustations on extracellular sheets. Millimeter- to centimeter-sized tubelike microfossils, common in Phanerozoic shelf and reef carbonates and significant in reef building as encrusters and binders.

Calcispheres Microfossils made of spherical calcium carbonate that are generally smooth or slightly pitted. Calcareous microfossils important in Mesozoic–Cenozoic biostratigraphy.

Calcisponges Now obsolete term referring to systematically different calcified chambered and nonchambered sponges of polyphyletic origin. Most chambered sponges are now regarded as demosponges.

Caprinidae A family of attached rudistid bivalves that inhabited reefs of Mid- to Late Cretaceous time. They had teeth and well-developed myophores, and pallial canals in one or both valves. Caprinid shells were dominantly aragonitic, with a relatively small amount of calcite.

Carbonate budget A formula that quantitatively accounts for all carbonate within a reef. The budget must account for (1) original carbonate (e.g., coral skeletons, coralline algae), (2) the rate at which it is reduced to sediment by bioerosion, and (3) how much of that sediment remains in the reef or is transported out. All the processes that contribute to the fabric of a reef are accounted for only after the budget is "balanced."

Carbonate buildup A thickened body of carbonate rock that may or may not have topographic relief. If the feature is large, it could be a carbonate platform or shelf.

Carbonate factories Sites of enormous calcium carbonate production by marine benthic organisms such as corals and calcareous algae. Reefs and associated lagoons are typical carbonate factories.

Catch-up reef A reef that under the influence of relative sea-level rise, initially did not keep up but was not inundated to the point of significant suppression of reef accretion, so that when sea-level rise slowed, the reef was able to catch up to sea level.

CCEs Acronym for calcium carbonate calcification events.

Cementation The chemical or biological precipitation of carbonate cement. This cement may fill pore spaces, bind up loose sediment into a solid substrate, or strengthen the rigidity of the reef by reinforcing points of contact within the primary framework.

CFCs Chlorofluorocarbons; any of several simple gaseous compounds that contain carbon, chlorine, fluorine, and sometimes hydrogen, which are used as refrigerants, cleaning solvents, and aerosol propellants; in the stratosphere, they promote the breakdown of ozone molecules and therefore are a major cause of ozone depletion.

Chaetetid sponges A polyphyletic group of coralline sponges present since Paleozoic times and characterized by a calcareous basal skeleton consisting of tubes with horizontal partitions.

Chancelloriids A group of Cambrian marine sessile fossils whose skeletons consist of rosettelike spiny hollow sclerites surrounding a sacklike body.

Chloralgal sediments Biogenic sediments dominated by the remains of calcareous green algae such as *Halimeda*; foramol constituents, particularly molluscan debris and foraminiferal shells, are typical secondary components; see also *photozoan association*.

Chlorophyll The green photosynthetic pigments found in cyanobacterial and plant cells.

Chlorophytes Green algae: fossil varieties had an aragonitic skeleton, e.g., receptaculitids, dasycladaceans.

Chlorozoan sediments Biogenic sediment type found around coral reefs and have coral and calcareous algal remains as the dominant constituents; coralline algae, foraminifers, mollusks, and urchin fragments are important secondary components; bryozoan, barnacle, and worm shell debris are typically scarce in chlorozoan sediments (see also *photozoan association*).

Chondrodontid bivalves A family of bivalves characterized by two slender, calcareous valves. Life orientation is with the hinge in or on top of the substrate and the valves opening toward the top. Chondrodontid bivalves often are found associated with rudist bivalves in the Caribbean region.

Chronostratigraphy The branch of stratigraphy that deals with the age of strata and their time relationships.

Climax [stage] The final (fourth) stage of reef community succession, characterized by the dominance of few taxa or a single growth habit.

Clionid Referring to a group of sponges of geological significance because they bore into skeletons of corals and other calcareous organisms. One of several biological agents of reef destruction and bioerosion.

Clonal Said of a colonial organism that asexually produces identical individuals or units. While these units usually function as individuals, they frequently involve cooperation (feeding, defense, etc.) for the benefit of the colony as a whole.

Cloudina A tubicolous calcareous marine microfossil with a tube consisting of stalked collarlike segments. Typical of Late Vendian time.

Coastal upwelling Upwelling that results from the interaction between wind systems that parallel the coastline and Coriolis effect that promote movement of surface waters offshore and their replacement by cooler, more nutrient-rich subsurface waters, generally originating from the uppermost thermocline–nutricline.

Colonization [stage] The second stage of reef community succession, characterized by the settlement of reef builders; the first stage of reef community succession, characterized by the initial colonization of substrate by reef builders.

Constratal Said of reef skeletal growth fabrics that grow at the same rate at which sediments accumulate, so that during life most of the skeleton remained buried in the substrate.

Constructional Pertaining to the ability of a reef organism to build a framework by organic growth.

Constructor guild Ecospace occupied by skeletal reef builders that provide most of the skeletal volume in a rigid framework. Members of the constructor guild usually are characterized by strong skeletonization and erect growth.

Coralline algae Calcifying members of the algal division Rhodophyta, informally known as red algae; most modern coralline algae precipitate calcite skeletons, though a few precipitate aragonite.

Coralline sponges Porifera (probably mostly Demospongea) with a secondary calcareous basal skeleton. Known since the Cambrian and comprising several important reef-building sponges, e.g., see sphinctozoid, inozoid, and chaetetid sponges.

Coralomorphs A group of coral-like, sessile marine Cambrian fossils whose calcareous skeletons resemble those of true corals.

Cribricyaths A group of sessile marine Early Cambrian calcareous microfossils that are hornlike asymmetrical, rounded, or subangular (often tetragonal) in cross-section.

Crinozoans Echinoderms with a stem, holdfast, and crown holding feeding arms. Includes crinoids, cystoids, and some other attached echinoderms.

Cryptic Referring to a habitat characterized by cavities beneath the sediment–water interface or on the underside of larger clasts or overhangs on the sea floor.

Cryptic biota Organisms within the ecosystem that live either inconspicuously or hidden on, in, and around the substrate. In the case of reefs, that substrate may be hard, rocky, or calcareous and composed of larger dead or living organisms. Some elements of the cryptos actively excavate cavities within the reef. Quantitatively, cryptic biotas may be more abundant than open surface communities.

Cryptobiont From *cryptos* = hidden. An organism inhabiting cavities within the reef rather than on the open surface. An obligate cryptobiont is an organism strictly adapted to life in reef cavities.

Cryptomicrobial Refers to reef fabric. A rock type that lacks distinctive macrofabrics of possible microbial origin.

Cyanobacteria Photosynthetic blue–green bacteria (formerly called blue–green "algae"). These were critical in raising atmospheric levels of oxygen during the early history of earth.

Dasycladacean algae "dasyclad algae" A group of calcareous green algae, known since the Cambrian and common in Mesozoic and Cenozoic lagoonal environments but occurring also within reefs. The centimeter-sized upright algae are characterized by a long stem cell bearing whorls of lateral branches. Calcification takes around the stem and between branches.

Dasycladacean limestone A limestone dominantly, or to a large degree, composed of skeletal fragments or particles of dasycladacean algae.

Debris potential The relative volume of loose skeletal material in and around a reef body. Debris may be coarse gravel or sand-sized particles.

Debris production The absolute volume of loose skeletal material in and around a reef body. Debris production depends on the potential of the skeletons and the size of a reef body.

Demosponge A sponge characterized by a skeleton composed of spongin, mixed spongin and silicious spicules, or silicious spicules. These sponges lived from the

Cambrian to Recent. Approximately 390 genera of demosponges are known from the fossil record.

Dendrolite A microframework found in reefs that is a nonlaminated, biomineralized deposit with a dominant dendritic or branching microfabric.

Destruction [stage] The second stage of reef community succession. Characterized by the hurricane-induced breaking of reef builders and their deposition as cobble layers and sand.

Deuterostomes Advanced metazoans. Includes echinoderms and chordate metazoans—two higher grades of animal life with a gut and comparable larval stages.

Displaced terrane A fault-bound tectonic block containing rocks, fossils, a stratigraphic succession, and other geologic attributes that contrast markedly with surrounding or adjacent tectonic blocks. Such blocks are said to be displaced or moved relative to the more stable craton.

Diversification [stage] The third stage of reef community succession, characterized by the high taxonomic diversity, niche partitioning, and trophic web complexity.

Drowned reef A reef whose accretion rate did not keep up with relative sea-level rise, also known as a "give-up reef."

Dysaerobic Environments with strongly reducing oxygen concentrations where a limited amount of oxygen is sufficient to sustain only some metazoans.

Echinoids Members of the phylum Echinodermata, class Echinoidea; specifically the sea urchins and sand dollars.

Ecological reef A term to describe structures that were rigid and stood above the sea floor. A modern or ancient reef produced by the organic growth of organisms and having wave resistance (see stratigraphic reef).

Ecospace A term referring to the ecological range of space occupied by a species or a fauna. It is a range beyond which the fauna cannot exist.

Encrustation The biological binding of primary framework by calcifying algae, foraminifers, or other organisms that can produce a sheetlike, skeletal structure. In special cases (e.g., algal ridges), encrusting algae can produce primary framework.

Encrusters Crustose coralline algae and other organisms that secrete skeletal crusts and serve to bind sediments together with reef framework.

Endosymbiotic An ecological condition in which two organisms live together for mutual benefit, one residing inside the others, body or tissues (e.g., a one-celled zooxanthella living inside the endodermal tissues of a coral host).

Equatorial upwelling A physical oceanographic process related to the Earth's rotation. The northeast trade winds, just north of the equator, move water northward away from the equator, and at the same time, the southeast trade winds, just south of the equator, move water southward away from the equator. As a result, subsurface water, which is slightly cooler and enriched in nutrients, is brought to the surface, stimulating phytoplankton growth.

Estrogenic Refers to chemicals that mimic hormones of the estrogen group.

Eukarya [=eukaryotes; Eucarya] Superkingdom of sexually reproducing organisms including all those with a cell nucleus, e.g., animals.

Euphotic zone Surface waters of the ocean where there is sufficient light (solar energy) for net primary productivity to occur by photosynthesis; the base of the euphotic zone is often defined for convenience as the depth of penetration of 1% of the solar energy present at the sea surface.

Eurybathic Said of organisms that can tolerate different conditions of bathymetry, such as those that can live in very different water depths (opposite is stenobathic).

Eurytrophic Said of eurytrophic organisms able to tolerate variable nutrient concentrations (opposite is stenotrophic).

Eustasy The position of global sea level.

Eustatic highstands In reference to the sea surface. This term refers to high stands of sea-level that can be determined to be global in nature and synchronous across a broad area.

Eutrophic Said of a marine environment oversaturated with nutrients. This term can be used to describe environments where nutrient flux is sufficient to support substantial phytoplankton densities that limit light penetration and therefore photosynthesis by benthic algae; benthic carbonate production is primarily by heterotrophic organisms (cf. oligotrophic).

Eutrophication Increased nutrient flux resulting in sufficiently high phytoplankton densities so that, even in shallow water, there is insufficient light to support benthic plants. Thus the benthos is dominated by heterotrophic microbes and animals.

Fabric The internal matrix of the reef. This includes primary framework (e.g., corals), secondary framework composed of toppled corals, encrusters, and marine cements, plus loose sediment, reef debris, and open void spaces.

Facies A mappable unit, either in outcrop or in plan view, that reflects a common set of environmental controls. In reefs, examples include the reef facies, the back reef facies, and the lagoonal facies.

Famennian A formal term for the last stage of the Upper Devonian system.

Fair-weather wave base The base of the bathymetric zone regularly affected by wave action. The fair-weather wave base can be very shallow in restricted lagoons or deeper on the open shelf (up to 15 m).

Fenestral fabric A rock fabric found in limestone when vugs and cavities develop in sediments and are preserved by large calcite cement filling minute void spaces. This occurs most frequently in the inner- and supratidal zones so that fenestral fabrics are characteristic of such settings.

Fixed nitrogen Nitrogen in compounds or ions, e.g., ammonia, ammonium, nitrite, and nitrate, available to organisms for growth.

Fixosessile Organisms fastened firmly to the sea bottom, via encrusting, stalks, stems, holdfasts, and so forth.

Fluviatile A synonym of *fluvial*. Geologists tend to use the term for the results of river features (e.g., fluviatile dam, or fluviatile sands).

Flux Rate of transfer of a substance or energy from one location to another.

Foraminifers Calcareous protists belonging to the class Foraminifera.

Forams Informal for foraminiferans, single-celled protists that capture food by pseudopods or by envelopment, passing food through their cell walls for digestion. Benthic forams in tropical belts have significantly sized exoskeletons and add substantial $CaCO_3$ to reefs.

Foramol sediments Biogenic sediments produced by the shells and skeletal debris of benthic foraminifers, mollusks, coralline red algae, sea urchin spines and plates, bryozoans, barnacles, and/or worm tubes; see also heterozoan association.

Forereef The seaward side of reef, including area and sediments of seaward slope.

Framework In a reef, the sum total of the material that provides rigidity to the structure. This includes *Primary framework*, such as in-place corals, and *secondary framework*, such as toppled corals, encrusting organisms, and marine cements. While many reefs have framework, the relative importance of in-place versus reworked skeletal material also must be considered.

Framework builders On Cenozoic reefs, those organisms such as (scleractinian) corals, which grow upward or outward in branching, massive or platy morphologies. Framework builders secrete substantial quantities of calcium carbonate, while trapping even greater quantities of sediment within and in the lee of the reef framework. Other types of organisms filled this role in pre-Cenozoic reefs.

Frasnian A formal term for the first stage of the Upper Devonian system.

Fusulinid foraminifer Any foraminifer belonging to the suborder Fusulinina, family Fusulindae, characterized by a multichambered elongate, calcareous microgranular test, commonly resembling the shape of a grain of wheat. Range, Ordovician to Triassic. Synonym: fusuline.

Ga Abbreviation for one billion years.

Give-up reef A reef, that under the influence of rising sea level, did not keep up and was inundated so that reef development was terminated; i.e. a drowned reef.

Globigerinid Any planktonic foraminifer belonging to the superfamily Globigerinacea, characterized by a perforate test with bilamellid septa and walls of radial calcite crystals. Range, Middle Jurassic to present. Var: globigerine.

Greenhouse oceans Refers to paleoceanographic conditions during the Cretaceous and Paleogene when atmospheric CO_2 levels were two to ten times higher than present (i.e., "greenhouse" conditions) and when temperatures of polar oceans and the deep sea were on the order of 10°C, i.e., much warmer than present, and (in the case of polar oceans) ice free.

Greenhouse world Global climatic situation of the world during certain periods of earth history. It defines the time when, at its maximum development, the tropics and subtropics were expanded and there was no permanent ice at the poles. During a greenhouse world time, cool–temperate marine biotas are found at the poles (see *icehouse world*).

Gross primary productivity (GPP) Total amount of energy fixed to organic carbon per unit time during photosynthesis.

Growth fabric A descriptive term for the presence of aggregated *in situ* corals (or other organisms) in growth position within a facies.

Guild An ecological grouping of species that exploit the same type of environmental resources in a similar way without regard to relationship or classification. They may overlap significantly in their niche requirements.The five guilds proposed for reefs are:

1. Constructors. Rigid, often high-growing organisms that build reefs and live in the water column obtaining their food directly on or above the surface of the reef.
2. Bafflers. Firmly attached organisms, often flexible, which present surface areas that reduce the velocity of a flowing current of water. These organisms also induce the deposition of sediment around and within the reef while at the same time capturing food.
3. Binders. Organisms that unite and consolidate reef material by encrusting the substrate as sheets, mats, and vines. Binders also may induce the precipitation

of chemical cements within pore spaces. The net effect is the formation of a rigid calcareous mass that resists the forces of the waves.

4. Destroyers. Bioeroding organisms that excavate, bore, rasp, and bite the calcareous framework.
5. Dwellers. Guild members that simply live on the reef or within the interstices. Dwellers make their life in a variety of ways as predators, scavengers, herbivores, and so forth, and compete for food and space. Upon death, their shells and skeletons consolidate the reef by filling in the cavities.

Guyot A flat-topped seamount (i.e., a submarine structure more than 915 m in elevation above the surrounding sea floor and usually of volcanic origin); the flat top indicates that the structure was subaerially eroded prior to permanent submergence.

Halimeda A genus of calcareous green algae characterized by an upright growth form consisting of segmented, calcified plates; *Halimeda* species are prolific producers of aragonite sediments.

***Halimeda* bioherms** Carbonate buildups consisting primarily of *Halimeda* skeletal debris (see *Halimeda*).

Halothermal circulation Density-driven subsurface ocean circulation caused primarily by salinity differences in ocean seawater.

Hardground Marine substrate that has been lithified, either by submarine or subaerial cementation; typically represents a period of interrupted sedimentation.

Hermatypic From "herma," mound. Said of constructional corals that both build reefs and possess symbiotic zooxanthellae.

Heterotrophic Referring to organisms that must consume organic matter to live and grow, e.g., many kinds of bacteria, protists, and most animals.

Heterotrophs Metazoan organisms that acquire their food directly by eating or assimilating organic material. See *heterozoans*; a term used for heterotrophic metazoans, especially reef biota, which live on a diet consisting of other organisms.

Heterotrophy An ecological strategy of heterotrophs.

Heterozoan association Term applied to biogenic sediments produced by the shells and skeletal debris of foraminifers, mollusks, coralline red algae, sea urchin spines and plates, bryozoans, barnacles, and worm tubes (see also *foramol sediments*).

Hexactinellids A group of sponges with a very regular skeleton composed of siliceous spicules (nonpreferred synonym of hyalosponge) that appeared in the Cambrian. They contributed to the formation of Mesozoic reefs (e.g., Late Jurassic reef mounds and mud mounds).

Hirnantian A formal term for the final stage of the Ordovician system.

Hydromechanical buildup A deposit built solely (or nearly so) by the physical action of waves or currents. As an example, a carbonate sand bar is more analogous to a similar siliciclastic structure than it is to a reef. While the origin of the sediment within it is biological, those processes had nothing to do with its ultimate deposition.

Hyoliths A group of Paleozoic marine benthic bivalved calcareous fossils consisting of bilaterally symmetrical shell and operculum.

Hypertrophic Referring to nutrient flux in surface waters that is so high as to support dense stands of phytoplankton. Hypertrophic water produces such an excess of organic matter that underlying benthic communities are limited to species tolerant of hypoxia and anoxia.

Hypertrophication Production of extreme nutrification in hypertrophic conditions.

Hypoxia [hypoxic] A condition where only minimal amounts of oxygen (usually less than 0.1 ml of oxygen per liter of seawater) are available for metabolism, thus affecting the abundance and diversity of life [=dysaerobic].

Icehouse world A time in geologic history when the world was cold at the poles, with permanent ice caps and a restricted tropical region.

Impact-tsunami event An extraterrestrial impact usually in the form of a meteorite or a meteor ranging in size from centimeters to kilometers in diameter. Such an impact in the ocean could have caused a sudden shift in ocean bottom topography, resulting in a tidal wave (tsunami) of exceptional magnitude.

Inequitable An ecological situation when a few or only one member of a fauna dominates an ecosystem.

Inozoid sponges Coralline sponges characterized by nonsegmented growth forms. Known since Cambrian time.

In-place An organism that is deposited in growth position.

In situ In contrast to in-place, an organism that remains within the general environment where it grew. It may be toppled and transported but must remain within the environment where it lived. As such, all in-place organisms are also *in situ*; however, the opposite need not be so.

Karstified A feature of limestone when acidic rainwater etches the surface to produce a surface of pinnacles, potholes, and sharp edges.

Keep-up reef A reef that under the influence of relative sea-level rise kept pace with sea level.

K-strategist A species possessing low population turnover and low dispersion ability that devotes a high proportion of its metabolic energy to production of its offspring.

Limestone A rock that consists predominantly of calcium carbonate.

Lithistid Said of a stonelike or stony sponge whose rigid skeletal framework consists of interlocking or fused siliceous spicules (desmas). Any demosponge belonging to the order Lithistida and characterized by the presence of desmas, interlocked and cemented to form a rigid framework.

Lithoherms Term used to describe steep-sided organomineralized mounds known today from Florida Straits and the margins of the Bahamas Platform at depths of 600–2000 m.

Lithophagid Literally "rock-eating," in reference to a type of bivalve that can bore into hard calcareous substrates and calcified organisms such as corals and sponges for protection. These organisms are principal agents of reef destruction.

Llandovery A formal term of a stage in the Lower Silurian system.

Lychniskid sponges A group of sponges characterized by a skeleton composed of siliceous spicules. Unlike hexactinellid sponges, the spicule junctions are reinforced by barlike structures.

Ma Abbreviation for one million years.

Malthusian overfishing Refers to such intensive fishing pressure that virtually all edible fish are caught and consumed for protein by local human populations.

Meridional transfer of heat Heat transfer across latitudes, for example, from the equator to the poles.

Mesotrophic [conditions] Said of a marine environment moderately saturated with respect to nutrients (cf. oligotrophic).

Metazoans Multicellular animals with complex walls and organs, e.g., sponges, corals and all higher animals.

Methanogenesis A metabolic process wherein CO_2 or more complex organic compounds are reduced to produce methane.

Methanogens Archaeans that produce methane (natural gas) as a by-product of their metabolism.

Microbes General term for millimeter-sized noncalcified and calcified eubacteria and cyanobacteria, algae, fungi, as well as protozoans and metazoans, commonly associated with organic crusts.

Micrite Fine-grained carbonate mud. Refers to such a carbonate rock.

Microbial Related to the activity of microbes, minute organisms usually not visible to the naked eye.

Microbial crust A calcareous crustlike sediment that owes its origin to the activity of bacteria, cyanobacteria, and other microbes.

Microbialite See *microbolite.*

Microbolite A rock type or sedimentary structure generated by the precipitation and sediment trapping and binding activity of calcifying microbes (also known as microbialite). It may be crustlike. Some reefs are composed almost completely of microbolite.

Microproblematica Micron- and millimeter-sized fossils that cannot readily be placed in established phyla or major groups. See *Tubiphytes* for one example.

Microsolenid coral A family of both solitary and colonial zooxanthellate corals, associated with Cretaceous deposits of rudists.

Millepora Stony cnidarians belonging to the order Milleporina; class Hydrozoa, commonly known as "fire corals," because of the strength of their stinging nematocysts. They are not true corals, however.

Mixotrophic Refers to organisms that can both feed and photosynthesize, often through a symbiotic association in which viable algal cells are maintained within the cells or tissue of a protozoan or animal host.

Modular Growth habit of an organism consisting of a set of morphologically and functionally identical and relatively independent units or individuals that were derived from the same zygote (see *clonal*).

Mud mounds Mound-shaped features on the seafloor composed mostly of fine-grained carbonate mud (micrite) and without important skeletal or framework components. Some are thought to have formed by organic processes by microbes, skeletal protists, or metazoan organisms.

Mutagenic Capable of causing changes in genetic material that result in mutations.

Nekton Organisms swimming in the watermass, e.g., fishes.

Nematocysts Stinging organs or structures characteristic of members of the phylum Cnidaria.

Net primary productivity (NPP) Total amount of energy fixed to organic carbon per unit time during photosynthesis, minus the total amount used in respiration.

Nonconstructional Being without organic framework and not capable of building a reef (see *constructional*).

Nonenzymatic A class of biomineralization wherein precipitation represents an incidental by-product of biological activity rather than a controlled secretion.

Nonzooxanthellate Said of corals and some other marine organisms lacking symbiotic, one-celled dinoflagellate cells (zooxanthellae) in their tissues.

Nucleation The initiation of a crystal's growth.

Nutricline Literally, a zone of rapidly changing nutrient concentrations; in the ocean, the nutricline refers to the region near the base and below the euphotic zone where there is insufficient light for the phytoplankton to fully utilize nutrients being excreted by other organisms, so nutrient concentrations rapidly increase with depth.

Nutrification An increase in nutrient flux sufficient to elicit a shift in benthic community structure, e.g., from coral–algal community to a sponge–algal community.

Nutrients Fixed nitrogen (NH_4^+, NO_2^-, NO_3^-) and phosphate ions (PO_4^{-3}) required by all organisms for growth and reproduction.

Octocoral Member of the phylum Cnideria, class Anthozoa, subclass Octocorallia, the group that includes the gorgonians, sea fans, and sea whips.

Oligotrophic Said of environments deficient or undersaturated in nutrients. Within the euphotic zone, oligotrophic is equivalent to deficient in fixed nitrogen and phosphorus; below the euphotic zone, oligotrophic means that food resources are limited, i.e., fixed inorganic nutrients may be plentiful, but without an energy source they are of no use to organisms; relating to tropical benthic environments, oligotrophic includes the range of conditions that support true reef framework, i.e., chlorophyll concentrations consistently less than 0.2 mg/m^{-3}.

Oncoids Large, pebble-sized, concentrically layered, spherical particles produced by microbial (usually cyanobacterial) precipitation of calcium carbonate films around small fragments.

Ooids Small, sand-sized, concentrically layered grains of microbially mediated and precipitated films of calcium carbonate (less often hematite, phosphate, etc.). Ooids are typically featured in tropical, high energy, shallow, sand shoals with above normal salinities.

Oolites (1) A sedimentary rock, usually a limestone, made up chiefly of tiny round ooliths cemented together. The rock was originally termed oolith. Syn: roestone; eggstone. (2) A term often used for oolith, or one of the ovoid particles of an oolite.

Organic matrix The organic matter in skeletal organisms that controls the growth of minerals in the skeleton.

Oxycline A subhorizontal boundary layer within the column of seawater where oxygen concentrations change drastically (often from oxic to anoxic).

Ozone O_3 form of oxygen; forms in the stratosphere under the influence of solar energy; ozone is an effective absorber of high-energy (UVB) radiation.

PaleoReefs A comprehensive database on Phanerozoic reefs developed at the University of Erlangen. Coded reef attributes and the paleopositions of reefs are linked to paleogeographic maps, permitting the spatial analysis of reef attributes in a given time slice. Synonyms: PaleoReef database, Paleoreef Maps.

Panthalassa An ancient ocean surrounding the supercontinent Pangea and extending across the area of the present-day Pacific Ocean.

Paucispecific Refers to a fauna containing very few species.

Phanerozoic A formal geologic era; the age of skeletal fossils including all rocks younger than 544 million years (the start of the Cambrian period).

Phosphate A mineral compound containing tetrahedral PO_4^{-3} groups.

Photoautotrophy Ability of an organism to produce food by photosynthesis in the presence of sunlight.

Photosymbionts Symbiotic algae or bacteria which supplement life activities of skeleton builders (today these are dinoflagellates such as *Symbiodinium*).

Photosymbiosis The coexistance of photosynthesizing microorganisms with heterotrophic organisms (usually invertebrates). The photosymbionts usually reside inside the host's body.

Photosynthate Product of photosynthesis; typically carbohydrates and lipids.

Photosynthesis Process by which plants fix solar energy into organic carbon.

Photosystem I First step in photosynthesis, in which chlorophyll traps solar energy by converting ADP to ATP.

Photosystem II The step in photosynthesis in which carbon is fixed to simple sugar.

Photozoans A term used for describing metazoans, especially tropical reef biota such as corals, forams, and bivalves today, possessing symbionts growing in their soft tissues and requiring light for their growth.

Photozoan association Biogenic sediment association consisting of predominantly coral and calcareous algal debris, the products of active biogeochemical and geochemical precipitation, specifically lime muds, ooids, and carbonate cements, and some components of the heterozoan association, i.e., shells and skeletal debris of benthic foraminifers, mollusks, coralline red algae, and sea urchin spines and plates.

Phylloid algae A nonsystematic category for partly calcified algae characterized by leaflike algal bodies with complex internal structures. Known from Devonian to Triassic. Abundant in Carboniferous and Permian reefs. Very rare in Triassic reefs.

Phytoplankton Algae that live as floaters and drifters in the water column or water bodies (i.e., lakes, bays, oceans, etc.).

Pisolite A sedimentary rock, usually a limestone, made up chiefly of concentric, round particles cemented together; a coarse-grained oolite. Synonym: peastone; pea grit.

Plankton Minute floating and weakly swimming organisms in the water column that form the basic part of the food chain for larger organisms.

Planktotrophic Feeding on plankton.

Poikiloaerobic Said of settings or environments with highly fluctuating oxygen concentrations.

Porostromate algae Tiny tubular fossils interpreted as the remains of microbes or algae (see *calcimicrobes*).

Porostromatolites Stromatolites produced by the secretion of $CaCO_3$ inside or around the cell walls of bacteria or protists [also = calcimicrobialites], e.g., including such fossil cyanobacteria as *Girvanella*, *Renalcis*, and *Epiphyton*.

Primary framework The part of the reef interior composed of in-place calcifiers. A *primary framework reef* is one in which the vast majority of the corals, stromatoporoids, or similar organisms are preserved in life position. In such as reef, structural rigidity could result from either the interlocking nature of the organisms, their attachment to one another, or both.

Primary productivity The rate at which energy is fixed into organic carbon (i.e., simple sugars) during photosynthesis.

Prokaryotes (also procaryotes) Primitive, single-celled organisms of small size, which lack a cell nucleus.

Protists Single-celled eukaryotes, many of which are significant carbonate sediment producers, e.g., forams, some of which may grow as multicellular or aggregate skeletal forms.

Protostomes Advanced invertebrates such as worms, arthropods, and mollusks with a linear body and gut.

Pseudoreefs Mound-shaped, irregular or tabular structures resembling reefs but cross-cutting bedding surfaces (e.g., including microbial mounds built around hydrothermal vents, seeps, fractures, etc.).

Pulchrilamina A problematic, finely laminated, marine encrusting organism common in Ordovician carbonate rocks, including reef deposits.

Radiocarbon age The age of a carbonate sample determined from its radioactive decay rate, based on a predetermined ratio of stable to unstable carbon. Ratio changes over time as the unstable carbon inverts to stable carbon. The age of a coral is determined by measuring this ratio and calculating the length of time needed for decay to occur.

Radiolitidae A family of rudist bivalves whose shells contain calcite as the dominant shell mineralogy. Radiolitids were part of reefal structures in the Late Cretaceous but also occurred as small, nonreef clusters in nontropical regions.

Rawtheyan The second last stage of the Ashgill, Late Ordovician. In North America, it includes Richmondian in part.

Receptaculitids A group of problematic Paleozoic marine sessile calcareous fossils with ovoid skeletons built of elements arranged in whorls around the central axis; each element consists of shaft and heads bearing four-ribbed stellate structure and plate. Said by some workers to be calcified, dasycladacean algae.

Recovery [core] The relative amount of solid material collected in a drill core. In a modern reef, this includes in-place or toppled corals, coral fragments, sediment, and rubble. In practice, some material is lost and recovery may not represent the actual amount of material preserved in the reef.

Recovery [stage] The final (third) stage in a reef community succession, characterized by the stabilization of rubble by binders and colonization by other reef organisms.

Red beds Sedimentary strata composed largely of reddish sandstone, siltstone, and shale with locally thin units of conglomerate, limestone, or marl that are predominantly red in color due to the presence of oxidized ferric hematite usually coating individual grains.

Reef Mound to irregular, to tabular shaped, biologically produced structure raised above the surrounding ocean substrate, usually by $CaCO_3$ secreting or sediment accreting organisms. Reefs have cavity spaces, cements, and vertical growth relief but do not necessarily require frame builders. Under this definition, reefs include a broad spectrum of features including bioherms, microbial mudmounds, sponge mounds, and so forth.

Reef accretion See *accretion*.

Reef crisis Time interval of a major decline in the global reef volume, usually but not always accompanied by a decline in reef diversity.

Reef flat (1) Any relatively flat area behind a reef; (2) relatively flat zone immediately behind reef crest characterized by water depths of 0–2 m; with scattered framework organisms and associated algae but generally lacking extensive organic framework in mature reef complexes.

Reef guilds Groups of reef organisms adapted for exploiting specific spatial resources in a similar way. Benthic sessile reef organisms can be assigned to constructor, baffler, or binder/encruster guilds representing major functional units. Criteria for guild assessment are growth habit and growth direction, life form, and growth form, skeletonization, skeletal volume and coloniality, skeletal density and areal cover, as well as burial and preservation of the organisms. The critical use of the guild concept provides a promising approach to understanding the role of reef organisms in spatial resource exploitation regardless of their geological age (see *guild*).

Reef mound Constructional reef type composed of about equal proportions of skeletal organisms and matrix.

Refugia (Sing. refugium) Havens or places where species survive climatic or environmental change. Refugia are usually small and protected geographic places or environmental habitats either isolated or left behind by major changes or shifts in Earth's global environment or where species have immigrated during times of stress.

Relic setting Pertaining to reefs and organisms, the geographic area occupied after a reef crisis. This setting is identical to the area occupied before the crisis.

Renalcids A group of problematic Paleozoic–Mesozoic marine sessile calcareous microfossils possessing simple tubular, chambered, dendritic, or fanlike skeletons with microgranular microstructure.

Requieniidae A family of rudist bivalves characterized by a coiled shell morphology. Members of this family were spread across back reef and shelf margin facies of the carbonate platforms.

Rhodolith A nodule composed of concentrically layered coralline red algae.

Rhodophytes Red algae: in the fossil record most of these had a calcite skeleton.

Rhuddanian A formal term for the first substage of the Lower Silurian, immediately following the Ordovician System.

Rotaliid foraminifers Members of the class Foraminifera, order Rotaliida, characterized by lamellar, perforate calcite tests.

Rubble Material larger than sand-size grains that is derived from the physical and biological breakdown of the reef.

Ruderal Referring to a species in the reef ecosystem whose life strategy is expressed in fast growth and small size.

Rudistid bivalve Member of the phylum Mollusca, class Bivalvia, order Rudistida; an unusual, highly specialized lineage of bivalves that flourished during the Cretaceous period, during which they were major limestone contributors to sedimentary limestone buildups.

Rugosans Solitary or colonial fossil corals of the order Rugosa with skeletons of calcite composition. Septal insertion serial in four quadrants. Range: Ordovician–Permian.

Runoff Fresh water, with suspended and dissolved sediments, nutrients, and pollutants, that moves from terrestrial environments to aquatic environments.

Saturation state The ratio of the ion activity of a dissolved solid to the thermodynamic solubility product of the solid.

Scleractinian A member of a living order of colonial and solitary corals characterized by an aragonitic skeleton and sixfold symmetry in the serial arrangement and insertion of their septa. This coral group is well-known builder of reefs and they first appeared in Middle Triassic time. Synonym: hertacoral.

Secondary framework That part of the reef interior that is comprised of dislodged calcifying organisms, encrusters and cement. A secondary framework reef is one in which less than 70–80% of the framework is composed of calcifiers in growth position.

Skeletal (or skeletogenic) Adjective for describing organisms able to secrete skeletons or shells.

Solenoporacean algae A group of calcareous red algae represented by millimeter- to centimeter-sized nodular, ramified, or encrusting growth forms. Composed of closely spaced parallel threads of vertically divergent calcified cells. Regarded as subgroup of the Floridae. Known from Cambrian to Early Tertiary. Important constituents of Paleozoic and Mesozoic open-shelf and reef environments.

Sphinctozoid sponges Segmented coralline sponges. The term "sphinctozoid" is used in order to underline the polyphyletic nature of the "Sphinctozoa." Known from the Cambrian to the present but most important as reef builders during Late Paleozoic and Triassic times.

Spongiomorphs Flat, domate, or branching fossil organisms with laminar or "spongy tissue," frequently found in Mesozoic reefs. Some are classified as hydrozoans, while others may be calcareous sponges. Polyphyletic and probably not a homogeneous taxonomic group.

Spongiostromate Used to describe irregular, encrusting, commonly clotted, microbial texture and early cementation or spongelike fabrics indicating the former activity of biofilms.

Spongiostromate crusts Fine-grained biogenic carbonate crusts of spongiostromate origin. More or less synonymous with "stromatolite."

Stabilization [stage] The initial (first) stage of a reef succession. This stage allows the stabilization of the substrate for further colonization by reef organisms.

Stratigraphic reef A term to identify stratigraphic thickenings that contrast with surrounding, otherwise bedded strata. The thicker sequence of reeflike strata may not represent actual topography at the time of deposition. Two common mechanisms that produce stratigraphic reefs are: (1) the successive stacking of tightly packed organisms at the same rate as surrounding sediments accrete; and (2) differential compaction of the sediment (see *ecological reef*).

Stratosphere Region of the atmosphere from 11 km to 50 km above the Earth's surface; region of the Earth's protective ozone layer which absorbs damaging UVB radiation.

"Stressed" reefs Reefs living near or at the limits of their threshold of tolerance for parameters of the environment such as abnormal salinity, reduced oxygenation, carbonate saturation, higher siltation, deeper waters, and so forth. Usually smaller, lower diversity reefs with restricted faunas that occur at the outer, marginal limits of reef growth.

Stress tolerant A species whose life strategy is oriented toward extremes of the environment. Usually expressed in slow growth and small size of organisms.

Stromatolite Organically produced sedimentary structure consisting of alternating layers of organic-rich and organic-poor sediment. The organic-rich layers usually have been formed by microbial activity of sticky threadlike cyanobacterial filaments and slime, which bind and trap the sediment of the organic-poor layers. Usually composed of calcium carbonate but more rarely composed of silica, phosphates, metallic oxides (e.g., hematite), manganese, or sulfides.

Subclimax [stage] The extrinsically stabilized stage of reef succession where physical factors affect the final stage of development.

Suprastratal Said of reef growth fabrics that project decimeters to meters above the substrate on which they grew.

Supertropical/Supertethyan Refers to a warmer, more saline ocean/climate zone that may have occurred episodically in the center of the Tethys seaway during the Cretaceous greenhouse world.

Symbionts Organisms that are part of a close ecological association, living usually within or around each other (see zooxanthellae of modern reef corals).

Symbiotic Said of two dissimilar organisms living together for mutual benefit (see *symbionts*).

Tabulates Paleozoic colonial corals (order Tabulata), with small corallites (normally <3 mm diameter), usually with short, reduced septa inserted in patterns of 4, 6, or 12.

Tannin Dark colored, dissolved organic matter usually derived from land.

Taphonomy In geology, the study of the processes of burial and preservation and how these processes preserve and help reconstruct the ancient community or ecosystem.

Tectonoeustatic Referring to a type of global sea-level change caused by tectonic factors. Seawater is displaced by an upward swelling mid-oceanic spreading center and that water invades former land areas.

Terrane A fault-bound block characterized by a stratigraphic, tectonic, and geologic history distinctly different from those of adjacent blocks or geologic areas.

Terrigenous Originating from land; of terrestrial origin.

Tethys/Tethyan An ancient east–west trending seaway in existence from Paleozoic to Cenozoic time. It was formed during and after breakup of the one-world continent called Pangea. The seaway separated two great Mesozoic supercontinents: Laurasia in the north and Gondwana in the south.

Thermocline Literally, a zone of rapidly changing temperature. In the ocean, the thermocline typically refers to the zone below the surface "mixed" layer, which is warmed by the sun and mixed by the winds. Within the thermocline, which is several hundred meters thick, temperatures rapidly drop from surface temperatures down to about 4°C, at roughly 1,000 m depth.

Thermohaline Refers to a condition characterized by density-driven, subsurface ocean circulation. This occurs primarily because of temperature differences.

Thrombolite A microbolite fabric usually in carbonate rocks, composed of irregular, clotted aggregation. Also used as a rock type based on the fabric. Differs from stromatolites by the lack of laminations (see *microbialite*).

Thrombolitic crusts Rinds of peloidal carbonate structures characterized by macroscopically clotted fabrics interpreted to be microbial in origin.

Thrombolitic fabric See *thrombolite*.

Tiering An ecological condition in which different species (usually suspension or filter-feeders) are juxtaposed vertically above the seafloor for the partitioning of food resources.

Tintinnid A ciliate protozoan belonging to the family Tintinnidae, ranging from the Jurassic to Recent.

Topographically induced upwelling Process by which subsurface waters, usually from the uppermost thermocline, are brought up into the zone of surface wind or current mixing by current flow that is disrupted by a topographic feature such as a bank or island.

Trophic Of or related to nutrition and food, e.g., autotrophic means self-feeding and refers to photosynthetic organisms; oligotrophic means "little nutrition."

Tubiphytes An enigmatic millimeter-sized fossil known from thin sections; very important in Permian and Middle Triassic reefs. It consists of irregular nodules with internal spar-filled tubes surrounded by a micrite envelope exhibiting a fine, network structure. Known from Carboniferous to Cretaceous time and interpreted as alga, sponge, or association of algae/bacteria and various organisms. The taxonomically correct name is *Shamovella.*

Turn-on/turnoff zone The range of environmental conditions that are marginal for coral reef growth, and within which biological activities and episodic events determine whether reef development will or will not occur. For example, a reef within the turn-on/turnoff zone might thrive as long as grazing sea urchins and fish keep algal growth in check, but it might be taken over by algae if grazing organisms are diminished through overfishing or disease.

Udoteacean algae Calcareous green algae. Formerly called Codiaceae. Characterized by erect, nodular or leaflike algal bodies and a differentiated internal structure. Present since the Cambrian. The breakdown of udoteacean algae (e.g., *Halimeda*) is a very important source for mud- and sand-sized carbonate sediment.

Upwelling Any process by which subsurface waters, usually from the uppermost thermocline, are brought up into the zone of surface wind or current mixing; because upwelled waters typically contain higher dissolved nutrient concentrations than surface waters, upwelling into the euphotic zone stimulates primary production by phytoplankton (and benthic algae on oceanic shelves or banks).

UVB Biologically damaging wavelengths (280–320 nm) of ultraviolet radiation.

Wenlockian A formal term for a Middle Silurian stage. It is followed by the Ludlovian–Pridolian stages.

Zooplankton Tiny marine organisms swimming and floating in the water column and providing a food source for larger organisms.

Zooxanthella (singular), **zooxanthellae** (plural) A dinoflagellate algal cell that lives symbiotically in the tissues of a protist or metazoan host. Today, zooxanthellae inhabit foraminifers, corals, sponges, and other reef organisms.

Zooxanthellate Adjective applied to organisms, primarily corals and other Cnidaria, that host dinoflagellate algae (zooxanthellae) as symbionts.

Index

MIX
Papier aus verantwortungsvollen Quellen
Paper from responsible sources
FSC® C105338

If you have any concerns about our products,
you can contact us on
ProductSafety@springernature.com

In case Publisher is established outside the EU,
the EU authorized representative is:
Springer Nature Customer Service Center GmbH
Europaplatz 3, 69115 Heidelberg, Germany

Printed by Libri Plureos GmbH
in Hamburg, Germany